A CHECK-LIST
OF THE
ORCHIDS OF BORNEO

A CHECK-LIST OF THE ORCHIDS OF BORNEO

JEFFREY J. WOOD

PHILLIP J. CRIBB

Assisted by Audrey Thorne and Sarah Thomas

ROYAL BOTANIC GARDENS
KEW

© Copyright The Trustees of The Royal Botanic Gardens, Kew 1994

First published 1994

ISBN 0 947643 59 1

General Editor: J.M. Lock

Cover Design By Media Resources, RBG, Kew

Typeset by Disc-to-Print Ltd., Olympia, London, U.K.

Printed and Bound in Great Britain by Whitstable Litho, Whitstable, Kent.

DEDICATION

Dedicated to the Foundation for the Preservation and Study of Wild Orchids, Zürich, Switzerland (Stiftung zum Schutze und zur Erhaltung Wildwachsender Orchideen, Zürich).

CONTENTS

LIST OF FIGURES

PREFACE

This checklist is part of the *Orchids of Borneo* project which is a collaborative research project aimed at elucidating the orchid flora of Borneo in a series of volumes, each of which includes accounts and detailed black and white drawings and colour photographs of 100 species. The first of these was published in 1991, the second in 1994 and further volumes will appear at regular intervals. The checklist is intended as the reference point for the series aiming to provide an up-to-date listing of the known taxa and their distribution both within and outside Borneo. It is also intended to provide baseline information useful to orchid conservation in Borneo, citing distributions, habitats where known and endemicity. Several new taxa are described here for the first time.

It is almost a century since Henry Ridley produced the first account of the orchids of Borneo in the Journal of the Linnean Society in 1896, and over fifty years since the last account was published by Masamune (1942). Ridley's treatment included 224 species in 62 genera while the numbers had risen to 1203 taxa (sp., subsp., var.) in 99 genera in Masamune's account. It was apparent, even to Ridley, that Borneo possesses one of the world's richest orchid floras in one of the largest remaining areas of tropical rain forest in the Old World. Since Masamune's account many new species have been recorded from Borneo and the identity of some of the species he listed has been reassessed, reducing some into synonymy. For example, Vermeulen (1991) in his recent account of Bornean bulbophyllums described thirty new species and reduced as many names into synonymy. The current account lists over 1400 species in 149 genera. Undoubtedly, there are more species still awaiting discovery in the remaining extensive forests of Borneo: judging by recent work on genera such as *Bulbophyllum*, *Dendrochilum* and *Paphiopedilum*, the number of orchids native to the island may be 20% higher than given here.

Vermeulen's *Bulbophyllum* account is the first volume to be published in a series, *Orchids of Borneo*, that will describe and illustrate all of the native orchids of the island. Therefore, an up-to-date checklist of Bornean orchids is timely. This account is intended to be the basis for the *Orchids of Borneo* series which was initiated by Anthony Lamb, Chan Chew Lun and Phyau Soon Shim in Sabah and is being undertaken with the collaboration of botanists from the Kew, Leiden and Singapore herbaria.

A further imperative for producing this checklist is the rapid deforestation that is currently underway in Borneo. This is not a phenomenon confined to the island but much of Borneo is still unexplored or under-explored biologically and it is certain that plant species are becoming endangered, and even extinct, before they have been described and named. The orchids are the most numerous and widespread family of flowering plants on the island. They also have a complex life-cycle which includes a fungal symbiosis for at least the start of that cycle and a specific pollination relationship to ensure seed production at the other end. When a habitat is disturbed the orchids are often the first plants to suffer and can, therefore, be used as a subtle gauge of the health of the environment. The fate of Borneo's orchids certainly mirrors that of the rest of its native flora and fauna. A good understanding of their diversity and distribution within Borneo will be useful to planners and biodiversity managers in selecting areas for protection and conservation.

Many of the orchids of Borneo are beautiful plants with attractive flowers and have potential in horticulture. In the past orchid collectors have often taken plants indiscriminately from the wild for sale abroad and also to the increasing home market.

Several local initiatives are currently underway to propagate native Bornean orchids in cultivation in order to take the pressure off the wild plants. We heartily condone these.

Orchids can be a powerful conservation tool in pinpointing areas of high biodiversity, in their use for education on conservation issues and as one group that might be used for sustainable exploitation programmes. We hope that this account will provide information that will help to protect some of the region's forests and will ensure that future generations can see these spectacular blooms in the wild.

Jeffrey Wood & Phillip Cribb

INTRODUCTION

HISTORY

The orchids of Borneo are still poorly understood, largely as a result of the sporadic activity of collectors in the island, but also because of the relative inaccessibility of much of the island. The Dutch, in particular, discouraged others from exploring in their part of Borneo. Nevertheless, the richness of Borneo's orchid flora has been apparent for many years and slowly we are gaining a better understanding of the species native to the island and their distribution therein.

The earliest recorded orchid collections in Borneo were made in the southern and western parts of the island by the Dutch but their activities were largely unrecorded at the time. The exploits of collectors in the northern part of the island are far better known, particularly those who explored the land of Rajah Brooke in Sarawak. Their exploits became the subject of fanciful accounts in books such as Ashmore Russan and Frederick Boyle's 'The Orchid Seekers in Borneo' which was serialised in a boy's magazine in late Victorian times. However, many of the collectors, notably Burbidge, Low, Hose and Beccari left their own accounts of their adventures and orchids feature prominently in these. It is remarkable that about nine-tenths of all the herbarium collections we have examined are from Sarawak and Sabah, both Kalimantan and Brunei being grossly under-collected even at the present day.

One of the earliest, and possibly most influential, orchid collectors in Borneo was **Hugh Low** (1824–1905). He was the son of Hugh Low, the founder of one of the best known London nurseries at Clapton that specialised in tropical orchids. Low went out to Sarawak in 1845 where he gained employment in the service of the first Rajah Brooke and became Colonial Treasurer for the Rajah on Labuan Island, just off the coast of Brunei. He is, perhaps, best remembered for his three ascents of Mount Kinabalu, the first in 1851 and the others in April and July of 1858 in the company of Spenser St.John, whose delightful book records both visits. On his first trip he was the first European to reach the summit of Mount Kinabalu, on 11th March 1851.

Low sent many living plants back to his family nursery in England, the most famous, probably, being the green and black-flowered orchid *Coelogyne pandurata*. He is commemorated by several orchids, including the beautiful yellow-flowered *Dendrobium lowii*, the strange *Dimorphorchis lowii*, which has two kinds of flowers in its inflorescence, and the well-known slipper orchid *Paphiopedilum lowii*, all collected by him in Borneo. Most of Low's collections were described by John Lindley (1799–1865), the foremost orchid taxonomist of the day. Lindley was at the time the Assistant Secretary to the Horticultural Society (later Royal Horticultural Society) of London.

Messrs. Veitch and Sons sent out **Thomas Lobb** (1820–1894) to the Far East for the first time in 1843 and he collected orchids for them in Java and the adjacent islands. He visited Borneo again in 1854 and 1856, collecting in Sarawak and Labuan but his attempt to climb Mount Kinabalu on the latter trip was thwarted by the locals at Kiau. He discovered *Bulbophyllum lobbii* in Borneo and it was named after him by John Lindley who described most of the discoveries from Borneo that entered Britain at the time.

The Italian naturalist **Odoardo Beccari** (1843–1920) arrived in Borneo in 1865 and spent nearly three years there exploring and collecting in Sarawak, Labuan and Brunei. He discovered the remarkable *Bulbophyllum beccarii* during his explorations. His book "Wanderings in the great forests of Borneo" (1904) gives a thrilling picture of Bornean forests before the massive destruction of recent years changed the face of the island.

Early collectors in the Dutch part of Borneo include **Peter Korthals** (1807–1892) who collected extensively in the Archipelago for the Dutch Government Commission for Natural Sciences. He visited South-east Borneo with Horner and S.Mueller from May to November 1836, collecting on the Barito River and inland from Banjarmasin and Martapura. He returned to the Netherlands in 1843 for a long retirement. Far sadder was the figure of **James Motley** who first arrived in Borneo in 1851, visiting Labuan, but then spent the period 1854 to 1859 in South-east Borneo, in the vicinity of Banjarmasin, where he was murdered for his troubles!

Joseph Hallier (1868–1932) participated in the Dutch expedition that explored West and Central Borneo from September 1893 to June 1894. This was a major exploration of the great Kapuas River and adjacent areas. Hallier's orchid collecting was not extensive but he did publish *Paphiopedilum amabile*, now considered a synonym of *P.bullenianum*.

Low, Lobb and Beccari's discoveries, in particular, stimulated the orchid nurseries in Britain to send out more collectors to Borneo. **Frederick Burbidge** (1847–1905), a Kew-trained gardener, and Peter Veitch ascended Mount Kinabalu, the highest mountain in Borneo, in 1877 and collected many living orchids there for Messrs. Veitch and Sons of Chelsea.

Frederick Sander, founder of the firm of Messrs. Sander and Sons of St. Albans, dominated the British orchid scene in the last two decades of the last century. His collector, **Ignatz Foerstermann**, visited Sarawak in the spring of 1886 and introduced the extraordinary *Paphiopedilum sanderianum* into cultivation. Its petals reach a metre long. *Coelogyne foerstermannii* is named in his honour.

John Whitehead, an ornithologist, ascended Mount Kinabalu in 1888 and was one of the earliest to reach the summit. He discovered the mountain's most famous orchid *Paphiopedilum rothschildianum* on this trip and introduced it into cultivation for Sander & Sons.

Henry Ridley (1855–1956) published the first list of Bornean orchids in 1896. He described 49 new species and the new genus *Porphyroglottis* in his treatment.

Despite Low's pioneering work on Mount Kinabalu, the plants of the mountain remained largely unknown until the present century. The first of the major collections on the mountain were made between December 1909 and late February 1910 by **Lillian Gibbs** (1870–1925) who explored Kinabalu and made about a thousand collections on the mountain. *Vanda hastifera* var. *gibbsiae* is named in her honour. The explorations of **Joseph Clemens** (1862–1936) and his wife, **Mary Strong Clemens**, are the most extensive and the most important for orchids. They first visited the mountain in October and November 1915, returning in November 1916 and again in November 1917. The results of their orchid collections, and a few more by George Haslam, were described by Oakes Ames and Charles Schweinfurth, in 1920, in "The orchids of Mount Kinabalu". 222 species in 52 genera are included in this account, 109 species and the genus *Nabaluia* being newly described there. Ames (1921) produced a checklist of Bornean orchids, based on published records, in which he listed 699 species in 85 genera. As a result he was moved to write that "The Orchidaceae....surpass numerically all the other families of spermatophytes native to Borneo". The Clemenses returned to Borneo in 1929, spending some months collecting in Sarawak. They returned to Kinabalu in 1931 and spent much of the next two years there collecting all over the mountain. The results of their collecting in this period have only just been fully published by Jeffrey Wood, and Reed and John Beaman (1993) in their account of the orchids of the mountain.

Another botanist who made a major contribution to our knowledge of Kinabalu's orchids was **Cedric Carr** (1892–1936), a New Zealander, who spent several months on the mountain in 1933. He published an account of his orchid collection in 1935 in which he listed 137 species in 40 genera with 39 of the species being newly described.

An updated checklist of Bornean orchids was produced by Masamune in 1942 which recorded 1203 species in 99 genera. Many of the additions to Ames' list of 1921 were the result of the publications of **Johannes Jacobus Smith** (1867–1947), the eminent Dutch botanist who was Director of the Herbarium Bogoriense in Java. Smith received both herbarium material and living plants from collectors in Borneo and from enthusiastic amateur orchid growers in Java. Over a period of nearly half a century he published descriptions of novelties from the Dutch part of Borneo (now Kalimantan). These included many orchids of horticultural merit, notably *Phalaenopsis gigantea, Paraphalaenopsis denevei* and *P.serpentilingua.*

Of the many botanists and collectors who sent orchids to Smith, a few merit special attention.

Anton Nieuwenhuis (1864– ?) made several expeditions to Borneo between 1893 and 1900. On the first of these, he joined **Hallier** (1868–1932) and **Molengraaf** in West and Central Borneo. In 1896 he undertook the first crossing of Borneo in the company of the Javanese collector, **Jaheri** (1857–1926). The expedition left Pontianak on the west coast in February, following the Mendalam River and then the Mahakam River, eventually arriving in Samarinda in May of the following year. His final expedition was another crossing of the island between 1898 and 1900, this time in the company of the Javanese collectors **Amdjah** and **Sakaran**. They departed from Pontianak in May 1898 and went up the Kapuas River, crossing the watershed to the Mahakam River in September. They spent several months in the headwaters of the Kayan and Mahakam Rivers before they finally travelled down the Mahakam and reached Samarinda in June 1899. The Javanese collectors returned to Bogor but Nieuwenhuis set out again for the headwaters of the Mahakam and spent many months exploring Central Borneo before returning to Samarinda in December 1900. His and Amdjah's many orchid discoveries were described by J.J.Smith.

Johannes Teijsmann (1808–1882) made extensive collections of orchids in western Borneo. He arrived in Java in 1830 and from 1831 until 1869 was the Curator of the Buitenzorg (now Bogor) Botanical Gardens. He travelled extensively in the Malay Archipelago visiting Borneo in 1874, arriving in Pontianak in July and exploring the Kapuas River and its tributaries until January 1875. He returned to West Borneo in July of that year and stayed until October.

Hans Winkler (1877–1945) undertook a major exploration of Central and West Borneo with **P.Dakkus** and **Rachmat** from October 1924 until March 1925. From Pontianak they travelled up the Kapuas River and several of its tributaries and then, in January 1925, climbed Bukit Raja. **Frederik Endert** (1891– ?), participated in the Central-East Borneo Expedition of 1925 that explored the Mahakam and other east coast rivers, the Kutai (Koetai) area and the mountains around Gunung Kemal (Kemul).

Smith also described novelties from several collections from Sarawak including those of the Clemenses, **John Hewitt** (1880–1961) and **John Moulton** (1886–1926). Hewitt visited Sarawak in 1905, 1907 and 1908, the last trip with Charles Brooke. He collected around Kuching, on the Limbang and Baram rivers, and visited Gunung Matang. Moulton collected in Sarawak in 1909, 1910, 1914 and 1920, ascending the Limbang, Seridan, Sarawak and Baram rivers and visiting Mount Batu Lawi and the Kelabit country on his travels. He also spent several weeks on Kinabalu in 1913.

Since the end of the Second World War, the collectors and botanists of the Herbaria in Kuching, Sandakan and Bogor together with a number of visiting botanists have collected orchids on their travels but many herbarium specimens have remained unnamed because of the lack of active taxonomic interest in Bornean orchids. The publication, by **Jeffrey Wood** (1984), of an account of the orchids of the Gunung Mulu National Park, resulting from the Royal Geographical Society's Expedition there in the 1970s, is a notable exception. The work of **Anthony Lamb**, **Chan Chew Lun** and **Phyau Soon Shim** in Sabah has stimulated greater interest in recent years. In collaboration with orchid taxonomists at Kew, Leiden, Singapore and elsewhere they have begun the first modern treatment of Bornean orchids, a series of volumes entitled *Orchids of Borneo*, in each of which, 100 species will be described in full and illustrated by line drawings and colour photographs. This has itself stimulated the production of this checklist and of the first detailed treatment of the orchids of Mount Kinabalu by Jeffrey Wood with Reed and John Beaman (1993). In that work they have accounted for 703 species in 121 genera that occur on the mountain which has, fortunately, been designated a National Park. There is still a great deal to be learned about the orchids of this large island and we offer this contribution as the starting point for what we hope will be a renewal of interest in the subject.

BORNEO — AN ORCHID-RICH ISLAND

Borneo is the third largest island in the world and the largest in the Malay Archipelago which stretches from Sumatra in the west to New Guinea in the east. The island is shaped like a lop-sided and angular pear with a dog's head in the north-east. It is 1300 kilometres long and 950 kilometres wide and about 740,000 square kilometres in extent. The island straddles the Equator and has a tropical climate except on its highest mountains.

The island is politically shared by three countries. The Malaysian states of Sabah and Sarawak lie in the north, the latter enclosing the small country of Brunei Darussalam that lies about two-thirds of the way along the north coast. The rest of the island, nearly three-quarters of the land surface, is Indonesian Kalimantan, which is itself divided into four large provinces: South, East, West and Central Kalimantan.

Geologically, the island of Borneo is part of the South-east Asian block and, in periods when the sea-level fell during the Ice Ages, was joined to the Malay Peninsula, Java and Sumatra but separated from the islands further east. Its flora and fauna reflect these links. A more detailed account of the geological history, present-day geology and soils of Borneo is given in the first volume of *Orchids of Borneo* by Chan et al.(in press). Ancient volcanic activity, tectonic movements and climatic changes have given Borneo a complex geology and wide range of vegetation types.

The major geological feature of the island is the complex of mountain chains that runs from the north-east corner diagonally across to the west coast. A rather isolated mountain block, the Meratus mountains, can be found in the south-east corner. For much of its length the main chain or Crocker Range runs parallel to the west coast, with the ranges at right angles to it running down, in places, almost to the east coast. This chain contains many peaks above 1500 m but is dominated by Mount Kinabalu, a massif granite outcrop rising to 4101 m, that can be seen from Kota Kinabalu on the west coast of Sabah. Kinabalu's rich variety of soils, climate and habitats has produced a commensurate wealth of orchids. Wood et al. (1993) list over 700 species in their recent account. These include some of the finest horticultural species such as *Paphiopedilum rothschildianum* and *Renanthera bella*.

CLIMATE

Situated on the Equator, Borneo's climate is essentially tropical with relatively uniform high temperatures and rainfall throughout the year, increasing during the two monsoon periods from November to January and from May to June. A marked seasonality, with significant dry periods extending to several weeks or more, has been detected in recent years, particularly in northern and eastern Borneo as a result of heavy logging and the El Niño effect. Most areas receive between 2000 mm and 4000 mm of rainfall a year but the drier areas of the east coast and the inland valleys of Sabah can receive as little as 1200 mm a year. The logging activities have had a serious effect on the lowland forests, with the hill and lowland dipterocarp forests being particularly badly damaged. Shifting agriculture and burning after logging have a devastating effect on the vegetation and prevent regeneration. Orchids do not survive such treatment.

The average temperatures fall with increasing altitude in the mountains and, on Mount Kinabalu, frosts are experienced at night in the summit area at certain times of the year. Clear vegetational zones can be seen with increasing altitude on Mount Kinabalu. On mountains closer to the sea, these zones occur at lower altitudes because of the cooling effect of the sea.

MAJOR VEGETATIONAL TYPES

Orchids are found in nearly every habitat in Borneo from the coastal mangrove swamps almost to the summit area of Mount Kinabalu. However, the distribution and number of species varies remarkably with habitat. For example, the mangrove forest, some forests on peat and the subalpine scrubby forest on Kinabalu are poor in species' diversity although they can be rich in numbers of plants. In contrast, some of the forests on ultramafic soils are exceedingly rich both in orchid species and in numbers of plants. Some of the more interesting habitats will be discussed here, but a more detailed treatment is given by Chan et al. (1994).

Mangrove and strand vegetation

Mangrove species form extensive forests on the coasts and deltas of Borneo's larger rivers and can extend many dozens of kilometres inland where the water is still brackish. Orchids are rare in this forest but more commom in the Nipa palm forest along the lower reaches of many of the rivers. Species of *Bulbophyllum*, *Eria javanica* and *Coelogyne rochussenii* are all found here.

Riverine and freshwater swamp forests

These forests can be extensive and contain large trees up to 30 m or more tall. The riverine forest is often a narrow belt nowadays in the more populated areas but the large trees can be rich in epiphytic orchids such as *Grammatophyllum speciosum*, *Paphiopedilum lowii* and numerous *Bulbophyllum* and *Coelogyne* species. In the upper canopy of the swamp forest, many *Bulbophyllum*, *Coelogyne*, *Dendrobium*, *Eria*, *Thrixspermum* and *Vanda* species can be found. In the open swampy areas the showy *Papilionanthe hookeriana* and *Thrixspermum amplexicaule* can be seen scrambling up the sedges for support.

Lowland evergreen forest

These forests are declining rapidly in Borneo because of logging. They are characterised by trees of great height and size, up to 30 m tall with massive straight boles and their lowermost branches 20 to 30 m above the ground. The best known trees of this type of forest are the dipterocarps, the conifer *Agathis*, and *Koompasia*, noted as the host for the native honey bees' pendent curving combs. The huge litter-gathering leopard orchid, *Grammatophyllum speciosum*, grows on the forks of the large emergents of the forest. Other common orchids here are *Bulbophyllum vaginatum*, *Coelogyne foerstermannii*, *Epigeneium treacherianum*, *Phalaenopsis cornucervi* and *P.pantherina*. In the hill dipterocarp forest, often along streams, from 600 m up to about 1000 m, may be found two of the showier moth orchids, *Phalaenopsis amabilis* and the rare *P.gigantea* and the strange vandaceous orchid *Dimorphorchis*. *Cymbidium*, *Dendrobium*, *Eria* and *Trichoglottis* species are also common.

Vanilla kinabaluensis can climb 30 m or more up trees in this forest. Other terrestrials, including *Calanthe triplicata*, *Claderia viridiflora* and *Nephelaphyllum pulchrum* are found here, while saprophytic orchids are not uncommon. In the secondary forests that develop after logging and felling for shifting cultivation, terrestrials such as *Arundina graminifolia*, *Bromheadia finlaysoniana*, *Corymborkis veratrifolia* and *Phaius tankervilleae* are in their element and can be common in places.

Arundina and *Phaius* are often collected by the local people and planted around their houses. The beautiful white-flowered *Phalaenopsis amabilis* suffers the same fate which has lead to its local extinction. Epiphytes that have a similar habitat preference include the fragrant *Aerides odorata*, *Cymbidium bicolor* subsp. *pubescens* and *C.finlaysonianum* and the Pigeon orchid, *Dendrobium crumenatum.*

Peat-swamp forest

Over a tenth of Borneo is covered in peat-swamp forest, although large areas in eastern Kalimantan were destroyed by fire during the long droughts of 1982 and 1983 and much has been logged. The rivers and streams of the peat-swamps have a characteristic dark tea-colour caused by the high levels of tannins in the water. Epiphytic orchids found here include *Acriopsis*, *Cymbidium* and *Grammatophyllum* species. In the drier marginal areas *Bromheadia*, *Claderia viridiflora*, *Malaxis* and *Plocoglottis* are common terrestrials.

Heath forests

Two types of heath forest, called respectively "kerangas" and "kerapah", are found in Borneo. These stunted open forests grow on infertile white sandy soils, the "kerangas" on the upper, drier slopes and the "kerapah" on the lower, poorly drained ones. Heath forest is found at all altitudes from the coast inland. Orchids grow here epiphytically but often near the ground, along with ant-plants, hoyas and dischidias. The heath forests can be rich in orchids, especially bulbophyllums, dendrobiums, dendrochilums and erias. They also boast a number of interesting rarities and endemics such as *Bulbophyllum beccarii*, *Cymbidium rectum* and the green-flowered *Coelogyne pandurata*, *C.peltastes* and *C.zurowetzii*. Terrestrials are also common here with *Bromheadia* and *Calanthe* species locally in abundance.

Forests on ultramafic substrates

Some of the most rewarding orchid areas are the forests and scrub found on ultramafic (also referred to as ultrabasic and serpentine) rocks and soils. These harbour a rich variety of orchids and many endemics, including some of Borneo's most beautiful species. The dominant trees in these areas are often gymnostomas (casuarinas) which fix nitrogen in these nutrient-poor soils. In more open areas *Leptospermum* bushes are common. Showy epiphytes that grow in ultramafic forest include *Arachnis longisepala*, *Dendrobium parthenium*, *D.sculptum*, *D.spectatissimum*, *Paraphalaenopsis labukensis* and *Renanthera bella*. It is equally rich in terrestrials with *Corybas serpentinus*, *Paphiopedilum dayanum*, *P.hookerae var.volonteanum*, *P.rothschildianum* and *Phaius reflexipetalus* all found here.

Forest on limestone

Forest that forms on limestone can be as species-rich as those on ultramafic rocks. The raised limestone hills of southern Sarawak have good examples of this type of vegetation. The forest is usually stunted and spare on the steep slopes. Epiphytic species such as *Phalaenopsis maculata*, *P.modesta* and *P.violacea* can be found here although all are severely endangered by over-collecting for horticulture. The limestone also supports some interesting terrestrials such as *Habenaria marmorophila* and the endemic slipper orchids, *Paphiopedilum sanderianum* and *P.stonei*. These last two are also the prey of collectors.

Tropical montane forest

In general, montane forest can be found in the mountains of Borneo from about 1200 to 2500 m but perhaps even lower down on mountains near the coast. The montane forests lack the dominant dipterocarps of the lowland and hill forests and these are replaced by oaks and chestnuts in the genera *Castanopsis, Lithocarpus, Quercus* and *Trigonobalanus*. Conifers such as *Agathis, Dacrycarpus, Dacrydium* and *Podocarpus* can be dominant in this forest. Epiphytic orchids are common in these forests and, on Kinabalu, Wood et al.(1993) demonstrate that these are the richest areas for orchids on the mountain. Epiphytic *Bulbophyllum, Coelogyne, Dendrobium* and *Eria* are all well represented here, while the terrestrial genera, *Calanthe, Corybas, Malaxis, Phaius* and various 'jewel orchids', such as *Anoectochilus* can be common. With increasing altitude, the number of species drops dramatically above 1600 m but orchids are still common and, in places, dominant. Thus, on Kinabalu, the terrestrial *Eria grandis* can form dense stands under *Dacrydium* and *Leptospermum* above 2600 m and several species of *Dendrochilum* are prolific.

Tropical sub-alpine and alpine vegetation

These are rare in Borneo and are more or less confined to above 3000 m on Mount Kinabalu. Orchids are common in the sub-alpine vegetation but species that are normally epiphytic become terrestrial or grow on the lower parts of the trunks of shrubs and dwarf trees. Numerous *Bulbophyllum, Coelogyne papillosa,* various *Dendrochilum, Eria grandis* and *Platanthera stapfii* are all characteristic of this zone.

No orchids have been recorded from the summit of Mount Kinabalu but *Coelogyne papillosa, Dendrochilum stachyodes* and *Eria grandis* frequent the cracks in the granite just below the summit dome.

Classification System of Bornean Orchids

(updated from Dressler, 1981 & 1990).

Subfamily Apostasioideae
Apostasia, Neuwiedia

Subfamily Cypripedioideae
Paphiopedilum

Subfamily Spiranthoideae

    Tribe Tropidieae
        Corymborkis, Tropidia

    Tribe Cranichideae
        Subtribe Goodyerinae
            Anoectochilus, Cheirostylis, Cystorchis, Dossinia,
            Erythrodes, Goodyera, Hetaeria, Hylophila,
            Kuhlhasseltia, Lepidogyne, Macodes, Myrmechis,
            Pristiglottis, Vrydagzynea, Zeuxine
        Subtribe Spiranthinae
            Spiranthes
        Subtribe Cryptostylidinae
            Cryptostylis

Subfamily Orchidoideae

    Tribe Diurideae
        Subtribe Acianthinae
            Corybas, Pantlingia

    Tribe Orchideae
        Subtribe Orchidinae
            Platanthera
        Subtribe Habenariinae
            Habenaria, Peristylus

Subfamily Epidendroideae

    Tribe Gastrodieae
        Subtribe Gastrodiinae
            Didymoplexiella, Didymoplexis,
            Gastrodia, Neoclemensia

Subtribe Epipogiinae
Epipogium, Stereosandra

Tribe Neottieae
Subtribe Limodorinae
Aphyllorchis

Tribe Nervilieae
Nervilia

Tribe Vanilleae
Subtribe Galeolinae
Cyrtosia, Erythrorchis, Galeola
Subtribe Vanillinae
Vanilla
Subtribe Lecanorchidinae
Lecanorchis

Tribe Malaxideae
Hippeophyllum, Liparis, Malaxis, Oberonia

Tribe Cymbidieae
Subtribe Bromheadiinae
Bromheadia
Subtribe Eulophiinae
Dipodium, Eulophia, Geodorum, Oeceoclades
Subtribe Thecostelinae
Thecopus, Thecostele
Subtribe Cyrtopodiinae
Chrysoglossum, Claderia, Collabium, Cymbidium, Grammatophyllum, Pilophyllum, Porphyroglottis
Subtribe Acriopsidinae
Acriopsis

Tribe Arethuseae
Subtribe Bletiinae
Acanthephippium, Ania, Calanthe, Mischobulbum, Nephelaphyllum, Pachystoma, Phaius, Plocoglottis, Spathoglottis, Tainia
Subtribe Arundinae
Arundina, Dilochia

Tribe Glomereae
Subtribe Glomerinae
Agrostophyllum
Subtribe Polystachyinae
Polystachya

Tribe Coelogyneae

    Subtribe Coelogyninae

        Chelonistele, Coelogyne, Dendrochilum, Entomophobia, Geesinkorchis, Nabaluia, Pholidota

Tribe Podochileae

    Subtribe Eriinae

        Ascidiera, Ceratostylis, Eria, Porpax, Sarcostoma, Trichotosia

    Subtribe Podochilinae

        Appendicula, Poaephyllum, Podochilus

    Subtribe Thelasiinae

        Octarrhena, Phreatia, Thelasis

Tribe Dendrobieae

    Subtribe Dendrobiinae

        Dendrobium, Diplocaulobium, Epigeneium, Flickingeria

    Subtribe Bulbophyllinae

        Bulbophyllum, Trias

Tribe Vandeae

    Subtribe Aeridinae

        Group 1

            Adenoncos, Doritis, Microsaccus, Taeniophyllum

        Group 2

            Abdominea, Arachnis, Bogoria, Ceratochilus, Cleisocentron, Cleisomeria, Cleisostoma, Cordiglottis, Dimorphorchis, Gastrochilus, Kingidium, Micropera, Ornithochilus, Pomatocalpa, Renanthera, Renantherella, Sarcoglyphis, Schoenorchis, Smitinandia, Staurochilus, Thrixspermum, Trichoglottis

        Group 3

            Aerides, Ascochilopsis, Brachypeza, Macropodanthus, Papilionanthe, Paraphalaenopsis, Phalaenopsis, Porphyrodesme, Pteroceras, Rhynchostylis, Robiquetia, Vanda

        Group 4

            Ascocentrum, Biermannia, Dyakia, Gastrochilus, Luisia

        Group 5

            Chamaeanthus, Chroniochilus, Grosourdya, Malleola, Microtatorchis, Pennilabium, Porrorhachis, Spongiola, Tuberolabium

KEY TO GENERA

(EXCLUDING SAPROPHYTES AND TRIBE VANDEAE, SUBTRIBE AERIDINAE)

1. Flowers with 2 or 3 fertile anthers...2

   Flowers with a single fertile anther 4

2. Perianth segments similar, the lip never deeply saccate 3

   Perianth segments very unequal, the lip deeply saccate, slipper- or pouch-shaped. Anthers 2, lateral.
   Staminode median, large and shield-shaped ....**Paphiopedilum**

3. Anthers 2, with or without a staminode. Inflorescence usually branched, curved and spreading, never erect..**Apostasia**

   Anthers 3, staminode absent. Inflorescence simple, erect. ....**Neuwiedia**

4. Anther erect or bending back, never short and operculate at apex of column. Leaves usually spirally arranged, convolute, not articulated at base 5

   Anther eventually bending downward over column apex to become operculate, or operculate at column apex but not bending downward. Leaves distichous, usually articulate at base 33

5. Plants exclusively terrestrial 31

6. Rostellum elongate, equalling the anther.
   Root-stem tuberoids (tubers) absent 7

   Rostellum usually shorter than the anther.
   Root-stem tuberoids (tubers) present or absent...25

7. Stems tough and rigid. Leaves plicate 8

   Stems weak and fleshy, often brittle, never tough and rigid.
   Leaves convolute or conduplicate 9

8. Lip widest at apex. Column long. Inflorescence often branched.....**Corymborkis**

   Lip widest at base. Column short. Inflorescence simple ....**Tropidia**

9. Pollinia sectile. Roots scattered along rhizome 10

   Pollinia not sectile. Roots in a close fascicle 24

10. Flowers resupinate 11

    Flowers non-resupinate 23

11. Spur or saccate base of lip containing neither glands nor hairs (hairs may occur near mid-lobe only) 12

Lip hairy within or having papillae or glands on either side near base or in spur or sac 13

12. Lip with spur which projects between lateral sepals ....**Erythrodes**

Lip saccate, entirely enclosed by lateral sepals ....**Hylophila**

13. Lip hairy within ....**Goodyera**

Lip otherwise 14

14. Apex of lip not abruptly widened into a distinct spathulate or transverse, bilobed blade...15

Apex of lip abruptly widened into a distinct spathulate or transverse, bilobed blade 17

15. Saccate base of lip with a transverse row of small calli.
Plants robust, up to 100 cm tall ....**Lepidogyne**

Saccate base of lip or spur containing stalked or sessile glands.
Plants much smaller..16

16. Hypochile swollen at base into twin lateral sacs each containing a sessile gland. Epichile with fleshy involute margins, forming a tube...........................**Cystorchis**

Hypochile otherwise, containing 2 stalked glands.
Epichile otherwise..**Vrydagzynea**

17. Claw of lip with a toothed or pectinate flange on either side ....**Anoectochilus**

Claw of lip otherwise..18

18. Leaves dark green with greenish-yellow, golden or pink median and secondary nerves...**Dossinia**

Leaves without such coloured secondary nerves, although median nerve sometimes coloured..19

19. Sepals connate for half their length to form a swollen tube ....**Cheirostylis**

Dorsal sepal and petals connivent, forming a hood, or free..............................20

20. Dorsal sepal and petals connivent, forming a hood 21

Dorsal sepal and petals free 22

21. Column without appendages ....**Kuhlhasseltia**

Column with 2 narrow wings which project into the base of the lip ...**Pristiglottis**

22. Lip with a long claw. Stigmas on short processes.
Inflorescence 1–2-flowered ....**Myrmechis**

Lip with a short claw. Stigmas sessile.
Inflorescence several-flowered ....**Zeuxine**

23. Lip and column twisted to one side..**Macodes**

Lip and column straight ....**Hetaeria**

24. Flowers non-resupinate, small, arranged spirally in a dense inflorescence
....**Spiranthes**

Flowers resupinate, large, arranged in all directions in a lax inflorescence
....**Cryptostylis**

25. Root-stem tuberoids (tubers) present 26

Root-stem tuberoids (tubers) absent..31

26. Lip without a spur...27

Lip spurred...28

27. Inflorescence produced with the leaf. Lip orbicular.
Column with a tooth-like process below...**Pantlingia**

Inflorescence produced before the leaf. Lip 3-lobed, the base embracing the
column.
Column without a tooth-like process ....**Nervilia**

28. Lip 2-spurred, tubular below. Flowers helmet-shaped...............................**Corybas**

Lip with 1 spur, not tubular below. Flowers otherwise 29

29. Stigmas each on a sigmatophore extending from the column, free from
hypochile ....**Habenaria**

Stigmas not freely extending in front of column, sometimes connate or
adpressed to lip hypochile..30

30. Lip simple, strap-shaped. Spur cylindric, rather long, not swollen at apex.
Stigmas joined to form a concave structure, free from hypochile ......**Platanthera**

Lip 3-lobed. Spur short, usually globular, saccate or fusiform.
Stigmas convex, cushion-like, connate with or adpressed to hypochile.**Peristylus**

31. Leaves fleshy, never plicate. Stems fleshy.
Pollinia soft and mealy, as monads ....**Vanilla**

Leaves plicate. Stem never fleshy, often rather tough, sometimes brittle.
Pollinia 2, cleft 32

32. Habit monopodial. Stems not distant on a creeping rhizome.
Leaves distichous, imbricate, ensiform.
Flowers pale yellowish with pink to crimson blotches...........................**Dipodium**

Habit sympodial. Stems placed distantly on a creeping rhizome.
Leaves elliptic, neither distichous or imbricate.
Flowers green ....**Claderia**

33. Plants terrestrial 34

    Plants epiphytic or lithophytic 61

34. Pollinia 2 or 4, naked, i.e. without caudicles; viscidia and stipes usually absent
   35

    Pollinia 2 to 8, with caudicles (sometimes reduced), or a stipes 38

35. Column-foot absent 36

    Column-foot prominent 37

36. Column long. Flowers usually resupinate.
    Lip without 2 large basal auricles, apex rarely pectinate...........**Liparis**
    (in part)
    Column short. Flowers non-resupinate.
    Lip with 2 large basal auricles, apex often pectinate ....**Malaxis**

37. Rhizomatous part of shoot (sometimes also the non- rhizomatous part) carrying
    one-noded pseudobulbs...**Epigeneium**
    (in part, sometimes *E.kinabaluense*)

    Non-rhizomatous part of shoot consisting of several internodes, wholly or partly
    fleshy, with or without pseudobulbs......................................**Dendrobium**
    (in part, some species in sections *Conostalix* & *Distichophyllum*)

38. Inflorescences numerous or not, borne along a slender leafy stem, lateral or
    terminal 39

    Inflorescences never numerous, usually solitary, never borne along a slender,
    leafy stem, usually lateral, sometimes axillary or terminal 44

39. Inflorescences lateral 40

    Inflorescences terminal 42

40. Pollinia 2. Inflorescences not numerous.
    Flowers c.4 cm across ....**Cymbidium**
    (in part, *C.elongatum*)
    Pollinia 6 or 8. Inflorescences numerous.
    Flowers much smaller 41

41. Pollinia 6. Leaf sheaths and flowers glabrous ....**Appendicula**
    (in part)
    Pollinia 8. Leaf sheaths and flowers usually covered in reddish brown hispid
    hairs ....**Trichotosia**
    (in part)

42. Pollinia 2 ....**Bromheadia**
    (in part, *B.borneensis*, *B.crassiflora*, *B.finlaysoniana*)

Pollinia 8..43

43. Flowers large, up to 8 cm across (sometimes peloric).
     Petals much broader than sepals. Inflorescence usually unbranched.
     Floral bracts small, acute, persistent ....**Arundina**

     Flowers much smaller. Petals similar to sepals. Inflorescence branching.
     Floral bracts conspicuous, concave, deciduous ....**Dilochia**

44. Pollinia 2..45

     Pollinia 4 or 8 47

45. Plants densely covered in yellowish brown hairs.
     Flowers non-resupinate ....**Pilophyllum**

     Plants glabrous. Flowers resupinate..46

46. Lip mobile. Column with 2 fleshy basal keels.
     Spur formed by column foot ....**Chrysoglossum**

     Lip immobile. Column without basal keels. Mentum formed by column foot,
     base of lateral sepals and base of lip ....**Collabium**

47. Pollinia 4..48

     Pollinia 8..53

48. Inflorescence arcuate, strongly decurved ....**Geodorum**

     Inflorescence otherwise..49

49. Lip not spurred 50

     Lip spurred...52

50. Lip divided into a distinct, somewhat saccate hypochile and a 2-lobed epichile.
     Pseudobulbs flattened, always 2-leaved..............................**Geesinkorchis**

     Lip otherwise. Pseudobulbs never flattened, sometimes elongated into a leafy
     stem, 1 – many-leaved 51

51. Lip convex, adnate to sides and apex of column foot to form a sac, usually with
     an elastic hinge that springs when touched.
     Pseudobulbs often elongated into a leafy stem ....**Plocoglottis**

     Lip never convex, free or fused at base to base of column, without an elastic
     hinge.
     Pseudobulbs short, often enclosed in sheathing leaf-bases ....**Cymbidium**
                    (in part, *C.borneense, C.ensifolium subsp. haematodes, C.lancifolium*)

52. Lip entire, or 3-lobed (mid-lobe not bilobulate) ....**Eulophia**
                                                     (in part, *E.graminea, E.spectabilis*)

     Lip '4-lobed', mid-lobe bilobulate...**Oeceoclades**

16

53. Pseudobulbs absent, replaced by a fleshy subterranean rhizome, swelling into a horizontal fusiform, sometimes v-shaped, tuber.
Leaves usually several, often withering before inflorescence appears.
Lip with a small basal pouch ....**Pachystoma**

Plants pseudobulbous...54

54. Lip spurred, or gibbous and partially adnate to and embracing column to form a tube 55

Lip not spurred 58

55. Pseudobulbs always unifoliate.
Plants remaining green when damaged..56

Pseudobulbs 2 to several-leaved.
Plants turning bluish-black when damaged. Lip spurred or gibbous.................57

56. Inflorescence lateral...**Ania**

Inflorescence terminal...**Nephelaphyllum**

57. Column margins fused with the base of the lip over nearly their entire length.
Lip spurred..**Calanthe**

Column margins fused with lip only at or near the base.
Lip shortly spurred or gibbous ....**Phaius**

58. Pseudobulbs unifoliate 59

Pseudobulbs with 2 or several leaves 60

59. Leaf base cordate in mature plants, petiole absent ....**Mischobulbum**

Leaf base ± decurrent along a petiole...**Tainia**

60. Flowers urn-shaped, sepals fleshy, fused to form a swollen tube, free at the apices.
Lip movably hinged to a column-foot, not clawed or callose....**Acanthephippium**

Flowers with free, usually spreading sepals.
Lip not movably hinged, mid-lobe very narrowly clawed, with 2 ovoid, often pubescent basal calli. Column-foot absent ....**Spathoglottis**

61. Pollinia 2 or 4, naked, i.e. without caudicles.......................................62

Pollinia 2 to 8, with distinct, though sometimes reduced, caudicles.................69

62. Column-foot absent. Leaves equitant, distichous, bilaterally flattened..............63

Column-foot prominent.
Leaves dorsiventral, or occasionally bilaterally flattened (in Dendrobium sections *Aporum*, *Oxystophyllum* and *Strongyle* only) 64

63. Groups of leaves close together. Column short.....................................**Oberonia**

Groups of leaves 4 cm apart. Column long....................................**Hippeophyllum**

64. Lip usually immobile, not hinged at base. Mentum often spur- like..................65

Lip movably hinged to column-foot. Mentum saccate 68

65. Rhizomatous part of shoot (sometimes also the non-rhizomatous part) bearing one-noded, 1- or 2, rarely 3-leaved pseudobulbs 66

Non-rhizomatous part of shoot (when present) consisting of several internodes, with or without 1- to several-noded pseudobulbs 67

66. Erect parts of shoot closely set, tufted, 15–25 cm high, consisting of a single internode tapering from a fleshy base into a slender neck, with 1 apical leaf and 1 – 2-flowered successive inflorescences. Flowers on very long pedicels, ephemeral. Sepals and petals narrowly caudate ....**Diplocaulobium**

Erect parts of shoot spreading or suberect, never tufted, consisting of several internodes bearing one-noded, 1, 2, or rarely 3-leaved pseudobulbs. Inflorescence 1- to several flowered. Flowers on shorter pedicels, long lived. Sepals and petals narrowly elliptic..**Epigeneium**

67. Stems superposed, the non-rhizomatous part of the shoot consisting of several quite long thin internodes, the uppermost pseudobulbous and unifoliate Flowers always ephemeral ....**Flickingeria**

Stems not superposed; either 1) rhizomatous, 2) erect and many-noded, 3) erect and 1-noded or several-noded from a many-noded rhizome, or 4) rhizome absent, new stems of many nodes arising from base of old ones. Leaves 1 to many. Flowers long-lived or ephemeral..........................**Dendrobium**
(in part)

68. Anther cap with a prolongation in front, of varying shape (deeply lacerate in *T.tothastes*). Column usually with insignificant stelidia ....**Trias**

Anther cap otherwise. Column stelidia usually prominent.............**Bulbophyllum**

69. Stems slender, leafy, without pseudobulbs...70

Stems pseudobulbous, pseudobulbs sometimes small and entirely enclosed by imbricate leaf sheaths 83

70. Pollinia 2 or 4...71

Pollinia 6 or 8 73

71. Pollinia 2. Leaves sometimes laterally flattened ....**Bromheadia**
(in part)

Pollinia 4. Leaves never laterally flattened 72

72. Stems slender, often branched, with many close, distichous leaves up to 1.2 cm long ....**Podochilus**

Stems very short, tufted, with 1 – 2 linear leaves 6 – 12 cm long.........**Sarcostoma**

73. Pollinia 6..**Appendicula**

    Pollinia 8...74

74. Inflorescence terminal, usually globose, surrounded by bracts.
    Flowers white or yellow ....**Agrostophyllum**

    Inflorescence lateral, terminal or subterminal, never of globose heads.
    Flowers variously coloured 75

75. Column-foot absent 76

    Column with a short or long foot 78

76. Leaves laterally compressed or terete, distichous.
    Flowers yellowish green ....**Octarrhena**

    Leaves dorsiventral, linear to linear-elliptic or strap-shaped 77

77. Inflorescence and sepals white-tomentose.
    Flowers arranged in whorls, non-resupinate ....**Ascidiera**

    Inflorescence and sepals glabrous.
    Flowers not arranged in whorls, resupinate ....**Thelasis**
    (in part, e.g. *T.carinata, T.micrantha*)

78. Leaf sheaths covered with reddish brown or rarely white, hispid hairs.
    Leaves never fleshy and subterete. ....**Trichotosia**

    Leaf sheaths glabrous. Leaves sometimes fleshy and subterete 79

79. Stems one-leaved ....**Ceratostylis**

    Stems few- to many-leaved 80

80. Stems short, entirely enclosed by imbricate leaf sheaths.
    Inflorescence a densely flowered raceme with small bracts....................**Phreatia**
    (in part, e.g. *P.amesii, P.densiflora, P.monticola, P.secunda*)

    Stems elongate, leafy throughout entire length 81

81. Inflorescence terminal or subterminal, usually densely many-flowered, densely
    hirsute. Floral bracts small...**Eria**
    (section *Mycaranthes*)

    Inflorescence axillary, few-flowered, glabrous 82

82. Floral bracts large and brightly coloured ....**Eria**
    (section *Cylindrolobus*)

    Floral bracts minute, green or brownish ....**Poaephyllum**

83. Pollinia 2..84

    Pollinia 4 or 8 87

84. Lip joined at its base with an outgrowth from the column and with column-foot
     to form a tube at right angles to base of column ....**Thecostele**

     Lip otherwise...85

85. Flowers non-resupinate. Lip convex when viewed from above, scoop-shaped
     when viewed from below, hairy, bee-like.
     Column with large, curved stelidia.
     Habit similar to *Grammatophyllum*..**Porphyroglottis**

     Flowers resupinate. Lip otherwise.
     Column lacking stelidia...86

86. Plants very large, with pseudobulbs up to 3 m or more long.
     Flowers up to 10 cm across. Sepals and petals up to 2.6 cm wide, with large
     irregular blotches. Stipes U-shaped ....**Grammatophyllum**

     Plants much smaller. Flowers up to 5.7 cm across.
     Sepals and petals narrow, without blotching. Stipes absent ....**Cymbidium**
                                        (all epiphytic species except *C. lancifolium*)

87. Pollinia 4...88

     Pollinia 8...98

88. Inflorescence terminal 89

     Inflorescence lateral 96

89. Flowers with a distinct column-foot, always non-resupinate ....**Polystachya**

     Flowers without a column-foot, resupinate, or, more rarely, non-resupinate....90

90. Pollinia attached to a stipes...**Geesinkorchis**
                                                                              (in part)

     Stipes absent..91

91. Basal half of the narrow, saccate lip adnate to basal half of column.
     Apical half of lip separated by a transverse, high, fleshy callus......**Entomophobia**

     Lip otherwise..92

92. Lip hypochile with long, slender lateral front lobes................................**Nabaluia**

     Lip hypochile without such lobes 93

93. Column usually with lateral arms (stelidia) ....**Dendrochilum**

     Column without lateral arms 94

94. Lip hypochile saccate, distinctly separate from epichile.
     Lip rarely 3-lobed ....**Pholidota**

Lip hypochile, although often concave, not sharply distinct from epichile.
Lip almost always 3-lobed 95

95. Side-lobes of lip (when present) narrow, borne from front part of hypochile at
right angles to the epichile.
Hypochile narrow, saccate...**Chelonistele**

Side-lobes of lip broad, widening gradually from base of lip.
Hypochile ± concave, broader and rarely saccate ....**Coelogyne**

96. Lip joined at its base with an outgrowth from the column and with column-foot
to form a tube at right angles to base of column ....**Thecopus**

Lip otherwise 97

97. Lateral sepals united into a synsepalum.
Stipes long, linear ....**Acriopsis**

Lateral sepals free. Stipes absent...**Cymbidium**
(in part, *C.lancifolium* only)
98. Sepals connate to varying degrees, forming a tube.
Pseudobulbs flattened ....**Porpax**

Sepals free 99

99. Column with a prominent foot. Rachis usually hirsute or woolly.
Pseudobulbs rarely flattened ....**Eria**
(in part)

Column absent or short. Rachis glabrous.
Pseudobulbs sometimes flattened..100

100. Column with a short foot. Anther cap horizontal on top of column, not beaked
....**Phreatia**
(in part, e.g. *P.listrophora, P.sulcata*)

Column-foot absent. Anther cap vertical behind column, beaked...........**Thelasis**
(in part, e.g. *T.capitata, T.carnosa, T.variabilis*)

KEY TO SAPROPHYTIC GENERA
(LEAFLESS TERRESTRIALS LACKING CHLOROPHYLL)

1. Flowers with sepals and petals fused (connate) to a varying degree, often appearing campanulate and always resupinate..2

    Flowers with free, spreading or connivent sepals and petals, not appearing campanulate, or lateral sepals connate; resupinate or non-resupinate 5

2. Petals fimbriate, bright orange...**Neoclemensia**

    Petals otherwise...3

3. Column with long decurved arms (stelidia), foot absent ....**Didymoplexiella**

    Column without long decurved arms, with a short foot......................................4

4. Pollinia 4. Dorsal sepal and petals adnate to form a single trifid segment forming a shallow cup or tube with the partially connate lateral sepals. Stigma near column apex ....**Didymoplexis**

    Pollinia 2. Sepals and petals connate to form a 5-lobed tube which is sometimes gibbous at the base, and which may be split between the lateral sepals. Stigma at base of column ....**Gastrodia**

5. Flowers always resupinate, lip lowermost 6

    Flowers non-resupinate or resupinate 11

6. Stem branching, tough and wiry. Sepals and petals surrounded by a shallow denticulate calyculus(cup ....**Lecanorchis**

    Stem simple, slender or fleshy. Sepals and petals not encircled by a shallow denticulate calyculus (cup.............7

7. Lip divided into a distinct hypochile and epichile. Hypochile with or without twin lateral sacs..8

    Lip not divided into a distinct hypochile and epichile...................................9

8. Hypochile swollen at base into twin lateral sacs, each containing a globular sessile gland. Epichile with fleshy involute margins, tube-like. Flowers pink to reddish, tipped with white...**Cystorchis**

    (*C.aphylla, C.salmoneus, C.saprophytica*)

    Hypochile without such sacs. Epichile 3-lobed. Flowers greenish white or creamy white and purple ....**Aphyllorchis**

9.   Flowers large, reddish brown. Lip 3-lobed, saccate ....**Eulophia**
     (*E. zollingeri* only)
     Flowers small, white, or white flushed with purple at apex.
     Lip entire, with or without a spur 10

10. Lip spurred, strap-shaped, margin not undulate................................**Platanthera**
     (*P.saprophytica* only)

     Lip not spurred, narrowly elliptic, margin undulate ....**Stereosandra**

11. Stem simple. Flowers non-resupinate. Lip with a short spur ....**Epipogium**

     Stem branching. Flowers resupinate or non-resupinate.
     Spur absent...12

12. Stems long and climbing. Flowers resupinate.
     Fruits dry, dehiscent 13

     Stems short, never climbing. Flowers non-resupinate.
     Sepals brownish mealy or blackish ramentaceous on reverse.
     Fruits succulent and indehiscent or dry and dehiscent 14

13. Rachis and flowers furfuraceous-pubescent. Stems stout.
     Column stout, arcuate, clavate ....**Galeola**

     Rachis and flowers glabrous. Column slender, erect.........................**Erythrorchis**

14. Plant robust, with several thick, fleshy stems borne from each rhizome.
     Sepals obtuse, concave, brownish mealy on reverse.
     Fruits succulent, indehiscent..**Cyrtosia**

     Plant slender, with a single narrow, wiry stem borne from each rhizome.
     Sepals acute, reflexed (*T.saprophytica*), or lateral sepals connate (*T.connata*),
     blackish ramentaceous on reverse. Fruits dry, dehiscent ....**Tropidia**
     (*T.connata* and *T.saprophytica* only)

KEY TO GENERA OF THE SUBTRIBE AERIDINAE

1. Pollinia 4...2

    Pollinia 2...27

2. Pollinia more or less equal, globular, free from each other (Group 1)...............3

    Pollinia appearing as 2 pollen masses, each completely divided into either
    unequal, or more or less equal, semiglobular free halves (Group 2) 6

3. Plants without leaves, or leaves reduced to minute brown scales.
    Stem minute. Roots terete or flattened, containing chlorophyll
    ....**Taeniophyllum**

    Plants with normal leaves. Roots lacking chlorophyll..4

4. Large terrestrial, *Phalaenopsis*-like. Leaves radical.
    Inflorescences long, erect, many-flowered...**Doritis**

    Small epiphytes. Leaves borne along a distinct stem.
    Inflorescence 1 – 4-flowered 5

5. Leaves dorsiventral.
    Inflorescence 1 – 4-flowered. Flowers green...**Adenoncos**

    Leaves bilaterally flattened/compressed.
    Inflorescence 2-flowered. Flowers white ....**Microsaccus**

6. Flowers without a distinct column-foot..7

    Flowers with a distinct, though sometimes short column-foot............................22

7. Leaves bilaterally flattened, distichous, with sheathing bases, resembling those
    of *Microsaccus*. Mid-lobe of lip expanded into a broadly
    oblong-elliptic, emarginate blade...**Ceratochilus**

    Leaves otherwise 8

8. Lip not adnate to column, movable. Sepals and petals narrow, usually rather
    spathulate. Spur short and conical...**Arachnis**

    Lip adnate to column, not movable 9

9. Spur with a distinct longitudinal internal median septum 10

    Spur without a longitudinal internal septum...3

10. Rostellum projection short or long, turned obliquely sideward and upward,
    supporting a thin linear stipes sometimes to 9 times as long as diameter of
    pollinia ....**Micropera**

    Rostellum projection and stipes otherwise..11

11. Column with a raised fleshy, laterally compressed rostellum which sits on top of the clinandrium and has a longitudinal furrow along its edge into which the stipes and dorsally placed pollinia recline ....**Sarcoglyphis**

Column without such a structure..12

12. Floral bracts, ovary and flowers densely pubescent.
Floral bracts large, longer than flowers...**Cleisomeria**

Floral bracts, ovary and flowers glabrous.
Floral bracts minute ....**Cleisostoma**

13. Back-wall of spur without calli and/or outgrowths...4

Back-wall of spur ornamented with calli and/or outgrowths...............................20

14. Hypochile of lip globose-saccate, the side-lobes reduced to low, often fleshy edges of the sac, mid-lobe fan-shaped ....**Gastrochilus**
(*G.patinatus* only, see also Group 4, couplet 43)

Hypochile otherwise 5

15. Midlobe of lip distinctly pectinate-fringed.
Stipes linear, about 4 times as long as diameter of pollinia.............**Ornithochilus**

Midlobe otherwise.
Stipes about twice as long as diameter of pollinia 6

16. Spur or sac separated from apical portion of lip by a fleshy transverse wall or ridge...17

Spur or sac not separated from apical portion of lip by a transverse wall or ridge 18

17. Flowers pale greenish-yellow or cinnamon-orange, spotted black.
Rostellum projection large, narrow at base, rising in front of column ....**Abdominea**

Flowers creamy-white with lilac-pink patch on lip.
Rostellum projection narrow, somewhat decurved ....**Smitinandia**

18. Lip as long as or longer than dorsal sepal.
Flowers small, white to pink, bluish or mauve.
Leaves narrowly lanceolate or terete..**Schoenorchis**

Lip much shorter than dorsal sepal.
Flowers red or yellow, showy 19

19. Leaves dorsiventral, bilobed.
Column ¼ the length of dorsal sepal ....**Renanthera**

Leaves more or less semi-terete, acute.
Column ⅔ the length of dorsal sepal. ....**Renantherella**

20. Lip with a tongue or valvate callus, often forked at the tip, projecting diagonally from deep inside the spur ....**Pomatocalpa**
Lip with an often hairy ligulate tongue placed close to the spur entrance or at the base of the lip..21

21. Inflorescence branched, scape long, several-flowered ....**Staurochilus**

Inflorescence unbranched, scape short, often several close together, one- to few-flowered ....**Trichoglottis**

22. Flowers large, showy, dimorphic, the basal two always strongly scented, differently coloured from those above..**Dimorphorchis**

Flowers much smaller, not dimorphic..23

23 Lip without a distinct spur or sac, but hypochile often somewhat concave. Flowers ephemeral..24

Lip with a distinct spur or sac. Flowers long-lasting..25

24. Leaves terete...**Cordiglottis**

Leaves dorsiventral ....**Thrixspermum**

25. Spur or sac with a median longitudinal septum.................................**Cleisostoma**

Spur or sac without a longitudinal septum 6

26. Lip epichile with two forward pointing teeth emerging from base ......**Kingidium**

Lip epichile otherwise 27

27. Stems very short. Inflorescences borne below the leaves.
Flowers small, greenish yellow and white, marked with crimson on the lip.
Lip deeply saccate...**Bogoria**

Stems long, usually pendent. Inflorescences axillary.
Flowers translucent lavender-blue.
Lip distinctly spurred ....**Cleisocentron**

28. Pollinia sulcate or porate..29

Pollinia entire..44

29. Pollinia sulcate, i.e. more or less, but not completely cleft or split (Group 3) 30

Pollinia porate (Group 4) 41

30. Column-foot absent or very indistinct 31

Column-foot distinct, though sometimes short 35

31. Flowers often large and showy, usually a few, well spaced on a raceme.
Stipes short and broad. Rostellum broad, entire, shelf-like ....**Vanda**

   Flowers small to medium-sized, crowded on to a usually densely many-flowered raceme or panicle. Stipes linear, spathulate, uncinnate, rarely hamate. Rostellum prominent, bifid or long and pointed 32

32. Leaves linear, acute, fleshy.
Inflorescence and flowers scarlet-red...**Porphyrodesme**

   Leaves broader, unequally bilobed.
Inflorescence and flowers otherwise...33

33. Plants small, stem and inflorescence less than 4 cm.
Rachis very fleshy, clavate. Flowers borne in succession, minute .....**Ascochilopsis**

   Plants larger. Rachis not clavate.
Flowers not borne in succession, small to medium sized 34

34. Stems short. Leaves borne close together, most often with many light-coloured nerves.
Lip entire or obscurely 3-lobed, deeply saccate or with a short backward-pointing spur without interior ornaments ....**Rhynchostylis**

   Stems rather long. Leaves distant, without pale nerves.
Lip 3-lobed, spur often apically inflated and occasionally with callosities or scales within ....**Robiquetia**

35. Leaves terete, sometimes up to 165 cm long..36

   Leaves dorsiventral, much shorter 37

36. Stems up to 2 m long.
Lip spurred, ecallose.
Column-foot entire..**Papilionanthe**

   Stems very short.
Lip not spurred, with a conduplicate, plate-like callus situated at the junction of the mid- and side lobes.
Column-foot 3-fingered..**Paraphalaenopsis**

37. Spur or sac, if present, developed from the hypochile.
Epichile dorsiventral...38

   Spur or sac borne centrally on lip.
Epichile reduced, fleshy 39

38. Spur absent, or rudimentary.
Lip with at least one forward-pointing forked appendage.
Flowers few, sometimes large and showy, distichous ....**Phalaenopsis**

   Spur well developed.
Forked appendages absent.
Flowers many, facing in every direction, developing simultaneously ....**Aerides**

39. Rostellum projection long, slender.
    Lip bent upward so as to make a right angle with column-foot, distinctly
    unguiculate.
    Flowers long-lasting, developing simultaneously..........................**Macropodanthus**

    Rostellum projection inconspicuous.
    Lip continuing the line of and usually flush with column-foot.
    Flowers usually ephemeral usually developing successively,
    a few open at a time 40

40. Column-foot longer than the column proper.......................................**Pteroceras**

    Column-foot short, column proper elongate ....**Brachypeza**

41. Leaves terete. Lip neither saccate nor spurred..**Luisia**

    Leaves dorsiventral. Lip saccate or spurred 42

42. Lip mobile on a short but distinct column-foot, spur or sac absent ...**Biermannia**

    Lip immobile. Column-foot absent. Spur or sac present 43

43. Lip with a globose-saccate hypochile, the epichile separated from it by a
    transverse ridge connecting front edges of side lobes ....**Gastrochilus**
    (see also Group 2, couplet 14)

    Lip with a rather long, cylindric or extinctoriform spur and a ligulate mid-lobe
   44

44. Spur without a prominent back-wall callus.
    Rostellum projection broadly triangular. Stipes without apical appendages.
    Pollinator guides absent ....**Ascocentrum**

    Spur with a prominent backwall callus.
    Rostellum projection elongate, attenuate, sigmoid.
    Stipes with apical appendages.
    A pair of crimson pollinator guides present at throat of spur....................**Dyakia**

45. Column with a distinct foot. Lip movable 46

    Column without a foot. Lip not movable 48

46. Lip without a spur, side lobes sometimes fimbriate........................**Chamaeanthus**

    Lip spurred or saccate 47

47. Spur-like conical portion of lip more or less solid.
    Peduncle short, glabrous...**Chroniochilus**

    Sac or spur thin-walled, without interior fleshiness.
    Peduncle longer, often prickly-hairy...**Grosourdya**

48. Lip with a bristle or tooth inside near the apex.
    Floral bracts conspicuous, triangular, leafy ....**Microtatorchis**

Lip without an apical bristle or tooth.
Floral bracts not leafy 49

49. Side lobes of lip very large, often fringed ....**Pennilabium**

Side lobes of lip small, never fringed...50

50. Mid-lobe of lip resembling a small spongy pouch, hollow above, solid towards
apex...**Spongiola**

Mid-lobe of lip otherwise...51

51. Lip not truly spurred, but with a spur-like tubular cavity.
Lateral sepals adpressed to lip..**Porrorhachis**

Lip spurred. Lateral sepals not adpressed to lip...52

52. Rachis slender, never thickened and sulcate, or clavate.
Column hammer-shaped.
Stipes linear-spathulate, much broadened at apex ....**Malleola**

Rachis fleshy, sulcate, or clavate. Column short and stout.
Stipes linear, much reduced ....**Tuberolabium**

THE CHECKLIST

Basionyms and Synonyms

All basionyms are cited throughout. Taxa described from Borneo but subsequently treated in synonymy, are listed below the currently accepted name. Non Bornean synonyms used as currently accepted names in earlier publications on Bornean orchids are also listed. Synonyms not cited in publications on Bornean orchids are omitted unless they have been widely used as accepted names in recent literature.

Types

Types of any status, whether holotype, isotype, syntype, lectotype or neotype, are cited for each currently accepted species and for synonyms where applicable. The location of much type material remains to be confirmed and a question mark appears before the herbarium code, or by itself, where this is the case. The majority of holotypes of Schlechter names formerly located at Berlin (B) were destroyed during bombing in World War Two. Isotypes of some of these taxa were, however, distributed to other herbaria and these are cited where known. Addresses of herbaria cited are given below.

Habitat and altitudinal range

Habitat data are compiled from herbarium labels, literature references and from personal knowledge. Many terms such as 'hill forest', 'montane forest', 'moss forest' and 'sand forest', etc. are very generalised and used rather loosely. The demarcation between 'lower' and 'upper' montane forest, for example, is very arbitrary and depends on factors such as aspect, location and local microclimate. More accurate and detailed habitat data is provided by several recent collectors. The distinctive kerangas vegetation associated with lowland and upland podsolic soils or the forest types developed on ultramafic (ultrabasic) rocks such as serpentine are good examples. For many taxa though, there remains little or no habitat information.

Altitudinal data is often equally vague, though much useful information has been gained from recent collections. Details from many earlier and a few more recent collectors are often unreliable or non-existent. Important exceptions among earlier collectors include the collections made in the 1930s by Joseph and Mary Strong Clemens (Mt.Kinabalu), C.E.Carr (various localities) and the 1932 Oxford University Expedition to Sarawak (Mt.Dulit). F.H.Endert, R.Schlechter and H.Winkler also provided valuable altitudinal data with their collections from the former Dutch colony of Kalimantan. Many collectors, however, provide a locality but no altitude. In such cases, when the locality is known, the altitude may sometimes be estimated. An example are the collections made between 1857 and 1858 at 'Bangarmassing' (modern Banjarmasin) in Kalimantan Selatan by J.Motley. These were most probably collected at or around sea level.

Distribution

Distribution within Borneo is indicated by B for Brunei, K for Kalimantan, SA for Sabah and SR for Sarawak. Distribution outside Borneo is given where applicable.

Herbarium Abbreviations

A: Herbarium, Arnold Arboretum, Harvard University, Cambridge, Massachusetts 02138, USA.

AAU : Herbarium Jutlandicum, Botanical Institute, University of Aarhus, Bygn. 137, Universitetsparken, DK-8000, Aarhus C, Denmark.

AMES : Orchid Herbarium of Oakes Ames, Botanical Museum, Harvard University, Cambridge, Massachusetts 02138, U.S.A.

B : Herbarium Botanischer Garten und Botanisches Museum Berlin-Dahlem, Königin-Luise-Strasse 6-8, D-1000 Berlin 33, Germany.

BM : Herbarium, Botany Department, The Natural History Museum, Cromwell Road, London SW7 5BD, England, U.K.

BO : Herbarium Bogoriense, Jalan Raya Juanda 22-24, Bogor, Java, Indonesia.

BRI : Herbarium, Plant Pathology Branch, Department of Primary Industries, Indooroopilly, Queensland 4068, Australia.

BRUN : Herbarium, Forestry Department, Bandar Seri Begawan, 2067, Brunei Darussalam.

C : Herbarium, Botanical Museum, University of Copenhagen, Gothersgade 130, DK-1123, Copenhagen, Denmark.

CAL : Central National Herbarium, P.O.Botanic Garden, Howrah, Calcutta 711 103, West Bengal, India.

CBG : Herbarium, Australian National Botanic Gardens, G.P.O.Box 1777, Canberra, A.C.T. 2601, Australia.

E : Herbarium, Royal Botanic Garden, Edinburgh EH3 5LR, Scotland, U.K.

FI : Herbarium Universitatis Florentinae, Museo Botanico, Via G.La Pira 4, I-50121, Firenze, Italy.

G: Herbarium, Conservatoire et Jardin botaniques de la Ville de Genève, Case Postale 60, CH-1292 Chambésy/Genève, Switzerland.

HBG : Herbarium , Institut für Allgemeine Botanik und Botanischer Garten Hamburg, Ohnhorststrasse 18, D-2000, Hamburg 52, Germany.

K : Herbarium, Royal Botanic Gardens, Kew, Richmond, Surrey TW9 3AB, England, U.K. (LINDL refers to the John Lindley herbarium housed in the Orchid Herbarium; WALL refers to the Nathaniel Wallich herbarium housed at Kew).

KEP: Herbarium, Forest Research Institute of Malaysia, PO Box 201, Kepong, 52109 Kuala Lumpur, Malaysia.

KYO: Herbarium, Botany Department, Faculty of Science, Kyoto University, Kyoto-shi, Kyoto 606, Japan.

L : Rijksherbarium, Postbus 9514, 2300 RA Leiden, The Netherlands.

LA : Herbarium, Biology Department, University of California, Los Angeles, California 90024-1606, U.S.A.

LAE : Papua New Guinea National Herbarium, Forest Research Institute, PO Box 314, Lae, Papua New Guinea.

LINN: Herbarium, Linnean Society of London, Burlington House, Piccadilly, London W1V 0LQ, England, U.K.

MEL : National Herbarium of Victoria, Royal Botanic Gardens, Birdwood Avenue, South Yarra, Victoria 3141, Australia.

MO : Herbarium, Missouri Botanical Garden, PO Box 299, Saint Louis, Missouri 63166-0299, U.S.A.

NY : Herbarium, New York Botanical Garden, Bronx, New York 10458-5126, U.S.A.

P : Herbier, Laboratoire de Phanérogamie, Muséum National d'Histoire Naturelle, 16 rue Buffon, F-75005, France.

S : Herbarium, Botany Department, Swedish Museum of Natural History, PO Box 50007, S-104 05 Stockholm, Sweden.

SAN : Herbarium, Forest Research Centre, Forest Department, PO Box 1407, 90008 Sandakan, Sabah, Malaysia.

SAR : Forest Herbarium, Department of Forestry, 93660 Kuching, Sarawak, Malaysia.

SEL : Herbarium, Marie Selby Botanical Gardens, 811 South Palm Avenue, Sarasota, Florida 34236, U.S.A.

SING : Herbarium, Parks and Recreation Department, Botanic Gardens, Cluny Road, Singapore 0511, Singapore.

SNP: Herbarium , Sabah National Parks, Mt. Kinabalu National Park, Sabah, Malaysia.

TUB: Herbarium, Institut für Biologie I, Lehrstuhl Spezielle Botanik, Eberhard – Karls – Universität Tübingen, Auf der Morgenstelle 1, D-7400 Tübingen 1, Germany.

UKMS : Herbarium Jabatan Biologi, Universiti Kebangsaan Malaysia, Kampus Sabah, Locked Bag No.62, 88996 Kota Kinabalu, Sabah, Malaysia.

UPS : Herbarium (Fytoteket), Uppsala University, P.O.Box 541, S-751 21, Uppsala, Sweden.

W : Herbarium, Department of Botany, Naturhistorisches Museum Wien, Burgring 7, A-1014 Wien, Austria.

New taxa

The following twenty four new taxa are described:

Ascochilopsis lobata J.J.Wood & A.Lamb

Dendrobium cymboglossum J.J.Wood & A.Lamb

Dendrobium limii J.J.Wood

Dendrobium nabawanense J.J.Wood & A.Lamb

Dendrobium trullatum J.J.Wood & A.Lamb

Dendrochilum auriculilobum J.J.Wood

Dendrochilum crassifolium Ames var. *murudense* J.J.Wood

Dendrochilum crassilabium J.J.Wood

Dendrochilum cruciforme J.J.Wood var. *cruciforme*

Dendrochilum cruciforme J.J.Wood var. *longicuspum* J.J.Wood

Dendrochilum cupulatum J.J.Wood

Dendrochilum devogelii J.J.Wood

Dendrochilum geesinkii J.J.Wood

Dendrochilum hosei J.J.Wood

Dendrochilum imitator J.J.Wood

Dendrochilum lacinilobum J.J.Wood & A.Lamb

Dendrochilum lumakuense J.J.Wood

Dendrochilum meijeri J.J.Wood

Dendrochilum ochrolabium J.J.Wood

Dendrochilum pachyphyllum J.J.Wood & A.Lamb

Dendrochilum pandurichilum J.J.Wood

Dendrochilum papillilabium J.J.Wood

Dendrochilum tenuitepalum J.J.Wood

Tropidia connata J.J.Wood & A.Lamb

New combinations

The following twelve new combinations are made:

Malaxis amabilis (J.J.Sm.) J.J.Wood

Malaxis partitiloba (J.J.Sm.) J.J.Wood

Malaxis repens (Rolfe) J.J.Wood

Malaxis sublobata (J.J.Sm.) J.J.Wood

Malaxis subtiliscapa (J.J.Sm.) J.J.Wood

Peristylus ovariophorus (Schltr.) J.J.Wood

Thrixspermum fimbriatum (Ridl.) J.J.Wood

Trias tothastes (J.J.Verm.) J.J.Wood

Trichotosia brevirachis (J.J.Sm.) J.J.Wood

Trichotosia conifera (J.J.Sm.) J.J.Wood

Trichotosia jejuna (J.J.Sm.) J.J.Wood

Trichotosia lawiensis (J.J.Sm.) J.J.Wood

New names

Two new names are proposed:

Eria magnibracteata J.J.Wood

Malaxis rajana J.J.Wood

New synonyms

Thirteen new synonyms are proposed:

Cleisostoma borneense J.J.Wood

Dendrochilum bicallosum J.J.Sm.

Dendrochilum bigibbosum J.J.Sm.

Dendrochilum hewittii J.J.Sm.

Dendrochilum lobongense Ames

Dendrochilum mantis J.J.Sm.

Dendrochilum minimum (Ridl.) Ames

Dendrochilum minus (Ridl.) Ames

Dendrochilum palawanense Ames

Kingidium deliciosum (Rchb.f.) H.R.Sweet var. *bellum* (Teijsm. & Binn.) O.Gruss & Röllke

Platyclinis minima Ridl.

Platyclinis minor Ridl.

Trichoglottis wenzelii Ames

ENUMERATION OF TAXA

SUBFAMILY APOSTASIOIDEAE

APOSTASIA Blume

Bijdr.:423 (1825).

de Vogel, E.F.(1969). Monograph of the tribe Apostasieae. (Orchidaceae). Blumea 17:313–350.

Erect, rhizomatous terrestials with nodular storage roots. Stems thin, often branched, leafy throughout. Leaves usually linear, acute, not plicate, the lowermost smaller or dying off. Inflorescences terminal, often branched, branches decurved or spreading, many flowered; bracts small. Flowers yellow or white; tepals free, almost equal, spreading, sometimes revolute; column straight or curved; fertile stamens 2, with short filaments adnate to the style to various degrees, anthers clasping the style; staminode present or absent, usually adnate to the style, free at the apex; stigma small; pollen grains as monads, not aggregated into pollinia; ovary narrowly cylindric, 3-angular in cross section.

Seven species distributed from the Himalayan region, India and Sri Lanka east to the Ryukyu Islands, New Guinea and Australia.

A.elliptica *J.J.Sm.* in Bull. Jard. Bot. Buitenzorg, ser. 3, 2:15 (1920). Type: Sumatra, Ophir District, north of Taloe, *Bünnemeijer* 107 (lecto. BO).

HABITAT: mixed dipterocarp forest on shale.
ALTITUDINAL RANGE: 500 m.
DISTRIBUTION: Borneo (B). Also Peninsular Malaysia, Sumatra.

A.nuda *R. Br.* in Wallich, Pl. As. Rar. 1: 76 (1830). Type: Peninsular Malaysia , *Porter* in *Wallich* 4449 (lecto. K, isolecto. K - WALL).
Apostasia brunonis Griff., Not. 3:234 (1851). Type: Burma, Mergui, *Griffith* 5604 (lecto. K, isolecto. P).
A.lobbii Rchb.f. in Flora 55:278 (1872). Type: Labuan, *Lobb* s.n. (lecto. K, isolecto. P).
Adactylus lobbii (Rchb.f.) Rolfe in Orchid Rev. 4:329 (1896).
A.nudus (R.Br.) Rolfe in Orchid Rev. 4:329 (1896).

HABITAT: lowland, hill and lower montane forest; mixed dipterocarp forest; sometimes on ultramafic substrate; rarely found in the lowlands.
ALTITUDINAL RANGE: sea level – 1300m
DISTRIBUTION : Borneo (B, K, SA, SR). Also Burma, Cambodia, S. Vietnam, Peninsular Malaysia, Sumatra, Bangka, W. Java.

A.odorata *Blume,* Bijdr.: 423 (1825). Type: Java, *Blume* s.n. (lecto. L).

HABITAT: Primary hill and lower montane forest, sometimes on ultramafic substrate.
ALTITUDINAL RANGE: (700–) 1000 – 1500 (–1700)m.
DISTRIBUTION: Borneo (SA). Also N.E. India (Assam), China (Yunnan), S.Thailand, N. & S. Vietnam, Peninsular Malaysia, W.Sumatra, Bangka, Belitung, W.Java, Sulawesi.

A.parvula *Schltr.* in Bull. Herb. Boissier 2, 6:295 (1906). Type: Kalimantan, *Schlechter* 13562 (holo. B, destroyed).

HABITAT: Primary forest.
ALTITUDINAL RANGE: unknown.
DISTRIBUTION Borneo (K). Endemic.

A.wallichii *R.Br.* in Wallich , Pl. As. Rar. 1:75, t.84 (1830). Type: Nepal, *Wallich* 4448 (lecto. K - WALL, isolecto. E,K).

A.gracilis Rolfe in J. Linn. Soc., Bot. 25: 242 (1889). Type: Kalimantan, *Korthals* s.n. (holo.K).

A.alba Rolfe in Orch.Rev. 4: 329 (1896). Types: Borneo, *Barber* s.n. (lecto. and isolecto.K); Labuan, *Motley* 95 (syntype K).

A.curvata J.J.Sm. in Mitt. Inst. Allg. Bot. Hamburg 7:11, t.1, f.1 (1927). Type: Kalimantan, Lebang Hara, *Winkler* 595 (holo. BO).

A.wallichii R.Br. var. *seraweiensis* J.J.Sm. loc. cit. :10 (1927). Type: Kalimantan, *Winkler* 268 (lecto. HBG, isolecto. BO).

HABITAT: Primary forest undergrowth; streamsides; dipterocarp forest; mixed dipterocarp, *Agathis, Gymnostoma* and *Fagaceae* forest on ultramafic substrate
ALTITUDINAL RANGE: (sea level–) 200 – 1200 (–1700)m.
DISTRIBUTION: Borneo (B, K, SA, SR). Also Sri Lanka, Nepal, N.E. India (Assam), Burma, Thailand, Cambodia, Vietnam, Peninsular Malaysia, Sumatra, Bangka, Java, Buru, Philippines, New Guinea, Australia (Queensland).

NEUWIEDIA Blume

Tijdschr. Natuurl. Gesch. Physiol. 1:140(1834).

de Vogel, E.F.(1969). Monograph of the tribe Apostasieae. (Orchidaceae). Blumea 17:313–350.

Erect or ascending, glabrous to hairy terrestrial herbs, without rhizomes. Stems usually simple, leafy. Leaves narrowly elliptic or linear, acute, plicate. Inflorescences terminal, racemose, mostly unbranched, many flowered; bracts conspicuous. Flowers yellow or white; tepals free, slightly convex, not spreading; column straight; stamens 3, filaments connate at base, partly adnate to style, anthers not clasping style, free from each other, dorsifixed; ovary ellipsoid, strongly contracted at apex, more or less triangular to rounded.

Eight species distributed throughout S.E.Asia east to New Guinea, the Solomon Islands and Vanuatu.

N.borneensis *de Vogel* in Blumea 17 (2): 325, fig. 7i (1969). Type: Sarawak, *Ridley* s.n. (holo. K).

HABITAT: dipterocarp forest; hill and lower montane forest on ultramafic substrate, in deep shade; de Vogel (1969) reports it as occuring in "sandy, wet, poor soil".
ALTITUDINAL RANGE: 300 – 1000m.
DISTRIBUTION: Borneo (B, SA, SR). Endemic.

N.elongata *de Vogel* in Blumea 17 (2): 331, fig. 7a–c (1969). Type: Kalimantan, *Kostermans* 12988 (holo. L, iso. BO, K).

HABITAT: *Agathis* forest on waterlogged, sandy, acid soil
ALTITUDINAL RANGE: 500m.
DISTRIBUTION: Borneo (K). Endemic.

N.inae *de Vogel* in Blumea 17(2):333, fig.8 (1969). Type: Kalimantan, Bukit Tilung, *Winkler* 1485 (holo. HBG).

HABITAT: "primary forest".
ALTITUDINAL RANGE: 700 – 800m.
DISTRIBUTION: Borneo (K). Endemic.

N.veratrifolia *Blume* in v.d. Hoeven and de Vriese, Tijd. Nat. Gesch. Phys.1: 142 (1834).Type: Java, *Blume* s.n. (lecto. L, isolecto. L).
N.lindleyi Rolfe in J. Linn. Soc., Bot. 25:232 (1889). Type: Peninsular Malaysia, Penang, Sungai Penang, *Curtis* 469 (lecto. SING, isolecto. BO,K).

HABITAT: field layer in primary and secondary forest; riversides; recorded on limestone and ultramafic substrate.
ALTITUDINAL RANGE: sea level – 1200m.
DISTRIBUTION: Borneo (B, K, SA, SR). Also Peninsular Malaysia, Singapore, Sumatra, Mentawai Islands, Bangka, Riau Archipelago, W.Java, Anambas Islands, Philippines, Sulawesi, Maluku, Aru Islands, New Guinea, Solomon Islands.

N.zollingeri *Rchb.f.* in Bonplandia 5:58 (1857).

var. **javanica** *(J.J.Sm.)de Vogel* in Blumea 17 (2):329 (1969). Type: Java, Djampang Tengah, Artana, *Smith* s.n. (lecto. L, isolecto. BO).
N.javanica J.J.Sm. in Bull. Jard. Bot. Buitenzorg, ser. 2, 14:5 (1914).

HABITAT: hill forest on ultramafic substrate.
ALTITUDINAL RANGE: 600 – 100m.
DISTRIBUTION: Borneo (SA). Also Sumatra, Java, Bali.

var. **singapureana** *(Baker) de Vogel* in Blumea 17 (2): 331 (1969).Type: Singapore, *Wallich* 5195 (holo. K - WALL).
Tupistra singapureana Baker in J. Linn. Soc., Bot. 14: 581 (1874).

HABITAT: mixed dipterocarp, Fagaceae, *Gymnostoma* and *Agathis* forest on ultramafic substrate; forest on poor or rich soils.
ALTITUDINAL RANGE: 600m.
DISTRIBUTION: Borneo (SA, ?SR). Also China (Kwangtung, Hainan), Hong Kong, Peninsular Malaysia, N. & S. Vietnam (Annam), Sumatra, Bangka, Belitung, Lingga Archipelago.

SUBFAMILY CYPRIPEDIOIDEAE

PAPHIOPEDILUM Pfitzer

Morph. Stud. über Orchideenbl.:11 (1886).

Cribb, P. (1987). The Genus *Paphiopedilum*.

Terrestrial or, rarely, epiphytic herbs. Rhizomes very short. Stem usually short, sometimes elongated, ascending and leafy. Leaves arranged in a fan, linear -ligulate to elliptic-oblong, fleshy or leathery, often tessellated above, sometimes purple stained below. Inflorescence erect or arcuate, usually hirsute, 1- to many-flowered; bracts lanceolate and acute to elliptic and obtuse. Flowers large, usually showy, usually unscented, long lived, often waxy-textured; lateral sepals united to form a synsepalum; petals deflexed and/or reflexed and often twisted; lip slipper- or helmet-shaped, deeply saccate; column bearing 2 ventral anthers and an apical, often shield-shaped staminode.

About 65 species distributed from India, southern China and S.E.Asia to the Philippines, the Malay Archipelago, New Guinea and the Solomon Islands.

P.bullenianum *(Rchb.f.) Pfitzer* in Bot. Jahrb. Syst. 19:40 (1894). Type: Borneo, hort. *Low* (holo. W).
   Cypripedium bullenianum Rchb.f. in Bot. Zeitung (Berlin) 23:99 (1865).
   C.bullenianum Rchb.f. var. *oculatum* Rchb.f. in Gard.Chron. 15:563 (1881). Type: Borneo, cult. *Bull* (holo. W).
   Paphiopedilum amabile Hallier f. in Natuurk. Tijdschr. Ned.-Indie 54: 450 (1895). Type: Kalimantan, *Hallier* s.n. (holo. BO).
   Cordula bulleniana (Rchb.f.) Rolfe in Orchid Rev. 20:2 (1912).
   C. amabilis (Hallier f.) Ames in Merr., Bibl. Enum. Born. Pl.:135 (1921).
   C.bulleniana (Rchb.f.) Rolfe var. *oculata* (Rchb.f.) Ames in Merr., Bibl. Enum. Born.Pl.:135(1921).
   Paphiopedilum linii Schoser in Die Orchidee 16:181 (1966). Type: Sarawak, cult. Tuebingen B.G. ex *Sheridan-Lea* s.n. (holo. TUB).

HABITAT: in moss and leaf litter amongst stilt-roots of mangroves; in leaf litter on steep slopes and wet moss covered rocks in shade of low forest.
ALTITUDINAL RANGE: sea level – 1900m.
DISTRIBUTION: Borneo (K, SR). Also Peninsular Malaysia, Sumatra.

P.x burbidgei *(Rchb.f.) Pfitzer* in Bot. Jahrb. Syst. 19:40 (1894). Type: Borneo, hort. *Veitch* (holo. W).
   Cypripedium x burbidgei Rchb.f. in Gard. Chron. 16: 38 (1881).
   Cordula burbidgei (Rchb.f.) Ames in Merr., Bibl. Enum. Born. Pl.:135 (1921).

HABITAT: as parents.
ALTITUDINAL RANGE: as parents.
DISTRIBUTION: Borneo (SA). Endemic.
A natural hybrid between *P. hookerae* var. *volonteanum* and *P.javanicum* var. *virens*.

P.dayanum *(Lindl.) Stein,* Orchideenbuch: 464 (1892). Type: Borneo, cult. Hort.Soc. (holo. K).
   Cypripedium spectabile Rchb.f. (sphalm. for *superbiens*) var. *dayanum* Lindl. in Gard. Chron.: 693 (1860).
   C. superbiens Rchb.f. var. *dayanum* (Lindl.) Rchb.f., Xenia Orch. 2:10 (1860).
   C. dayanum (Lindl.) Rchb.f. in Bot. Zeitung (Berlin) 20:214 (1862).

C. ernestianum Hort., J. Hort. Soc. London : 375, fig.67 (1887). Type: not located.
C. petri Rchb.f. in Gard. Chron. n.s. 13:680 (1880). Type: Sabah, hort. *Veitch* (holo. W).
Paphiopedilum dayanum (Lindl.) Stein var. *petri* (Rchb.f.) Pfitzer in Engler, Pflanzenr. Orch. Pleon.:86 (1903).
Cordula dayana (Lindl.) Rolfe in Orchid Rev. 20:2 (1912).

HABITAT: leaf litter under bamboo and at base of trees on steep ridges; on serpentine rock outcrops.
ALTITUDINAL RANGE: 300 – 1500m.
DISTRIBUTION: Borneo (SA). Endemic.

P.hookerae (*Rchb.f.*) *Stein* Orchideenbuch: 470 (1892). Type: Borneo, *Low* s.n. (holo. W).
Cypripedium hookerae Rchb.f. in Bot.Mag. 89:t.5362 (1863).
Cordula hookerae (Rchb.f.) Rolfe in Orchid Rev. 20:2 (1912).

var. **hookerae**

HABITAT: edges of steep limestone and sandstone cliffs under open tree canopy: in leaf litter at base of trees and in damp crevices.
ALTITUDINAL RANGE: 100 – 600m
DISTRIBUTION: Borneo (K, SR). Endemic.

var.**volonteanum** (*Sander ex Rolfe*) *Kerch.*, Orch.:456 (1894). Type: Borneo, *Low* (holo. K).
Cypripedium hookerae Rchb.f. var. *volonteanum* Sander ex Rolfe in Gard. Chron. ser. 3, 8:66 (1890).
Paphiopedilum volonteanum (Sander ex Rolfe)Pfitzer in Engler, Pflanzenr. Orch. Pleon.:80 (1903).
Cordula hookerae (Rchb.f.) Rolfe var. *volonteana* (Sander ex Rolfe) Ames in Merr., Bibl. Enum. Born. Pl. :136 (1921).

HABITAT: hill and lower montane forest; screes under the shade of bushes; in leaf litter along ridge tops; under *Gymnostoma*.
ALTITUDINAL RANGE: 200 – 2600m.
DISTRIBUTION: Borneo (SA). Endemic.

P.javanicum (*Reinw.ex Lindl.*) *Pfitzer* in Pringsh. Jahrb. Wiss. Bot. 19:165 (1888).

var. **virens** (*Rchb.f.*) *Stein*, Orchideenbuch: 471 (1892). Type: Borneo, cult. *Day* (holo. W).
Cypripedium virens Rchb.f. in Bot. Zeitung (Berlin) 21:128 (1863).
C.javanicum Reinw. ex Lindl. var *virens* (Rchb.f.) J.J.Veitch, Man. 4:35 (1881).
Paphiopedilum virens (Rchb.f.) Pfitzer in Bot. Jahrb. Syst. 19:41 (1896).
Cordula virens (Rchb.f.) Rolfe in Orchid Rev. 20:2 (1912).
Paphiopedilum purpurascens Fowlie in Orchid Digest 38:155 (1974). Type: cult. Los Angeles Arb.,*Hilberg* H66B1 (holo. UCLA).

HABITAT: hill and lower montane forest; steep, boulder strewn slopes, often above rivers and streams, growing in boulder crevices in deep shade.
ALTITUDINAL RANGE: 900 – 1700m.
DISTRIBUTION: Borneo (SA, ?SR). Endemic.

P.kolopakingii *Fowlie* in Orchid Digest 48: 41 (1984). Type: Kalimantan, hort.*Kolopaking* LKW 82 K1 (holo. UCLA).

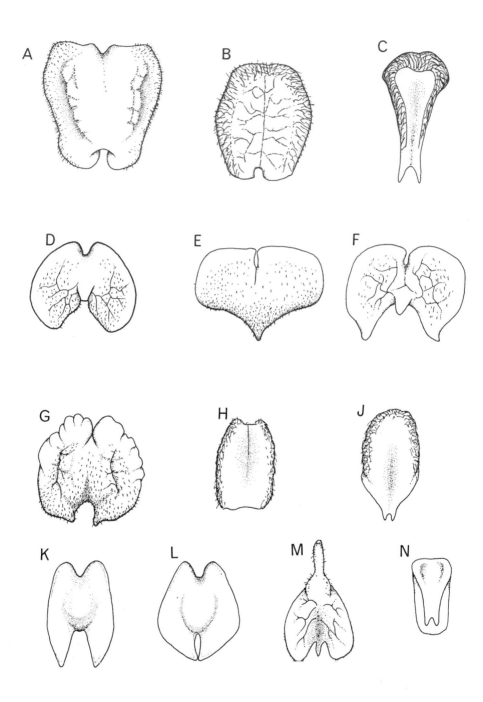

FIG. 1. *Paphiopedilum* staminodes. **A**, *P. philippinense*, x 4; **B**, *P. stonei*, x 2; **C**, *P. rothschildianum*, x 2; **D**, *P. javanicum* var. *virens*, x 3; **E**, *P. dayanum*, x 3; **F**, *P. lawrenceanum*, x 2; **G**, *P. hookerae* var. *volonteanum*, x 1; **H**, *P. kolopakingii*, x 2; **J**, *P. supardii*, x 3; **K & L**, *P. bullenianum*, x 3; **M**, *P. lowii*, x 2; **N**, *P. sanderianum*, x 1. Drawn by Sarah Thomas.

HABITAT: growing between stones over steep river gorges.
ALTITUDINAL RANGE: 600 – 700m.
DISTRIBUTION: Borneo (K). Endemic.

P.lawrenceanum (*Rchb.f.*) *Pfitzer* in Jahrb. Wiss. Bot. 19: 163 (1888). Type: Borneo, cult. Sander, *Burbidge* s.n. (holo. W).
Cypripedium lawrenceanum Rchb.f. in Gard. Chron. n.s. 10: 748 (1878).
Cordula lawrenceana (Rchb.f.) Rolfe in Orchid Rev. 20:2 (1912).
Paphiopedilum barbatum (Lindl.) Pfitzer ssp. *lawrenceanum* (Rchb.f.) M.W.Wood in Orchid Rev. 84:352 (1976).

HABITAT: in leaf litter in primary forest, less commonly on mossy limestone rocks.
ALTITUDINAL RANGE: 300 – 500m.
DISTRIBUTION: Borneo (?SA, SR). Endemic.

P.lowii (*Lindl.*) *Stein*, Orchideenbuch: 476 (1892). Type: Borneo, *Low* (holo. K).
Cypripedium lowii Lindl. in Gard. Chron.:765 (1847).
Cordula lowii (Lindl.) Rolfe in Orchid Rev. 20:2 (1912). (as *C.lowiana*).

HABITAT: riverine, lower montane forest; moss- or humus-filled hollows of rocks and boulders, especially limestone.
ALTITUDINAL RANGE; 200 – 1700m.
DISTRIBUTION; Borneo (SA, SR). Also Peninsular Malaysia, Sumatra, Java, Sulawesi.

P.philippinense (*Rchb.f.*) *Stein*, Orchideenbuch: 480 (1892). Type: without provenance (holo. W ?lost).
Cypripedium philippinense Rchb.f. in Bonplandia 10:335 (1862).

HABITAT: limestone cliffs and boulders, often in quite open places.
ALTITUDINAL RANGE: sea level – 500m.
DISTRIBUTION: Borneo (SA). Also Philippines.

P.rothschildianum (*Rchb.f.*) *Stein*, Orchideenbuch: 482 (1892). Type: hort. *Sander* (holo. W).
Cypripedium rothschildianum Rchb.f. in Gard. Chron. ser. 3, 3:457 (1888).
C.elliottianum O'Brien in Gard. Chron. ser. 3, 4:501 (1888). Type: hort. *Sander* (holo. K).
Paphiopedilum elliottianum (O'Brien) Stein, Orchideenbuch: 466 (1892).
P.rothschildianum (Rchb.f.) Stein var. *elliottianum* (O'Brien) Pfitzer in Engler, Pflanzenr. Orch. Pleon.: 59 (1903).
Cordula rothschildiana (Rchb.f.) Ames in Merr., Bibl. Enum. Born. Pl.:137 (1921).

HABITAT: hill forest; ledges on steep ultramafic slopes and cliffs in the open and in shade.
ALTITUDINAL RANGE: 500 – 1800m.
DISTRIBUTION: Borneo (SA). Endemic.

P.sanderianum (*Rchb.f.*) *Stein*, Orchideenbuch: 482 (1892). Type: Malay Archipelago, hort. *Sander* (holo. W).
Cypripedium sanderianum Rchb.f. in Gard. Chron. n.s.: 554 (1886).

HABITAT: vertical south-east facing limestone cliffs.
ALTITUDINAL RANGE: 100 – 900m.
DISTRIBUTION: Borneo (SR). Endemic.

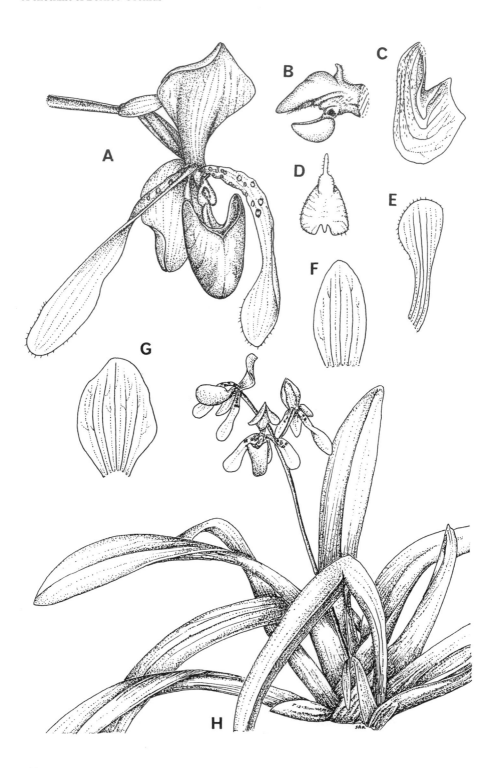

P.stonei (*Hook.*) *Stein*, Orchideenbuch: 487 (1892). Type: Borneo, hort. *Low* (holo. K).

   Cypripedium stonei Hook. in Bot.Mag. 88,t. 5349 (1862).

   C.stonei Hook. var. *platytaenium* Rchb.f. in Gard. Chron.:1118 (1867). Type: Borneo, cult. *Day* (holo. K).

   Paphiopedilum stonei (Hook.) Stein var. *platytaenium* (Rchb.f.) Pfitzer in Engler, Pflanzenr. Orch. Pleon.: 64 (1903).

   Cordula stonei (Hook.) Rolfe in Orchid Rev. 20:2 (1912).

   C.stonei (Hook.) Rolfe var. *platytaenia* (Rchb.f.) Ames in Merr., Bibl. Enum. Born. Pl.: 137 (1921).

HABITAT: limestone cliffs, lightly shaded by the crowns of trees growing at the base of the cliffs.
ALTITUDINAL RANGE: sea level – 500m.
DISTRIBUTION: Borneo (SR). Endemic.

P.supardii *Braem* & *Loeb* in Die Orchidee 36 (4): separate & 142 (1985). Type: Kalimantan, *Supardi* in *Braem* GB585 (holo. herb. Braem).

HABITAT: rock crevices on limestone.
ALTITUDINAL RANGE: 600 – 1000m.
DISTRIBUTION: Borneo (K). Endemic.

FIG. 2. *Paphiopedilum lowii.* **A**, flower, x ¾; **B**, column (side view), x 1½; **C**, lip (longitudinal section), x ¾; **D**, staminode, x 1½; **E**, petal apex, x ½; **F**, synsepalum, x ¾; **G**, dorsal sepal, x ¾; **H**, habit, x ⅕. Drawn by Sarah Thomas.

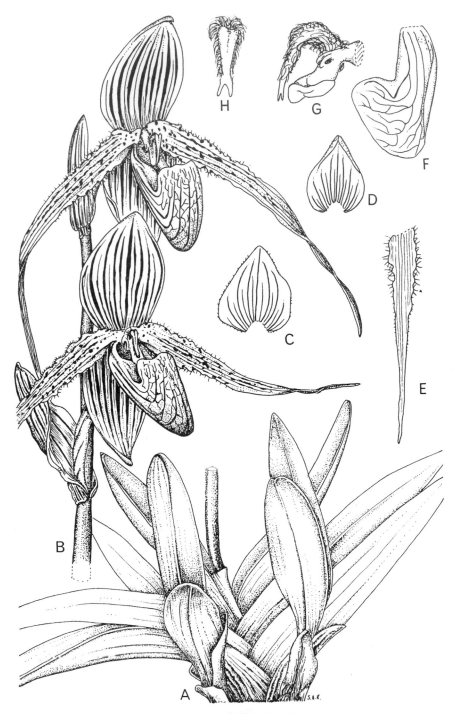

FIG. 3. *Paphiopedilum rothschildianum*. **A**, habit, x ⅕; **B**, inflorescence, x ⅓; **C**, dorsal sepal, x ½; **D**, synsepalum, x ½; **E**, petal, distal portion, x ½; **F**, lip (longitudinal section), x 1½; **G**, column (side view), x 1½; **H**, staminode, x ⅔. Drawn by Sarah Thomas.

SUBFAMILY SPIRANTHOIDEAE

TRIBE TROPIDIEAE

CORYMBORKIS Thouars

Hist. Orch.(Orch. Iles Afr.), t.37 & 38 (1822).

Rasmussen, F. (1977). The genus *Corymborkis* Thou., Orchidaceae: A taxonomic revision. Bot.Tidsskr. 71:161–192.

Terrestrial herbs with subterranean rhizomes and persistent roots. Stems erect, slender, tough, unbranched. Leaves narrowly elliptic to ovate, plicate, scattered along upper part of stem, sheathing at base, thin, but tough-textured. Inflorescences lateral, paniculate, corymbose, many flowered. Flowers white to greenish-white; sepals and petals long and spathulate, of equal length; lip equal to sepals and petals, narrow, apex expanded and flabellate, disc with 2 obscure longitudinal keels; column straight, long and slender, apex dilated, rostellum erect; pollinia 2.

A pantropical genus containing 5 species.

C.veratrifolia (*Reinw.*) *Blume*, Fl. Javae, ser. 2, 1, Orch. (Col. Orch. Arch. Ind.): 125, Pl.42 & 43 (1859). Type: Java, *Lobb* 162 (holoneo K, isoneo BM, C (photograph)).
  Hysteria veratrifolia Reinw., Syll. Pl. Nov. 2:5 (1825–26).

HABITAT: primary hill and lower montane forest; secondary forest in shade; old rubber plantations.
  ALTITUDINAL RANGE: sea level – 1500m.
  DISTRIBUTION: Borneo (B, K, SA, SR). Widespread from India and Sri Lanka to Samoa.

TROPIDIA Lindl.

Bot.Reg. 19, sub t. 1618 (1833).

Muluorchis J.J.Wood in Kew Bull. 39:73 (1984).

Terrestrial herbs. Stems erect, leafy, rarely leafless. Leaves distichous, tough, plicate, sheathing at the base. Inflorescences terminal, simple, with 1 or 2 flowers opening in succession; floral bracts large, imbricate, sheathing. Flowers often opening only partially; sepals and petals free, or lateral sepals connate, together with lateral sepals enclosing base of lip; lip entire, concave, saccate or spurred; column short, rostellum long, erect. Pollinia 2.

A pantropical genus containing about 20 species, two of which are saprophytic.

T.connata *J.J.Wood* et *A.Lamb* **sp.nov.**, saprophyton *T.saprophytico* J.J.Sm. affine, sed rhachidi fractiflexa, sepalis lateralibus connatis labelloque calcarato distinguitur. Typus: East Malaysia, Sabah, Sipitang District, Gunung Lumaku, 27.6.1992, *Lamb, Surat & Lim* in *Lamb* AL 1512/92 (holotypus K, herbarium and spirit material).

Erect, clump-forming saprophytic herb. Stems 10 to 15 per clump, 15 – 20 cm high, wiry, simple or branched, internodes 0.8 – 2.2 cm long, 1 mm wide, off white to cream coloured, glabrous above, sparsely ramentaceous on lower internodes, bearing 5–10 ovate, acute to acuminate, adpressed, sparsely ramentaceous, brown sterile sheaths each 3–5 mm long. Inflorescence 10 to 15-flowered, 1 or 2 flowers open at a time; rachis 3–9 cm long, fractiflex, sparsely ramentaceous; floral bracts 3 mm long, ovate, acute, off white tipped brown, sparsely ramentaceous. Flowers non-resupinate, 1 cm across, white, apex of lip yellow. Pedicel with ovary 2 mm long, white, sparsely ramentaceous. Sepals and petals spreading, the sepals sparsely ramentaceous on reverse. Dorsal sepal 5 x 1.5 mm, narrowly elliptic, acute. Lateral sepals connate into a 4 – 5 x 2.5 mm ovate-elliptic, cymbiform synsepalum, its apex minutely bifid. Petals 6 x 1.5 mm, narrowly elliptic, acute. Lip 5 mm long, 2 mm wide when flattened, entire, elliptic, acute, concave, rather fleshy, margin minutely erose, curving back to a horizontal position, spur saccate, obtuse, enclosed by synsepalum, c.1.1 mm long. Column 1 mm long, rostellum c.1.5 mm long. Anther rostrate. Pollinia 2. Fig. 4.

This curious saprophyte was found growing between 400 and 600 metres in mixed hill dipterocarp forest with isolated patches of podsol forest on sandstone. It is easily distinguished from *T.saprophytica* J.J.Sm., the only other saprophytic species, by the fractiflex rachis, connate lateral sepals forming a synsepalum, and shortly spurred lip. The flowers recall those of the non-saprophytic *T.angulosa* (Lindl.) Blume, but are much smaller and have a shorter spur.

T.curculigoides *Lindl.*, Gen. & Sp. Orch. Pl.:497 (1840). Type:India (Assam), *Wallich* 7386A (holo. K-WALL).

HABITAT: lowland and hill dipterocarp forest, sometimes on ultramafic substrate; in shade.
ALTITUDINAL RANGE: 700 – 1300m.
DISTRIBUTION: Borneo (SA, SR). Also India, Burma, China (Hainan), Taiwan, Thailand, Vietnam, Cambodia, Peninsular Malaysia, Java, Timor.

T.graminea *Blume*, Fl. Javae, ser. 2, 1, Orch. (Coll. Orch. Arch. Ind.): 124, Pl. 41, fig.3 (1859). Type: Java, *Blume* (holo. L).

HABITAT: ridge forest on sandstone soil; mixed hill dipterocarp, oak and chestnut forest on sandstone ridges.
ALTITUDINAL RANGE: 200–1400m.
DISTRIBUTION: Borneo (B, SA, SR). Also Peninsular Malaysia, Java, probably Sumatra.

T.pedunculata *Blume*, Fl.Javae ser. 2, 1, Orch. (Coll. Orch. Arch. Ind.):122, Pl.40, (1858). Types: Sumatra, *Korthals* (syn. L, isosyn. K); *Praetorius* (syntype: BO).

HABITAT: hill dipterocarp forest; lower montane forest; dipterocarp, Fagaceae, *Gymnostoma, Agathis* forest on ultramafic substrate.
ALTITUDINAL RANGE: 100 – 1200m.
DISTRIBUTION: Borneo (SA,SR). Also Peninsular Malaysia, Thailand, Laos, Sumatra, Java, Timor, Tanimbar Islands.

T.saprophytica *J.J.Sm.* in Mitt. Inst. Allg. Bot. Hamburg 7: 27, t.3, fig.16 (1927). Type Kalimantan, Sungai Raun, *Winkler* 1546 (holo. HBG, drawing at BO).
Muluorchis ramosa J.J.Wood in Kew Bull. 39: 74, fig.1 (1984).
Type: Sarawak, Hansen 437 (holo. C, iso. K).

HABITAT: lower and upper montane forest; mixed dipterocarp forest; riverine alluvial soils.
ALTITUDINAL RANGE: 300–1900m.
DISTRIBUTION: Borneo (K, SA, SR). Endemic.

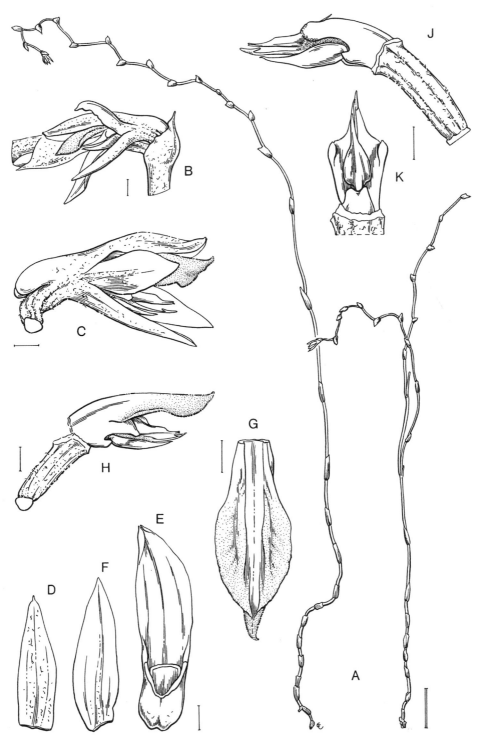

FIG. 4. *Tropidia connata*. **A**, habit; **B**, part of inflorescence; **C**, flower (side view); **D**, dorsal sepal; **E**, synsepalum and spur, lip detached (front view); **F**, petal; **G**, lip (front view); **H**, pedicel with ovary, lip and column (side view); **J**, pedicel with ovary and column (side view); **K**, column and anther-cap (back view). **A-K** from *Lamb, Surat* and *Lim* in *Lamb* AL 1512/92. Scale: single bar = 1 mm; double bar = 1 cm. Drawn by Eleanor Catherine.

49

SUBTRIBE GOODYERINAE

ANOECTOCHILUS Blume

Bijdr.:411, t.15 (1825).

Terrestrial herbs. Leaves green, or with colourful silvery or red nerves, stalked. Inflorescence with a rather short peduncle, few flowered. Flowers resupinate, rather showy, usually white, pink flushed; dorsal sepal and petals connivent, forming a hood; lip in contact with base of column, either with a projecting spur or a saccate base enclosed by the sepals, inside which are 2 large, sessile glands; middle part of lip narrowed into a channelled claw, the edges of which are involute, with a toothed or fringed flange on either side, sometimes with distinct side lobes at base of claw, apex of claw widened into a transverse bilobed blade; column 2 winged in front, either small, or prolonged downwards as free parallel plates into spur; anther acute, short or elongate; stigmas 2; pollinia 2.

When considered in the broad sense to include *Odontochilus* Blume, about 40 species can be recognised distributed from Sri Lanka and India to Japan, south to Malaysia and Indonesia, and eastwards to the Pacific islands.

A.integrilabris *Carr* in Gard. Bull. Straits Settlem. 8, 3: 186 (1935). Type: Sabah, Mt. Kinabalu, Minitinduk Gorge, *Carr* C.3162, (SFN26634) (holo. SING., iso. AMES).

HABITAT: young secondary hill forest.
ALTITUDINAL RANGE: 800m.
DISTRIBUTION: Borneo (SA). Endemic.

A.longicalcaratus *J.J.Sm.* in Bull. Jard. Bot. Buitenzorg, ser. 3, 5: 18 (1922). Types: Sumatra, *Jacobson* 760, *Bünnemeijer* 765, 4036, 5350, 5445, *Groeneveldt* 1762, *Rothert* s.n. (syn. L).

HABITAT: hill and lower montane forest; in dense shade.
ALTITUDINAL RANGE: 600–1800m.
DISTRIBUTION: Borneo (SA). Also Sumatra.

A.reinwardtii *Blume*, Fl. Javae, ser. 2, 1, Orch., (Coll. Orch. Arch. Ind.): 48, Pl.12, fig.2, Pl.12B, fig. 14–29 (1858). Type: Java, *Blume* (holo. L).

HABITAT: unknown.
ALTITUDINAL RANGE: 900m.
DISTRIBUTION: recorded from Mt. Kinabalu, Sabah by Ames (1920), but all subsequent collections from there have proved to be *A.longicalcaratus* J.J.Sm. Also Sumatra, Java and possibly Bunguran (Natuna) Islands and Maluku (Ambon).

[A.spicatus *(Blume) Miq.*, Fl. Ind. Bat. 3: 734 (1859). Type: Kalimantan, *Muller* s.n. (holo. L).

Cystopus spicatus Blume, Fl. Javae, ser. 2,1, Orch. (Coll. Orch. Arch. Ind.): 84, Pl.31, fig 1–10 (1858).

HABITAT: unknown
ALTITUDINAL RANGE: unknown.
DISTRIBUTION: Borneo (K). Endemic.]

See *Pristiglottis spicata*, p.61.

CHEIROSTYLIS Blume

Bijdr.: 413, fig.16 (1825).

Small terrestrial herbs. Stem delicate, leafy, arising from a fleshy, creeping base. Inflorescence erect, often pubescent, few flowered. Flowers resupinate; sepals joined for half their length to form a swollen tube; dorsal sepal joined to petals; petals often oblique; lip joined to base of column, with a short saccate hypochile containing a few papillae each side, claw entire, epichile transversely widened, 2-lobed, lobes toothed or erose; column short, thickened, with 2 slender vertical stelidia, and 2 lateral stigmas under the rostellum, which has 2, often slender vertical arms; pollinia 2.

Between 20 and 25 species mainly found in S.E. Asia, but distributed from East Africa to Japan, the Philippines, New Guinea and Vanuatu.

C.montana *Blume*, Bijdr.: 413, fig.16 (1825). Type: Java, Parang, *Blume* 1245 (holo. L).

HABITAT: mixed hill dipterocarp forest on sandstone ridges.
ALTITUDINAL RANGE: 200 – 300m.
DISTRIBUTION: Borneo (K, SA). Also Thailand, Java, Sumbawa.

C.spathulata *J.J.Sm.* in Bull. Jard. Bot. Buitenzorg, ser. 3, 9: 32, t.5, II (1927). Type: Java, Soerabaja, Gunung Kembangan bei Gresik, *Dorgelo* 2072 (holo. L).

HABITAT: lowland forest.
ALTITUDINAL RANGE: lowlands.
DISTRIBUTION: Borneo (SA). Also Thailand, Vietnam, Java.

CYSTORCHIS Blume

Fl.Javae ser. 2, 1, Orch.(Col. Orch. Arch. Ind.):87 (1858).

Terrestrial herbs. Stem with a few leaves or, in saprophytic species, replaced by brown scale-leaves. Leaves green, purplish, occasionally variegated. Inflorescence lax or rather dense, few to many flowered. Flowers small, resupinate; dorsal sepal and petals connivent, forming a hood; lateral sepals partially or entirely enclosing base of lip; lip divided into a hypochile and epichile; hypochile swollen at base into twin lateral sacs, each usually containing a globular sessile gland; epichile with fleshy involute margins, forming a tube; column very short; rostellum rather large (absent in one species); stigma on front of column, large, rounded; pollinia 2.

About 20 species distributed in India, China, Malaysia, Indonesia, the Philippines, New Guinea and Vanuatu.

C.aphylla *Ridl.* in J. Linn. Soc., Bot. 32: 400 (1896). Type: Peninsular Malaysia, *Ridley* s.n. (holo. SING).

HABITAT: hill and montane dipterocarp forest, sometimes on ultramafic substrate.
ALTITUDINAL RANGE: 200–1300m.
DISTRIBUTION: Borneo (K, SA, SR). Also Peninsular Malaysia, Thailand, Sumatra, Java, Buru, Philippines.

C.macrophysa *Schltr.* in Feddes Repert.9: 429 (1911). Type: Sarawak, *Beccari* 2636 (holo. B, destroyed, iso. FI).

HABITAT: lowland rainforest; alluvial forest; mixed dipterocarp forest on sandstone
ALTITUDINAL RANGE: sea level – 600m.
DISTRIBUTION: Borneo (SR). Endemic.

C.saccosepala *J.J.Sm.* in Mitt. Inst. Allg. Bot. Hamburg 7: 22, t.2, fig.11 (1927). Type: Kalimantan, *Winkler* 521 (holo. HBG, iso. BO).

var. **saccosepala**.

HABITAT: primary forest.
ALTITUDINAL RANGE: 900m.
DISTRIBUTION: Borneo (K,SA). Endemic.

var. **menabaiensis** *J.J.Sm.*, loc.cit.: 23, t.3, fig.12 (1927). Type: Kalimantan, *Winkler* 823 (holo. HBG, iso. BO).

HABITAT: unknown.
ALTITUDINAL RANGE: unknown.
DISTRIBUTION: Borneo (K). Endemic.

C.salmoneus *J.J.Wood* in Lindleyana 5, 2: 81, fig.1 (1990). Type: Sabah, Sipitang District, *Wood* 625 (holo. K).

HABITAT: lower montane forest of *Lithocarpus, Castanopsis, Agathis*, etc. with a field layer of small rattan palms and gingers.
ALTITUDINAL RANGE: 1300m.
DISTRIBUTION: Borneo (SA).Endemic.

C.saprophytica *J.J.Sm.* in Mitt. Inst. Allg. Bot. Hamburg 7: 23, t.3, fig.13 (1927). Type: Kalimantan, *Winkler* 465 (holo. HBG).

HABITAT: Primary forest; ridge forest of oak, chestnut and *Agathis*, on sandstone.
ALTITUDINAL RANGE: 700–1400m.
DISTRIBUTION: Borneo (K, SA, SR).Endemic.

C.variegata *Blume* Fl.Javae, ser. 2, 1, Orch. (Coll. Orch. Arch. Ind.): 89, Pl.24, fig.3 & Pl. 36c (1858). Type: Java, *van Hasselt* s.n. (holo. L).

var.**variegata**.

HABITAT: lowland and hill dipterocarp and lower montane forest, on sandstone.
ALTITUDINAL RANGE: 200 – 1500m.
DISTRIBUTION: Borneo (B, K, SA, SR). Also Peninsular Malaysia, Sumatra, Java, Vanuatu.

var. **purpurea** *Ridl.* in J. Linn. Soc., Bot.32: 399 (1896). Types: Singapore, *Ridley* s.n., Peninsular Malaysia, *Ridley* s.n. (syntypes SING).

HABITAT: primary lowland and hill forest; damp streamside forest; peat swamp forest.
ALTITUDINAL RANGE: sea level – 700m.
DISTRIBUTION: Borneo (SA, SR). Also Peninsular Malaysia, Singapore, Sumatra, Java.

DOSSINIA E.Morren

Ann. Soc. Roy. Agric. Gand 4: 171(1848).

Terrestrial herbs. Leaves basal, 3 – 5, dark green with greenish-yellow, golden or pink nerves. Inflorescence subdense or dense, many-flowered; rachis and floral bracts salmon-pink. Flowers small, resupinate; sepals hirsute; dorsal sepal and petals connivent, forming a hood; lateral sepals free; lip divided into a hypochile and epichile; hypochile saccate, with 2 basal glands and a thick median callus; epichile abruptly expanded into a transverse, obtusely bilobed blade; column short, with 2 longitudinal lamellae in front; pollinia 2.

A monotypic genus endemic to Borneo.

D.marmorata *E.Morren* in Ann. Soc. Roy. Agric. Gand 4: 171, t.193 (1848). Type: 'Java', *Verschaffelt* (holo.: BR).

HABITAT: lowland and hill forest on limestone, growing in leaf litter or moss on or between rocks and on ledges.
ALTITUDINAL RANGE: sea level – 400m.
DISTRIBUTION: Borneo (K, SA, SR). Endemic.

ERYTHRODES Blume

Bijdr.: 410, t.72 (1825).

Terrestrial herbs. Leaves ovate, oblique, green, petioles sheathing. Inflorescence lax, racemose, peduncle quite long, rachis and ovary pubescent. Flowers resupinate; dorsal sepal and petals connivent, forming a hood; lip spurred, concave, apex reflexed, spur 2-lobed, projecting between lateral sepals, internal warts and hairs absent; column short; stigma hollow, at foot of rostellum; pollinia 2.

Between 60 and 100 species, primarily tropical American, with about 15 – 20 represented in the East Asiatic - Pacific area.

E.glandulosa (*Lindl.*) *Ames* in Merr., Bibl. Enum. Born. Pl.: 140 (1921). Type: Borneo, *Lobb* s.n. (holo. K - LINDL).
Physurus glandulosus Lindl. in J.Linn. Soc., Bot. 1:180 (1857).

HABITAT: in humus in forest shade.
ALTITUDINAL RANGE: unknown.
DISTRIBUTION: Borneo (unspecified). Endemic.

E.latifolia *Blume*, Bijdr.:411, fig.72(1825). Type: Java, Salak, *Blume* s.n. (holo. L).
E.humilis (Blume) J.J.Sm. in Bull. Dép. Agric. Indes. Néerl. 13:11 (1907). Type: West Java, Bantam Province, *Blume* s.n. (holo. L).

HABITAT: hill forest; mixed oak-chestnut forest on sandstone ridges; riverine forest.
ALTITUDINAL RANGE: 900 – 1400m.
DISTRIBUTION: Borneo (SA). Also Peninsular Malaysia, Sumatra, Java.

E triloba *Carr* in Gard. Bull. Straits Settlem., 8:181 (1935). Types: Sabah, Mt. Kinabalu, Tenompok, by bridle path to Ranau, *Carr* C.3564 (SFN 27859) (syn. SING, isosyn. AMES, K); Tenompok, *Clemens* 27689 & 28420 (syn. SING).

HABITAT: lower montane forest.
ALTITUDINAL RANGE: 1400 – 1500m.
DISTRIBUTION: Borneo (SA). Endemic.

GOODYERA R.Br.

in W.Aiton & W.T.Aiton, Hort.Kew, ed.2, 5:197 (1813).

Terrestrial, rarely epiphytic, herbs. Leaves ovate, often asymmetric, sometimes variegated, petioles sheathing stem. Stem, peduncle and ovaries finely pubescent. Inflorescence erect, 1 to many flowered. Flowers usually not opening widely, resupinate, white, pale green, pink to purplish; sepals parallel to floral axis, or the lateral sepals spreading, dorsal sepal connivent with petals, forming a hood; petals sometimes joined near their tips; lip concave or saccate, glabrous or papillose, with bristly hairs inside, narrowed to an acute, entire, often reflexed tip; column short, without basal appendages; rostellum usually long, deeply cleft; stigma not divided, large, ventral; pollinia 2.

A cosmopolitan genus of about 50 species, mostly distributed in warmer climates.

G.bifida *(Blume) Blume*, Fl.Javae, ser. 2,1, Orch. (Coll. Orch. Arch. Ind.):40 (1858). Type: Java, Gede, *Blume* s.n. (holo. L).
  Neottia bifida Blume, Bijdr.:408 (1825).

HABITAT: lower montane forest.
ALTITUDINAL RANGE: 1500 – 2100m.
DISTRIBUTION: Borneo (SA). Also Peninsular Malaysia, Sumatra, Java.

G.colorata *(Blume) Blume*, Fl. Javae, ser. 2, 1, Orch. (Coll. Orch. Arch. Ind.): 37 (1858). Type: Java, Salak Mountains, Tjapus River, *Blume* s.n. (holo. L).
  Neottia colorata Blume, Bijdr.: 409 (1825).

HABITAT: mixed oak-chestnut lower montane forest on sandstone ridges.
ALTITUDINAL RANGE: 1200 – 1500m.
DISTRIBUTION: Borneo (SA). Also Peninsular Malaysia, Sumatra, Java.

G.hylophiloides *Carr* in Gard. Bull. Straits Settlem., 8,3: 195 (1935). Types: Sabah, Mt. Kinabalu, above Tenompok, *Carr* C.3256 (SFN 26939) (syn. SING, isosyn. AMES,K); Silau basin and trail to Lumu Lumu, *Clemens* 29993 (syn. BM); Dallas, *Clemens* s.n.(syn. BM).

HABITAT: hill and lower montane forest; ridge forest; recorded from sandstone soil.
ALTITUDINAL RANGE: 400 – 2300m.
DISTRIBUTION: Borneo (SA, SR). Endemic.

G.kinabaluensis *Rolfe* in Gibbs in J. Linn. Soc., Bot., 42: 159 (1914). Types: Sabah, Mt. Kinabalu, Gurulau Spur, above Kiau ridge, *Gibbs* 3997, 4003 (syn. unlocated).

HABITAT: hill and lower montane forest; recorded from sandstone soil.
ALTITUDINAL RANGE: 800 – 1700m.
DISTRIBUTION: Borneo (SA). Endemic.

G.procera (*Ker Gawl.*) *Hook.*, Exot. Fl. 1, 3, t.39 (1823). Type: Nepal, (holo. BM-LAMBERT).
Neottia procera Ker Gawl. in Bot. Reg., 8: t.639 (1822).

HABITAT: upper montane forest; forest trails; moss-covered rocks.
ALTITUDINAL RANGE: 2100 – 2400m.
DISTRIBUTION: Borneo (SA). Widespread from India, Sri Lanka and Burma east through Indochina and Thailand to Indonesia, north to China, Taiwan, Japan and the Philippines.

G.pusilla *Blume*, Fl. Javae, ser. 2, 1, Orch. (Coll. Orch. Arch. Ind.): 36, Pl.9b, fig.3 (1858). Type: West Java, Pangarangu, *Blume* s.n. (holo. L).

HABITAT: in shade on sandstone soil.
ALTITUDINAL RANGE: 1000 – 1200m.
DISTRIBUTION: Borneo (SA). Also Java.

G.reticulata (*Blume*) *Blume*, Fl. Javae, ser. 2, 1, Orch. (Coll. Orch. Arch. Ind.): 35, Pl.96, fig.1 (1858). Type: West Java, *Blume* s.n. (holo. L, iso. K).
Neottia reticulata Blume, Bijdr.: 408 (1825).

HABITAT: mixed oak-chestnut lower montane forest on sandstone ridges.
ALTITUDINAL RANGE: 900m.
DISTRIBUTION: Borneo (SA). Also Java.

G.rostellata *Ames & C.Schweinf.*, Orch. 6:12 (1920). Type: Sabah, Mt. Kinabalu, Kiau, *Clemens* 401 (holo. AMES).

HABITAT: hill and lower montane forest.
ALTITUDINAL RANGE: 900 – 2700m.
DISTRIBUTION: Borneo (SA). Endemic.

G.rostrata *Ridl.* in J. Straits Branch Roy. Asiat. Soc. 49:40 (1908). Type: Sarawak, Lingga, *Hewitt* s.n. (holo. SING).

HABITAT: unknown.
ALTITUDINAL RANGE: unknown.
DISTRIBUTION: Borneo (SR). Endemic.

G.rubicunda (*Blume*) *Lindl.* in Bot. Reg. 25, misc. 61 (1839). Type: Java, *Blume* s.n. (holo. BO).
Neottia rubicunda Blume, Bijdr.: 408 (1825).

HABITAT: primary hill and lower montane forest; river banks; in leaf litter among limestone boulders.
ALTITUDINAL RANGE: 500 – 1500m.
DISTRIBUTION: Borneo (K, SA). Also Peninsular Malaysia, Indonesia, Philippines, New Guinea, Australia (Queensland), north to Taiwan and the Ryukyu Islands.

G.ustulata *Carr* in Gard. Bull. Straits Settlem. 8:194 (1935). Types: Sabah, Ulu Kagitang, *Carr* 3236 (SFN 27359) (syn. SING); Dallas, *Clemens* 26911 (syn. SING) & 26690a (syn. BM).

HABITAT: hill forest; secondary forest; ridge forest; in leaf litter among limestone boulders; also recorded from hill dipterocarp forest on sandstone soil.
ALTITUDINAL RANGE: 500 – 1100m.
DISTRIBUTION: Borneo (SA). Endemic.

G.viridiflora *(Blume) Blume*, Fl. Javae, ser. 2, 1, Orch. (Coll. Orch. Arch. Ind.):41, Pl.9c, fig.2 (1858). Type: Java, *Blume* s.n. (holo. L).
Neottia viridiflora Blume, Bijdr.: 415 (1825).
Goodyera rosans J.J.Sm. in Bull. Jard. Bot. Buitenzorg, ser. 3,9:35, t.3, f.4 (1927). Type: Java, Buitenzorg, Tjipoeti bei Tjiampea, *Bakhuizen van den Brink* s.n., cult. *Smith* (holo. L).

HABITAT: unknown.
ALTITUDINAL RANGE: 400m.
DISTRIBUTION: Borneo (SA). Widespread from N.India and the Himalayas south to Thailand and Peninsular Malaysia, east through Indonesia to New Guinea and the Pacific islands, north to the Philippines, Taiwan, Ryukyu Islands and China.

HETAERIA Blume

Bijdr.: 409 (1825).

Terrestrial herbs. Rhizome decumbent, rooting at nodes. Plants 25–60cm. tall when in flower. Leaves broad, ovate to elliptic, usually asymmetric, with sheathing petioles, green. Inflorescence a lax or dense, many flowered raceme. Flowers small, non-resupinate, lip uppermost; dorsal sepal and petals connivent, forming a hood; lateral sepals enclosing lip base; lip ventricose at base, convex, apex shallow and narrow, entire or with divaricate apical lobes, containing papillae or glands above base on both sides; column short, with 2 parallel, keel-like lamellae in front; stigma bilobed; pollinia 2.

Around 20 species distributed in the Old World tropics, extending from India to Fiji, the majority of species native to Malaysia.

H.alta *Ridl.* in J. Linn. Soc., Bot., 32:404 (1896). Type: Peninsular Malaysia, Perak, Hermitage Hill, Kwala Kangsa Valley, *Ridley* s.n. (holo. SING).

HABITAT: in humus in lowland and hill forest; recorded from limestone.
ALTITUDINAL RANGE: 100 – 500m.
DISTRIBUTION: Borneo (SA). Also Peninsular Malaysia, Peninsular Thailand.

H.angustifolia *Carr* in Gard. Bull. Straits Settlem. 8:190 (1935). Type: Sabah, Mt. Kinabalu, main spur above Koung, *Carr* 3770 (holo. SING, iso. K).

HABITAT: hill dipterocarp forest; interface between old secondary and primary forest; lower montane forest of *Lithocarpus, Castanopsis, Agathis alba*, etc. with field layer of small rattans and gingers.
ALTITUDINAL RANGE: 700 – 1000m.
DISTRIBUTION: Borneo (SA). Endemic.

H.anomala *(Lindl.) Hook.f.*, Fl. Brit. Ind. 6:116 (1890). Type: India (Assam), Tingree, *Griffith* s.n. (holo. K-LINDL).
Aetheria anomula Lindl. in J.Linn. Soc., Bot., 1:185 (1857).
Hetaeria biloba (Ridl.) Seidenf. & J.J.Wood in Orch. Pen. Mal. & Sing.: 95 (1992). Type: Peninsular Malaysia, Telom, on ridge above Batang Padang Valley, *Ridley* s.n. (holo. SING).
Zeuxine biloba Ridl. in J. Fed. Malay States Mus. 4:73 (1909).
Hetaeria grandiflora Ridl. in J. Straits Branch Roy. Asiat. Soc. 87:98 (1923). Type: Sumatra, Berastagi hill woods on the way to Sibayak, *Ridley* s.n. (holo. SING, iso. K).

H.rotundiloba J.J.Sm. in Svensk Bot. Tidskr. 20:470 (1926). Type: Sulawesi, Balaang Mongondou, *Kaudern* 129 (holo. L).

HABITAT: lower and upper montane forest; oak-laurel forest; among rocks; in humus; beside rivers; old landslide areas on ultramafic substrate.
ALTITUDINAL RANGE: 1700 – 2000m.
DISTRIBUTION: Borneo (SA). Also India (Assam), Burma, Peninsular Malaysia, Thailand, Sulawesi, Sumatra.

H.obliqua *Blume,* Fl. Javae, ser. 2, 1, Orch. (Coll. Orch. Arch. Ind.): 104, Pl.34, fig.1 (1858). Type: Kalimantan, Babay, *Korthals* s.n. (holo. L).

HABITAT: in leaf litter on limestone and sandstone rocks, in shade.
ALTITUDINAL RANGE: 300 – 400m.
DISTRIBUTION: Borneo (B, K, SA, SR). Peninsular Malaysia, Thailand (Terutao Island), Sumatra.

H.oblongifolia *(Blume) Blume,* Fl. Javae, ser. 2, 1, Orch. (Coll. Orch. Arch. Ind.): 102, Pl.32, fig.3 (1858). Type: Java, Tjanjor, *Blume* s.n. (holo. L).
Etaeria oblongifolia Blume, Bijdr.: 410 (1825).

HABITAT: primary ridge forest.
ALTITUDINAL RANGE: lowlands.
DISTRIBUTION: Borneo (K, SA). Widespread from the Andaman Islands east through Thailand, Vietnam, Peninsular Malaysia, Indonesia, New Guinea and the Pacific islands, north to the Philippines, Japan and Ryukyu Islands.

HYLOPHILA Lindl.

Bot. Reg. 19: sub t. 1618 (1833)

Dicerostylis Blume, Fl.Javae, ser. 2, 1, Orch. (Coll. Orch. Arch.Ind.) : 98 (1858).

Terrestrial herbs. Rhizomes fleshy, creeping. Stems erect, leafy. Leaves ovate to elliptic, often asymmetric, with sheathing petioles. Inflorescence erect, many flowered, dense. Flowers small, resupinate, usually not opening widely; dorsal sepal and petals connivent, forming a hood; lateral sepals oblique, enclosing whole of lip; lip saccate, hairy within, near the narrow reflexed blade; column short; anther long, acute; rostellum long, deeply divided; stigma convex, ventral, sometimes with a short, horn-like appendage spreading on either side.

About 6 species distributed from Thailand through Malaysia, Indonesia and the Philippines eastwards to New Guinea and the Solomon Islands.

H.cheangii *Holttum* in Gard. Bull. Singapore 25:106 (1969). Type: Peninsular Malaysia, Bukit Fraser, *Cheang* s.n. (holo. K).

HABITAT: unknown.
ALTITUDINAL RANGE: 900m.
DISTRIBUTION: Borneo (SR). Also Peninsular Malaysia.

H.lanceolata *(Blume) Miq.,* Fl. Ind. Bat. 3:746 (1859). Type: W.Java, *Blume* s.n. (holo. BO).
Dicerostylis lanceolata Blume, Fl. Javae, ser. 2, 1, Orch. (Coll. Orch. Arch. Ind.): 116, Pl.38, fig.1 (1858).
D.kinabaluensis Carr in Gard. Bull. Straits Settlem. 8: 192 (1935). Type: Sabah, Mt.

Kinabalu, near Bundu Tuhan, *Carr* 3614 (SFN 27861) (holo. SING, iso. AMES, K).

HABITAT: damp shady areas in hill and lower montane forest; on mossy rocks.
ALTITUDINAL RANGE: 900 – 1500m.
DISTRIBUTION: Borneo (K, SA). Also Thailand, Sumatra, Java, Flores, Philippines.

H.mollis *Lindl.*, Gen. Sp. Orch. Pl.: 490 (1840). Type: Singapore, *Wallich* s.n. (holo. K -LINDL).

HABITAT: under *Gymnostoma sumatrana* beside stream in heath forest.
ALTITUDINAL RANGE: unknown.
DISTRIBUTION: Borneo (K, SR). Also Peninsular Malaysia, Singapore, Sumatra.

KUHLHASSELTIA J.J.Sm.

Icon. Bogor. 4:1 (1910).

Terrestrial herbs. Rhizomes decumbent, rooting at nodes. Leaves small, petioles sheathing, blade often purple on reverse, margin sometimes undulate. Inflorescence lax or dense, few to several flowered. Flowers small, resupinate; dorsal sepal and petals connivent, forming a hood; lateral sepals enclosing lip base; lip saccate at base and containing 2 glands, narrowed above into a claw with inflexed margins and a short spathulate apical blade; column without appendages; stigma entire; pollinia 2.

About 6 species distributed in Malaysia, Indonesia, the Philippines and New Guinea.

K.javanica *J.J.Sm.* in Icon. Bogor. 4:t.301 (1910). Type: Java, Gunung Boender, *Smith* s.n. (holo. L).
 K.kinabaluensis Ames & C.Schweinf., Orch. 6:14 (1920). Type: Sabah, Mt. Kinabalu, Marai Parai Spur, *Clemens* 398 (holo. AMES).

HABITAT: lower montane forest; in moss and leaf litter; recorded from sandstone and ultramafic substrate.
ALTITUDINAL RANGE: 1400 – 2000m.
DISTRIBUTION: Borneo (SA, SR). Also Java.

K.rajana *J.J.Sm.* in Mitt. Inst. Allg. Bot. Hamburg 7:26, t.3, fig.15 (1927). Type: Kalimantan, Bukit Raja, *Winkler* 1013 (holo. HBG, drawing at BO).

HABITAT: lower montane forest.
ALTITUDINAL RANGE: 1400m.
DISTRIBUTION: Borneo (K). Endemic.

LEPIDOGYNE Blume

Fl. Javae, ser. 2, 1, Orch. (Coll. Orch. Arch. Ind): 93, Pl.25 (1858).

Tall terrestrial herbs. Leaves grouped at base, many, narrowly elliptic, acuminate, with sheathing base. Inflorescence densely many flowered, long, peduncle up to 30cm. long, rachis up to 40cm. long, pubescent; floral bracts narrow, longer than flowers. Flowers reddish-brown; dorsal sepal and petals connivent, forming a hood; lateral sepals enclosing base of lip; lip with swollen concave base containing a transverse row of small calli, blade 3-lobed, side lobes small, erect, mid-lobe long and narrow; column short; anther rostrate; rostellum very long, deeply cleft; stigma transverse, covered by a plate springing from its lower edge; pollinia 2, long, clavate, deeply divided.

Three species distributed in Malaysia, Indonesia, the Philippines and New Guinea.

L.longifolia *(Blume) Blume*, Fl. Javae, ser. 2, 1, Orch. (Coll. Orch. Arch. Ind.): 94, Pl.25 (1858). Type: Java, Salak, *Blume* s.n. (holo. L).
Neottia longifolia Blume, Bijdr.: 406 (1825).

HABITAT: hill and lower montane forest; oak-laurel forest; recorded from ultramafic substrate.
ALTITUDINAL RANGE: 900 – 1500m.
DISTRIBUTION: Borneo (SA). Peninsular Malaysia east to the Philippines and probably New Guinea.

MACODES (Blume) Lindl.

Gen.Sp.Orch. Pl.:496(1840).

Terrestrial herbs. Rhizome creeping, fleshy. Stems erect, short or long. Leaves rather fleshy, ovate to elliptic-subcircular, margins sometimes undulate, petioles sheathing, blade with coloured veins. Inflorescence laxly few to many flowered, peduncle and bracts pubescent. Flowers rather small, non-resupinate, lip uppermost, asymmetric owing to twisting of lip and column; sepals pubescent on exterior; lateral sepals enclosing base of lip; lip bipartite or 3-lobed, hypochile saccate at base, sac containing 2 glands, epichile twisted to one side; column short, twisted, with 2 thin close parallel wings on the front, the wings descending down into the saccate hypochile; anther acute; rostellum cleft; stigma entire, large; pollinia 2.

About 10 species distributed from Malaysia to the Pacific islands.

M.angustilabris *J.J.Sm.* in Bull. Jard. Bot. Buitenzorg, Ser. 3, 11:90 (1931). Type: Kalimantan, Long Liah Leng, *Endert* 3026 (holo. L, iso. BO).

HABITAT: wet places; rocks.
ALTITUDINAL RANGE: 200 – 300m.
DISTRIBUTION: Borneo (K). Endemic.

M.lowii *(E.J.Lowe) J.J.Wood* in Orchid Digest 48, 4:155 (1984). Type: Beautiful Leaved Plants, fig.40, leaf 'from a plant supplied by Mr.Howard', lecto.
Anoectochilus lowii E.J.Lowe in Beautiful Leaved Plants: 81–82, fig.40 (1868).

HABITAT: hill forest under *Gymnostoma*; among rocks in light shade, on ultramafic substrate.
ALTITUDINAL RANGE: 400 – 1100m.
DISTRIBUTION: Borneo (SA). Endemic.

M.petola *(Blume) Lindl.*, Gen. Sp. Orch. Pl.:497 (1840). Type: Java, locality unknown, *Blume* s.n. (holo. ?L).
Neottia petola Blume, Bijdr.:407, t.2 (1825).

HABITAT: mossy rocks in hill and lower montane forest, on ultramafic substrate.
ALTITUDINAL RANGE: 100 – 1500m.
DISTRIBUTION: Borneo (K, SA). Also Peninsular Malaysia, Sumatra, Java, Philippines.

MYRMECHIS (Lindl.) Blume

Fl. Javae, ser. 2, 1, Orch. (Coll. Orch. Arch. Ind.): 76 (1858).

Small terrestrial herbs. Rhizomes decumbent. Leaves ovate, spread out along stem and with several clustered at base of inflorescence. Inflorescence 2–3 flowered. Flowers resupinate, white; dorsal sepal and petals connivent, forming a hood; lateral sepal with an oblique, concave base; lip long clawed, spathulate, base saccate, containing 2 bilobed glands, apex shallowly bilobed; column short; rostellum short, bifid; stigmas 2; pollinia 2.

Five or six species distributed in India, China and Japan, south to Malaysia, Indonesia and the Philippines.

M.kinabaluensis *Carr* in Gard. Bull. Straits Settlem. 8:188 (1935). Types: Sabah, Mt. Kinabalu, above Kamborangah, *Carr* 353a (SFN 27595) (syn. SING, isosyn. AMES); Mesilau, *Clemens* 29217 (syn. BM).

HABITAT: in very wet, mossy, rather stunted upper montane forest; mossy rock faces.
ALTITUDINAL RANGE: 2100 – 2800m.
DISTRIBUTION: Borneo (SA). Endemic.

PRISTIGLOTTIS Cretz. & J.J.Sm.

Acta Fauna Fl. Universali, ser. 2, Bot. 1, no.14: 4 (1934).

Cystopus Blume, Fl.Javae, ser. 2, 1, Orch. (Coll. Orch. Arch. Ind.): 82 (1858), non Leveille

Terrestrial herbs. Leaves rather small, green or slightly coloured. Inflorescence short, with a few rather large flowers. Flowers resupinate, white; dorsal sepal and petals connivent, forming a hood; lateral sepals spreading; lip with a short saccate base enclosed by lateral sepals, containing 2 sessile glands, narrowed to a long channelled claw which expands into a bilobed blade; column with 2 narrow wings which project into base of lip; rostellum and anther long; stigma entire, ventral; pollinia 2.

About 15 species distributed from India and China southeastwards through Malaysia and Indonesia to the Pacific islands.

P.hasseltii *(Blume) Cretz. & J.J.Sm.* in Acta Fauna Fl. Universali ser. 2, Bot., 1, no.14: 4 (1934). Type: Java, Bantam Province, *van Hasselt* s.n. (holo. L).
Cystopus hasseltii Blume, Fl. Javae, ser. 2, 1, Orch. (Coll. Orch. Arch. Ind.): 86, Pl.30, fig.4, Pl.36B (1858).

HABITAT: mossy stream banks; in leaf litter on sandy soil under oaks, etc. in lower montane forest.
ALTITUDINAL RANGE: 1200 – 1600m.
DISTRIBUTION: Borneo (SA, SR). Also Java.

P.hydrocephala *J.J.Sm.* in Bull. Dép. Agric. Indes Néerl., 22:9 (1909). Type: Kalimantan, Amai Ambit, *Hallier* 3285 (holo. L, iso. BO).

HABITAT: unknown.
ALTITUDINAL RANGE: unknown.
DISTRIBUTION: Borneo (K). Endemic.

P.occulta *(Blume) Cretz. & J.J.Sm.*, in Acta Fauna Fl. Universali, ser. 2, Bot., 1, no.14: 4 (1934). Type: Java, Bantam Province, *Blume* (holo. L).
Cystopus occultus Blume, Fl. Javae, ser. 2, 1, Orch. (Coll. Orch. Arch. Ind.): 85, Pl.30, fig.2 (1858).

HABITAT: in moss on boulders; moss forest.
ALTITUDINAL RANGE: 1500m.
DISTRIBUTION: Borneo (SR). Also Java.

P.serriformis *(J.J.Sm.) Cretz. & J.J.Sm.* in Acta Fauna Fl. Universali, ser. 2, Bot.,1, no.14: 6 (1934). Type: Kalimantan, Bukit Raja, *Winkler* 995 (holo. HBG, drawing at BO).
Cystopus serriformis J.J.Sm. in Mitt. Inst. Allg. Bot. Hamburg 7:25, t.3, fig.14 (1927).

HABITAT: unknown.
ALTITUDINAL RANGE: 1400m.
DISTRIBUTION: Borneo (K). Endemic.

P.spicata *(Blume) Cretz. & J.J.Sm.* in Acta Fauna Fl. Universali, ser. 2, Bot., 1 (14): 6 (1934). Type: Borneo, *Muller* s.n. (holo. L).
Cystopus spicatus Blume, Fl. Javae, ser. 2, 1, Orch. (Coll. Orch. Arch. Ind.): 84, Pl.31, fig.1 (1858).
Anoectochilus spicatus (Blume) Miq., Fl. Ind. Bat. 3: 734 (1859).

HABITAT: unknown.
ALTITUDINAL RANGE: unknown.
DISTRIBUTION: Borneo (K). Endemic.

VRYDAGZYNEA Blume

Fl. Javae, ser. 2, 1, Orch. (Coll. Orch. Arch. Ind.): 71 (1858).

Terrestrial herbs. Rhizome decumbent, rooting at nodes. Stems weak, fleshy. Leaves few, green, sometimes with a median white stripe. Inflorescence usually short and densely many flowered. Flowers small, resupinate, not opening widely; dorsal sepal and petals connivent, forming a hood; lip parallel to column, entire, hypochile bearing a prominent spur projecting between lateral sepals and containing 2 stalked glands; column very short; stigma bilobed; pollinia 2.

Between 20 and 40 species, according to opinion, distributed from northern India to Taiwan and through Malaysia, Indonesia eastwards to New Guinea, Australia and the Pacific islands.

V.albida *(Blume) Blume,* Fl. Javae, ser. 2, 1, Orch. (Coll. Orch. Arch. Ind.): 75, Pl.19, fig.2 (1858). Type: Java, Salak, Seribu, etc. *Blume* s.n. (holo. L).
 Etaeria albida Blume, Bijdr.: 410 (1825).

HABITAT: stream banks in hill and lower montane forest; recorded from limestone soils.
ALTITUDINAL RANGE: 300 – 1500m.
DISTRIBUTION: Borneo (SA, SR). Widespread from India, Thailand, Vietnam, Peninsular Malaysia to Sumatra and Java east to the Philippines and, possibly, New Guinea.

V.angustisepala *J.J.Sm.* in Mitt. Inst. Allg. Bot. Hamburg 7: 20, t.2, fig.9 (1927). Type: Kalimantan, Bidang Menabai, *Winkler* 1095 (holo. HBG, drawing at BO).

HABITAT: primary hill forest.
ALTITUDINAL RANGE: 700m.
DISTRIBUTION: Borneo (K). Endemic.

V.argentistriata *Carr* in Gard. Bull. Straits Settlem. 8:183 (1935). Type: Sabah, Mt. Kinabalu, near Bundu Tuhan, *Carr* 3713 (SFN 28051) (holo. SING, iso. K).

HABITAT: hill forest.
ALTITUDINAL RANGE: 900m.
DISTRIBUTION: Borneo (SA). Endemic.

V.beccarii *Schltr.* in Feddes Repert. 9:429 (1911). Type: Sarawak, Kuching, *Beccari* 905 (holo. B).

HABITAT: unknown.
ALTITUDINAL RANGE: sea level.
DISTRIBUTION: Borneo (SR). Endemic.

V.bicostata *Carr* in Gard. Bull. Straits Settlem. 8:185 (1935). Type: Mt. Kinabalu, Upper Kinunut valley, *Carr* 3344 (SFN 27134) (holo. SING, iso. K).

HABITAT: hill and lower montane forest.
ALTITUDINAL RANGE: 1000 – 1500m.
DISTRIBUTION: Borneo (SA). Endemic.

V.bractescens *Ridl.* in Kew Bull.: 87 (1926). Type: Sumatra, Siberut Island, *Boden-Kloss* 11443 (holo. SING, iso. K).

HABITAT: hill forest.
ALTITUDINAL RANGE: 1200m.
DISTRIBUTION: Borneo (SA, SR). Also Sumatra.

V.elata *Schltr.* in Feddes Repert. 9:430 (1911). Type: Sarawak, Kuching, *Beccari* 2673 (holo. B, iso. FI).

HABITAT: mixed lowland and hill dipterocarp forest on sandstone.
ALTITUDINAL RANGE: sea level – 1200m.
DISTRIBUTION· Borneo (SA,SR). Endemic.

V.endertii *J.J.Sm.* in Bull. Jard. Bot. Buitenzorg, ser. 3, 11:85 (1931). Type: Kalimantan, near Long Liah Leng, *Endert* 2992 (holo. L, iso. BO).

HABITAT: primary forest.
ALTITUDINAL RANGE: 200 – 300m.
DISTRIBUTION: Borneo (K). Endemic.

V.grandis *Ames & C.Schweinf.*, Orch. 6:16 (1920). Type: Sabah, Mt. Kinabalu, Kiau, *Clemens* 340 (holo. AMES).

HABITAT: hill and lower montane forest.
ALTITUDINAL RANGE: 900 – 1500m.
DISTRIBUTION: Borneo (SA). Endemic.

V.lancifolia *Ridl.* in J. Linn. Soc., Bot. 32: 398 (1896). Types: Singapore, Bukit Timah, *Ridley* s.n. (syn. SING); Peninsular Malaysia, Johor, Gunung Panti, *Ridley* s.n. (syn. SING).

HABITAT: among leaf litter in podsol forest.
ALTITUDINAL RANGE: 500 – 600m.
DISTRIBUTION: Borneo (SA). Also Peninsular Malaysia, Singapore, Thailand.

V.nuda *Blume,* Fl. Javae, ser. 2, 1, Orch.(Coll. Orch. Arch. Ind.): 74, Pl.20, fig.3 (1858). Type: Java, Bantam Province, Harriang, *van Hassell* (holo. L).

HABITAT: leaf litter on rocky cliffs; dipterocarp forest on sandstone.
ALTITUDINAL RANGE: 200 – 300m.
DISTRIBUTION: Borneo (SA). Also Java.

V.pauciflora *J.J.Sm.* in Mitt. Inst. Allg. Bot. Hamburg 7:19, t.2, fig.8 (1927). Type: Kalimantan, Bidang Menabai, *Winkler* 822 (holo. HBG, drawing at BO).

var. **pauciflora.**

HABITAT: primary hill forest.
ALTITUDINAL RANGE: 700m.
DISTRIBUTION: Borneo (K).Endemic.

var.**unistriata** J.J.Sm., loc. cit.: 20 (1927). Type: Kalimantan, Bukit Obat, *Winkler* 1325 (holo. HBG).

HABITAT: lowland primary forest.
ALTITUDINAL FOREST: 100 – 200m.
DISTRIBUTION: Borneo (K). Endemic.

V.semicordata *J.J.Sm.* in Bull. Jard. Bot. Buitenzorg, ser. 3, 11:86 (1931). Type: Kalimantan, Gunung Kemoel, *Endert* 4347 (holo. L, iso. BO).

HABITAT: lower montane forest.
ALTITUDINAL RANGE: 1600m.
DISTRIBUTION: Borneo (K). Endemic.

V.tilungensis *J.J.Sm.* in Mitt. Inst. Allg. Bot. Hamburg 7:21, t.2, fig.10 (1927). Type: Kalimantan, Bukit Tilung, *Winkler* 1462 (holo. HBG, drawing at BO).

HABITAT: lowland primary forest.
ALTITUDINAL RANGE: 100 – 200m.
DISTRIBUTION: Borneo (K). Endemic.

V.tristriata *Ridl.* in J. Linn. Soc., Bot. 32:398 (1896). Type: Singapore, Chan Chu Kang, *Ridley* s.n. (holo. SING).

HABITAT: on streamside rocks in secondary forest.
ALTITUDINAL RANGE: 300m.
DISTRIBUTION: Borneo (SR). Also Peninsular Malaysia, Singapore, Thailand.

ZEUXINE Lindl.

Coll. Bot., app., 18, as *Zeuxina* (1826).

Terrestrial herbs. Rhizome decumbent, rooting at nodes. Stems weak, fleshy. Leaves ovate, elliptic or linear, sessile on a broad sheath, or petiolate, sometimes with a coloured median nerve. Inflorescence few to many-flowered, lax or dense. Flowers small, resupinate, not opening widely; dorsal sepal and petals connivent, forming a hood; lateral sepals enclosing base of lip; lip with a saccate base, usually with inflexed margins and containing 2 glands, more or less sulcate on lower surface, blade transversely widened and bilobed, small, connected to the saccate base by a short neck or an elongated claw; column short, with or without appendages in front; stigma bilobed; pollinia 2.

Between 40 and 50 species distributed throughout the Old World tropics of Africa, Asia and the Pacific islands. Kores (1989:32) quotes a figure of 70 species. The circumscription of the genus and the species delimitation are uncertain.

Z.flava *(Wall.ex Lindl.) Benth. ex Hook.f.*, Fl. Brit. Ind. 6:108 (1890). Type: Nepal, Gukkurrum, Mayo, *Wallich* 7380A (holo. K-LINDL).
  Monochilus flavum Wall.ex Lindl., Gen. Sp. Orch. Pl.:487 (1840).

  HABITAT: secondary forest, associated with gingers and grasses.
  ALTITUDINAL RANGE: 100 – 200m.
  DISTRIBUTION: Borneo (SA). Also India, Nepal, Bhutan, Burma, Thailand.

Z.gracilis *(Breda) Blume*, Fl.Javae, ser. 2, 1, Orch. (Coll. Orch. Arch. Ind.): 69, Pl.18, fig.2, t.23D (1858). Type: Java, Bantam Province, *van Hasselt* s.n. (holo. L).
  Psychechilos gracile Breda, Gen. Sp. Orch. Asclep., t.9 (1827).

  HABITAT: lower montane oak-laurel forest; mixed hill dipterocarp forest; low stature forest;, secondary forest; sometimes on ultramafic substrate.
  ALTITUDINAL RANGE: 200 – 1900m.
  DISTRIBUTION: Borneo (SA). Also Peninsular Malaysia, islands off Sumatra, Java, Krakatau.

Z.kutaiensis *J.J.Sm.* in Bull. Jard. Bot. Buitenzorg, ser. 3, 11:89 (1931). Type: Kalimantan, near Moeara Antjaloeng, *Endert* 2132 (holo. L, iso. BO).

  HABITAT: 'on a rather high bank wall'.
  ALTITUDINAL RANGE: lowlands.
  DISTRIBUTION: Borneo (K). Endemic.

Z.linguella *Carr* in Gard. Bull. Straits Settlem. 8:71 (1935). Type: Sarawak, Mt.Dulit, *Synge* S.302 (holo. SING, iso. K).

  HABITAT: 'on banks of rocky stream in primary forest'.
  ALTITUDINAL RANGE: under 300m.
  DISTRIBUTION: Borneo (SR). Endemic.

Z.papillosa *Carr* in Gard. Bull. Straits Settlem. 8:189 (1935). Type: Sabah, Mt. Kinabalu, beside Kadamaian River, *Carr* C.3159 (SFN 26671) (holo. SING, iso. ?).

  HABITAT: hill forest.
  ALTITUDINAL RANGE: 900m.
  DISTRIBUTION: Borneo (SA). Endemic.

Z.petakensis *J.J.Sm.* in Bull. Jard. Bot. Buitenzorg, ser. 3, 87 (1931). Type: Kalimantan, near Long Petak, *Endert* 3260 (holo. L, iso. BO).

HABITAT: marshy areas in primary forest.
ALTITUDINAL RANGE: 400 – 500m.
DISTRIBUTION: Borneo (K). Endemic.

Z.purpurascens *Blume,* Fl.Javae, ser. 2,1, Orch. (Coll. Orch. Arch. Ind.): 71, Pl.18, fig.3, Pl.23E (1853). Types: Sumatra, Borneo, *Blume* s.n. (holo. L).

HABITAT: near streams in rocky lowland primary forest.
ALTITUDINAL RANGE: under 300m.
DISTRIBUTION: Borneo (SA, SR). Also Sumatra, Java.

Z.strateumatica *(L.) Schltr.,* in Bot. Jahrb. Syst. 45:394 (1911). Type: Sri Lanka, 'habitat in Zeylona' (holo. LINN).
Orchis strateumatica L., Sp. Pl.: 943 (1753).

HABITAT: open grassy areas; waste places; cultivated land; rarely in hill forest.
ALTITUDINAL RANGE: sea level – 900m.
DISTRIBUTION: Borneo (SA). Widespread from Afghanistan to Japan, south to New Guinea.

Z.violascens *Ridl.,* Mat. Fl. Mal. Pen. 1:218 (1907). Type: Peninsular Malaysia, Tanjong Kupang, *Ridley* s.n. (holo. SING).

HABITAT: peat swamp forest.
ALTITUDINAL RANGE: unknown.
DISTRIBUTION: Borneo (SR). Also Peninsular Malaysia, Sumatra.

Z.viridiflora *(J.J.Sm.) Schltr.* in Bull. Herb. Boissier, ser. 2, 6:298 (1906). Type: Sulawesi, Gorontalo, *Smith* s.n. (holo. L).
Haplochilus viridiflorus J.J.Sm., Icon. Bogor. 2:21, t.IVB (1903).

HABITAT: unknown.
ALTITUDINAL RANGE: unknown.
DISTRIBUTION: Borneo (unspecified). Also Java, Sulawesi.

SUBTRIBE SPIRANTHINAE

SPIRANTHES Rich.

Mém. Mus. Hist. Nat. 4:50 (1818).

Terrestrial herbs. Roots fleshy, fasciculate. Stems erect, rather short, leafy. Leaves narrow, conduplicate, fleshy. Inflorescence, more or less densely spicate, flowers arranged spirally on rachis. Flowers small, tubular, white or pink; sepals free, dorsal sepal porrect, connivent with petals, forming a hood; lateral sepals oblique, more or less spreading; lip concave at base, containing 2 appendages, obscurely 3-lobed, undulate at apex; column short; stigma convex; pollinia 2.

A cosmopolitan genus of over 50 species, particularly well represented in the Americas.

S.sinensis *(Pers.) Ames*, Orch. 2:53 (1908). Type: China, 'prope Cantonem Sinarum' (holo. UPS).
Neottia sinensis Pers., Syn. Pl. 2:511 (1807).

HABITAT: open grassy places; roadside banks; waste ground; rarely in hill and lower montane forest.
ALTITUDINAL RANGE: 900 – 1500m.
DISTRIBUTION: Borneo (SA, but probably throughout whole island). Widespread throughout Asia east to Japan and southeast to Australia, New Zealand and the Pacific islands.

SUBTRIBE CRYPTOSTYLIDINAE

CRYPTOSTYLIS R.Br.

Prodr. :317 (1810).

Terrestrial herbs. Rhizome short, thick. Stem erect, slender. Leaves 1–4, ovate, petiolate, with dark reticulate veins, arising from ground level. Inflorescence terminal, arising separately on rhizome, lax or subdense, many flowered. Flowers non-resupinate, lip uppermost; sepals and petals free, sepals usually longer, both very narrow, spreading; dorsal sepal revolute; lip entire, not spurred, strongly concave at base and enclosing column, tapered at apex; column very short, with lateral auricles; pollinia 4.

About 20 species distributed from S.E.Asia to Australia, New Guinea and the Pacific islands.

C.acutata *J.J.Sm.* in Bull. Jard. Bot. Buitenzorg, ser. 3, 3:243 (1921). Type: Java, Salak, Gunung Gadjah, *Bakhuizen van den Brink* 4029 (holo. L).

var.**acutata**

HABITAT: lower montane dipterocarp forest; oak-laurel forest; low stature forest on ultramafic substrate; often beside paths, or on logs and mossy banks.
ALTITUDINAL RANGE: 1100 – 2700m.
DISTRIBUTION: Borneo (SA). Also Sumatra, Java.

var.**borneensis** *J.J.Sm.* in Mitt. Inst. Allg. Bot. Hamburg, 7:17 (1927). Type: Kalimantan, Bukit Mulu, *Winkler* 515 (holo. HBG).

HABITAT: unknown.
ALTITUDINAL RANGE: 1000m.
DISTRIBUTION: Borneo (K). Endemic.

C.arachnites *(Blume) Hassk.*, Cat. Bog.: 48 (1844). Type: Java, Salak, Seribu, *Blume* s.n. (holo. BO).
Zosterostylis arachnites Blume, Bijdr.: 419, t.32 (1825).

HABITAT: hill and lower montane dipterocarp forest; low primary forest on ultramafic substrate; low oak-laurel forest; upper montane forest; often growing in moss at base of trees or beside paths.
ALTITUDINAL RANGE: 800 – 2200m.
DISTRIBUTION: Borneo (K, SA, SR). Widespread from Sri Lanka and India through Thailand east to Indonesia, the Philippines and the Pacific islands.

C.clemensii *(Ames & C.Schweinf.) J.J.Sm.* in Mitt. Inst. Allg. Bot. Hamburg 7:17 (1927). Type: Sabah, Mt. Kinabalu, Marai Parai, *Clemens* 399 (holo. AMES).
Chlorosa clemensii Ames & C.Schweinf., Orch. 6: 9, Pl.80 (1920).
Cryptostylis tridentata Carr in Gard. Bull. Straits Settlem. 8:174 (1935). Type: Sabah, Mt. Kinabalu, Penibukan Ridge, *Carr* C.3089 (SFN 26556) (holo. SING, iso. K).

HABITAT: lower montane forest on ultramafic substrate.
ALTITUDINAL RANGE: 1100 – 1700m.
DISTRIBUTION: Borneo (SA). Endemic.

SUBFAMILY ORCHIDOIDEAE

TRIBE DIURIDEAE

SUBTRIBE ACIANTHINAE

CORYBAS Salisb.

Parad. Lond., t.83 (1807).

Dwarf terrestrial, lithophytic, rarely epiphytic herbs arising from small tuberoids. Stem erect, very short and slender. Leaf solitary, ovate to orbicular or cordate, borne horizontally above the ground, often with red or white veins. Inflorescence terminal, erect, 1 flowered. Flower sessile or shortly stalked, 1–2 cm. high, helmet-shaped, large for size of plant, resupinate; dorsal sepal large, hooded; lateral sepals and petals linear, often filiform; lip large, recurved, entire, tubular below, with 2 short basal spurs, embracing column at base, apex rounded, often fimbriate; column small, erect; pollinia 2, 2-lobed, granular.

About 100 species distributed from India, S.China, Taiwan and S.E.Asia eastwards to New Guinea, Australia, New Zealand and the S.W.Pacific.

C.bryophilus *J.J.Sm.* in Mitt. Inst. Allg. Bot. Hamburg 7:15, t.1, fig.5 (1927). Type: Kalimantan, between Bukit Raja and Bidang Menabai, *Winkler* 1057 (holo. HBG).

HABITAT: mossy places in lower montane forest.
ALTITUDINAL RANGE: 1000m.
DISTRIBUTION: Borneo (K). Endemic.

C.carinatus *(J.J.Sm.) Schltr.* in Feddes Repert. 19:19 (1924). Type: Java, Gunung Salak, *Smith* s.n. (holo. BO; iso. L).
   Corysanthes carinata J.J.Sm. in Bull. Dép. Agric. Indes Néerl., 13:8 (1907).
   Corybas johannis-winkleri J.J.Sm. in Mitt. Inst. Allg. Bot. Hamburg 7:14, t.1, fig.4 (1927). Type: Kalimantan, Bukit Mulu, *Winkler* 498 (holo. HBG, iso. BO).
   C.johannis-winkleri J.J.Sm. var. *interruptus* J.J.Sm., loc. cit.: 14 (1927). Type: Kalimantan, between Bukit Raja and Bidang Menabai, *Winkler* 1056 (holo. HBG).

HABITAT: lower montane forest; moss carpets on the lips of podsolized ridges; amongst well-drained but damp crumbly humus.
ALTITUDINAL RANGE: 400 – 2000m.
DISTRIBUTION: Borneo (K, SA, SR). Also Peninsular Malaysia, Sumatra, Java.

C.crenulatus *J.J.Sm.* in Mitt. Inst. Allg. Bot. Hamburg, 7:16, t.2, fig.6 (1927). Type: Kalimantan, Bukit Mulu, *Winkler* 494 (holo. HBG).

HABITAT: in thin moss on limestone.
ALTITUDINAL RANGE: 400 – 1000m.
DISTRIBUTION: Borneo (K, SR). Endemic.

C.geminigibbus *J.J.Sm.* in Mitt. Inst. Allg. Bot. Hamburg 37:13, t.1, fig.3 (1927). Type: Kalimantan, Bukit Mulu, *Winkler* 496 (holo. HBG).

HABITAT: lower montane forest.
ALTITUDINAL RANGE: 1200m.

DISTRIBUTION: Borneo (K). Also Peninsular Malaysia.

C.kinabaluensis *Carr* in Gard. Bull. Straits Settlem., 8: 173 (1935). Type: Sabah, Mt. Kinabalu, Penibukan, *Carr* 3067 (SFN 26395) (holo. SING, iso. K).

HABITAT: lower montane forest; in mossy areas on ultramafic substrate.
ALTITUDINAL RANGE: 1200m.
DISTRIBUTION: Borneo (SA). Endemic.

C.muluensis *J.Dransf.* in Kew Bull. 41 (3):588 (1986). Type: Sarawak, Fourth Division, Gunung Mulu summit ridge, *Lewis* 352 (holo. K).

HABITAT: mossy banks in moss forest.
ALTITUDINAL RANGE: 2000m.
DISTRIBUTION: Borneo (SR). Endemic.

C.pictus *(Blume) Rchb.f.*, Beitr. Syst. Pflanz.: 67 (1871). Type: Java, Gunung Salak, *Blume* s.n. (holo. L).
Calcearia picta Blume, Bijdr.: 418, t.33 (1825).

HABITAT: mossy, well-drained banks, sometimes on bases of mossy tree trunks, rarely on rocks, in primary hill and lower montane forest; infrequently on ultramafics.
ALTITUDINAL RANGE: 700 – 1800m.
DISTRIBUTION: Borneo (SA). Also Sumatra, Java.

C.piliferus *J.Dransf.* in Kew Bull. 41(3):588 (1986). Type: Sabah, Nabawan, *Dransfield* 5129 (holo. K).

HABITAT: low, well drained mossy banks in 'kerangas' forest dominated by *Dacrydium*.
ALTITUDINAL RANGE: c.300m.
DISTRIBUTION: Borneo (SA). Endemic.

C.serpentinus *J.Dransf.* in Kew Bull. 41(3):604 (1986). Type: Sabah, Lahad Datu, Bukit Silam, *Dransfield* 5847 (holo. K).

HABITAT: thin soil and moss carpets on serpentine rock on ridgetops.
ALTITUDINAL RANGE: 700 – 800m.
DISTRIBUTION: Borneo (SA). Endemic.

PANTLINGIA Prain

J.Asiat. Soc. Bengal, Pt.2, Nat.Hist. 65 (2):107 (1896).

Stigmatodactylus Maxim. ex Makino, Illus. Fl. Japan 1, nr.7, tab.43 (1891), nom. nud.

Delicate small tuberous terrestrial herbs. Stems erect, 1-leaved. Leaf positioned halfway up stem well above ground level, cordate, sessile, green. Inflorescences terminal, few-flowered, lax, racemose; floral bracts leafy. Flowers resupinate, green with purple callus on lip. Sepals and petals linear-ligulate, acute to acuminate. Lip orbicular, slightly retuse, not embracing column, spur absent, callus basal, simple to complicated, often with lateral processes. Column slender, curved, with a tooth-like process below, sometimes with a second process beneath stigma. Pollinia 2.

About ten species distributed in the northern parts of eastern Asia, ranging from the Himalayas to Japan, extending south east to New Guinea and New Caledonia. One species each in Java, Borneo and Sulawesi.

P.lamrii *J.J.Wood & C.L.Chan* in Wood, Beaman & Beaman, Plants of Mt. Kinabalu 2, Orchids: 274, fig.46 (1993). Type: Sabah, Mt. Kinabalu, Pinosuk Plateau, path to Pig Hill, *Atwood & Beaman* s.n. (holo. K).

HABITAT: mossy stream banks in lower montane forest.
ALTITUDINAL RANGE: 1500 – 2900m.
DISTRIBUTION: Borneo (SA). Endemic.

TRIBE ORCHIDEAE

SUBTRIBE ORCHIDINAE

PLATANTHERA Rich.

Mém. Mus. Hist. Nat.4:48 (1818).

Terrestrial herbs arising from tubers or a short rhizome. Stem erect, bearing a few leaves, often near the ground, sheathed at the base. Leaves thin, usually broad, elliptic, not jointed at base, uppermost bract-like. Inflorescence terminal, usually many flowered, lax or dense. Flowers white or green, sometimes fragrant; dorsal sepal and petals usually forming a hood over column; lateral sepals spreading or reflexed; lip spurred, blade simple, horizontal to pendent; column short; stigmas 2, flat, joined across below rostellum, not freely extending in front of column; pollinia 2, separate, the caudicles enclosed in tubes (canals or thecae) separated by a rostellum.

About 85 species in Europe, temperate and tropical Asia, New Guinea, North Africa, North and Central America. A couple of species are found inside the Arctic Circle.

P.angustata *(Blume) Lindl.*, Gen. Sp. Orch. Pl.: 290 (1835). Type: Java, Salak and Burangrang, *Blume* s.n. (holo. BO).
  Mecosa angustata Blume, Bijdr.:404, fig.1(1825).
  Habenaria angustata (Blume) Kuntze, Rev. Gen. Pl.2: 664 (1891).

HABITAT: in moss under trees; scrub; upper montane forest.
ALTITUDINAL RANGE: 2200m.
DISTRIBUTION: Borneo (SA). Widespread from Peninsular Malaysia, Thailand and Vietnam east to Indonesia, the Philippines and New Guinea, north to Taiwan, Hong Kong and the Ryukyu Islands.

P.borneensis *(Ridl.) J.J.Wood* in Wood, Beaman & Beaman, Plants of Mt. Kinabalu, 2, Orchids: 291 (1993). Type: Sabah, Mt. Kinabalu, *Haviland* s.n. (holo. SING).
  Habenaria borneensis Ridl. in Trans. Linn. Soc. London, Bot. ser. 2, 4: 240 (1894).

HABITAT: upper montane forest.
ALTITUDINAL RANGE: 3000m.
DISTRIBUTION: Borneo (SA). Endemic.

P.crassinervia *(Ames & C.Schweinf.) J.J.Sm.* in Mitt. Inst. Allg. Bot. Hamburg 7:12(1927). Type: Sabah, Mt. Kinabalu, Pakka, *Clemens* 221 (holo. AMES).
  Habenaria crassinervia Ames & C.Schweinf., Orch. 6:8 (1920).

HABITAT: upper montane forest.
ALTITUDINAL RANGE: 3000m.
DISTRIBUTION: Borneo (SA). Endemic.

P.gibbsiae *Rolfe* in J. Linn. Soc., Bot. 42:160 (1914). Type: Sabah, Mt. Kinabalu, below Pakapaka, *Gibbs* 4258 (holo. K).
  Habenaria gibbsiae (Rolfe) Ames & C.Schweinf., Orch. 6:8 (1920).

HABITAT: open scrub; upper montane forest, sometimes on ultramafic substrate.

ALTITUDINAL RANGE: 2700 – 3400m.
DISTRIBUTION: Borneo (SA). Endemic.

P.kinabaluensis *Kraenzl. ex Rolfe* in J. Linn. Soc., Bot. 42:160 (1914). Type: Sabah, Mt. Kinabalu, *Haviland* s.n. (holo. K).
Habenaria kinabaluensis (Kraenzl. ex Rolfe) Ames & C.Schweinf., Orch. 6:9 (1920).

HABITAT: upper montane forest; often under *Leptospermum recurvum;* sometimes on ultramafic substrate.
ALTITUDINAL RANGE: 1600 – 3200m.
DISTRIBUTION: Borneo (SA). Endemic.

P.saprophytica *J.J.Sm.* in Mitt. Inst. Allg. Bot. Hamburg 7:12, t.1, fig.2 (1927). Type: Kalimantan, Bukit Mulu, *Winkler* 525 (holo. HBG, drawing at BO).

HABITAT: lower montane forest; often on ultramafic substrate.
ALTITUDINAL RANGE: 1000 – 1700m.
DISTRIBUTION: Borneo (K, SA). Endemic.

P.stapfii *Kraenzl. ex Rolfe* in J. Linn. Soc., Bot. 42:160 (1914). Type: Sabah, Mt. Kinabalu, Marai Parai, *Haviland* 1158 (holo. K).
Habenaria stapfii (Kraenzl. ex Rolfe) Ames & C.Schweinf., Orch. 6:9 (1920).

HABITAT: lower montane scrub and grassland; recorded from rocky landslips on ultramafic substrate.
ALTITUDINAL RANGE: 1500 – 2000m.
DISTRIBUTION: Borneo (SA). Endemic.

SUBTRIBE HABENARIINAE

HABENARIA Willd.

Sp.Pl. 4:44 (1805).

Terrestrial herbs arising from tubers. Stem erect, few to many leaved, sheathed below. Leaves thin, narrowly elliptic to orbicular, not jointed at base, uppermost bract-like. Inflorescence terminal, usually quite long, with many small to fairly large flowers, lax or dense. Flowers white, green or pink, sometimes fragrant; dorsal sepal and petals usually forming a hood over column; lateral sepals spreading or reflexed; lip spurred, blade usually 3-lobed, sometimes finely divided and fringed; column short, consisting mainly of the anther, usually with a small auricle either side; stigmas 2, usually separate, the caudicles enclosed in long or short, often prominent, tubes (canals or thecae) separated by a rostellum.

A genus of between 600 and 800 species with a cosmopolitan distribution, but particularly well represented in Africa.

H.damaiensis *J.J.Sm.* in Bull. Jard. Bot. Buitenzorg, ser. 3, 8:35 (1926). Type: Kalimantan, Soengai Damai, *Jaheri* 1338 (holo. L, iso. BO).

HABITAT: hill and lower montane forest.
ALTITUDINAL RANGE: 900 – 1500m.
DISTRIBUTION: Borneo (K, SA). Endemic.

H.elatius *Ridl.* in Sarawak Mus. J. 1:37 (1912). Type: Sarawak, Bungo Range, *Brookes* 44 (holo. SAR, iso. K).

HABITAT: unknown.
ALTITUDINAL RANGE: unknown.
DISTRIBUTION: Borneo (SR). Endemic.

H.hewittii *Ridl.* in J. Straits Branch Roy. Asiat. Soc. 54:55 (1910). Type: Sarawak, *Hewitt* 1908 (holo. SING).

HABITAT: unknown.
ALTITUDINAL RANGE: unknown.
DISTRIBUTION: Borneo (SR). Endemic.

H.hystrix *Ames*, Orch.2:35(1908). Type: Philippines, Luzon, *Meyer* s.n. (holo. AMES).

HABITAT: unknown.
ALTITUDINAL RANGE: unknown.
DISTRIBUTION: Borneo (SA). Also Philippines.

H.koordersii *J.J.Sm.* in Bull. Jard. Bot. Buitenzorg, ser. 2, 9:2 (1913). Types: Java, Wilis, G.Pitjis, *Koorders* 23268B (syn. L); Telomojo, *Koorders* 35864B (syn. L).

HABITAT: montane oak forest.
ALTITUDINAL RANGE: 1200 – 1500m.
DISTRIBUTION: Borneo (SA). Also Java.

H.marmorophylla *Ridl.* in J. Linn. Soc., Bot. 31:304 (1895). Type: Sarawak, Buseau, *Haviland* s.n. (holo. SING).
  H.havilandii Kraenzl., Orch. Gen. Sp. 1:427 (1898). Type: Sarawak, Sarawak River, *Haviland* s.n. (holo. B, iso. K).

HABITAT: limestone rocks in dense shade.
ALTITUDINAL RANGE: sea level – 500m.
DISTRIBUTION: Borneo (K, SR). Endemic.

H.medusa *Kraenzl.* in Bot. Jahrb. Syst. 16:203 (1893). Type: Java, *Blume* s.n. (holo. L).

HABITAT: unknown
ALTITUDINAL RANGE: unknown
DISTRIBUTION: Borneo (SA). Also Sumatra, Java, Sulawesi.

H.rumphii *(Brongn.) Lindl.,* Gen. Sp. Orch. Pl.: 230 (1835). Type: Maluku, Ambon, *Rumphius, D'Urville* (holo. P).
Platanthera rumphii Brongn. in Duperr., Voy. Coquill. Phan.: 194, t.38A (1829).

HABITAT: unknown.
ALTITUDINAL RANGE: unknown.
DISTRIBUTION: Borneo (K). Widespread from Thailand and Indonesia to Australia, New Guinea and possibly the Philippines.

H.setifolia *Carr* in Gard. Bull. Straits Settlem. 8:171 (1935). Type: Sabah, Mt. Kinabalu, Tenompok, *Clemens* 28323 (holo. BM, iso. AMES).

HABITAT: hill and lower montane forest; riversides.
ALTITUDINAL RANGE: 1000 – 1500m.
DISTRIBUTION: Borneo (SA). Endemic.

H.singapurensis *Ridl.* in J. Linn. Soc., Bot. 32:410 (1896). Type: Singapore, Choa Chu Kang, *Ridley* s.n. (holo. SING).

HABITAT: lower montane forest.
ALTITUDINAL RANGE: 1200 – 1400m.
DISTRIBUTION: Borneo (SA). Also Peninsular Malaysia, Singapore.

PERISTYLUS Blume

Bijdr.: 404, t.30 (1825).

Terrestrial herbs arising from tubers. Stem erect, few to several leaved, either basal, spaced along stem or grouped near the centre. Leaves thin, linear to broadly elliptic, not jointed at base, uppermost often bract-like. Inflorescence terminal, few to many flowered, lax or dense. Flowers small, white or green; pedicel with ovary porrect, close to rachis; dorsal sepal and petals forming a hood over column; petals usually a little broader than lateral sepals; lip 3-lobed, spur shorter than ovary, often reduced to a globular sac shorter than sepals; column short; stigmas 2, usually rather short, convex, cushion-like, often adnate to lip edge, but in some species appearing flat and clavate, protruding and adnate to hypochile; pollinia 2, the caudicles very short, without tubes (canals or thecae), attached to short rostellar sidelobules, rostellar midlobule small, more or less hidden between the anthers.

Some 135 species are listed in Index Kewensis, although half this figure is probably more correct. They are distributed from India and Sri Lanka to China, Taiwan and the Ryukyu Islands, south through Thailand, Indochina, Malaysia and Indonesia east to New Guinea, Australia and the Pacific islands.

P.brevicalcar *Carr* in Gard. Bull. Straits Settlem. 8:167 (1935). Type: Sabah, Mt. Kinabalu, near Bundu Tuhan, *Carr* C.3353 (holo. SING).
Habenaria brevicalcar (Carr) Masam., Enum. Phan. Born.: 188 (1942).

HABITAT: hill forest.
ALTITUDINAL RANGE: 900m.
DISTRIBUTION: Borneo (SA). Endemic.

P.candidus *J.J.Sm.*, Orch. Java: 36, fig.18 (1905). Type: Java, Salak, collector unknown (holo. L).
Habenaria candida (J.J.Sm.) Masam., Enum. Phan. Born. 188 (1942).

HABITAT: rubber plantations; grassy places; hill forest.
ALTITUDINAL RANGE: 200 – 900m.
DISTRIBUTION: Borneo (K, SA, SR). Also Peninsular Malaysia, Cambodia, Vietnam, Sumatra, Lingga Archipelago, Java, Buru, Sulawesi, Ambon.

P.ciliatus *Carr* in Gard. Bull. Straits Settlem. 8: 169 (1935). Type: Sabah, Mt. Kinabalu, head of Kulapis river, *Carr* 3368 (SFN 27111) (syn. SING, isosyn. AMES); Tenompok, *Clemens* 29816 (syn. SING).
Habenaria ciliata (Carr) Masam., Enum. Phan. Born.:189 (1942).

HABITAT: lower montane forest.
ALTITUDINAL RANGE: 300 – 1500m.
DISTRIBUTION: Borneo (SA). Endemic.

P.goodyeroides *(D.Don) Lindl.*, Gen. Sp. Orch. Pl.: 299 (1835). Type: Nepal, *Wallich* s.n. (holo. K-LINDL).
Habenaria goodyeroides D.Don, Prodr. Fl. Nepal: 25 (1825).

HABITAT: streamsides; hill forest; recorded from sandstone soils.
ALTITUDINAL RANGE: 600 – 900m.
DISTRIBUTION: Borneo (K, SA). Widespread from N.W.Himalaya to China and the Philippines and through Malaysia and Indonesia to New Guinea.

P.gracilis *Blume*, Bijdr.: 404 (1825). Type: Java, Buitenzorg Province, Seribu, *Blume* s.n. (holo. L).
Habenaria bambusetorum Kraenzl., Orch. Gen. Sp.1: 384 (1898). Type: Java, *Zollinger* 1151 (holo. L).

HABITAT: hill forest, sometimes on ultramafic substrate.
ALTITUDINAL RANGE: 900m.
DISTRIBUTION: Borneo (SA). Also N.India, Peninsular Malaysia, Thailand, Sumatra, Java and Krakatau.

P.grandis *Blume*, Bijdr.: 405, fig.30 (1825). Type: Java, Salak, *Blume* s.n. (holo. L).
Habenaria grandis auct., non Benth. ex Ridl. in Forbes, Nat. Wand. East. Arch.: 519 (1885), nom. nud.
H.gigas Hook.f. var *papuana*. (J.J.Sm.) Ames & C.Schweinf., Orch 6: 8 (1920).

HABITAT: hill and lower montane forest; recorded from sandstone soils.
ALTITUDINAL RANGE: 900 – 1000m.
DISTRIBUTION: Borneo (SA). Also Peninsular Malaysia, Indonesia, New Guinea.

P.hallieri *J.J.Sm.* in Bull. Dép. Agric. Indes Néerl. 22:1 (1909). Types: Kalimantan, Soeka Lanting, *Hallier* 37 (syn. BO, L); Soengei Kelasar, *Hallier* 1542 (syn. BO, L).
Habenaria hallieri (J.J.Sm.) Schltr. in Feddes Repert. 9:338 (1911).

HABITAT: open grassy areas; rough mossy ground; landslips; mossy riverbed rocks and boulders; secondary vegetation; lower montane forest; able to withstand inundation.
ALTITUDINAL RANGE: 400 – 1600m.
DISTRIBUTION: Borneo (B, K, SA, SR). Endemic.

P.kinabaluensis *Carr* in Gard.Bull. Straits Settlem. 8:170 (1935). Type: Sabah, Mt. Kinabalu, between Dallas and Tenompok, *Carr* C.3389 (SFN27308) (holo. SING).

HABITAT: hill and lower montane forest; 'by a stream in very young secondary forest'.
ALTITUDINAL RANGE: 1200m.
DISTRIBUTION: Borneo (SA). Endemic.

P.lacertiferus *(Lindl.) J.J.Sm.* in Bull. Jard. Bot. Buitenzorg, ser. 3, 9:23 (1927). Type: Burma, Tavoy, *Wallich* s.n. (holo. K-WALL).
Coeloglossum lacertiferum Lindl., Gen. Sp. Orch. Pl.: 302 (1835).

HABITAT: unknown.
ALTITUDINAL RANGE: unknown.
DISTRIBUTION: Borneo (SR). Also N.India, Burma, Thailand, Peninsular Malaysia, Sumatra, Java, north to China, Hong Kong, Taiwan, Ryukyu Islands.

P.ovariophorus *(Schltr.) J.J.Wood*, **comb.nov.**
Habenaria ovariophora Schltr. in Feddes Repert. 9:434 (1911). Type: Sarawak, Kuching, *Schlechter* 15838 (holo. B, destroyed).

HABITAT: between grasses at forest margins.
ALTITUDINAL RANGE: lowlands.
DISTRIBUTION: Borneo (SR). Endemic.

P.spathulatus *J.J.Sm.* in Bull. Dép. Agric. Indes Néerl. 22:5 (1909). Type: Kalimantan, between Dawar and Goenoeng Damoes, *Hallier* 481 (holo. L, iso. BO).
Habenaria spathulata (J.J.Sm.) Ames in Merr., Bibl. Enum. Born. Pl.: 138 (1921).

HABITAT: unknown.
ALTITUDINAL RANGE: unknown.
DISTRIBUTION: Borneo (K), Endemic.

P.unguiculatus *J.J.Sm.* in Bull. Dép. Agric. Indes Néerl. 22:4 (1909). Type: Kalimantan, Liliboelan Tepoetsi, *Jaheri* 934 (holo. BO).
Habenaria unguiculata (J.J.Sm.) Ames in Merr., Bibl. Enum. Born. Pl.: 139 (1921).

HABITAT: unknown.
ALTITUDINAL RANGE: unknown.
DISTRIBUTION: Borneo (K). Endemic.

SUBFAMILY EPIDENDROIDEAE

TRIBE GASTRODIEAE

SUBTRIBE GASTRODIINAE

DIDYMOPLEXIELLA Garay

Arch. Jard. Bot. Rio de Janeiro 13:33 (1955).

Leafless saprophytic herbs. Rhizome small, fleshy. Stems simple, erect, slender. Inflorescence terminal, racemose. Flowers small, resupinate; sepals and petals connate near the base, lateral sepals a little more than to the middle; petals connate halfway with dorsal sepal; lip entire and retuse at apex or 3-lobed, callose; column with long decurved, falcate wings (stelidia), foot absent; stigma near column apex; pollinia 4.

Six species distributed in Thailand, Peninsular Malaysia and Borneo.

D.borneensis *(Schltr.) Garay* in Arch. Jard. Bot. Rio de Janeiro 13:33 (1954). Type: Sarawak, Gunung Matang, *Beccari* 1479 (holo. B, destroyed, iso. FI).
Leucolaena borneensis Schltr. in Feddes Repert. 9:428 (1911).

HABITAT: limestone hills.
ALTITUDINAL RANGE: unknown.
DISTRIBUTION: Borneo (SR). Endemic.

D.forcipata *(J.J.Sm.) Garay* in Arch. Jard. Bot. Rio de Janeiro 13:33 (1954). Type: Kalimantan, Sungei Raun, *Winkler* 1545 (holo. HBG, drawing at BO).
Didymoplexis forcipata J.J.Sm. in Mitt. Inst. Allg. Bot. Hamburg 7:18, t.2, fig.7 (1927).

HABITAT: swampy lowland primary forest.
ALTITUDINAL RANGE: 300m.
DISTRIBUTION: Borneo (K). Endemic.

D.kinabaluensis *(Carr) Seidenf.* in Dansk Bot. Ark. 32:175 (1978). Type: Sabah, Mt. Kinabalu, Menetendok Gorge, *Carr* 3155 (SFN 26606) (holo. SING, iso. K).
Didymoplexis kinabaluensis Carr in Gard. Bull. Straits Settlem. 8:178 (1935).

HABITAT: hill forest.
ALTITUDINAL RANGE: 900 – 1300m.
DISTRIBUTION: Borneo (SA). Endemic.

D.ornata *(Ridl.)Garay* in Arch. Jard. Bot. Rio de Janeiro 13:33 (1954). Type: Peninsular Malaysia, Bukit Sadanen, *Ridley* s.n. (holo. SING).
Leucolaena ornata Ridl. in J. Linn. Soc., Bot. 28:340, pl.43 (1891).

HABITAT: unknown.
ALTITUDINAL RANGE: unknown.
DISTRIBUTION: Borneo (unspecified). Also Peninsular Malaysia, Singapore, Thailand, Sumatra.

DIDYMOPLEXIS Griff.

Calcutta J. Nat. Hist. 4:383, t.17 (1844).

Leafless saprophytic herbs. Rhizome small, fleshy. Stems simple, erect, slender. Inflorescence terminal, racemose. Flowers small, resupinate; pedicels elongating rapidly after fruit is set; dorsal sepal and petals adnate to form a single trifid segment forming a shallow cup or tube with the partially connate lateral sepals; lip entire or 3-lobed, adnate to column foot, free from sepals and petals, callose or papillose; column long, slender, with a short foot, apical wings (stelidia) absent; stigma near column apex; pollinia 4.

About 20 species distributed in Africa, Madagascar and Asia eastwards to new Guinea, Australia and the Pacific islands.

D.latilabris *Schltr.* in Bull. Herb. Boissier ser. 2, 6:300 (1906). Type: Kalimantan, Long Dett, *Schlechter* 13532 (holo. B, destroyed).

HABITAT: forest humus.
ALTITUDINAL RANGE: unknown.
DISTRIBUTION: Borneo (K). Endemic.

D.pallens *Griff.* in Calcutta J. Nat. Hist. 4:383, t.17 (1844). Type: India, Calcutta, Serampore, *Griffith* s.n. (holo. CAL).

HABITAT: unknown.
ALTITUDINAL RANGE: unknown.
DISTRIBUTION: Borneo (unspecified). Widespread through India, Malaysia and Indonesia to the Philippines and Australia.

D.striata *J.J.Sm.* in Icon. Bogor. 2:17 t.CIVB (1902). Type: West Java, Pasir Njarungsum am Salak (holo. BO).

HABITAT: lowland forest.
ALTITUDINAL RANGE: 100m.
DISTRIBUTION: Borneo (SR). Also Java.

GASTRODIA R.Br.

Prodr.:330 (1810).

Leafless saprophytic herbs. Rhizome tuberous, ± horizontal. Stems tall or short, simple, erect, often elongating after fertilisation. Inflorescence terminal, racemose, 1 to many flowered. Flowers campanulate; sepals and petals connate to form a 5-lobed tube which may or may not be gibbous at the base, and which may be split between the lateral sepals; petals smaller than sepals; lip shorter than sepals, entire or 3-lobed, disc with 2 keels and sometimes glandular basal calli; column quite long, foot short; stigma at base of column; pollinia 2.

About 20 species widespread from E. and S.E.Asia to Australia, New Zealand and the Pacific islands.

G.grandilabris *Carr* in Gard. Bull. Straits Settlem. 8: 179 (1935). Type: Sabah, Mt. Kinabalu, Tenompok - Ranau path, *Carr* 3264 (SFN 26970) (holo. SING, iso. K).

HABITAT: lower montane forest.
ALTITUDINAL RANGE: 1500 – 1600m.
DISTRIBUTION: Borneo (SA). Endemic.

G.javanica *(Blume) Lindl.*, Gen. Sp. Orch. Pl.: 384 (1840). Type: Java, Seribu, *Blume* s.n. (holo. BO).
Epiphanes javanica Blume, Bijdr.: 421, fig.4 (1825).

HABITAT: recorded from under a bamboo garden hedge.
ALTITUDINAL RANGE: lowlands.
DISTRIBUTION: Borneo (SA, SR). Also China?, Ryukyu Islands, Taiwan, Peninsular Malaysia, Thailand, Java, Philippines.

NEOCLEMENSIA Carr

Gard. Bull. Straits Settlem. 8:180 (1935).

Leafless saprophytic herb. Rhizome cylindrical, villose. Stem simple. Inflorescence erect, terminal, laxly 2–3 flowered; floral bracts adpressed to pedicel. Flowers medium-sized, resupinate, campanulate, white with orange petals and greenish-brown lip; sepals adnate, forming a tube, free, recurved and papillose at apex; petals adnate at base to lateral sepals, much shorter than sepals, linear-spathulate, fimbriate; lip adnate to base of column, entire, clawed, claw oblong or subquadrate and bearing 2 subglobose, papillose- vesiculose apical calli, disc transversely rugulose, with a short median bilobed keel; column stout, with acute stelidia; pollinia 2.

A monotypic genus endemic to Borneo.

N.spathulata *Carr* in Gard. Bull. Straits Settlem. 8: 180 (1935). Type: Sabah, Mt. Kinabalu, Penibukan Ridge, *Clemens* s.n. (holo. SING, iso. K).

HABITAT: hill forest.
ALTITUDINAL RANGE: 1100m.
DISTRIBUTION: Borneo (SA). Endemic.

SUBTRIBE EPIPOGIINAE

EPIPOGIUM J. G. Gmel. ex Borkh.

Tent. Disp. Pl.German. :139 (1792).

Leafless saprophytic herbs. Rhizome tuberous. Stems erect, fleshy. Inflorescence terminal, racemose, brittle, ephemeral. Flowers non-resupinate, nodding at first. Sepals and petals about equal, narrow, acute, free; lip sessile, concave, entire, with minutely papillose ridges, shortly spurred; column short; pollinia 2, each with a filiform caudicle.

Two species widely distributed in tropical Africa and from western Europe to Japan, S.E.Asia to Australia and the Pacific islands.

E.roseum *(D.Don) Lindl.*, J. Linn. Soc., Bot.1: 177 (1857). Type: Nepal, *Wallich* s.n. (holo. BM, iso. K).
Limodorum roseum D.Don, Prodr. Fl. Nepal.: 30 (1825).

HABITAT: hill dipterocarp forest.
ALTITUDINAL RANGE: 700 – 1000m.
DISTRIBUTION: Borneo (K, SA). Very widespread from Africa, mainland Asia, eastwards through Malaysia, Indonesia and New Guinea to Australia, the Solomon Islands and New Caledonia.

STEREOSANDRA Blume

Mus. Bot. Lugd. Bat. 2:179 (1856).

Leafless saprophytic herbs. Rhizome tuberous. Stems erect. Inflorescence terminal, few to many flowered. Flowers resupinate; sepals and petals about equal, narrow; lip narrow, entire, with 2 basal glands; column short; anther erect, on a broad filament rising from the back of the column; stigma forming an erect bilobed structure with the rostellum on front of column; pollinia 2, powdery, with a caudicle.

About 5 species distributed from S.E.Asia to New Guinea and the Solomon Islands.

S.javanica *Blume*, Mus. Bot. Lugd. Bat., 2:176 (1856). Type: West Java, Bantam Province, *Blume* s.n. (holo. BO).

HABITAT: open lowland mixed dipterocarp forest on sandstone.
ALTITUDINAL RANGE: 300 – 500m.
DISTRIBUTION: Borneo (SA, SR). Also Taiwan, Ryukyu Islands, Peninsular Malaysia, Thailand, Sumatra, Java, Philippines, New Guinea, Solomon Islands.

TRIBE NEOTTIEAE

SUBTRIBE LIMODORINAE

APHYLLORCHIS Blume

Bijdr.:t.77 (1825).

Leafless saprophytic herbs. Rhizome short, rather thin, erect, with thick spreading roots. Stem erect, slender. Inflorescence terminal, many flowered. Flowers resupinate; sepals and petals about equal, free; lip divided into a short narrow basal hypochile and a 3-lobed apical epichile; column long; anther erect, dorsal; pollinia 2, powdery.

Around 30 species have been proposed, although the true figure is probably much lower. Distributed from Sri Lanka, India and the western Himalaya to China, through Indochina, Malaysia, Indonesia, Taiwan and the Philippines, eastwards to New Guinea and Australia.

A.kemulensis *J.J.Sm.* in Bull. Jard. Bot. Buitenzorg, ser. 3, 11:84 (1931). Type: Kalimantan, near Gunung Kemoel, *Endert* 3652 (holo. L, iso. BO).

HABITAT: lower montane forest on mountain ridges.
ALTITUDINAL RANGE: 1200m.
DISTRIBUTION: Borneo (K). Endemic.

A.montana *Rchb.f.* in Linnaea 41:57 (1876). Types: Sri Lanka, Ambagumowa District, *Thwaites* C.P.3189; *Mrs.Walker* s.n. (syn. W).
A.borneensis Schltr. in Bull. Herb. Boissier 2, 6:299 (1906). Type: Kalimantan, Long Sele, *Schlechter* 13520 (holo. B).

HABITAT: lowland mixed dipterocarp forest; hill and lower montane forest.
ALTITUDINAL RANGE: sea level – 1500m.
DISTRIBUTION: Borneo (K, SA, SR). Also Sri Lanka, India and the Himalayan region to China, Taiwan and Ryukyu Islands, south to Thailand, Peninsular Malaysia and the Philippines.

A.pallida *Blume*, Bijdr.: t.77 (1825). Type: Java, *Blume* s.n. (holo. BO).

HABITAT: in rich humus among rocks in primary lowland, hill and lower montane forest; freshwater swamp forest; ridge forest dominated by *Quercus*.
ALTITUDINAL RANGE: sea level – 1600m.
DISTRIBUTION: Borneo (B, K, SA, SR). Widespread from Peninsular Malaysia and Thailand to Sumatra, Java and east to the Philippines.

A.spiculaea *Rchb.f.* in Linnaea 41:58 (1877). Type: Borneo, *Lobb* s.n. (holo. W).

HABITAT: unknown.
ALTITUDINAL RANGE: unknown.
DISTRIBUTION: Borneo (unspecified). Endemic.

A.striata *(Ridl.) Ridl.*, Mat. Fl. Mal. Pen. 1:205 (1907). Type: Peninsular Malaysia, Tahan woods, *Ridley* s.n. (holo. SING).
Pogonia striata Ridl. in Trans. Linn. Soc. London, Bot. ser. 2, 3: 377 (1893).

HABITAT: unknown.
ALTITUDINAL RANGE: unknown.
DISTRIBUTION: Borneo (unspecified). Also Peninsular Malaysia.

TRIBE NERVILIEAE

NERVILIA Comm. ex Gaudich.

Freycinet, Voy. Uranie: 421, t.35 (1829 "1826").

Terrestrial herbs arising from a rounded, reduced corm bearing short roots. Leaf solitary, appearing after flowers mature, on a separate stalk, erect to horizontal, plicate, often reniform or ovate to cordate, rarely lobed, glabrous or hirsute. Inflorescence erect, produced before the leaf appears, elongating after fertilization, terminating in a 1 to few-flowered raceme. Flowers small to large, often pendulous; sepals and petals similar, free, long and narrow; lip entire or 3-lobed, usually without a spur, base embracing column; column long, apex widened, without wings or foot; pollinia 2.

About 80 species distributed in the tropics and subtropics of Africa, Asia, New Guinea, Australia and the Pacific islands.

N.borneensis *(J.J.Sm.) Schltr.* in Bot. Jahrb. Syst.45: 402 (1911). Type: Kalimantan, Goenoeng Damoes, *Hallier* 980 (holo. BO).
Pogonia borneensis J.J.Sm. in Bull. Dép. Agric. Indes Néerl. 22:8 (1909).

HABITAT: unknown.
ALTITUDINAL RANGE: unknown.
DISTRIBUTION: Borneo (K). Endemic.

N.dilatata *(Blume) Schltr.* in Bot. Jahrb. Syst.45: 402 (1911). Type: Kalimantan, *Blume* s.n. (holo. BO).
Pogonia dilatata Blume, Fl. Javae, ser. 2, 1, Orch. (Coll. Orch. Arch. Ind.): 151, Pl.10, fig.4 (1858).

HABITAT: lowland forest on limestone.
ALTITUDINAL RANGE: 300 – 400m.
DISTRIBUTION: Borneo (K, SR). Endemic.

N.punctata *(Blume) Makino* in Bot. Mag. (Tokyo) 16: 199 (1902). Type: West Java, *Blume* s.n. (holo. L).
Pogonia punctata Blume, Mus. Bot. Lugd. Bat. 1:32 (1849).

HABITAT: secondary forest; hill forest.
ALTITUDINAL RANGE: 1100m.
DISTRIBUTION: Borneo (SA). Also Peninsular Malaysia, Thailand, Sumatra, Java.

TRIBE VANILLEAE

SUBTRIBE GALEOLINAE

CYRTOSIA Blume

Bijdr.: 396, t.6 (1825).

Leafless saprophytic herbs. Rhizome stout, erect, bearing several fleshy roots. Stems several from each rhizome, simple or branched. Inflorescence terminal. Flowers not fully opening; sepals and petals connivent; lip spurless, entire, erect, fused with base of column; column slightly curved, shortly clawed below, flabellate above, foot absent; anther cap terminal; pollinia 2; fruits succulent, indehiscent, seeds exalate.

Five species distributed from Sri Lanka, Thailand and Indochina to Malaysia and Indonesia, with one species native to Japan.

C.javanica *Blume*, Bijdr.: 396 (1825). Type: Java, *Blume* s.n. (holo. BO).
Galeola javanica (Blume) Benth. & Hook. f., Gen. Plant. 3:590 (1883).

HABITAT: secondary forest; lower montane forest.
ALTITUDINAL RANGE: 1000 – 2700m.
DISTRIBUTION: Borneo (SA). Also Java, Sulawesi.

ERYTHRORCHIS Blume

Rumphia 1:200, t.70 (1835).

Leafless saprophytic herbs. Stems climbing, flexuous, much-branched, with scale-leaves. Inflorescence simple or paniculate; rachis glabrous, densely many flowered. Flowers thin, glabrous; sepals and petals connivent, not opening widely; lip with numerous transversely parallel ridges on each side of the median ridge; column with a short, descending foot, tapering into the thick, median ridge of the lip; pollinia 2, solid; fruit a long, dry dehiscent capsule; seeds winged.

Three species distributed in Taiwan, Japan and the Ryukyu Islands, and from India (Assam), Burma, Thailand and Indochina to Malaysia, Indonesia, the Philippines and Australia.

E.altissima *(Blume) Blume* in Rumphia 1:200 (1837). Type: Java, Seribu, *Blume* s.n. (holo. BO).
Cyrtosia altissima Blume, Bijdr.: 396 (1825).
Galeola altissima (Blume) Rchb.f., Xenia Orch. 2:77 (1865).

HABITAT: unknown.
ALTITUDINAL RANGE: 500m.
DISTRIBUTION: Borneo (SA, SR). Also Peninsular Malaysia, Java, Philippines.

GALEOLA Lour.

Fl.Cochinch. 2:520 (1790).

Saprophytic herbs with scale-leaves. Stems stout, long-climbing, with a root and a scale-leaf at each node. Inflorescences terminal, and also lateral in the axils of the upper scale-leaves of climbing stems; rachis pubescent-furfuraceous. Flowers fleshy, yellowish or brownish, pubescent-furfuraceous; sepals and petals about equal, free, usually hardly spreading; base of lip surrounding the column, the blade concave to saccate, with longitudinal ridges; column stout, rather short, arcuate, clavate; pollinia 2, cleft, granular; fruit a long dry capsule, dehiscent, opening with two unequal valves; seeds winged, rather large.

About 10 species distributed in Madagascar eastwards and from India (Sikkim) to China, south-eastwards through Thailand to Malaysia, Indonesia, the Philippines, New Guinea and the Solomon Islands.

G.nudifolia *Lour.*, Fl. Cochinch. 2:521 (1790). Type: 'in sylvis Cochinchinae' (holo. BM).

HABITAT: lowland primary and secondary forest; hill dipterocarp forest.
ALTITUDINAL RANGE: sea level – 800m.
DISTRIBUTION: Borneo (SA). Widespread through Burma and Thailand to Indochina and China, south through Malaysia and Indonesia to the Philippines.

85

SUBTRIBE VANILLINAE

VANILLA Mill.

Gard. Dict. abr. ed.4 (1754).

Scandent or climbing plants, bearing a leaf and a root at each node. Stems long, branched, green. Leaves large, fleshy, sessile or shortly petiolate, or replaced by small scales. Inflorescences lateral, arising from leaf axils, short, few to many flowered. Flowers rather large, fugaceous; sepals and petals about equal, free, spreading; lip tubular, entire or obscurely lobed, its claw adnate to column, usually with hairy appendages inside; column long, foot absent; anther pointing downwards on the front of the column; rostellum broad; pollinia granular; fruit a long, fleshy, cylindrical 'pod', often fragrant when mature; seeds large, exalate.

About 100 species widely distributed in the Old and New World tropics.

V.abundiflora *J.J.Sm.* in Bull.Jard. Bot. Buitenzorg, ser. 3, 2:21 (1920). Types: Kalimantan, Soengei Djongkang, Tengaroeng, *van Gelder* 1 (syn. L); Soengei Landak *Teysmann* s.n. (syn. L); Kapoeas, *Teysmann* 10898 (syn. L).

HABITAT: lowland swamp forest.
ALTITUDINAL RANGE: lowlands.
DISTRIBUTION: Borneo (K). Endemic.

V.albida *Blume*, Cat. Gen. Buitenzorg: 100 (1823). Type: Java, *Blume* s.n. (holo. BO).

HABITAT: hill forest.
ALTITUDINAL RANGE: 900m.
DISTRIBUTION: Borneo (SA). Also Peninsular Malaysia, Thailand, Sumatra, Java.

V.borneensis *Rolfe* in J. Linn. Soc., Bot. 32:460 (1896). Type: Kalimantan, Banjarmassing, *Motley* 1248 (holo. K).

HABITAT: unknown.
ALTITUDINAL RANGE: unknown.
DISTRIBUTION: Borneo (K). Endemic.

V.griffithii *Rchb.f.* in Bonplandia 2:88 (1854). Type: origin unknown (holo. W).

HABITAT: unknown.
ALTITUDINAL RANGE: unknown.
DISTRIBUTION: Borneo (B, SA). Also Peninsular Malaysia, Karimun Besar Island, Sumatra.

V.havilandii *Rolfe* in Kew Bull.: 236 (1918). Type: Sarawak, Kuching, *Haviland* s.n. (syn. K); Matang, *Ridley* s.n. (syn. K).

HABITAT: secondary forest; riverine forest; lowland dipterocarp forest.
ALTITUDINAL RANGE: 200 – 500m.
DISTRIBUTION: Borneo (SA, SR). Endemic.

V.kinabaluensis *Carr* in Gard. Bull. Straits Settlem. 8:176 (1935). Types: Sabah, Mt. Kinabalu, Kadamaian (Tampassuk) river, *Carr* C.3157 (syn. SING, isosyn. K); Mt. Kinabalu, Dallas, *Clemens* 26300 (syn. SING) and 26725 (syn. SING, isosyn. K).

HABITAT: hill dipterocarp forest.
ALTITUDINAL RANGE; 700 – 900m.
DISTRIBUTION: Borneo (SA). Also Peninsular Malaysia.

V.pilifera *Holttum* in Gard. Bull. Singapore, 13, 2:253 (1951). Type: Peninsular Malaysia, Johore, Kota Tinggi, *le Doux* s.n. (holo. SING).

HABITAT: hill forest on sandstone.
ALTITUDINAL RANGE: 200 – 300m.
DISTRIBUTION: Borneo (SA). Also Peninsular Malaysia, Thailand.

V.sumatrana *J.J.Sm.* in Bull. Jard. Bot. Buitenzorg, ser. 3, 2:22 (1920). Type: Sumatra, Benkoelen, Boekit Barisan, Rimbo Pengadang, *Ajoeb* 210 (holo. L).

HABITAT: lowland and hill forest.
ALTITUDINAL RANGE: lowlands – 400m.
DISTRIBUTION: Borneo (SA). Also Sumatra.

SUBTRIBE LECANORCHIDINAE

LECANORCHIS Blume

Mus. Bot. Lugd. Bat. 2:188 (1856).

Leafless saprophytic herbs. Rhizomes branched, fleshy. Stems simple or branched. Inflorescence terminal, erect, racemose, lax, few- to many-flowered. Flowers small, resupinate; sepals and petals free, subequal, usually spreading and encircled by a shallow, denticulate cup or calyculus; lip adnate to base of column, 3-lobed, rarely entire, not spurred, disc papillose or hirsute; column slender, elongate, clavate, partially adnate to claw of lip; pollinia 2.

About 20 species distributed in Thailand, Malaysia, Indonesia, north to Japan including Ryukyu Islands, Taiwan and the Philippines eastwards to New Guinea.

L.malaccensis *Ridl.* in Trans. Linn. Soc. London, Bot. 3:377, t.65 (1893). Type: Peninsular Malaysia, Tahan woods, *Ridley* s.n. (holo. SING).

HABITAT: unknown.
ALTITUDINAL RANGE: unknown.
DISTRIBUTION: Borneo (SA, SR). Also Peninsular Malaysia, Thailand, Sumatra.

L.multiflora *J.J.Sm.* in Bull. Jard. Bot. Buitenzorg, ser. 2, 26:8 (1918). Types: Java, *Backer* 7158, 6275, 10124, 14020 (syn. L).

HABITAT; lowland and hill forest; lower montane oak - chestnut forest; sandy forest; *Casuarina* forest; in humus among rocks.
ALTITUDINAL RANGE: sea level – 1500m.
DISTRIBUTION: Borneo (B, K, SA, SR). Also Peninsular Malaysia, Thailand, Sumatra, Java.

TRIBE MALAXIDEAE

HIPPEOPHYLLUM Schltr.

in K. Schum. & Lauterb., Nachtr. Fl. Schutzgeb.

Südsee:107 (1905).

Epiphytic herbs. Rhizome long, creeping and rigid. Stems short, lacking pseudobulbs, usually spaced 2–6cm. apart on the rhizome. Leaves distichous, bilaterally flattened, sheathing at base. Inflorescence terminal, densely many-flowered. Flowers small, non-resupinate; sepals and petals free, with the sepals broader than the petals; lip usually trilobed, sometimes with a bilobed callus at the base; column slender; pollinia 4.

About 5 species distributed from Peninsular Malaysia to the Solomon Islands.

H.scortechinii *(Hook.f.) Schltr.* in Bot. Jahrb. Syst. 45, Beibl. 104:13 (1911). Types: Peninsular Malaysia, Perak, *Scortechini* s.n., *King's collector* s.n. (syn. K).
 Oberonia scortechinii Hook.f., Fl. Brit. Ind. 5:683 (1890).

HABITAT: unknown.
ALTITUDINAL RANGE: unknown.
DISTRIBUTION: Borneo (K, SA, SR). Also Peninsular Malaysia, Sumatra, Java.

LIPARIS Rich.

in Mém. Mus. Hist. Nat. 4:43, 52 (1818).

Epiphytes, lithophytes or terrestrials. Pseudobulbs ovoid to cylindrical, fleshy, 1- to several-leaved at apex. Leaves plicate or conduplicate, narrowly elliptic or ovate, often fleshy, membranaceous, sheathing at base. Inflorescence terminal, a several to many flowered raceme, erect, lax or dense. Flowers resupinate, small to medium-sized, sometimes self-fertilising, green, purple, yellow-green or dull orange; sepals and petals free, often reflexed; sepals ovate to narrowly elliptic; petals usually narrow, linear; lip entire or 3-lobed, apex often deflexed, usually with a fleshy basal callus; column long, arcuate, sometimes swollen at base, narrowly winged above, foot absent; pollinia 4.

About 250 species with a cosmopolitan distribution, the greatest number being in Asia and New Guinea.

L.anopheles *J.J.Wood* in Nordic J. Bot. 11(1):85, fig.1 (1991). Type: Sabah, Tambunan District, Mt. Trus Madi, *Surat* in *Wood* 871 (holo. K).

HABITAT: lower montane forest with *Agathis*.
ALTITUDINAL RANGE: 1500 – 1600m.
DISTRIBUTION: Borneo (SA, SR). Endemic.

L.araneola *Ridl.* in J. Linn. Soc., Bot. 31:265 (1896). Type: Kalimantan, Pontianak, cult. *Singapore* (holo. SING).

HABITAT: unknown.
ALTITUDINAL RANGE: unknown.

DISTRIBUTION: Borneo (K). Endemic.

L.atrosanguinea *Ridl.* in J. Straits Branch Roy. Asiat. Soc. 39:71 (1903). Type: Peninsular Malaysia, Perak, Larut Hills, *Curtis & Derry* s.n. (holo. SING).

HABITAT: lower montane forest; *Leptospermum* scrub; mostly on ultramafic substrate.
ALTITUDINAL RANGE: 1200 – 1800m.
DISTRIBUTION: Borneo (SA, SR). Also Peninsular Malaysia, Thailand, Vietnam, Sumatra.

L.aurantiorbiculata *J.J.Wood & A.Lamb*, in Wood, Beaman & Beaman, Plants of Mt. Kinabalu 2, Orchids: 243, fig 41 (1993).Type: Sabah, Mt. Kinabalu, Pinosuk Plateau, *Lamb* AL22/82 (holo. K).

HABITAT: lower montane forest.
ALTITUDINAL RANGE: 1200 – 1700m.
DISTRIBUTION: Borneo (SA). Endemic.

L.bicuspidata *J.J.Sm.* in Icon. Bogor. 2:45, t.109, fig.C (1903). Type: Kalimantan, *Nieuwenhuis* s.n. (holo. BO).

HABITAT: hill forest; streamside forest.
ALTITUDINAL RANGE: 300 – 600m.
DISTRIBUTION: Borneo (B, K, SA). Endemic.

L.bootanensis *Griff.*, Itin.Not.: 98 (1848, description) & Not. 3:278 (1851, name). Type: Bhutan, Durunga, *Griffith* 1460 (holo. K).

HABITAT: hill and lower montane forest, often on ultramafic substrate.
ALTITUDINAL RANGE: 1200 – 1500m.
DISTRIBUTION: Borneo (K, SA). Also China, Japan, N.India, Bhutan, Burma, Peninsular Malaysia, Thailand, Vietnam, Java, Sumbawa, Philippines.

L.brookesii *Ridl.* in J. Straits Branch Roy. Asiat. Soc. 54:47 (1910). Type: Sarawak, Bidi, *Brookes* s.n. (holo. SING).

HABITAT: limestone rocks.
ALTITUDINAL RANGE: unknown.
DISTRIBUTION: Borneo (SR). Endemic.

L.caespitosa *(Thouars) Lindl.* in Bot. Reg. 11, sub t.882 (1825). Type: Mauritius and Bourbon, *du Petit Thouars*, Hist. Orch., t.89.
Malaxis caespitosa Thouars, Hist. Orch., t.90 (1822).

HABITAT: hill and lower montane forest; dipterocarp, *Fagaceae, Gymnostoma* and *Agathis* forest on ultramafic substrate; kerangas forest with field layer of *Cyperus, Araceae* and *Bromheadia*, with patches of open, somewhat higher forest including *Eugenia, Ficus* and *Tristania*.
ALTITUDINAL RANGE: 400 – 1600m.
DISTRIBUTION: Borneo (K, SA, SR). Widespread from Africa to the Pacific islands.

L.compressa *(Blume) Lindl.*, Gen. Sp. Orch. Pl.: 32 (1830). Type: Java, Buitenzorg, Bantam and Tjanjor, *Blume* s.n. (holo. BO).
Malaxis compressa Blume, Bijdr.: 390, fig 54 (1825).

var. **compressa**

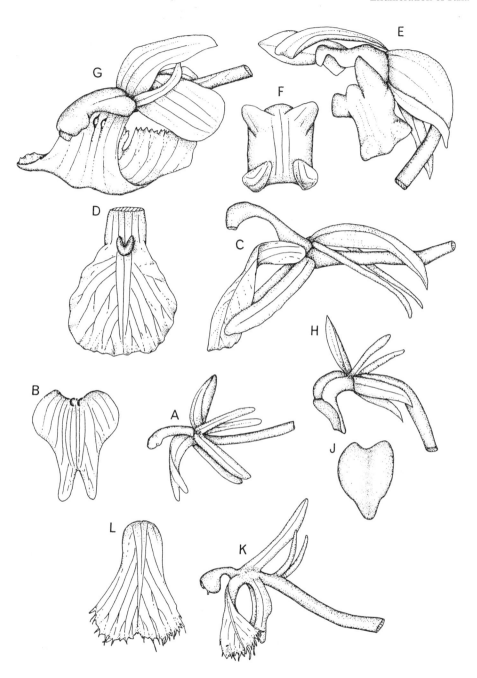

FIG. 5. *Liparis.* **A**, *L. bicuspidata*, flower (side view), x 2½; **B**, *L. bicuspidata*, lip, x 4; **C**, *L. compressa* var. *compressa*, flower (side view), x 4; **D**, *L. compressa*, lip, x 4; **E**, *L. gibbosa*, flower (side view), x 10; **F**, *L. gibbosa*, lip (front view), x 10; **G**, *L. grandiflora*, flower (oblique view), x 2½; **H**, *L. viridiflora*, flower (side view), x 4; **J**, *L. viridiflora*, lip, x 5; **K**, *L. wrayi*, flower (side view), x 4; **L**, *L. wrayi*, lip, x 4. Drawn by Sarah Thomas.

HABITAT: mixed hill dipterocarp forest; lower montane forest; sometimes on ultramafic substrate.
ALTITUDINAL RANGE: 300 – 1700m.
DISTRIBUTION: Borneo (K, SA, SR). Also Peninsular Malaysia, Sumatra, Java, Sulawesi, Philippines.

var.**maxima** J.J.Sm. in Bull. Jard. Bot. Buitenzorg ser. 3, 11:123 (1931). Type: Kalimantan, Gunung Kemoel, *Endert* 4513 (holo. L, iso. BO).

HABITAT: lower montane forest.
ALTITUDINAL RANGE: 1600m.
DISTRIBUTION: Borneo (K). Endemic.

L.condylobulbon *Rchb.f.* in Hamburger Garten-Blumenzeitung 18:34 (1862). Type: Java, cult.*Schiller* (holo. W).
L.confusa J.J.Sm., Orch. Java :275, fig.211 (1905). Types: Java, Vogelberg, Tjampea, am Tjiliwoeng bei Buitenzorg, *Hallier* s.n. (syn. L); Tjipakoe, *Hallier* s.n. (syn. L); Garoet, *Ader* s.n. (syn. L); Palaboean Ratoe, *Koorders* s.n. (syn. L); Pangentjongan, *Koorders* s.n. (syn. L) Tengger, bei Poespo, *Kobus* s.n. (syn. L); Wilis bei Ngebel, *Koorders* s.n. (syn. L).
L.confusa J.J.Sm. var. *latifolia* J.J.Sm. in Nova Guinea 12:226 (1915). Type: Kalimantan, cult. Buitenzorg, *Nieuwenhuis* s.n. (holo. L).

HABITAT: podsol forest with bamboo understorey; recorded growing on *Ficus;* streamside forest; hill dipterocarp forest on sandstone soil; montane *Agathis/Lithocarpus* forest.
ALTITUDINAL RANGE: 100 –1600m.
DISTRIBUTION: Borneo (K, SA, SR). Also Taiwan, Sumatra, Java, Sulawesi, Maluku(Ambon), Philippines, New Guinea, Solomon Islands, Samoa, Fiji.

L.dolichostachys *Schltr.* in Bull. Herb. Boissier, ser. 2, 6:307 (1906). Type: Kalimantan, Long Wahau, *Schlechter* 13540 (holo. B,destroyed).

HABITAT: swampy forest.
ALTITUDINAL RANGE: unknown.
DISTRIBUTION: Borneo (K). Endemic.

L.elegans *Lindl.,* Gen. Sp. Orch. Pl.: 30 (1830). Type: Peninsular Malaysia, Penang, *Wallich* 1943 (holo. K-LINDL).
L.stricta J.J.Sm. in Bull. Dép. Agric. Indes Néerl. 5:3 (1907). Type: Kalimantan, Bukit Kasian, *Nieuwenhuis* s.n. (holo. L).

HABITAT: hill and lower montane forest; moss forest with *Gymnostoma* and undergrowth of climbing bamboo on serpentine rock.
ALTITUDINAL RANGE: 400 – 2600m.
DISTRIBUTION: Borneo (K, SA, SR). Also Peninsular Malaysia, Sumatra, Anambas Islands, Mentawai and Riau Islands, Philippines.

L.endertii *J.J.Sm.* in Bull. Jard. Bot. Buitenzorg, ser. 3, 11:122 (1931). Type: Kalimantan, Gunung Kemoel, *Endert* 4270 (holo. L, iso. BO).

HABITAT: hill and lower montane forest.
ALTITUDINAL RANGE: 1200 – 1800m.
DISTRIBUTION: Borneo (K,SA). Endemic.

L.ferruginea *Lindl.* in Gard.Chron.: 55 (1848). Type: Peninsular Malaysia, Penang, *Loddiges* s.n. (holo. K - LINDL).

HABITAT: unknown.
ALTITUDINAL RANGE: unknown.
DISTRIBUTION: Borneo (SA, questionable). Also China (Hainan), Peninsular Malaysia, Thailand, Cambodia, Sumatra, Java.

L.gibbosa *Finet* in Bull. Soc. Bot. France 55:342, t.11, figs.36–44 (1908). Type: Java, *Blume* s.n. (holo. P).
L.disticha auct, non (Thouars) Lindl.

HABITAT: lower montane forest; ridge oak/chestnut forest with *Tristania*, *Agathis alba*, small rattans, etc.; recorded on *Ficus*.
ALTITUDINAL RANGE: sea level – 1900m.
DISTRIBUTION: Borneo (K, SA, SR). Also Burma, Peninsular Malaysia, Thailand, Laos, Sumatra, Java, New Guinea, Solomon Islands, Vanuatu, New Caledonia, Fiji, Samoa and probably elsewhere.

L.glaucescens *J.J.Sm.* in Icon. Bogor. 2:47, t.109, fig.D (1903). Type: Kalimantan, Kelam, *Hallier* s.n. (holo. BO).

HABITAT: on rocks in primary forest.
ALTITUDINAL RANGE: 1000m.
DISTRIBUTION: Borneo (K). Endemic.

L.grandiflora *Ridl.* in J. Bot.: 333 (1884). Type: Sarawak, Mindai-Pramassan, *Grabowsky* s.n. (holo. BM).

HABITAT: mixed hill dipterocarp forest.
ALTITUDINAL RANGE: unknown.
DISTRIBUTION: Borneo (K,SR). Endemic.

L.grandis *Ames & C.Schweinf.*, Orch. 6:87 (1920). Type: Sabah, Mt. Kinabalu, *Haslam* s.n. (holo. AMES, iso. K).

HABITAT: lower montane forest on ultramafic substrate.
ALTITUDINAL RANGE: 1200 – 1500m.
DISTRIBUTION: Borneo (SA). Endemic.

L.kamborangensis *Ames & C.Schweinf.*, Orch. 6:89 (1920). Type : Sabah, Mt. Kinabalu, Kamborangah, *Clemens* 220 (holo. AMES).

HABITAT: lower and upper montane forest; ridge moss forest; wet mossy rock faces; on ultramafic substrate.
ALTITUDINAL RANGE: 1400 – 3150m.
DISTRIBUTION: Borneo (SA). Endemic.

L.kemulensis *J.J.Sm.* in Bull. Jard. Bot. Buitenzorg, ser. 3, 12:149 (1932). Type: Kalimantan, Gunung Kemoel, *Endert* 4167 (holo. L).
L.amesiana J.J.Sm. in Bull. Jard. Bot. Buitenzorg, ser. 3, 11:124 (1931), non Schltr.

HABITAT: primary forest.
ALTITUDINAL RANGE: 1500m.
DISTRIBUTION: Borneo (K). Endemic.

L.kinabaluensis *J.J.Wood* in Lindleyana 5(2):84, fig.2 (1990). Type: Sabah, Mt. Kinabalu, Penibukan Ridge, east of Dahobang River, *Clemens* 50099 (holo. K).

HABITAT: hill forest on ultramafic substrate; lower montane forest.
ALTITUDINAL RANGE: 1000 – 1800m.

DISTRIBUTION: Borneo (SA). Endemic.

L.lacerata *Ridl.* in J. Linn. Soc., Bot. 22:284 (1886). Type: Sarawak, Lawas River, *Burbidge* s.n. (holo. BM).

HABITAT: lowland forest; kerangas forest; heath forest on podsolic soil; peat swamp forest; hill mixed dipterocarp forest.
ALTITUDINAL RANGE: sea level – 600m.
DISTRIBUTION: Borneo (B, K, SA, SR). Also Burma, Peninsular Malaysia, Sumatra, Mentawai Islands (Siberut).

L.latifolia *(Blume) Lindl.*, Gen. Sp. Orch. Pl.: 30 (1830). Type: Java, Bantam Province, Buitenzorg Province, Pantjar, *Blume* s.n. (holo. BO).
Malaxis latifolia Blume, Bijdr.: 393 (1825).

HABITAT: on cultivated *Citrus* spp.; hill and lower montane forest.
ALTITUDINAL RANGE: 200 – 1500m.
DISTRIBUTION: Borneo (SA, SR). Also China (Hainan), Peninsular Malaysia, Thailand, Sumatra, Java, Timor, New Guinea.

L.lingulata *Ames & C.Schweinf.*, Orch. 6:90 (1920). Type: Sabah, Mt. Kinabalu, Kiau, *Clemens* 324 (holo. AMES, iso. K,SING).

HABITAT: hill and lower montane forest.
ALTITUDINAL RANGE: 800 – 1500m.
DISTRIBUTION: Borneo (SA). Endemic.

L.lobongensis *Ames*, Orch.6:92 (1920). Type: Sabah, Mt. Kinabalu, Lobong, *Clemens* 219 (holo. AMES).

HABITAT: lower montane forest, sometimes on ultramafic substrate.
ALTITUDINAL RANGE: 800 – 2300m.
DISTRIBUTION: Borneo (SA). Endemic.

L.longissima *J.J.Sm.* in Bull. Jard. Bot. Buitenzorg, ser. 2, 13:6 (1914). Type: Kalimantan, Gunung Labang, *Amdjah* s.n., cult. Bogor (syn. BO), Gunung Djempanga, *Amdjah* s.n., cult. Bogor (syn. BO).

HABITAT: secondary forest.
ALTITUDINAL RANGE: below 300m.
DISTRIBUTION: Borneo (K, SR). Endemic.

L.lycopodioides *J.J.Sm.* in Bull. Jard. Bot. Buitenzorg, ser. 3, 11:121 (1931). Type: Kalimantan, Gunung Kemoel, *Endert* 4270 (holo. L).

HABITAT: lower montane forest.
ALTITUDINAL RANGE: 1800m.
DISTRIBUTION: Borneo (K). Endemic.

L.montana *(Blume) Lindl.*, Gen. Sp. Orch. Pl.:29 (1830).

var.**maxima** *Ridl.* in J. Linn. Soc., Bot. 31:264 (1896). Type: Sarawak, Trusan River, *Haviland* s.n. (holo. SING, iso. SAR).

HABITAT: unknown.
ALTITUDINAL RANGE: unknown.
DISTRIBUTION: Borneo (SR). Endemic; var. *montana* is recorded from Java only.

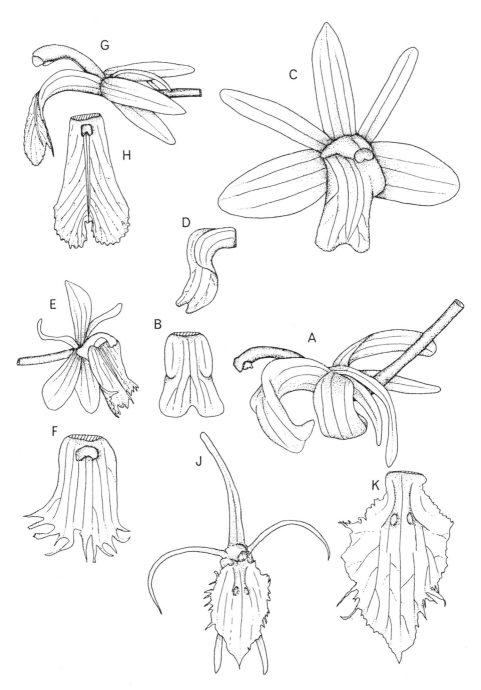

FIG. 6. *Liparis.* **A**, *L. caespitosa,* flower (side view), x 2½; **B**, *L. caespitosa,* lip, x 2½; **C**, *L. elegans,* flower (front view), x 6; **D**, *L. elegans,* lip (side view), x 6; **E**, *L. lacerata,* flower (side view), x 4; **F**, *L. lacerata,* lip, x 4; **G**, *L. latifolia,* flower (side view), x 2½; **H**, *L. latifolia,* lip, x 2½; **J**, *L. purpureoviridis,* flower (front view), x 3; **K**, *L. purpureoviridis,* lip, x 3. Drawn by Sarah Thomas.

L.mucronata *(Blume) Lindl.*, Gen. Sp. Orch. Pl.: 32 (1830). Type: Java, Bantam and Buitenzorg, *Blume* s.n. (holo. BO).
Malaxis mucronata Blume, Bijdr.: 391 (1825).
Liparis divergens J.J.Sm., Icon.Bogor. 2:48, t.109E, fig. 1–4 (1903). Types: Sumatra, Deli, *Heldt* s.n.; Java, Sukabumi, *Smith* s.n. (syn. BO).

HABITAT: hill forest on ultramafic substrate, with *Gymnostoma sumatrana*, small rattans, etc.; lower montane forest.
ALTITUDINAL RANGE: 800 – 1400m.
DISTRIBUTION: Borneo (SA). Also Sumatra, Java, Bali.

L.pandurata *Ames*, Orch. 6:94 (1920). Type: Sabah, Mt. Kinabalu, Lobong, *Clemens* 117 (holo. AMES).

HABITAT: swamp forest; oak-laurel forest; lower montane forest.
ALTITUDINAL RANGE: 1200 – 2600m.
DISTRIBUTION: Borneo (SA). Endemic.

L.parviflora *(Blume) Lindl.*, Gen. Sp. Orch. Pl.: 31 (1830). Type: Java, Salak, *Blume* s.n.(holo. BO).
Malaxis parviflora Blume, Bijdr.: 392 (1825).
Liparis flaccida Rchb.f. in Linnaea 41:45 (1877).Type: 'Ins. Sundaic', *collector unknown* (holo. W).

HABITAT: rubber plantations; riverine forest; hill and lower montane forest.
ALTITUDINAL RANGE: sea level – 2000m.
DISTRIBUTION: Borneo (K, SA, SR). Also Peninsular Malaysia, Thailand, Sumatra, Java, Bali, Krakatau, Sulawesi, Philippines.

L.purpureoviridis *Burkill ex Ridl.*, Fl. Mal. Pen. 4:21 (1924). Type: Peninsular Malaysia, Pahang, Bukit Fraser, *Burkill & Holttum* 8422 (holo. SING, iso. K).

HABITAT: on rocks in moss forest.
ALTITUDINAL RANGE: 1300 – 1400m.
DISTRIBUTION: Borneo (SR). Also Peninsular Malaysia.

L.rheedii *(Blume) Lindl.*, Gen. Sp. Orch. Pl.: 34 (1830). Type: Java, Seribu, *Blume* s.n.(holo. BO).
Malaxis rheedii Blume, Bijdr.: 389, t.54 (1825).

HABITAT: hill forest.
ALTITUDINAL RANGE: 800m.
DISTRIBUTION: Borneo (K,SA). Also Peninsular Malaysia, Thailand, Sumatra, Java, Sumbawa, Sulawesi, New Guinea.

L.rhodochila *Rolfe* in Kew Bull.: 412 (1908). Type: Java, locality unknown, ex cult. *Lawrence*, imported by *Moore* (holo. K).

HABITAT: hill forest.
ALTITUDINAL RANGE: 1000m.
DISTRIBUTION: Borneo (SA). Also Java.

L.rhombea *J.J.Sm.* in Bull. Dép. Agric. Indes Néerl. 43:35 (1910). Type: Java, Salak, cult. *Joseph* (holo. L).

HABITAT: riverine lower montane forest.
ALTITUDINAL RANGE: 1200 – 1300m.
DISTRIBUTION: Borneo (B, SA). Also Peninsular Malaysia, Thailand, Java.

L.tenella *J.J.Sm.* in Bull. Jard. Bot. Buitenzorg, ser. 3, 12:149 (1931). Type: Kalimantan, Long Liang Leng, *Endert* 3078 (holo. L, iso. BO).

L.tenuis J.J.Sm. in Bull. Jard. Bot. Buitenzorg, ser. 3, 11:121 (1931), non Rolfe ex Downie.

HABITAT: lowland primary forest.
ALTITUDINAL RANGE: 300m.
DISTRIBUTION: Borneo (K). Endemic.

L.tricallosa *Rchb.f.* in Gard. Chron. ser. 2, 11:684 (1879). Type: Borneo, *Burbidge* s.n., cult.*Bull* (holo. W, iso. BM).

HABITAT: peat swamp forest.
ALTITUDINAL RANGE: sea level.
DISTRIBUTION: Borneo (SR). Peninsular Malaysia, Sumatra, Philippines.

L.viridiflora *(Blume) Lindl.*, Gen. Sp. Orch. Pl., 31 (1830). Type: Java, Tjanjor Province, Solassie, *Blume* s.n. (holo. BO).

Malaxis viridiflora Blume, Bijdr.: 392 (1825).

Liparis longipes Lindl. in Wall., Pl. As. Rar. 1:31, t.35 (1830). Type: India, East Bengal, *Wallich* s.n. (holo. K- LINDL).

HABITAT: hill and lower montane forest.
ALTITUDINAL RANGE: 300 – 2000m.
DISTRIBUTION: Borneo (SA, SR). Widespread in S.E.Asia east to Fiji and Samoa.

L.wrayi *Hook.f.*, Fl. Brit. Ind. 6: 181 (1891). Type: Peninsular Malaysia, Perak, *Wray* 3713 (holo. K).

L.pectinifera Ridl. in J. Bot. 36:210 (1898). Type: Peninsular Malaysia, locality unknown, *Ridley* s.n. (holo. not located).

HABITAT: mixed dipterocarp forest on sandy soil; lower montane forest.
ALTITUDINAL RANGE: 100 – 1400m.
DISTRIBUTION: Borneo (B, SA). Also Burma, Peninsular Malaysia, Thailand, Sumatra, Java, New Guinea.

MALAXIS Sol. ex Sw.

Prodr. 8:119 (1788).

Terrestrial or, rarely, epiphytic herbs. Rhizomes creeping. Stems erect, leafy, basally extended into a small jointed pseudobulb. Leaves 1 to several, plicate or conduplicate, ovate or narrowly elliptic, acute, thin textured or fleshy, more or less petiolate and articulate to a tubular sheath. Inflorescence terminal, erect, racemose or subumbellate, laxly or densely many-flowered. Flowers small, non-resupinate; sepals and petals free, subequal, spreading, rarely reflexed; lateral sepals often connate at base; petals usually narrower than sepals, circinate; lip free from column, sessile, entire or lobed, apex often toothed, cordate or auriculate at base, enveloping base of column, often with a nectar secreting hollow or fovea at base; column short, foot absent; pollinia 4.

About 300 species with a cosmopolitan distribution, most numerous from Asia to New Guinea.

M.amabilis *(J.J.Sm.) J.J.Wood*, **comb.nov.**
Microstylis amabilis J.J.Sm. in Bull. Jard. Bot. Buitenzorg, ser. 3, 8:40 (1926). Type: Kalimantan, Liang Karing, *Sakeran* s.n., cult. Bogor (holo. BO).

HABITAT: unknown.
ALTITUDINAL RANGE: unknown.
DISTRIBUTION: Borneo (K). Endemic.

M.amplectens *(J.J.Sm.) Ames & C.Schweinf.*, Orch. 6:73 (1920). Type: West Java, Salak, *collector unknown* (holo. BO).
Microstylis amplectens J.J.Sm. in Icon. Bogor. 2:39, t.108E (1903).

HABITAT: lowland and hill dipterocarp forest; kerangas forest; riverine alluvial forest with residual limestone karst.
ALTITUDINAL RANGE: sea level – 1000m.
DISTRIBUTION: Borneo (SA, SR). Also Java, Bali.

M.andersonii *(Ridl.) Ames* in Merr., Bibl. Enum. Born. Pl.: 150 (1921). Type: Sarawak, Bau, *Anderson* 42 (holo. K).
Microstylis andersonii Ridl. in Kew Bull.: 210 (1914).

HABITAT: limestone rocks.
ALTITUDINAL RANGE: under 300m.
DISTRIBUTION: Borneo (SR). Endemic.

M.aurata *(Ridl.) Ames* in Merr., Bibl. Enum. Born. Pl.: 150 (1921). Type: Sarawak, Quop, *Hewitt* s.n.(holo. SING, iso. K).
Microstylis aurata Ridl. in J. Straits Branch Roy. Asiat. Soc. 49:27 (1908).

HABITAT: unknown.
ALTITUDINAL RANGE: unknown.
DISTRIBUTION: Borneo (SR). Endemic.

M.bidentifera *(J.J.Sm.) Ames & C.Schweinf.*, Orch. 6:73 (1920). Type: Kalimantan, between Muara Uja and Kundim Baru, *Winkler* 2675 (holo. HBG).
Microstylis bidentifera J.J.Sm. in Bot. Jahrb. Syst. 48:97 (1912).

HABITAT: limestone rocks.
ALTITUDINAL RANGE: unknown.
DISTRIBUTION: Borneo (K). Endemic.

M.blumei *(Boerl.& J.J.Sm.) Bakh.f.* in Blumea 12:68 (1963). Type: Java, Tjiapus, cult.Bogor (holo. BO).
Microstylis blumei Boerl. & J.J.Sm. in Icon. Bogor. 2:34, t.108B (1903).

HABITAT: lowland forest.
ALTITUDINAL RANGE: sea level – 500m.
DISTRIBUTION: Borneo (SR). Also Java.

M.burbidgei *(Rchb.f.ex Ridl.) Kuntze*, Rev. Gen. Pl.2:673 (1891). Type: Labuan, *Burbidge* s.n. (holo. K).
Microstylis burbidgei Rchb.f. ex Ridl. in J. Linn. Soc., Bot. 24:336 (1888).

HABITAT: lowland forest; lower montane forest; mossy rocks.
ALTITUDINAL RANGE: around sea level – 1600m.
DISTRIBUTION: Borneo (K, SA). Endemic.

M.calophylla *(Rchb.f.) Kuntze*, Rev. Gen. Pl.2: 673 (1891). Types: Peninsular Malaysia, ex *Groenewegen, Makoy, Veitch & Bull*, cult. Hamburg Bot. Gard. (syn. W).
Microstylis calophylla Rchb.f. in Gard. Chron.12:718 (1879).

HABITAT: hill forest on ultramafic substrate.
ALTITUDINAL RANGE: 1100m.
DISTRIBUTION: Borneo (SA). Also India (Sikkim), Burma, Peninsular Malaysia, Thailand, Cambodia.

M.chlorophrys *(Rchb.f.) Kuntze*, Rev. Gen. Pl.2:673 (1891). Type: Borneo, *Bull* s.n. (holo. W).
Microstylis chlorophrys Rchb.f. in Gard. Chron.15:266 (1881).

HABITAT: unknown.
ALTITUDINAL RANGE: unknown.
DISTRIBUTION: Borneo (unspecified). Endemic.

M.clemensii *(J.J.Sm.) J.J.Wood* in Jermy (ed.) Studies Flora Gunung Mulu Nat. Park: 29 (1984). Type: Sarawak, Bidi cave, *Clemens* 20715 (holo. L, iso. K).
Microstylis clemensii J.J.Sm. in Brittonia 1:108 (1931).

HABITAT: hill forest on limestone.
ALTITUDINAL RANGE: 800 – 900m.
DISTRIBUTION: Borneo (SR). Endemic.

M.commelinifolia *(Zoll. & Moritzi) Kuntze*, Rev. Gen. Pl.2:673 (1891). Type: Java (not located).
Microstylis commelinifolia Zoll. & Moritzi in Natuur-Geneesk. Arch. Ned.-Indië, 1:402 (1844).

HABITAT: limestone cliffs; hill and lower montane forest, sometimes on ultramafic substrate; open swamp forest on ultramafic substrate; kerangas forest with dominant *Dacrydium* and *Tristania*; coastal mixed dipterocarp forest on sandy clay soil.
ALTITUDINAL RANGE: sea level – 1400m.
DISTRIBUTION: Borneo (B, SA, SR). Also Java.

M.damusica *(J.J.Sm.) Ames* in Merr., Bibl. Enum. Born. Pl.:151 (1921). Type: Kalimantan, Gunung Damoes, *Hallier* 419 (holo. L, iso. BO).
Microstylis damusica J.J.Sm. in Bull. Dép. Agric. Indes Néerl. 20:20 (1909).

HABITAT: open swamp forest on ultramafic substrate; on boulders in river beds.
ALTITUDINAL RANGE: 300 – 700m.
DISTRIBUTION: Borneo (K, SA). Endemic.

M.elegans *(J.J.Sm.) Ames* in Merr., Bibl. Enum. Born. Pl.: 151 (1921). Type: Sarawak, *Haviland* s.n. (holo. SING).
Microstylis elegans J.J.Sm. in J. Linn. Soc., Bot. 31:264 (1896).

HABITAT: unknown.
ALTITUDINAL RANGE: unknown.
DISTRIBUTION: Borneo (SR). Endemic.

[**M.graciliscapa** *Ames & C.Schweinf.*, Orch. 6:73, pl.88 (1920). Type: Sabah, Mt. Kinabalu, Marai Parai, *Clemens* 258 (holo. AMES).
Microstylis graciliscapa (Ames & C.Schweinf.) J.J.Sm. in Bull. Jard. Bot. Buitenzorg, ser. 3, 11:119 (1931).

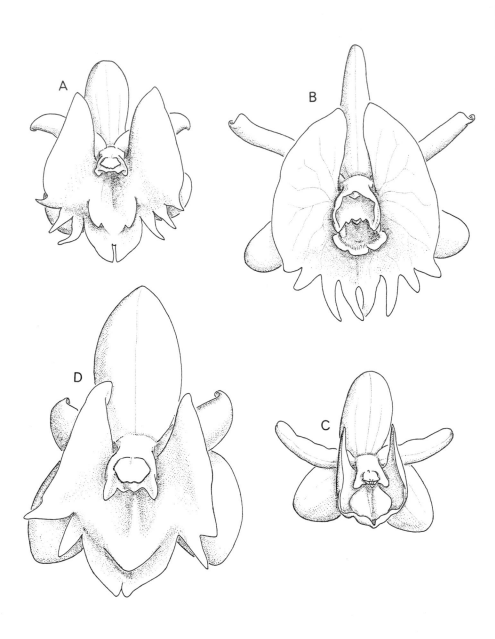

FIG. 7. *Malaxis.* **A**, *M. amplectens*, x 9; **B**, *M. blumei*, x 9; **C**, *M. burbidgei*, x 9; **D**, *M. commelinifolia*, x 18. Drawn by Mutsuko Nakajima.

HABITAT: primary riverine forest; lower montane forest on ultramafic substrate.
ALTITUDINAL RANGE: 400 – 2400m.
DISTRIBUTION: Borneo (K, SA). Endemic.]

See *M.subtiliscapa*, p.104.

M.kinabaluensis *(Rolfe) Ames & C.Schweinf.*, Orch. 6:73 (1920). Type: Sabah, Mt. Kinabalu, Penibukan Ridge, *Gibbs* 4065 (holo. K).
Microstylis kinabaluensis Rolfe in J. Linn. Soc., Bot. 42:146 (1914).

HABITAT: oak-laurel forest; lower montane forest with *Drimys*, rattans,etc.
ALTITUDINAL RANGE: 1200 – 2000m.
DISTRIBUTION: Borneo (K, SA). Endemic.

M.laciniosa *(Ridl.) Ames* in Merr., Bibl. Enum. Born. Pl.:151 (1921). Type: Sarawak, Trusan River, *Haviland* s.n. (holo. SING, iso. K).
Microstylis laciniosa Ridl. in J. Linn. Soc., Bot. 31:264 (1896).

HABITAT: unknown.
ALTITUDINAL RANGE: 100 – 200m.
DISTRIBUTION: Borneo (SR). Endemic.

M.latifolia *Sm.* in Rees, Cyclop.: 22 (1812). Type: Nepal, Narainhetty, *Buchanan* s.n. (holo. LINN).

HABITAT: lowland and hill forest; 'white sand forest'.
ALTITUDINAL RANGE: sea level – 1500m.
DISTRIBUTION: Borneo (B, K, SA, SR). Widespread in China, India, S.E.Asia, New Guinea, Australia.

M.lowii *(E.Morren) Ames* in Merr., Bibl. Enum. Born. Pl.: 151 (1921). Type: Borneo, *Low* s.n., introduced by Jacob-Makoy & Co. (holo. not located).
Microstylis lowii E.Morren in Belgique Hort. 34:281, t.14, fig.2 (1884).

HABITAT: hill and lower montane forest on sandy soil with scattered limestone rocks, or ultramafic substrate; mixed dipterocarp forest on sandstone; limestone cliffs.
ALTITUDINAL RANGE: sea level – 2100m.
DISTRIBUTION: Borneo (SA, SR). Endemic.

M.maculata *(Ridl.) Ames* in Merr., Bibl. Enum. Born. Pl.: 151 (1921). Type: Sarawak, Dulit, *Hose* s.n.(holo. SING).
Microstylis maculata Ridl. in J. Linn. Soc., Bot. 31:263 (1896).

HABITAT: unknown.
ALTITUDINAL RANGE: unknown.
DISTRIBUTION: Borneo (SR). Endemic.

M.metallica *(Rchb.f.) Kuntze,* Rev. Gen. Pl.2:673 (1891). Type: Borneo, *Bull* s.n.(holo. W).
Microstylis metallica Rchb.f. in Gard. Chron. 12:750 (1879).

HABITAT: hill forest on ultramafic substrate; moss forest with *Gymnostoma* and field layer of climbing bamboo; rocks and boulders in river beds, above high water level.
ALTITUDINAL RANGE: 400 – 1200m.
DISTRIBUTION: Borneo (SA). Endemic.

M.micrantha *(Hook.f.) Kuntze*, Rev. Gen. Pl.2:673 (1891). Type: Peninsular Malaysia, Perak, *Scortechini* s.n. (holo. K).
   Microstylis micrantha Hook.f., Icon. Plant., t.1834 (1889).

HABITAT: rocky places in lowland forest; limestone rocks.
ALTITUDINAL RANGE: sea level – 200m.
DISTRIBUTION: Borneo (SR). Also Peninsular Malaysia and probably Sumatra.

M.multiflora *Ames & C.Schweinf.*, Orch. 6:75, Pl.88 (1920). Type: Sabah, Mt. Kinabalu, Kiau, *Clemens* 86 (holo. AMES).
   Microstylis multiflora (Ames & C.Schweinf.) Masam., Enum. Phan. Born.:197 (1942).

HABITAT: hill forest.
ALTITUDINAL RANGE: 900 – 1200m.
DISTRIBUTION: Borneo (SA). Endemic.

M.partitiloba *(J.J.Sm.) J.J.Wood*, **comb.nov.**
   Microstylis partitiloba J.J.Sm. in Mitt. Inst. Allg. Bot. Hamburg, 7:41, t.6, fig.31 (1927). Type: Kalimantan, Bukit Raja, *Winkler* 875 (holo. HBG).

HABITAT: lower montane forest.
ALTITUDINAL RANGE: 1100m.
DISTRIBUTION: Borneo (K). Endemic.

M.perakensis *(Ridl.) Holttum* in Gard. Bull. Singapore 11:283 (1947). Type: Peninsular Malaysia, Perak, Batu Kuran, *Curtis* s.n. (holo. SING).
   Microstylis perakensis Ridl. in J. Linn. Soc., Bot. 32:222 (1896).
   M.longidens J.J.Sm. in Bull. Jard. Bot. Buitenzorg, ser. 2, 26:25 (1918). Type: Java, Priangan, *Backer* 12 (holo. L).

HABITAT: limestone boulders; riverine alluvial forest; hill forest.
ALTITUDINAL RANGE: sea level – 800m.
DISTRIBUTION: Borneo (SA, SR). Also Peninsular Malaysia, Thailand, Java.

M.pleistantha *(Schltr.) Ames* in Merr., Bibl. Enum. Born. Pl.: 151 (1921). Type: Sarawak, Bitulu, Tubao, *Beccari* 3771 (holo. B, destroyed, iso. FI).
   Microstylis pleistantha Schltr. in Feddes Repert. 9:340 (1911).

HABITAT: unknown.
ALTITUDINAL RANGE: unknown.
DISTRIBUTION: Borneo (SR). Endemic.

M.punctata *J.J.Wood* in Kew Mag. 4 (2):76, Pl.78 (1987). Type: Sabah, Mt. Kinabalu, between Kiau and Dahobang River, *Sands* 4011 (holo. K).

HABITAT: mixed hill dipterocarp forest often with palm understorey; moss forest; oak-chestnut forest; lower montane forest, often on ultramafic substrate.
ALITUDINAL RANGE: 800 – 1800m.
DISTRIBUTION: Borneo (SA). Endemic.

M.rajana *J.J.Wood*, **nom.nov.**
   Microstylis laxa J.J.Sm. in Mitt. Inst. Allg. Bot. Hamburg 7:40, t.6, fig.30 (1927), non Ridl. Type: Kalimantan, Bukit Raja, *Winkler* 1017 (holo. HBG).

HABITAT: lower montane forest.
ALTITUDINAL RANGE: 1400m.
DISTRIBUTION: Borneo (K). Endemic.

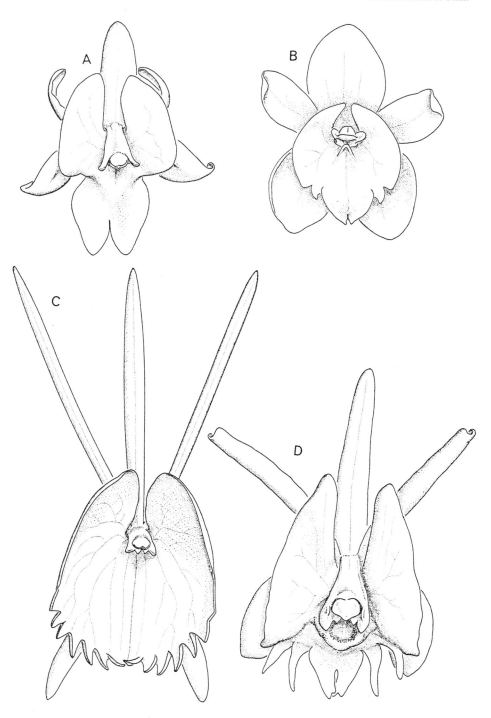

FIG. 8. *Malaxis.* **A**, *M. damusica*, x 9; **B**, *M. kinabaluensis*, x 9; **C**, *M. metallica*, x 6; **D**, *M. perakensis*, x 9. Drawn by Mutsuko Nakajima.

M.ramosa *(J.J.Sm.) Ames* in Merr., Bibl. Enum. Born. Pl.: 152 (1921). Type: Kalimantan, Kapuas, *Nieuwenhuis* s.n. (holo. BO).
Microstylis ramosa J.J.Sm. in Icon. Bogor. 2:40, t.108F (1903).

HABITAT: unknown.
ALTITUDINAL RANGE: unknown.
DISTRIBUTION: Borneo (K). Endemic.

M.repens *(Rolfe) J.J.Wood,* **comb.nov.**
Microstylis repens Rolfe in Kew Bull.: 127 (1899). Type: Sulawesi, Minahassa, Lokon, Gunung Klabat, *Koorders* 29537 (holo. K).

HABITAT: lower montane forest.
ALTITUDINAL RANGE: 1300 – 1400m
DISTRIBUTION: Borneo (SR). Also Sulawesi.

M.sublobata *(J.J.Sm.) J.J.Wood,* **comb.nov.**
Microstylis sublobata J.J.Sm. in Bull. Jard. Bot. Buitenzorg, ser. 3, 11:118 (1931). Type: Kalimantan, Gunung Kemoel, *Endert* 4342 (holo. L, iso. BO).

HABITAT: lower montane forest.
ALTITUDINAL RANGE: 1600m.
DISTRIBUTION: Borneo (K). Endemic.

M.subtiliscapa *(J.J.Sm.) J.J.Wood,* **comb.nov.**
Microstylis subtiliscapa J.J.Sm. in Bull. Jard. Bot. Buitenzorg, ser. 3, 12:149 (1932).
Malaxis graciliscapa Ames & C.Schweinf., Orch. 6:73 (1920). Type: Sabah, Mt. Kinabalu, Marai Parai, Clemens 258 (holo. AMES).
Microstylis graciliscapa (Ames & C.Schweinf.) J.J.Sm. in Bull. Jard. Bot. Buitenzorg, ser. 3, 11:119 (1931), non Schltr.

HABITAT: hill and lower montane forest.
ALTITUDINAL RANGE: 400 – 1700m.
DISTRIBUTION: Borneo (K, SA). Endemic.

M.variabilis *Ames & C.Schweinf.,* Orch. 6:77, pl.88 (1920). Type: Sabah, Mt. Kinabalu, Kiau, *Haslam* s.n.(holo. AMES).
Microstylis variabilis (Ames & C.Schweinf.), Masam., Enum.Phan. Born.: 198(1942).

HABITAT: hill forest.
ALTITUDINAL RANGE: 900m.
DISTRIBUTION: Borneo (SA,SR). Endemic.

M.venosa *(J.J.Sm.) Ames* in Merr., Bibl. Enum. Born. Pl.: 152 (1921). Type: Kalimantan, *Nieuwenhuis* s.n. (holo. BO).
Microstylis venosa J.J.Sm. in Icon. Bogor. 2:34, t.108A (1903).

HABITAT: unknown.
ALTITUDINAL RANGE: unknown.
DISTRIBUTION. Borneo (K). Endemic.

OBERONIA Lindl.

Gen. Sp. Orch. Pl.:15 (1830).

Epiphytic herbs. Stems short or long, leafy, often pendulous, lacking pseudobulbs. Leaves equitant, distichous, bilaterally flattened, sometimes jointed at base of blade. Inflorescence terminal, densely many-flowered, the flowers arranged in whorls, the lowermost usually opening last. Flowers minute, 1–2mm. long, non-resupinate, white, pale green, yellow, orange, red or brown; sepals and petals free, subequal, erect or reflexed; sepals broader than petals; petals often erose; lip sessile, entire or 3-lobed and erose, with 2 basal auricles that may or may not embrace base of column; column very short, without a foot; pollinia 4.

About 330 species distributed from tropical Africa to the Pacific islands and most numerous in mainland Asia.

O.affinis *Ames & C.Schweinf.*, Orch. 6:79, Pl.89 (1920). Type: Sabah, Mt. Kinabalu, Lobong Cave, *Clemens* 102 (holo. AMES, iso. BO, K, SING).

HABITAT: lower montane oak-laurel forest, sometimes on ultramafic substrate.
ALTITUDINAL RANGE: 1200 – 2100m.
DISTRIBUTION: Borneo (SA). Endemic.

O.anceps *Lindl.*, Sert. Orch. sub t.8 (1838). Type: Burma, *Griffith* 1097 (holo. K-LINDL).

HABITAT: mixed lowland dipterocarp forest.
ALTITUDINAL RANGE: sea level and lowlands.
DISTRIBUTION: Borneo (B, K). Also Burma, Peninsular Malaysia, Thailand, Vietnam, Sumatra, Java, Sulawesi, Maluku (Ambon), Philippines.

O.beccarii *Finet* in Bull. Soc. Bot. France 55:336, t.10, fig.20–28 (1908). Type: Sarawak, *Beccari* 2676 (holo. P, iso. FI, K).

HABITAT: unknown.
ALTITUDINAL: unknown.
DISTRIBUTION: Borneo (SR). Endemic.

O.borneensis *Schltr.* in Feddes Repert. 8:563 (1910). Type: Borneo, Landakan (= Sandakan), *Routledge* s.n. (holo. B, destroyed).

HABITAT: unknown.
ALTITUDINAL RANGE: unknown.
DISTRIBUTION: Borneo (unspecified). Endemic.

O.ciliolata *Hook.f.*, Fl.Brit.Ind. 6:181 (1890). Type: Singapore, Krangi, *Ridley* 375 (holo. K).
O.spathulata auct., non J.J.Sm.

HABITAT: moss forest.
ALTITUDINAL RANGE: 1200 – 1300m.
DISTRIBUTION: Borneo (SA, SR). Also Peninsular Malaysia, Singapore, Thailand, Sumatra, Java.

O.djongkongensis *J.J.Sm.* in Mitt. Inst. Allg. Bot. Hamburg 7:43, t.6, fig.33 (1927). Type: Kalimantan, Djongkong, *Winkler* 1304 (holo. HBG, drawing at BO).

HABITAT: coastal forest; hill forest; lower montane forest.

ALTITUDINAL RANGE: sea level – 1500m.
DISTRIBUTION: Borneo (K, SA). Endemic.

O.equitans *(G.Forst.) Mutel* in Mém. Soc. Roy. Centr. Agric. Sci. Arts Dépt. Nord :84 (1837). Type: Tahiti, *Forster* 170 (lecto. BM, isolecto. P).
Epidendrum equitans G.Forst., Fl. Ins. Austr. Prodr.:60 (1786).

HABITAT: unknown.
ALTITUDINAL RANGE: 300 – 400m.
DISTRIBUTION: Borneo (SA). *O.equitans* is widely distributed in the Pacific islands.

NOTE: The specimen *Carr* SFN 26267 (SING) is fruiting only, but agrees with *O.equitans* vegetatively.

O.filaris *Ridl.* in J. Straits Branch Roy. Asiat. Soc. 50:126 (1908). Type: Sarawak, *Hewitt* 19 (holo. SING, iso. K).

HABITAT: unknown.
ALTITUDINAL RANGE: unknown.
DISTRIBUTION: Borneo (SR). Endemic.

O.hosei *Rendle* in J. Bot. 39:173 (1901). Type: Sarawak, Baram, *Hose* 30 (holo. BM).

HABITAT: unknown.
ALTITUDINAL RANGE: unknown.
DISTRIBUTION: Borneo (SR). Endemic.

O.insectifera *Hook.f.*, Icon. Plant. t.2004 (1890). Type: Peninsular Malaysia, Perak, Larut, *King's collector* 2793 (holo. K).

HABITAT: primary hill forest
ALTITUDINAL RANGE: 200 – 300m.
DISTRIBUTION: Borneo (SA, SR). Also Peninsular Malaysia.

O.integerrima *Guillaumin* in Bull. Mus. Hist. Nat. (Paris), ser. 2, 26:692 (1954). Type: Vietnam, Annam, Da Lat, *Sigaldi* 158 (holo. P).

HABITAT: unknown.
ALTITUDINAL RANGE: unknown.
DISTRIBUTION: Borneo (SA). Also China (Yunnan), Vietnam.

O.iridifolia *(Roxb.) Lindl.*, Gen. Sp. Orch. Pl.: 15 (1830). Type: Bangladesh, Sylhet, *Roxburgh* s.n. (holo. not located).
Cymbidium iridifolium Roxb., Fl. Ind. 3:458 (1832).

HABITAT: unknown.
ALTITUDINAL RANGE: unknown.
DISTRIBUTION: Borneo (SR). Widespread from N.W.Himalaya and India (Deccan) to Burma, Laos and Thailand, southwards to Peninsular Malaysia, eastwards from Sumatra and Java to Sulawesi, north to the Philippines, and China (Hainan), extending to the Pacific islands.

O.kinabaluensis *Ames & C.Schweinf.*, Orch. 6:81, pl.89 (1920). Type: Sabah, Mt. Kinabalu, Kiau, *Clemens* 329 (holo. AMES).

HABITAT: hill and lower montane forest, sometimes on ultramafic substrate.
ALTITUDINAL RANGE: 900 – 1500m.
DISTRIBUTION: Borneo (SA). Endemic.

O.linearifolia *Ames*, Orch. 7:113, Pl.114, fig.1 & 2 (1922). Type: Sarawak, *Native collector* 921 (holo. AMES).

HABITAT: unknown.
ALTITUDINAL RANGE: unknown.
DISTRIBUTION: Borneo (SR). Endemic.

O.lobbiana *Lindl.*, Fol. Orch. Ober.: 7 (1859). Type: Borneo, *Lobb* s.n. (holo. K-LINDL).

HABITAT: unknown.
ALTITUDINAL RANGE: lowlands.
DISTRIBUTION: Borneo (K). Endemic.

O.longhutensis *J.J.Sm.* in Bull. Jard. Bot. Buitenzorg, ser. 3, 11: 119 (1931). Type: Kalimantan, Long Hoet, *Endert* 2836 (holo. L, iso. BO).

HABITAT: hill forest; riverine forest.
ALTITUDINAL RANGE: 300m.
DISTRIBUTION: Borneo (K). Endemic.

O.longifolia *Ridl.* in J. Straits Branch Roy. Asiat. Soc., 50:127 (1908). Types: Sarawak, Bukit Tendang, Busau, *Ridley* s.n., Quop, *Hewitt* s.n. (syn. SING).

HABITAT: unknown.
ALTITUDINAL RANGE: unknown.
DISTRIBUTION: Borneo (SR). Endemic.

O.macrostachys *Ridl.* in J. Linn. Soc., Bot. 31:263 (1896). Type: Sarawak, Braang, *Haviland* s.n. (holo. SING).

HABITAT: hill forest.
ALTITUDINAL RANGE: 400 – 500m.
DISTRIBUTION: Borneo (SA, SR). Endemic.

O.melinantha *Schltr.* in Bull. Herb. Boissier ser. 2, 6:305 (1906). Type: Kalimantan, Moera-Kelindjau, *Schlechter* 13364 (holo. B, destroyed).

HABITAT: unknown.
ALTITUDINAL RANGE: unknown.
DISTRIBUTION: Borneo (K). Endemic.

O.miniata *Lindl.* in Bot. Reg. 29, misc. no.8 (1843). Type: Singapore, *Loddiges* s.n. (holo. K-LINDL).

HABITAT: riverine forest.
ALTITUDINAL RANGE: 200 – 300m.
DISTRIBUTION: Borneo (SA). Also Peninsular Malaysia, Thailand, Sumatra, Java.

O.multiflora *Ridl.* in J. Linn. Soc., Bot. 31:262 (1896). Type: Sarawak, Trusan River, *Haviland* s.n. (holo. SING).

HABITAT: unknown.
ALTITUDINAL RANGE: unknown.
DISTRIBUTION: Borneo (SR). Endemic.

O.neglecta *Schltr.* in Feddes Repert. 9:26 (1910). Type: Sarawak, Kuching race course, *Schlechter* 15842 (holo. B, destroyed).

HABITAT: unknown.
ALTITUDINAL RANGE: sea level.
DISTRIBUTION: Borneo (SR). Endemic.

O.patentifolia *Ames & C.Schweinf.*, Orch. 6: 83, Pl.90, fig.1 (1920). Type: Sabah, Mt. Kinabalu, Lobong Cave, *Clemens* 104 (holo. AMES).

HABITAT: hill and lower montane forest on ultramafic and other substrates.
ALTITUDINAL RANGE: 600 – 1500m.
DISTRIBUTION: Borneo (SA). Endemic.

O.rubra *Ridl.* in J. Straits Branch Roy. Asiat. Soc. 50:127 (1908). Type: Sarawak, Matang Estate, *Hewitt* s.n. (holo. SING, iso. K).

HABITAT: coffee plantations; lower montane forest.
ALTITUDINAL RANGE: lowlands – 1400m.
DISTRIBUTION: Borneo (SA, SR). Endemic.

O.sarawakensis *Schltr.* in Feddes Repert. 9:342 (1911). Type: Sarawak, Gunung Matang, *Beccari* 1543 (holo. B, destroyed, iso. FI).

HABITAT: unknown.
ALTITUDINAL RANGE: unknown.
DISTRIBUTION: Borneo (SR). Endemic.

O.sinuosa *Ridl.* in J. Linn. Soc., Bot. 31:263 (1896). Types: Sarawak, Kuching, *Hose* s.n., *Haviland* s.n. (syn. SING).

HABITAT: unknown.
ALTITUDINAL RANGE: unknown.
DISTRIBUTION: Borneo (SR). Endemic.

O.spiralis *Lindl.*, Fol. Orch. Oberonia 7, no. 40 (1859). Type: Java, *Kuhl & van Hasselt* s.n. (K-LINDL, drawing by Rchb.f.)
    O.spathulata Lindl., Gen. Sp. Orch. Pl.:16 (1830), nom. illeg.
    O.padangensis Schltr. in Feddes Repert. 9:26 (1910). Type: Sumatra, Padang-Pandjang, *Schlechter* 16019 (holo. B, destroyed, iso. K).

HABITAT: riverine forest
ALTITUDINAL RANGE: 300 m.
DISTRIBUTION: Borneo (K). Also Peninsular Malaysia, Thailand, Sumatra, Lingga Archipelago, Java, Bunguran Islands, Sulawesi and possibly New Guinea.

O.stenophylla *Ridl.* in J. Linn. Soc., Bot., 32:218 (1896). Type: Peninsular Malaysia, Johor, Hulu Sembrong, *Lake & Kelsall* s.n. (holo. SING).

HABITAT: podsol forest.
ALTITUDINAL RANGE: 400 – 500m.
DISTRIBUTION: Borneo (SA). Also Peninsular Malaysia, Singapore.

O.sumatrana *Schltr.* in Bull. Herb. Boissier, ser. 2, 6:306 (1906).

    var. **endertii** *J.J.Sm.* in Bull. Jard. Bot. Buitenzorg, ser. 3, 11:119 (1931). Type: Kalimantan, Moeara Kaman, *Endert* 1503 (holo. L, iso. BO).

HABITAT: epiphytic on *Lagerstroemia*.
ALTITUDINAL RANGE: sea level.
DISTRIBUTION: Borneo (K). Endemic.

O.triangularis *Ames & C.Schweinf.*, Orch. 6:85, Pl.90, fig.2 (1920). Type: Sabah, Mt. Kinabalu, Lobong Cave, *Clemens* 104B (holo. AMES).

HABITAT: lower montane forest.
ALTITUDINAL RANGE: 300 – 1500m.
DISTRIBUTION: Borneo (SA). Endemic.

TRIBE CYMBIDIEAE

SUBTRIBE BROMHEADIINAE

BROMHEADIA Lindl.

Bot. Reg. 27: misc. 89, no.184 (1841).

Terrestrial or epiphytic herbs. Stems slender, usually leafy, pseudobulbs absent. Leaves normal and oblong or laterally compressed. Inflorescence terminal, occasionally also lateral, producing a succession of flowers one or few at a time in two opposite ranks; bracts regularly alternate. Flowers resupinate, small to medium-sized; sepals and petals similar, spreading; lip straight, parallel with column, side lobes erect and touching column, disc between them thickened, hirsute, mid-lobe ovate, recurved, with a median warty or papillose area; column long and slender, without a foot, winged above; pollinia 2, on a short, broad stipes.

Around 17 species distributed from Sri Lanka, Thailand and Indochina through Malaysia and Indonesia to the Philippines, New Guinea and Australia.

B.alticola *Ridl.* in J. Linn. Soc., Bot. 28:338, Pl.42 (1891). Types: Singapore, Bukit Timah, Jurong & Bukit Mandai, *Ridley* s.n. (syn. SING); Peninsular Malaysia, Selangor, Kuala Lumpur, *Ridley* s.n. (syn. SING).

HABITAT: unknown.
ALTITUDINAL RANGE: unknown.
DISTRIBUTION: Borneo (SR). Also Thailand, Peninsular Malaysia, Singapore and possibly the Philippines.

B.aporoides *Rchb.f.* in Otia Bot. Hamb.1:44 (1878). Type: Burma, Moulmein, *Parish* s.n. (holo. W).

HABITAT: unknown.
ALTITUDINAL RANGE: lowlands.
DISTRIBUTION: Borneo (K). Also Burma, Thailand, Laos, Singapore.

B.borneensis *J.J.Sm.* in Bull. Jard. Bot. Buitenzorg, ser. 2, 25:18 (1917). Type: Kalimantan, Gunung Samonggaris, *Amdjah* 980 (holo. L, iso. K).

HABITAT: swamp forest; lowland dipterocarp forest; hill forest; podsol forest; ridge top kerangas.
ALTITUDINAL RANGE: lowlands – 700m.
DISTRIBUTION: Borneo (B, K, SA, SR). Also Peninsular Malaysia, Singapore.

B.brevifolia *Ridl.* in J. Linn. Soc., Bot. 32:340 (1896). Type: Peninsular Malaysia, Perak, Hermitage Hill, *Ridley* s.n. (holo. :SING).

HABITAT: 'sand forest'; lower montane forest.
ALTITUDINAL RANGE: 800 – 1100m.
DISTRIBUTION: Borneo (SA, SR). Also Peninsular Malaysia.

B.crassiflora *J.J.Sm.* in Bull. Jard. Bot. Buitenzorg, ser. 3, 11: 149 (1931). Type: Kalimantan, West Koetai, Gunung Kemoel, *Endert* 4273 (holo. L).

HABITAT: lower montane forest.
ALTITUDINAL RANGE: 1300 – 1600m.

DISTRIBUTION: Borneo (K,SA). Endemic.

B.divaricata *Ames & C.Schweinf.*, Orch. 6:155 (1920). Type: Sabah, Mt. Kinabalu, above Marai Parai Spur, *Clemens* 389 (holo. AMES, iso. BO, K, SING).

HABITAT: lower and upper montane forest on ultramafic substrate.
ALTITUDINAL RANGE: 2100 – 2700m.
DISTRIBUTION: Borneo (SA). Endemic.

B.finlaysoniana *(Lind.) Miq.*, Fl.Ind.Bat. 3:709 (1859). Types: Singapore, *Finlayson* s.n. (syn. K-LINDL); *Wallich* 7561 (syn. K-LINDL).
Grammatophyllum finlaysonianum Lindl., Gen. Sp. Orch. Pl. 173 (1833).

HABITAT: peat swamp forest; secondary forest; 'heath woodland'; low kerangas with *Baeckea* and *Ploiarium* dominant; high kerangas; hill and lower montane forest.
ALTITUDINAL RANGE: sea level – 1500m.
DISTRIBUTION: Borneo (B, K, SA, SR). Also Thailand, Indochina, Peninsular Malaysia, Singapore, Sumatra, Mentawai Islands, Lingga Archipelago, Riau, Bangka, Belitung, Anambas, Maluku, Philippines, New Guinea.

B.scirpoidea *Ridl.* in J. Bot. 38:71 (1900). Types: Peninsular Malaysia, Perak, near Ipoh, *Ridley* s.n. (syn. SING); Pahang, Tahan Valley, *Ridley* s.n. (syn. SING).

HABITAT: hill and lower montane forest.
ALTITUDINAL RANGE: 500 – 1500m.
DISTRIBUTION: Borneo (SA). Also Peninsular Malaysia.

B.tenuis *J.J.Sm.* in Brittonia 1:110 (1931). Types: Sarawak, Mt. Poi, *Clemens* 20308 (syn. L), 20184 (syn. L, SAR) & 22588 (syn. L).

HABITAT: lower montane forest.
ALTITUDINAL RANGE: 1300 – 1800m.
DISTRIBUTION: Borneo (SA, SR). Endemic.

B.truncata *Seidenf.* in Opera Bot. 72:14, fig.5 (1983). Type: Thailand, Doi Suthep, *Seidenfaden & Smitinand* GT2691 (holo. C).

HABITAT: unknown.
ALTITUDINAL RANGE: below 900m.
DISTRIBUTION: Borneo (SA). Also Thailand, Peninsular Malaysia, Singapore, Sumatra.

SUBTRIBE EULOPHIINAE

DIPODIUM R.Br.

Prodr.:330 (1810).

Terrestrial or climbing herbs of diverse habit: either (1) erect-stemmed, sympodial and green-leaved, (2) climbing, leafy, monopodial and green-leaved or (3) leafless, saprophytic (habit 1 and 2 in Borneo). Leaves, when present, distichous, imbricate, ensiform, plicate. Inflorescence lateral or terminal, racemose, ± erect, few to many-flowered. Flowers medium-sized, resupinate; sepals and petals free, spreading, usually spotted and blotched on reverse; lip 3-lobed, close and parallel to column, side lobes narrow, more or less embracing column to form a tube, mid-lobe much larger, ovate, convex, pubescent; column short, thick, foot absent; pollinia 2, cleft, each with a rather long stipes.

About 20 species distributed from tropical Asia to New Guinea, Australia and the Pacific islands.

D.paludosum *(Griff.) Rchb.f.*, Xenia Orch. 2:15 (1862). Type: Peninsular Malaysia, Melaka, Ayer Punnus, *Griffith* 5317 (holo. K).
Grammatophyllum paludosum Griff., Not. 3:844 (1851).

HABITAT: swampy kerangas.
ALTITUDINAL RANGE: lowlands.
DISTRIBUTION: Borneo (K, SA, SR). Also Thailand, Cambodia, Vietnam, Peninsular Malaysia, Singapore, Sumatra, Mentawai, Bangka, Philippines.

D.pictum *(Lindl.) Rchb.f.*, Xenia Orch. 2:15 (1862). Type: cult. ex 'Java', *Lobb* s.n. (holo. K-LINDL, iso. K,W).
Wailesia picta Lindl. in J. Hort. Soc. London 4:261 (1849).

HABITAT: kerangas forest; lower montane forest, with *Agathis;* peat swamp forest.
ALTITUDINAL RANGE: 300 – 1100m.
DISTRIBUTION: Borneo (B, K, SA, SR). Also Peninsular Malaysia, Sumatra, Java, Philippines.

D.purpureum *J.J.Sm.* in Icon. Bogor. 4, t.304 (1910). Type: Kalimantan, Bukit Mili, *Sakeran* 1854, Cult. Bogor (holo. BO).

HABITAT: unknown.
ALTITUDINAL RANGE: 100 – 200m.
DISTRIBUTION: Borneo (K). Endemic.

D.scandens *(Blume) J.J.Sm.*, Orch. Java. 6:488 (1905). Type: Java, *Blume* s.n. (holo. ?L).
Leopardanthus scandens Blume, Rumphia 4:47 (1848).

HABITAT: swamp forest; lowland dipterocarp forest; hill forest on ultramafic substrate.
ALTITUDINAL RANGE: sea level – 1400m.
DISTRIBUTION: Borneo (B, SA, SR). Also Peninsular Malaysia, Java.

EULOPHIA R.Br. ex Lindl.

Bot. Reg. 8:t.686 (1823).

Terrestrial, sometimes saprophytic, herbs. Rhizome short. Stems pseudobulbous, squat, with either (1) terminal leaves, (2) slender, leafy and sheathed in basal cataphylls or (3) leafless. Leaves, when present, narrowly elliptic, plicate. Inflorescence lateral, arising from near base of stem, erect, racemose, few- to many-flowered. Flowers small to large, resupinate; sepals free, similar; petals smaller than sepals; lip entire or 3-lobed, spurred at base; column slender; pollinia 4.

A pantropical genus of about 200 species, particularly well represented in Africa.

E.borneensis *Ridl.* in J. Linn. Soc., Bot.: 31:289 (1896). Type: Sarawak, *Haviland* s.n. (holo. SING).

HABITAT: unknown.
ALTITUDINAL RANGE: unknown.
DISTRIBUTION: Borneo (SR). Endemic.

E.chrysoglossoides *Schltr.* in Bull. Herb. Boissier ser. 2, 6:453 (1906). Type: Sabah, locality unknown, *Little* s.n. (holo. G).

HABITAT: unknown.
ALTITUDINAL RANGE: unknown.
DISTRIBUTION: Borneo (SA). Endemic.

E.graminea *Lindl.*, Gen. Sp. Orch. Pl.: 182 (1833). Type: Singapore, *Wallich* 7372 (holo. K-LINDL).

HABITAT: sandy places near the sea; open grassy places; roadsides; waste ground.
ALTITUDINAL RANGE: sea level – 500m.
DISTRIBUTION: Borneo (B, K, SA). Also Sri Lanka, India, Nepal, Burma, Nicobar Islands, Thailand, Laos, Vietnam, Peninsular Malaysia, Singapore, Sumatra, Java, Natuna, Taiwan, China, Hong Kong, Ryukyu Islands.

E.spectabilis *(Dennst.) Suresh* in Regnum Veg. 119:300 (1988). Type: India, Malabar coast, *coll.?* (lecto. Rheede's t.36 in Hort. Malab. 11 (1692)).
Wolfia spectabilis Dennst., Schlüssel Hortus Malab., 11, 25, 38 (1818).
Eulophia squalida Lindl. in Bot. Reg. 27, misc.77 (1841). Type: Philippines, *Cuming* s.n. (holo. K-LINDL).

HABITAT: open grassy places; roadsides; lowland forest; peat swamp forest.
ALTITUDINAL RANGE: sea level – 900m.
DISTRIBUTION: Borneo (B, K, SA, SR). Widespread from Sri Lanka, India and the Himalayas to Indochina, Malaysia, Indonesia, the Philippines, New Guinea east to Fiji and Tonga.

E.zollingeri *(Rchb.f.) J.J.Sm.*, Orch. Java: 228 (1905). Type: Java, Lampong Province and Gebbok Klakka, *Zollinger* 585 (holo. W).
Cyrtopera zollingeri Rchb.f. in Bonplandia 5:38 (1857).

HABITAT: hill and lower montane forest; secondary forest.
ALTITUDINAL RANGE: 500 – 1500m.
DISTRIBUTION: Borneo (SA). Widespread from India, Malaysia, Indonesia and the Philippines, north to Taiwan and Japan, east to New Guinea and Australia.

GEODORUM Jacks.

Bot. Repos. 10: pt.626 (1811).

Terrestrial herbs. Stems short, later swelling at base to form almost round, subterranean pseudobulbs. Leaves terminal, few, plicate, the uppermost the largest, broad, petiolate. Inflorescence separate from the leaf-bearing stem, erect; rachis decurved, nodding, straightening prior to seed dispersal, densely several- to many-flowered. Flowers not opening widely, bell-shaped; sepals and petals similar; lip entire or obscurely trilobed, forming with the column foot a short saccate base, not spurred; column short with a distinct foot; pollinia 4.

Between 5 and 10 species distributed from tropical Asia eastward through S.E.Asia to New Guinea, Australia and the Pacific islands.

G.densiflorum *(Lam.) Schltr.* in Feddes Repert 4:259 (1919). Type: India, Malabar coast, *collector unknown* (lecto.: Rheede's t.35 in Hort Malab. 11 (1692)).
Limodorum densiflorum Lam., Encycl. Meth. Bot. 3:516 (1792).
Geodorum nutans sensu Masamune, Enum. Phan. Born.:187 (1942), non (C. Presl) Ames.

HABITAT: secondary hill forest; roadsides; recorded from ultramafic substrate.
ALTITUDINAL RANGE: sea level – 1800m.
DISTRIBUTION: Borneo (SA). Widespread from southern China, Burma, India and Sri Lanka east to New Guinea, Solomon Islands, Vanuatu, New Caledonia, Fiji, Samoa, Tonga, Niue and Australia.

OECEOCLADES Lindl.

Bot. Reg. 18: sub t.1522 (1832).

Terrestrial herbs. Pseudobulbs short and one-noded to long, slender and few-noded. Leaves 1–3, terminal, conduplicate or plicate. Inflorescence lateral, arising from near the base of the pseudobulb, erect, racemose, many-flowered. Flowers with narrow, acute sepals; petals usually shorter and broader than sepals; lip trilobed or 4-lobed, with a short spur; column slender; pollinia 4.

About twenty species in the Old World Tropics, mainly in Africa and Madagascar, extending east to the Pacific islands.

O.pulchra *(Thouars) P.J.Cribb & M.A.Clem.* in Clements, Austr. Orch. Res. 1:99 (1989). Type: Mascarene Islands, Reunion, *Thouars* s.n. (holo. P).
Limodorum pulchrum Thouars, Hist. Orch., Pl.43 & 44 (1822).
Eulophia macrostachya Lindl., Gen. Sp. Orch. Pl.:183 (1833). Type: Sri Lanka, *Macrae* s.n. (holo. K-LINDL).

HABITAT: on sandy soil in light shade; hill forest.
ALTITUDINAL RANGE: sea level – 1100m.
DISTRIBUTION: Borneo (K, SA). Widespread in Africa, Madagascar, Mascarene Islands, India, Sri Lanka, Malaysia, Indonesia, Philippines, New Guinea and Australia.

SUBTRIBE THECOSTELINAE

THECOPUS Seidenf.

Opera Bot. 72:101 (1983).

Epiphytic herbs. Pseudobulbs of one node, one-leaved. Leaves oblong-elliptic, petiolate. Inflorescences lateral, arising from near base of pseudobulbs, racemose, several- to many-flowered, pendulous. Flowers large, resupinate; sepals and petals free; petals narrower than sepals; lip 3-lobed, joined at its base with an outgrowth from the column and with column foot to form a tube at right angles to base of column, side lobes erect, tips curving inwards, mid-lobe acute, convex, hirsute, disc with or without keels; column arcuate, with spreading apical arms, foot hollow, with an entrance near the articulate lip base; anther cap conical; pollinia 4, unequal, in pairs on a long narrow stipes; viscidium obscure.

Two species distributed in Thailand, Peninsular Malaysia and Borneo.

T.maingayi *(Hook.f.) Seidenf.* in Opera Bot. 72:101 (1983). Type: Peninsular Malaysia, Malacca, *Maingay* 3155 (holo. K).
Thecostele maingayi Hook.f., Fl. Brit. Ind. 6:20 (1890).

HABITAT: rather open, swamp forest on ultramafic substrate.
ALTITUDINAL RANGE: lowlands – 700m.
DISTRIBUTION: Borneo (SA). Also Thailand, Peninsular Malaysia.

T.secunda *(Ridl.) Seidenf.* in Opera Bot. 72:101 (1983). Type: Sarawak, *Everett* s.n. (holo. SING).
Thecostele secunda Ridl. in J. Linn. Soc., Bot. 31:299 (1896).

HABITAT: podsol forest.
ALTITUDINAL RANGE: 400 – 500m.
DISTRIBUTION: Borneo (K, SA, SR). Also Peninsular Malaysia.

THECOSTELE Rchb.f.

Bonplandia 5:37 (1857).

Epiphytic herbs. Pseudobulbs of one node, one-leaved. Leaves oblong-elliptic or narrowly elliptic, shortly petiolate. Inflorescences lateral, arising from near base of pseudobulbs, racemose, many-flowered, pendulous. Flowers rather small, resupinate; sepals and petals free, spreading; petals narrower than sepals; lip 3-lobed, joined to the column in the same manner as in *Thecopus*, shape similar to *Thecopus*, mid-lobe broad, convex, decurved, finely hairy, disc with 2 short keels between side lobes; column curved into an S shape, with a narrow nectary opening at base of lip on upper side of foot; anther cap semiglobular; pollinia 2, cleft; stipes small; viscidium semicircular.

A monotypic genus distributed in Burma, Thailand, Indochina, Malaysia and Indonesia to the Philippines.

T.alata *(Roxb.) C.S.P.Parish & Rchb.f.* in Trans. Linn. Soc. London, Bot. 30:135 & 144, t.29 (1874). Type: Bangladesh, Chittagong, *Roxburgh* s.n. (holo. Roxburgh drawing, K).

Cymbidium alatum Roxb., Fl. Ind. 2, ed. 3:459 (1832).

HABITAT: riverine forest; hill forest.
ALTITUDINAL RANGE: sea level – 500m.
DISTRIBUTION: Borneo (B, K, SA, SR). Also India, Bangladesh, Burma, Thailand, Laos, Vietnam, Peninsular Malaysia, Sumatra, Java, Philippines.

SUBTRIBE CYRTOPODIINAE

CHRYSOGLOSSUM Blume

Bijdr.:337 (1825).

Terrestrial, rarely epiphytic, herbs. Rhizome creeping. Pseudobulbs articulated, one-leaved. Leaves convolute, linear-elliptic to narrowly elliptic, acuminate, petiolate, sometimes variegated. Inflorescence erect, racemose, many flowered. Flowers resupinate; sepals and petals free, similar; lateral sepals inserted on column-foot; lip 3-lobed, mobile, fleshy, hypochile pleated at base, disc with 3 keels, epichile recurved, concave; column erect, with a distinct foot, having a saccate spur, and a small lobe in front; pollinia 2, waxy.

Four species distributed from Sri Lanka and India eastwards to New Guinea and the Pacific islands.

C.reticulatum *Carr* in Gard. Bull. Straits Settlem. 8:197 (1935). Type: Sabah, Mt. Kinabalu, main spur west of Tenompok Pass, *Carr* C.3314 (SFN 27060) (lecto. K, isolecto. SING).

HABITAT: lower montane forest.
ALTITUDINAL RANGE: 1300 – 1800m.
DISTRIBUTION: Borneo (SA, SR). Endemic.

CLADERIA Hook.f.

Fl. Brit. Ind. 5:810 (1890).

Terrestrial herbs. Rhizome creeping, long, slender, bearing leafy shoots 20cm. or more apart, climbing a little way up the bases of tree trunks where the inflorescence is produced. Leaves 4 – 6 per shoot, close together, elliptic, plicate. Inflorescence racemose, bearing a succession of flowers one or two at a time; peduncle up to 15cm. long; rachis slowly elongating, pubescent; floral bracts stiff, pubescent. Flowers resupinate, green, lip paler green with darker green veins. Sepals and petals free, the sepals pubescent. Dorsal sepal erect or curved over column. Lateral sepals spreading horizontally, curved towards each other. Lip 3–lobed, saccate; side lobes large, obtuse, erect; mid-lobe wider than long, apex reflexed, retuse; disc with 2 low rounded hairy keels between side lobes. Column long, arcuate. Pollinia 2, deeply cleft.

Two species distributed in Thailand, Peninsular Malaysia, Sumatra, Bangka, Mentawai, Sulawesi and New Guinea.

C.viridiflora *Hook.f.*, Fl. Brit. Ind. 5:810 (1890). Types: Peninsular Malaysia, Perak, Sunga Ryah, *King's collector* 774 (syn. K); Malacca, *Maingay* 3294 (syn. K).

HABITAT: primary hill forest on sandstone; 'sand forest'; mixed dipterocarp forest.
ALTITUDINAL RANGE: sea level – 800m.
DISTRIBUTION: Borneo (B, K, SA, SR). Also Thailand, Peninsular Malaysia, Sumatra, Bangka, Mentawai, Sulawesi.

117

COLLABIUM Blume

Bijdr.:357 (1825).

Terrestrial, rarely epiphytic herbs. Rhizome creeping. Pseudobulbs alternately 1 – 6 bearing a leaf and 1 bearing an inflorescence, more or less articulated at junction with petiole or scape. Leaves convolute, narrowly elliptic, acute to acuminate, sometimes mottled. Inflorescence erect, racemose, few- to many-flowered. Flowers resupinate; lateral sepals inserted on column foot; lip immobile, entire or 3-lobed, hypochile more or less parallel to column, narrow at base, disc 2- or 3-keeled; column-foot with bases of lateral sepals and claw of lip forming a saccate, cylindrical or somewhat flattened mentum; pollinia 2, waxy.

Eleven species distributed in China and Taiwan south to Burma, Indochina and Peninsular Malaysia, eastwards through Sumatra, Java, Borneo and Sulawesi to New Guinea, Vanuatu and Fiji.

C.bicameratum *(J.J.Sm.) J.J.Wood* in Orch. Borneo 1:101 (1994). Type: Kalimantan, West Koetai, Gunung Kemoel, *Endert* 4351 (lecto. L).
    Chrysoglossum bicameratum J.J.Sm.in Bull. Jard. Bot. Buitenzorg, ser. 3, 11:91 (1931).

HABITAT: lower montane forest.
ALTITUDINAL RANGE: 1600 – 1800m.
DISTRIBUTION: Borneo (K, SA). Endemic.

C.simplex *Rchb.f.* in Gard. Chron. n.s. 15:462 (1881). Types: Borneo, cult. *Veitch* 288 and *Bull* s.n. (holo. W).
    Chrysoglossum simplex (Rchb.f.) J.J.Sm., Orch. Java: 177 (1905).

HABITAT: lower montane forest.
ALTITUDINAL RANGE: unknown.
DISTRIBUTION: Borneo (SA). Also Peninsular Malaysia, Sumatra, Java.

CYMBIDIUM Sw.

Nova Acta Regiae Soc. Sci. Upsal., ser. 2, 6: 70 (1799).

Du Puy, D. & Cribb, P. (1988). The Genus *Cymbidium.* Christopher Helm/Timber Press.

Epiphytic, lithophytic or terrestrial herbs. Pseudobulbs short, rarely elongate, occasionally absent and replaced by a slender stem, ovoid to spindle-shaped, often enclosed in sheathing leaf bases. Leaves up to a dozen, distichous, linear-elliptic or narrowly ligulate to elliptic, acuminate to strongly bilobed at apex, articulate to sheaths. Inflorescences racemose, dense or lax, erect, porrect or pendulous, 2- to many-flowered; peduncle loosely covered by inflated cymbiform sheaths. Flowers resupinate, often large & showy, up to 12 cm. in diameter (Bornean species smaller), sometimes fragrant. Sepals and petals free, spreading or erect. Lip 3-lobed, free or fused at base to base of column; side lobes erect around column; mid-lobe often recurved; disc with usually 2 pubescent parallel ridges. Column long, somewhat arcuate. Pollinia usually 2, deeply cleft, sometimes 4 in 2 unequal pairs.

Du Puy & Cribb (1988) recognise 44 species widely distributed from N.W. Himalaya to Japan, south through Indochina, Malaysia & Indonesia to the Philippines, New Guinea and Australia.

C.atropurpureum *(Lindl.) Rolfe* in Orchid Rev. 11:190 (1903). Type: ?Java, cult. *Rollissons* (neo. K).
C.pendulum (Roxb.) Sw. var. *atropurpureum* Lindl. in Gard. Chron.: 287 (1854).

HABITAT: lowland forest; lower montane forest.
ALTITUDINAL RANGE: sea level – 1200m.
DISTRIBUTION: Borneo (K, SA). Also Thailand, Peninsular Malaysia, Sumatra, Java, Philippines.

C.bicolor *Lindl.*, Gen. Sp. Orch. Pl.:164 (1833).

subsp. **pubescens** *(Lindl.) Du Puy & P.J.Cribb*, Genus Cymbidium:73 (1988). Type: Singapore, *Cuming* s.n., cult. *Loddiges* (holo. K-LINDL).
C.pubescens Lindl. in Bot. Reg. 26: misc.75 (1840).

HABITAT: riverine forest; hill forest; roadside secondary vegetation.
ALTITUDINAL RANGE: 400 – 1100m.
DISTRIBUTION: Borneo (SA, SR). Also Peninsular Malaysia, Java, Sumatra, Sulawesi, Philippines.

C.borneense *J.J.Wood* in Kew Bull. 38:69, t.1 (1983). Type: Sarawak, Gunung Mulu National Park, Melinau Gorge area, *Lewis* 314 (holo. K, iso. SAR).

HABITAT: lowland forest on limestone; hill forest on ultramafic substrate, with *Gymnostoma sumatrana*.
ALTITUDINAL RANGE: 100 – 1200m.
DISTRIBUTION: Borneo (SA, SR). Endemic.

C.chloranthum *Lindl.* in Bot. Reg. 29:68 (1843). Type: cult. *Loddiges* (holo. K-LINDL).
C.sanguinolentum Teijsm. & Binn. in Natuurk. Tijdschr. Ned.-Indië 24:318 (1862). Type: Java, Gunung Salak, *Teijsmann & Binnendijk* 902 (holo. BO).
C.pulchellum Schltr. in Feddes Repert. 8:570 – 571 (1910). Type: Sarawak, Kuching, *Schlechter* 15846 (holo. B, destroyed).

HABITAT: riverine forest; hill forest.
ALTITUDINAL RANGE: 400 – 900m.
DISTRIBUTION: Borneo (SA, SR). Also Peninsular Malaysia, Sumatra, Java.

C.dayanum *Rchb.f.* in Gard.Chron.:710 (1869). Type: India, Assam, cult. *Day* (holo. W, ?iso. K).
C.angustifolium Ames & C.Schweinf., Orch. 6:212 (1920). Type: Sabah, Mt. Kinabalu, Kiau, *Clemens* 74 (holo. AMES).

HABITAT: hill forest; *Agathis, Podocarpus, Lithocarpus* forest; lower montane forest; often on mossy banks and stumps.
ALTITUDINAL RANGE: 600 – 1700m.
DISTRIBUTION: Borneo (SA). Also India, China, Taiwan, Japan, Ryukyu Islands, Thailand, Cambodia, Peninsular Malaysia, Sumatra, Philippines.

C.elongatum *J.J.Wood, Du Puy & Shim* in Du Puy & Cribb, Genus Cymbidium: 103, fig.22.1, photos 76 & 77 (1988). Type: Sabah, Mt. Kinabalu, Marai Parai, *Collenette* A47 (holo. BM, iso. K).

HABITAT: scrubby *Leptospermum* woodland; open marshy areas among montane scrub, stunted trees, rattans and sedges on sandstone or ultramafic substrate.
ALTITUDINAL RANGE: 1200 – 1800m.
DISTRIBUTION: Borneo (SA, SR). Endemic.

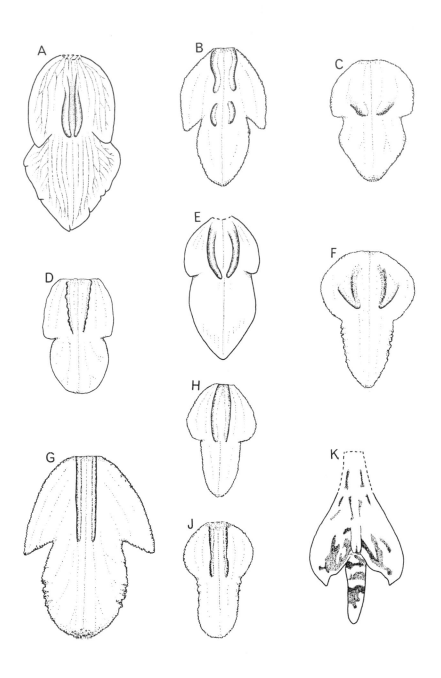

FIG. 9. *Cymbidium* lips. **A**, *C. atropurpureum*, x 2; **B**, *C. bicolor* subsp. *pubescens*, x 2; **C**, *C. borneense*, x 2; **D**, *C. chloranthum*, x 2; **E**, *C. elongatum*, x 2; **F**, *C. ensifolium* subsp. *haematodes*, x 2½; **G**, *C. finlaysonianum*, x 2; **H**, *C. lancifolium*, x 2; **J**, *C. rectum*, x 2; **K**, *C. sigmoideum*, x 2. Drawn by Sarah Thomas.

C.ensifolium *(L.) Sw.* in Nova Acta Regiae Soc. Sci. Upsal.6:77 (1799).

subsp. **haematodes** *(Lindl.) Du Puy & P.J.Cribb*, Genus Cymbidium: 161 (1988). Type: Sri Lanka, *Macrae* 12 (holo. K-LINDL).
    C.haematodes Lindl., Gen. Sp. Orch. Pl.: 162 (1833).
    C.ensifolium (L.) Sw. var. *haematodes* (Lindl.) Trimen, Cat.: 89 (1885).

HABITAT: hill and lower montane forest; recorded from ultramafic substrate.
ALTITUDINAL RANGE: 500 – 1700m.
DISTRIBUTION: Borneo (SA). Also Sri Lanka, India, Thailand, Peninsular Malaysia, Sumatra, Java, New Guinea.

C.finlaysonianum *Lindl.*, Gen. Sp. Orch. Pl.: 164 (1833). Type: Vietnam, Turon (Tourane?) Bay, *Finlayson* in *Wallich* 7358 (holo. K-LINDL).

HABITAT: open lowland primary or secondary forest; mangrove forest; rubber plantations; mixed hill dipterocarp forest; coastal rocks.
ALTITUDINAL RANGE: sea level – 1200m.
DISTRIBUTION: Borneo (K, SA, SR). Also S.Vietnam, Cambodia, Thailand, Peninsular Malaysia, Sumatra, Java, Sulawesi, Philippines.

C.lancifolium *Hook.*, Exot. Bot. 1:t.51 (1823). Type: Nepal, *Wallich* s.n., cult. *Shepherd* (holo. K).

HABITAT: lowland and hill dipterocarp forest; lower montane forest.
ALTITUDINAL RANGE: 300 – 1500m.
DISTRIBUTION: Borneo (K, SA, SR). Widespread in India, Nepal, Bhutan, China, Hong Kong, Taiwan, Ryukyu Islands, Japan, Burma, Thailand, Indochina, Peninsular Malaysia, Sumatra, Java, Maluku, New Guinea.

C.rectum *Ridl.* in J. Straits Branch Roy. Asiat. Soc. 82:198 (1920). Type: Peninsular Malaysia, Negeri Sembilan, *Genyns-Williams* s.n. (holo. SING).

HABITAT: open damp forest; podsol forest with *Baeckea frutescens.*
ALTITUDINAL RANGE: 400 – 500m.
DISTRIBUTION: Borneo (SA). Also Peninsular Malaysia.

C.sigmoideum *J.J.Sm.* in Bull. Dép. Agric. Indes Néerl. 13:52 (1907). Type: Java, Loemadjang, *Connell* s.n. (holo. L).
    C.kinabaluense K.M.Wong & C.L.Chan in Sandakania 2:86, fig.1 & 2 (1993). Type: Sabah, Mt. Kinabalu, *Chan & Nais* s.n. (holo. SAN, iso. SNP).

HABITAT: lower montane forest.
ALTITUDINAL RANGE: 1700m.
DISTRIBUTION: Borneo (SA). Also Sumatra, Java.

GRAMMATOPHYLLUM Blume

Bijdr.:377, t.20 (1825).

Large, robust epiphytic, rarely lithophytic herbs. Roots long, spreading and erect, acuminate. Pseudobulbs medium sized to very large, ovoid to elongate, clustered, few- to many-leaved. Leaves distichous, plicate, long and narrow, flexible. Inflorescences lateral, simple, erect or pendulous, a many-flowered raceme. Flowers resupinate, medium sized to large, often showy, yellow, marked with brown or maroon; sepals and petals similar, spreading, obtuse; lip much smaller, 3-lobed, fused at base to base of column, usually hairy, disc 2- or 3-ridged; column short, clavate, foot absent, but often with a concave basal outgrowth; pollinia 2, deeply cleft; stipes deeply bilobulate; viscidium fat.

About 12 species distributed in S.E. Asia eastwards to New Guinea and the Solomon Islands.

G.kinabaluense *Ames & C.Schweinf.*, Orch. 6:210 (1920). Type: Sabah, Mt. Kinabalu, Kiau, *Clemens* 55 (holo. AMES, iso. BO, K, SING).

HABITAT: hill forest.
ALTITUDINAL RANGE: 500 – 900m.
DISTRIBUTION: Borneo (SA). Endemic.

G.papuanum *J.J.Sm.* in Bull. Dép. Agric. Indes Néerl. 45:11 (1911). Type: Irian Jaya, north coast, *Gjellerup* 246 (holo. L).

HABITAT: unknown.
ALTITUDINAL RANGE: unknown.
DISTRIBUTION: Borneo (SA). Also New Guinea.

G.rumphianum *Miq.* in Ann. Mus. Bot. Lugduno-Batavi. 4:219, t.8, 9 (1869). Type: Cult. Bogor (holo.: probably Ann. Mus. Bot. Lugduno-Batavi. 4, t.8 & 9, 1869).

HABITAT: unknown.
ALTITUDINAL RANGE: unknown.
DISTRIBUTION: Borneo (unspecified). Also Maluku.

G.scriptum *Blume* in Rumphia 4:48 (1836). Type: Maluku, Amboina, *Rumphius* s.n. (holo. L,lost, lecto. illustration in Rumphius, Herb. Amb. 6:98, t.42).

HABITAT: lowland forest; coastal forest.
ALTITUDINAL RANGE: sea level.
DISTRIBUTION: Borneo (SA). Also Sulawesi, Maluku, New Guinea, Solomon Islands.

G.speciosum *Blume*, Bijdr.:378, Pl.20 (1825). Type: Java, Buitenzorg, *Blume* s.n. (holo. L).

HABITAT: podsol forest; lowland and hill forest; peat swamp forest.
ALTITUDINAL RANGE: sea level – 900m.
DISTRIBUTION: Borneo (B, K, SA, SR). Also Burma, Thailand, Laos, Vietnam, Peninsular Malaysia, Sumatra, Riau, Bangka, Java, Philippines.

PILOPHYLLUM Schltr.

Die Orchideen: 131 (1914).

Terrestrial herbs. Rhizomes short, creeping. Pseudobulbs alternately 1–4 bearing a solitary leaf and 1 bearing an inflorescence, clearly articulated at junction with petiole or scape. Leaf large, convolute, erect, ovate, acuminate, brown-hairy. Inflorescence erect, racemose, laxly many-flowered; peduncle, rachis and bracts brown-hairy. Flowers non-resupinate, few open at a time, glabrous, sepals and petals yellow with dark red stripes, lip white, spotted purple; sepals and petals free, spreading; lateral sepals and petals falcate; lip 3-lobed, side lobes oblong, spreading, mid-lobe narrowed from a broad base and expanded abruptly into a reniform, apiculate epichile; column with lateral appendages on its front edge, with a distinct foot; pollinia 2.

A monotypic genus from Peninsular Malaysia, Java, Borneo, Seram, the Philippines, New Guinea and the Solomon Islands.

P.villosum *(Blume) Schltr.*, Die Orchideen :131 (1914). Type: Java, Gede, *Blume* s.n. (holo. L).

> *Chrysoglossum villosum* Blume, Bijdr.:338 (1825).

HABITAT: lower montane forest, especially on ridges.
ALTITUDINAL RANGE: 700 – 1500m.
DISTRIBUTION: Borneo (SA). As for the genus.

PORPHYROGLOTTIS Ridl.

J. Linn. Soc., Bot. 31: 290 (1896).

Epiphytic herb. Pseudobulbs clustered, elongate, cane-like, leafy. Leaves distichous, long and narrow, linear, acuminate. Inflorescences lateral, erect to porrect, simple or sometimes with a lateral branch, slender, up to 150cm or more long, bearing a succession of many flowers, only 2 or 3 open at once. Flowers non-resupinate, pale pinkish, lip purple-brown outside, with a large yellow apical patch; sepals and petals reflexed, lying against pedicel with ovary, oblong-elliptic, obtuse; lip attached to column-foot by a strap-like hinge, mobile, obovate, entire, sides deflexed so that the whole is convex when viewed from above, scoop-shaped when viewed from below, hairy, bee-like; column 1.8 cm long, curved, sulcate, with large curved spreading arms at about the middle; stigma round, much broader than anther; pollinia 2, large, cleft, attached by flattened caudicles to a broad, round viscidium.

A monotypic genus from Peninsular Malaysia, Sumatra and Borneo.

P.maxwelliae *Ridl.* in J. Linn. Soc., Bot. 31:290, Pl.15 (1896). Type: Sarawak, *Maxwell* s.n. (holo. SING).

HABITAT: podsol forest, often epiphytic on *Tristaniopsis*.
ALTITUDINAL RANGE: 400 – 500m.
DISTRIBUTION: Borneo (K, SA, SR). As for the genus.

SUBTRIBE ACRIOPSIDINAE

ACRIOPSIS Reinw. ex Blume

Bijdr.:376 (1825).

Minderhoud, M.E. & de Vogel, E.F. (1986). A taxonomic revision of the genus *Acriopsis*. Reinwardt ex Blume (Acriopsidinae, Orchidaceae). Orch. Mon.1:1 – 16.

Epiphytic herbs. Rhizome creeping, branched. Roots slender, branched, with tufts of erect, acuminate branches at base of pseudobulb. Pseudobulbs crowded, ovoid, covered at the base by thin, silvery bracts. Leaves apical, (1–)2–3(–4), oblong or linear, the midrib sunken above, prominent below, petiolate. Inflorescence arising from the base of the pseudobulb on a short, rooting rhizome, a raceme or panicle, usually many-flowered; peduncle relatively long, terete; floral bracts persistent. Flowers small, more or less twisted but not resupinate, widely open; sepals lanceolate, concave at the apex; petals spreading, oblong to obovate, about as long as the sepals; lip trilobed, pandurate or entire, the disc with 2 keels; column more or less straight or slightly S-shaped, with 2 long parallel, porrect or decurved, elongate stelidia; pollinia 4.

Six species distributed from India (Sikkim), Burma, Thailand and Indochina eastwards through Malaysia and Indonesia to the Philippines, New Guinea, Solomon Islands and Australia.

A.densiflora *Lindl.* in Bot. Reg. 33, sub t.20 (1847). Type: Borneo, *Low* s.n. (holo. K-LINDL).
A.purpurea Ridl. in Trans. Linn. Soc. London, Bot., ser. 2, 3: 406 (1893). Type: Peninsular Malaysia, Pekan, *Ridley* s.n. (holo. SING).

var. **densiflora**

HABITAT: unknown.
ALTITUDINAL RANGE: lowlands.
DISTRIBUTION: Borneo (B, SR). Also Peninsular Malaysia,Sumatra.

var. **borneensis** *(Ridl.) Minderh. & de Vogel* in Orch. Mon. 1: 7 (1986). Type: Sarawak, Matang, *Ridley* s.n. (holo. SING).
A.borneensis Ridl. in J. Straits Branch Roy. Asiat. Soc., 44: 193 (1905).

HABITAT: high kerangas with dominant *Dacrydium* and *Tristania*.
ALTITUDINAL RANGE: 200 – 500m.
DISTRIBUTION: Borneo (SA, SR). Endemic.

A.gracilis *Minderh. & de Vogel* in Orch. Mon. 1:7, fig.4, Pl.1a (1986). Type: Sabah, Sook, *Lamb* s.n. (holo. L).

HABITAT: open stunted kerangas
ALTITUDINAL RANGE: 300 – 600m.
DISTRIBUTION: Borneo (SA). Endemic.

A.indica *Wight*, Icones 5:20, t.1748 – 1 (1852). Type: Burma, Moulmein, *Parish* 76 (neo. K), original probably lost.

HABITAT: hill and lower montane forest; podsol forest.

ALTITUDINAL RANGE: 400 – 1500m.
DISTRIBUTION: Borneo (K, SA). Also ?India, Burma, Thailand, Indochina, Peninsular Malaysia, Java, Timor, Philippines.

A.javanica *Reinw. ex Blume*, Bijdr.:377 (1825). Type: Java, Salak, *Blume* s.n., HLB 902-322-65 (lecto. L).

var. **javanica**

HABITAT: lowland, hill and lower montane forest; secondary forest; moss forest.
ALTITUDINAL RANGE: sea level – 1200m.
DISTRIBUTION: Borneo (B, K, SA, SR). Widespread from Thailand and Peninsular Malaysia east through Indonesia to the Philippines, New Guinea, Solomon Islands and Australia.

var.**auriculata** *Minderh. & de Vogel* in Orch. Mon. 1:14, fig.6 (1986). Type: Sarawak, Kuching, *Haviland* s.n. (holo. K).

HABITAT: lowland primary and secondary forest.
ALTITUDINAL RANGE: sea level – 600m.
DISTRIBUTION: Borneo (B, K, SA, SR). Also Burma, North Vietnam, Peninsular Malaysia, Sumatra, Java.

A.ridleyi *Hook.f.*, Fl. Brit. Ind. 6:79 (1890). Type: Singapore, Bukit Mandai, *Ridley* s.n. (holo. SING).

HABITAT: coastal podsol forest; 'heath woodland'; recorded from low stature hill forest on ultramafic substrate.
ALTITUDINAL RANGE: sea level – 800m.
DISTRIBUTION: Borneo (K, SA, SR). Also Peninsular Malaysia, Singapore.

TRIBE ARETHUSEAE

SUBTRIBE BLETIINAE

ACANTHEPHIPPIUM Blume

Bijdr.:353 (1825).

Terrestrial herbs. Pseudobulbs few noded, rather long, fleshy, subcylindrical, covered by sheaths when young, 2- to 3-leaved at apex. Leaves suberect, large, plicate. Inflorescence lateral, shorter than leaves, fleshy, few flowered. Flowers erect, rather large, urn-shaped; sepals fleshy, connate to form a swollen tube, free at the apices; lateral sepals forming an obtuse mentum with column foot; petals narrower, enclosed together with lip in sepal tube; lip 3-lobed, movably hinged to column-foot; column long, foot curved; pollinia 8, unequal.

About 15 species distributed from tropical Asia to the Pacific islands.

A.curtisii *Rchb.f.* in Gard. Chron. 15:169 (1881). Type: 'Malayan Archipelago', *Veitch* s.n. (holo. W).

HABITAT: unknown.
ALTITUDINAL RANGE: unknown.
DISTRIBUTION: Borneo (SR). Endemic.

A.eburneum *Kraenzl.* in Gard. Chron. 20:266 (1896). Type: 'introduced by *Mr.P.Wolter*, Magdeburg - Wilhelmstadt (holo. B).
A.lycaste Ridl. in Sarawak Mus. J.1:35 (1912). Type: Sarawak, Kuching, *Ridley* s.n. (holo. SING, iso. K).

HABITAT: unknown.
ALTITUDINAL RANGE: unknown.
DISTRIBUTION: Borneo (SR). Endemic.

A.javanicum *Blume,* Bijdr.: 353, Pl.47 (1825). Type: Java, Salak, *Blume* s.n. (holo. BO).

HABITAT: hill and lower montane forest; streamsides.
ALTITUDINAL RANGE: 300 – 1800m.
DISTRIBUTION: Borneo (SA, SR). Also Peninsular Malaysia, Sumatra, Java.

A.lilacinum *J.J.Wood & C.L.Chan* in Orch. Borneo 1:53, fig.1, Pl.1A (1994). Type: Sabah, Crocker Range, Sinsuron Road, *Lamb* K51 (holo. K).

HABITAT: hill and lower montane forest.
ALTITUDINAL RANGE: 300 – 1300m.
DISTRIBUTION: Borneo (SA). Endemic.

ANIA Lindl.

Gen. Sp. Orch. Pl.: 129 (1831).

Turner, H. (1992). A revision of the orchid genera *Ania* Lindley, *Hancockia* Rolfe, *Mischobulbum* Schltr. and *Tainia* Blume. Orch. Mon. 6:43–100.

Ascotainia Ridl., Mat. Fl. Mal. Pen.1:115 (1907).

Terrestrial herbs. Shoots arising from base of last pseudobulb; sterile shoots with 1 terminal leaf. Pseudobulb with 1 or several internodes, usually erect, conical, rarely ovoid to ellipsoid. Leaf petiolate, elliptic to slightly obovate, acute to acuminate.inflorescence lateral, erect, arising from base of pseudobulb of previous shoot. Flowers resupinate, glabrous (except in *A.ponggolensis*); sepals and petals free, elliptic to obovate; lateral sepals slightly decurrent along column foot; lip entire or 3-lobed, usually distinctly spurred, disc with 3 – 7 keels; column alate; pollinia 8, in 4 pairs.

Eight species distributed in India, Burma, China, Hong Kong, Vietnam, Thailand and Peninsular Malaysia eastwards to the Philippines and New Guinea.

A.borneensis *(Rolfe) Senghas* in Schltr., Die Orchideen ed.3,1: 863 (1984). Type: Sabah, Mt. Kinabalu, Kiau, *Gibbs* 3958 (holo. K, iso. BM).
  Ascotainia borneensis Rolfe in Gibbs in J. Linn. Soc., Bot.42:154 (1914).
  Tainia rolfei P.F.Hunt in Kew Bull. 26:182 (1971).

HABITAT: in shady places among limestone rocks; in leaf litter under bamboo in secondary hill forest.
ALTITUDINAL RANGE: 900 – 1200m.
DISTRIBUTION: Borneo (SA). Also Sumatra, Java, Maluku (Ambon), New Guinea.

A.ponggolensis *H.Turner* in Orch. Mon. 6:58, fig.30, Pl.5d (1992). Type: Sabah, Batu Ponggol, *Lamb* AL204/84 (holo. K, iso. L).

HABITAT: in shady places among limestone rocks.
ALTITUDINAL RANGE: 900m.
DISTRIBUTION: Borneo (SA). Endemic.

CALANTHE R.Br.

Bot. Reg. 7: sub t.573 (1821).

Terrestrial, rarely epiphytic, herbs. Roots thick, long and hairy. Stems pseudobulbous, short or long, usually 2- to several-leaved, often clustered, pseudobulbs small, ovoid, covered by leaf bases. Leaves persistent, rarely deciduous, narrowly elliptic or linear-elliptic, plicate, petiolate, up to 1m. long. Inflorescence axillary, terminal or arising from base of leafy pseudobulb, erect, long or short, few-to many-flowered, racemose; floral bracts persistent or deciduous. Flowers resupinate, small to rather large and showy, usually turning blue-black when damaged; sepals and petals similar, free, spreading, rarely connivent; lip entire, 3- or 4-lobed, adnate to column at base, usually spurred, mid-lobe often deeply bifid, disc usually with basal verrucose calli or lamellae; column short, rarely long, fleshy, truncate, without a foot; pollinia 8.

About 100 species from tropical, subtropical and warm temperate regions, from

Africa to Asia and the Pacific islands, with a single species in the tropical Americas.

C. aureiflora *J.J.Sm.* in Bull. Jard. Bot. Buitenzorg, ser. 3, 5:67 (1922). Types: Sumatra, Goenoeng Malintang, *Bünnemeijer* 3947, 4116, 4254, Goenoeng Sago, *Bünnemeijer* 4363 (syn. L).

> HABITAT: unknown.
> ALTITUDINAL RANGE: unknown.
> DISTRIBUTION: Borneo (SA). Also Sumatra.

> NOTE: The specimen *Lamb* AL 26/82 (K) agrees with *C.aureiflora* except for the broader petals and flabellate lip mid-lobe.

C.crenulata *J.J.Sm.* in Bot. Jahrb. Syst. 48: 97 (1912). Type: Kalimantan, between Muara Uja and Kundim baru, *Winkler* 2676 (holo. ? HBG).

> HABITAT: hill and lower montane forest, often on ultramafic substrate.
> ALTITUDINAL RANGE: sea level – 1200m.
> DISTRIBUTION: Borneo (K, SA). Endemic.

C.flava *(Blume) E.Morren* in Belgique Hort. 2:238, t.46 (1834). Type: Java, Gede, *Blume* s.n. (holo. BO).
> *Amblyglottis flava* Blume, Bijdr.: 370, t.64 (1825).
> *Calanthe parviflora* Lindl. in Paxton's Fl. Gard. 3:37, t.61 (1852). Type: Java, *Lobb* 334 (holo. K-LINDL).

> HABITAT: lower montane forest.
> ALTITUDINAL RANGE: 2100m.
> DISTRIBUTION: Borneo (SA). Also Sumatra, Java.

C.gibbsiae *Rolfe* in J. Linn. Soc., Bot. 42:156 (1914). Type: Sabah, above Tenom, *Gibbs* 2870 (holo. K).

> HABITAT: lowland and hill forest on limestone; lower montane forest.
> ALTITUDINAL RANGE: 100 – 1700m.
> DISTRIBUTION: Borneo (K, SA). Endemic.

C.kemulensis *J.J.Sm.* in Bull. Jard. Bot. Buitenzorg, ser. 3, 11:117 (1931). Type: Kalimantan, G.Kemoel, *Endert* 4179 (holo. L).

> HABITAT: lower montane forest.
> ALTITUDINAL RANGE: 1500m.
> DISTRIBUTION: Borneo (K). Endemic.

C.kinabaluensis *Rolfe* in J. Linn. Soc., Bot.42: 156 (1914). Type: Sabah, Mt. Kinabalu, Lobang, *Gibbs* 4108 (holo. K).
> *C.cuneata* Ames & C.Schweinf., Orch. 6:159 (1920). Type: Sabah, Mt. Kinabalu, *Haslam* s.n. (holo. AMES).

> HABITAT: lower montane oak-laurel forest; rarely hill forest.
> ALTITUDINAL RANGE: 900 – 2100m.
> DISTRIBUTION: Borneo (SA). Endemic.

FIG. 10. *Calanthe* lips. **A**, *C.salaccensis*, with column and part of spur, x 2; **B**, *C. kinabaluensis*, x 3½; **C**, *C. pulchra*, x 5; **D**, *C. undulata*, x 5; **E**, *C. vestita*, x 2; **F**, *C. gibbsiae*, x 2½; **G**, *C. triplicata* (variant), x 1¼; **H**, *C. salaccensis*, longitudinal section, x 2; **J**, *C. sylvatica*, x 2; **K**, *C. transiens*, x 2; **L**, *C. crenulata*, x 2½. Drawn by Sarah Thomas.

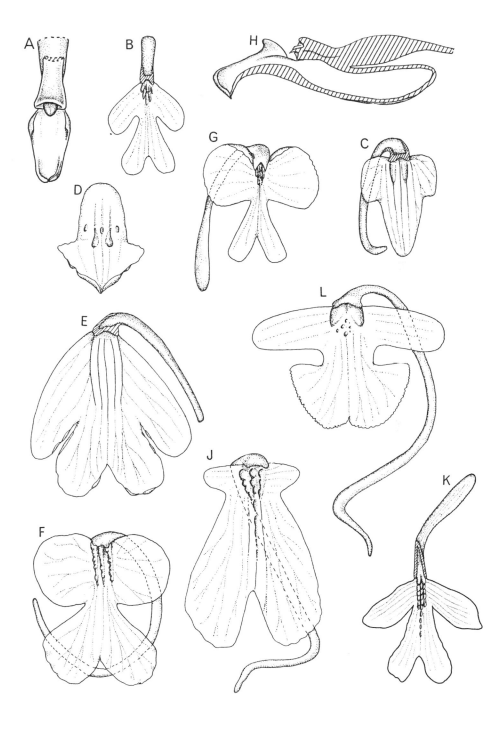

C. lyroglossa *Rchb.f.* in Otia Bot. Hamburg 1:53 (1878). Type: Philippines, Luzon, Mt.Mahahai, *collector unknown* (holo. HBG).

HABITAT: lower montane forest.
ALTITUDINAL RANGE: 1300m.
DISTRIBUTION: Borneo (SA). *C.lyroglossa* is recorded from Burma, Peninsular Malaysia, Thailand and Indochina to the Philippines, north to Japan.

NOTE: The specimen *Lamb* SAN 89675 (SAN) agrees with *C.lyroglossa* except that it has pronounced lip side lobes.

C.ovalifolia *Ridl.* in Trans. Linn. Soc. London, Bot. 4:239 (1894). Type: Sabah, Mt. Kinabalu, Penokok River, *Haviland* s.n. (holo. SING).

HABITAT: hill forest.
ALTITUDINAL RANGE: 900m.
DISTRIBUTION: Borneo (SA). Endemic.

C.pulchra *(Blume) Lindl.*, Gen. Sp. Orch. Pl.: 250 (1833). Type: Java, Seribu, *Blume* s.n.(holo. BO).
Amblyglottis pulchra Blume, Bijdr.: 371 (1825).

HABITAT: hill and lower montane forest; streamsides; rocky places; heath forest; sometimes on ultramafic substrate.
ALTITUDINAL RANGE: 600 – 1700m.
DISTRIBUTION: Borneo (K, SA, SR). Also Peninsular Malaysia, Sumatra, Java, Philippines. Recently recorded from Thailand.

C.rajana *J.J.Sm.* in Mitt. Inst. Allg. Bot. Hamburg 7:39, t.6, fig.29 (1927). Type: Kalimantan, Bukit Raja, *Winkler* 944 (holo. HBG, drawing at BO).

HABITAT: lower montane forest.
ALTITUDINAL RANGE: 1200 – 1300m.
DISTRIBUTION: Borneo (K). Endemic.

C.salaccensis *J.J.Sm.* in Bull. Dép. Agric. Indes Néerl. 43: 21 (1910). Type: Java, Salak, Gunung Malang, *Joseph* s.n. (holo. BO).

HABITAT: podsol forest with *Dacrydium*, etc.
ALTITUDINAL RANGE: 400 – 500m.
DISTRIBUTION: Borneo (SA). Also Peninsular Malaysia, Java, Bangka.

C.shelfordii *Ridl.* in J. Straits Branch Roy. Asiat. Soc. 33: 23 (1900). Type: Sarawak, Mt.Penrissen, *Shelford* s.n. (holo. SING).

HABITAT: lower montane forest.
ALTITUDINAL RANGE: 1400 – 1500m.
DISTRIBUTION: Borneo (SR). Endemic.

C.speciosa *(Blume) Lindl.*, Gen. Sp. Orch. Pl.: 250 (1833). Type: Java, Bantam & Buitenzorg, Provinces, *Blume* s.n. (holo. BO).
Amblyglottis speciosa Blume, Bijdr.: 371 (1825).

HABITAT: lowland and hill dipterocarp forest; riversides; near waterfalls; lower montane oak-laurel forest, on sandstone.
ALTITUDINAL RANGE: 100 – 1900m.
DISTRIBUTION: Borneo (SA, SR). Also Peninsular Malaysia, Sumatra, Java.

C.sylvatica *(Thouars) Lindl.*, Gen. Sp. Orch. Pl.: 250 (1833). Type: Mascarene Islands, *du Petit Thouars* s.n. (holo. P).
    Centrosis sylvatica Thouars, Orch. Iles Aust. Afr. t.35, 36 (1822).

HABITAT: mixed dipterocarp forest, in deep shade, on sandstone, limestone or basalt.
ALTITUDINAL RANGE: 400 – 1500m.
DISTRIBUTION: Borneo (K, SA, SR). Widespread in Africa, Madagascar, Mascarene Islands and tropical Asia.

C.tenuis *Ames & C.Schweinf.*, Orch. 6: 161 (1920). Type: Sabah, Mt. Kinabalu, *Haslam* s.n. (holo. AMES).

HABITAT: unknown.
ALTITUDINAL RANGE: unknown.
DISTRIBUTION: Borneo (SA). Endemic.

C.transiens *J.J.Sm.* in Bull. Jard. Bot. Buitenzorg, ser. 3, 5: 70 (1922). Types: Sumatra, Korintji Peak, *Bünnemeijer* 9661, 9747, 9895, 9901, 10469 (syn. L).

HABITAT: beside streams in upper montane forest on ultramafic substrate.
ALTITUDINAL RANGE: 1700 – 2700m.
DISTRIBUTION: Borneo (SA). Also Sumatra.

C.triplicata *(Willemet) Ames* in Philipp. J. Sci. 2: 326 (1907). Type: Maluku, Amboina, *Rumphius* s.n. (holo. L, lost, lecto. illustration in Rumphius, Herb. Amb. 6: t.52, f.2).
    Orchis triplicata Willemet in Ann. Bot. (Usteri) 18:52 (1796).
    Calanthe veratrifolia (Willd.) Ker Gawl. in Bot. Reg. 9, t.720 (1823). Type: 'India orientali' (holo. B).
    C.furcata Bateman ex Lindl. in Bot. Reg. 24, misc. 28 (1838). Type: Philippines, Luzon, *Cuming* s.n. (holo. K).
    C.veratrifolia (Willd.) Ker Gawl. var *incurvicalcar* J.J.Sm. in Bull. Jard Bot. Buitenzorg, ser. 3, 11:115 (1931). Types: Kalimantan, Kombeng, *Endert* 5232 (syn. BO, L); Lakoem, *Endert* 1763 (syn. BO, L).
    C.veratrifolia (Willd.) Ker Gawl. var. *densissima* J.J.Sm. in Bull. Jard. Bot. Buitenzorg, ser. 3, 11:116 (1931). Types: Kalimantan, Liang Gagang, *Hallier* 2621 (syn. L); Kiau, *Endert* 4588 (syn. BO, L).

HABITAT: lowland alluvial rainforest; mixed lowland and hill dipterocarp forest; lower montane forest; limestone rocks and scree; streamsides; also recorded from ultramafic substrate.
ALTITUDINAL RANGE: sea level – 1500m.
DISTRIBUTION: Borneo (K, SA, SR). Widespread from Madagascar and Mascarene Islands to India, China and Japan, through Malaysia and Indonesia east to Australia and the Pacific islands.

C.truncicola *Schltr.* in Bot. Jahrb. Syst. 45: 26 (1911). Type: Sumatra, Bukit Djarat, *Schlechter* 16001 (holo. B, destroyed, iso. K).

HABITAT: among bamboo and scrub in lower montane forest on ultramafic substrate.
ALTITUDINAL RANGE: 800 – 1500m.
DISTRIBUTION: Borneo (SA). Also Sumatra.

C.undulata *J.J.Sm.* in Icon. Bogor. 2:67, t.112, fig.B (1903). Type: Kalimantan, *Nieuwenhuis* s.n. (holo. BO).

HABITAT: podsol forest of *Dacrydium*, etc. with understorey of *Rhododendron malayanum* on very wet sandy soil.
ALTITUDINAL RANGE: 400 – 1000m.
DISTRIBUTION: Borneo (K, SA). Endemic.

C.vestita *Lindl.*, Gen. Sp. Orch. Pl.: 250 (1833). Type: Burma, Tavoy, *Wallich* s.n. (holo. K - WALL).
C. vestita Lindl. var. *igneo-oculata* Hort. ex Rchb.f. in Gard. Chron. 5:534 (1876). Type: Borneo, ex cult. *Sir Trevor Lawrence* (holo. W).
C. vestita Lindl. var. *oculata-gigantea* Rchb.f. ex B.S.Williams, Orchid Growers Man. ed.6:166 (1885). Type: ? Borneo, ex *Spiers* (holo. not located).
C. vestita Lindl. var. *fournieri* Rolfe in Gard. Chron. ser. 3, 11:488 (1892). Type: Borneo, ex cult. *Fournier* (holo. K).

HABITAT: forest on limestone; low stature hill forest on ultramafic substrate.
ALTITUDINAL RANGE: 600 – 900m.
DISTRIBUTION: Borneo (K, SA, SR). Also Burma, Peninsular Malaysia, Thailand, Vietnam, Java, Sulawesi, Seram.

C.zollingeri *Rchb.f.* in Bonplandia 5:38 (1857). Type: Java, Litjin Prov. Banjuwangie, *Zollinger* 2858 (holo. W).

HABITAT: riverine forest.
ALTITUDINAL RANGE: 100 – 200m.
DISTRIBUTION: Borneo (SA). Also Sumatra, Java.

MISCHOBULBUM Schltr.

Feddes Repert. Beih.1:98 (1911).

Turner, H. (1992). A revision of the orchid genera *Ania* Lindley, *Hancockia* Rolfe, *Mischobulbum* Schltr. and *Tainia* Blume. Orch.Mon. 6:43–100.

Terrestrial herbs. Pseudobulbs slender, unifoliate. Leaf apical, broadly ovate. Inflorescence arising from the base of a recently matured pseudobulb, with few to many, well-spaced, rather large flowers. Flowers resupinate; sepals and petals free, elliptic to lanceolate; lateral sepals joined to the column-foot to form a mentum; lip entire, with prominent keels on the upper surface; column with a foot; pollinia 8.

About seven species distributed from China and India to Peninsular Malaysia east to New Guinea and the Solomon Islands.

M.scapigerum *(Hook.f.) Schltr.* in Feddes Repert. Beih. 1:98 (1911). Type: Borneo, cult. *Low* (holo. K).
Nephelaphyllum scapigerum Hook.f. in Bot. Mag. ser. 3, 19: t.5390 (1863).
Tainia scapigera (Hook.f.) J.J.Sm. in Bull. Jard. Bot. Buitenzorg, ser. 2, 8:6 (1912).

HABITAT: among rocks in hill dipterocarp forest, on sandstone ridges.
ALTITUDINAL RANGE: 200 – 900m.
DISTRIBUTION: Borneo (B, SA). Endemic.

NEPHELAPHYLLUM Blume

Bijdr.:372, t.22 (1825).

Terrestrial herbs. Rhizome creeping, rooting at nodes. Pseudobulbs slender, fusiform, rarely distinct from the short petioles. Leaves slightly fleshy, more or less cordate, often marbled light and dark green and flushed purple. Inflorescence terminal, lax or dense, few-flowered. Flowers small to medium-sized, non-resupinate; sepals and petals subequal, acute, often reflexed; lip entire or slightly 3-lobed, shortly spurred, disc keeled; column usually rather short, slightly winged, without a foot; anther bicornute; pollinia 8.

Some 8 to 12 species distributed in China and from Indochina through Thailand and Peninsular Malaysia east through Indonesia and north to the Philippines.

N.aureum *J.J.Wood* in Orch. Borneo 1: 183, fig.53, Pl.11E (1994). Type: Sabah, Gunung Alab, *Lamb* 2309 in SAN 93500 (holo. K).

HABITAT: hill and lower montane forest on sandstone ridges.
ALTITUDINAL RANGE: 900 – 1400m.
DISTRIBUTION: Borneo (SA, SR). Endemic.

N.beccarii *Schltr.* in Feddes Repert. 9:338 (1911). Types: Sarawak, Gunung Matang, *Beccari* 1682 & 1945 (syn. FI); Mt. Poe, *Hewitt* s.n.; Sinjii Mountains, *Brooks* s.n. (syn. B, destroyed).
   Tainia beccarii (Schltr.) Gagnep. in Bull. Mus. Natl. Hist. Nat., ser. 2, 4: 706 (1932).

HABITAT: unknown.
ALTITUDINAL RANGE: 700 – 800m.
DISTRIBUTION: Borneo (SR). Endemic.

[**N.borneense** *Schltr.* in Bull Herb. Boissier 2, 6:301 (1906). Type: Kalimantan, Long Sele, *Schlechter* 13526 (holo. B,destroyed).

HABITAT: in leaf litter.
ALTITUDINAL RANGE: unknown.
DISTRIBUTION: Borneo (K). Endemic.]

   See *N.pulchrum*, p.134.

N.flabellatum *Ames & C.Schweinf.*, Orch. 6:19 (1920). Type: Sabah, Mt. Kinabalu, Marai Parai Spur, *Clemens* s.n. (holo. AMES).
   Tainia flabellata (Ames & C.Schweinf.) Gagnep. in Bull. Mus. Natl. Hist. Nat., ser. 2, 4: 706 (1932).

HABITAT: leaf litter in deep shade in hill and lower montane forest on ultramafic substrate.
ALTITUDINAL RANGE: 600 – 1400m.
DISTRIBUTION: Borneo (SA). Endemic.

N.gracile *Schltr.* in Bull. Herb. Boissier 2, 6:301 (1906). Type: Kalimantan, Long Sele, *Schlechter* 13523 (holo. B, destroyed).
   Tainia gracilis (Schltr.) Gagnep. in Bull. Mus. Natl. Hist. Nat., ser. 2,4: 706 (1932).

HABITAT: mossy places.
ALTITUDINAL RANGE: unknown.

DISTRIBUTION: Borneo (K) Endemic..

N.pulchrum *Blume,* Bijdr., 373 (1825). Type: Java, Salak, *Blume* s.n. (holo.? L).
N.latilabre Ridl. in Stapf. in Trans. Linn. Soc. London, Bot. ser. 2, 4: 238 (1894). Type: Sabah, Mt. Kinabalu, *Haviland* 1165 (holo. SING).
N.borneense Schltr. in Bull. Herb. Boissier 2, 6: 301 (1906).Type: Kalimantan, Long Sele, *Schlechter* 13526 (holo. B, destroyed).
Tainia borneensis (Schltr.) Gagnep. in Bull. Mus. Natl. Hist. Nat. ser. 2, 4:706 (1932).
T.latilabra (Ridl.) Gagnep. in Bull. Mus. Natl. Hist. Nat. ser. 2, 4: 706 (1932).

HABITAT: damp shady places in mixed dipterocarp forest; lower montane forest; oak-laurel forest.
ALTITUDINAL RANGE: 500 – 2000m.
DISTRIBUTION: Borneo (K, SA, SR). Also Bhutan south to Burma, Peninsular Malaysia, Thailand, Vietnam, Sumatra, Java, Philippines.

N.tenuiflorum *Blume,* Bijdr.: 373 (1825). Type: Java, Salak, *Blume* s.n. (holo.?L).

HABITAT: unknown.
ALTITUDINAL RANGE: unknown.
DISTRIBUTION: Borneo (SR). Also Peninsular Malaysia, Thailand, Vietnam, Sumatra, Java, Bengkoeloe.

N.trapoides *J.J.Sm.* in Mitt. Inst. Allg. Bot. Hamburg 7:37, t.6, fig.27 (1927). Types: Kalimantan, Bukit Raja, *Winkler* 948 (syn. HBG) & 1014 (syn. BO, HBG); Gunung Kenepai, *Hallier* 2453 (syn. HBG).

HABITAT: mossy places in lower montane forest.
ALTITUDINAL RANGE: 1200m.
DISTRIBUTION: Borneo (K, SR). Endemic.

N.verruculosum *Carr* in Gard. Bull. Straits Settlem. 8:200 (1935).Type: Sabah, Mt. Kinabalu, Menetendok/Kinataki Divide, *Carr* C.3160 (SFN26607) (holo. SING).

HABITAT: hill forest.
ALTITUDINAL RANGE: 900m.
DISTRIBUTION: Borneo (SA). Endemic.

PACHYSTOMA Blume

Bijdr.:376 (1825).

Terrestrial herbs. Rhizome subterranean, fleshy, swelling into a horizontal fusiform, sometimes v-shaped tuber, 1- to few-leaved. Leaves narrow, grass-like, often withering before inflorescence appears. Inflorescence long, many-flowered. Flowers resupinate, not widely opening, often nodding, hairy; lateral sepals distinctly swollen at base, resembling a small mentum; lip 3-lobed, immobile, slightly saccate at base, disc with 5 tuberculate longitudinal keels; column slender, finely pubescent; pollinia 8.

Probably comprising only one variable species widely distributed from India, north to China, through S.E.Asia to New Guinea, Australia and New Caledonia.

P.pubescens *Blume*, Bijdr.: 376, Pl.29 (1825). Type: Java, Tjiradjas, Krawang, *Blume* s.n. (holo. BO).

Pachystoma pantanum (Blume) Miq., Fl. Ind. Bat. 3:675 (1859). Type: Borneo, *Korthals* s.n. (holo. BO, iso. K).

HABITAT: unknown.
ALTITUDINAL RANGE: unknown.
DISTRIBUTION: Borneo (K, SR). As for genus.

PHAIUS Lour.

Fl. Cochinch. 2: 517, 529 (1790).

Terrestrial herbs. Stems clustered, short, long, cylindrical or pseudobulbous, few leaved. Pseudobulbs, when present, conical or ovoid. Leaves medium to large, petiolate, narrowly obovate or narrowly elliptic, acuminate, plicate. Inflorescence lateral from base of pseudobulb, or axillary halfway up stem, lax or rather dense, racemose. Flowers usually large, often showy, resupinate, turning blue-black when old or damaged; sepals and petals similar, free, rather fleshy, spreading or reflexed; lip free, entire or obscurely 3-lobed, erect, sessile and partially adnate to and embracing column to form a tube, gibbous or shortly spurred; column long or short, fleshy, curved, with an inflexed foot; pollinia 8.

About 30 species distributed in the Old World tropics of Africa, Madagascar, S. and S.E.Asia, Australia, New Guinea and the Pacific islands.

P.baconii *J.J.Wood & Shim* in Orch. Borneo 1:221, fig.70, Pl.14D (1994). Type: Sabah, Mt. Kinabalu, Penibukan Ridge, *Lamb* AL 39/83 (holo. K).

HABITAT: lower montane forest on ultramafic substrate; in leaf litter over rocks in bamboo thickets.
ALTITUDINAL RANGE: 1200 – 1500m.
DISTRIBUTION: Borneo (SA). Endemic.

P.borneensis *J.J.Sm.* in Icon. Bogor. 2:61, t.3, fig.C (1903). Type: Kalimantan, Bukit Kasian, *Nieuwenhuis* s.n. (holo. BO, iso. L).

HABITAT: hill forest on limestone;lower montane forest.
ALTITUDINAL RANGE: near sea level – 1500m.
DISTRIBUTION: Borneo (K, SA). Endemic.

P.callosus *(Blume) Lindl.*, Gen. Sp. Orch. Pl.: 128 (1831). Type: Java, *Blume* s.n. (holo. BO).

Limodorum callosum Blume, Bijdr.: 374 (1825).

HABITAT: mossy boulders and leaf litter in montane forest.
ALTITUDINAL RANGE: 1800m.
DISTRIBUTION: Borneo (SR). Also Peninsular Malaysia, Sumatra, Java, Sulawesi.

P.pauciflorus *(Blume) Blume*, Mus. Bot. Lugd. Bat. 2:181 (1856).

subsp. **sabahensis** J.J.Wood & A.Lamb in Wood, Beaman & Beaman, Plants of Mt. Kinabalu 2, Orchids: 282, fig.47 (1993). Type: Sabah, Mt. Kinabalu, Hempuen Hill, *Lamb & Surat* in *Lamb* AL 1320/91 (holo. K).

HABITAT: hill forest with *Gymnostoma sumatrana* on ultramafic substrate.
ALTITUDINAL RANGE: 500 – 1500m.
DISTRIBUTION: Borneo (SA). Endemic.

P.reflexipetalus *J.J.Wood & Shim* in Orch. Borneo. 1:231, fig. 69, Pl.14F (1994). Type: Sabah, Mt. Kinabalu, Penataran Ridge, *Lamb* AL 25/82 (holo. K).

HABITAT: hill forest on ultramafic substrate.
ALTITUDINAL RANGE: 700 – 1100m.
DISTRIBUTION: Borneo (SA). Endemic.

P.subtrilobus *Ames & C.Schweinf.*, Orch. 6:157 (1920). Type: Sabah, Mt. Kinabalu, *Haslam* s.n. (holo. AMES).

HABITAT: lower montane *Agathis, Podocarpus*, oak forest; swampy forest.
ALTITUDINAL RANGE: 1300 – 1500m.
DISTRIBUTION: Borneo (SA). Endemic.

P.tankervilleae *(Banks ex L'Her.) Blume*, Mus. Bot. Lugd. Bat. 2, 12: 177 (1856). Type: China, based on a Banks drawing 'icon absque sermone'.
Limodorum tankervilleae Banks ex L'Her., Sert. Ang.: 28 (1789).
Phaius blumei Lindl., Gen. Sp. Orch. Pl.: 127 (1831). Type: Java, cult., *Blume* s.n. (holo. ? BO or L).

HABITAT: open grassy places; secondary forest; lower montane forest.
ALTITUDINAL RANGE: 300 – 1500m.
DISTRIBUTION: Borneo (K, SA, SR). Widespread from India, through South and S.E. Asia east to Australia and the Pacific islands

PLOCOGLOTTIS Blume

Bijdr.:380 (1825).

Terrestrial, rarely epiphytic, herbs. Stems slenderly pseudobulbous, 1-leaved, or elongated into a leafy stem. Leaves plicate, narrowly elliptic, sheathing at base, sometimes mottled yellow or, rarely, variegated with silver lines. Inflorescences lateral, from base of pseudobulb, erect; peduncle long; rachis bearing a succession of many flowers, a few open at a time; floral bracts small to long acuminate, persistent. Flowers small to medium-sized, resupinate; sepals and petals similar, free, spreading, sepals pubescent on outer surface; lateral sepals adnate to column-foot forming a very short mentum; lip short, fleshy, entire to obscurely lobed, usually convex, adnate to sides and apex of column foot to form a sac and usually with an elastic hinge that springs when lip is touched, apex often narrow and strongly reflexed; column rather short; pollinia 4.

About 35 species distributed from Peninsular Malaysia eastwards to New Guinea and the Solomon Islands.

P.acuminata *Blume*, Mus. Bot. Lugd. Bat. 1:46 (1849). Type: Java, Sumatra, *Blume* s.n. (holo. ? L).

HABITAT: lowland dipterocarp forest; secondary forest.
ALTITUDINAL RANGE: sea level – 1000m.
DISTRIBUTION: Borneo (B, K, SA, SR). Also Sumatra, Java, Philippines.

P.angulata *J.J.Sm.* in Mitt. Inst. Allg. Bot. Hamburg 7: 38, t.6, fig.28 (1927). Type: Kalimantan, Nanga Arung, *Winkler* 1566 (holo. HBG, drawing at BO).

HABITAT: lowland primary forest.
ALTITUDINAL RANGE: 100m.
DISTRIBUTION: Borneo (K). Endemic.

P.borneensis *Ridl.* in J. Straits Branch Roy. Asiat. Soc. 49: 33 (1908). Type: Sarawak, Lundu and Tambusan, *Ridley* s.n. (holo. SING).

HABITAT: hill forest; damp riverside forest; swamp forest; river banks; on sandy soils.
ALTITUDINAL RANGE: sea level – 1200m.
DISTRIBUTION: Borneo (B,SA,SR). Endemic.

P.dilatata *Blume*, Mus.Bot. Lugd. Bat.1: 47 (1849). Type: West Java, *Blume* s.n. (holo. ?L).

HABITAT: unknown.
ALTITUDINAL RANGE: unknown.
DISTRIBUTION: Borneo (K). Also Java.

P.gigantea *(Hook.f.) J.J.Sm.* in Feddes Repert. 32:228 (1933). Types: Peninsular Malaysia, Perak, Ulu Bubong, *King's collector* 10277 (syn. K); Perak, Assam Kumbong, *Wray* 2932 (syn. K).
Calanthe gigantea Hook.f., Fl. Brit. Ind. 5:856 (1890).

HABITAT: hill forest.
ALTITUDINAL RANGE: 900m.
DISTRIBUTION: Borneo (SA). Also Peninsular Malaysia, Sumatra, Mentawai Islands (Siberut Island).

P.hirta *Ridl.* in J. Straits Branch Roy. Asiat. Soc. 50:137 (1908). Type: Sarawak, Bidi, *Brookes* s.n. (holo. SING).

HABITAT: lowland forest; limestone boulder scree.
ALTITUDINAL RANGE: sea level – 600m.
DISTRIBUTION: Borneo (SA,SR). Endemic.

P.javanica *Blume*, Bijdr.: 381, fig.21 (1825). Type: Java, Salak, Pantjar, *Blume* s.n. (holo. ?L).

HABITAT: lowland forest.
ALTITUDINAL RANGE: sea level – 100m.
DISTRIBUTION: Borneo (K). Also Andaman Islands, Peninsular Malaysia, Thailand, Sumatra, Java.

P.lowii *Rchb.f.* in Gard. Chron.: 434 (1865). Type: Borneo, *Low* s.n. (holo. W).
P.porphyrophylla Ridl. in Trans. Linn. Soc. London, Bot., ser. 2, 3: 368 (1893). Type: Peninsular Malaysia, Pekan, *Ridley* s.n. (holo. SING).

HABITAT: heath forest; lowland and hill forest; recorded from ultramafic substrate.
ALTITUDINAL RANGE: sea level – 1000m.
DISTRIBUTION: Borneo (SA, SR). Also Andaman Islands, Peninsular Malaysia, Thailand, Sumatra, Maluku.

P.parviflora *Ridl.* in J. Straits Branch Roy. Asiat. Soc. 49:35 (1907). Type: Sarawak, Mt.Lingga, *Hewitt* s.n. (holo. SING).

>   HABITAT: unknown.
>   ALTITUDINAL RANGE: unknown.
>   DISTRIBUTION: Borneo (SR). Endemic.

SPATHOGLOTTIS Blume

Bijdr.:400 (1825).

Terrestrial herbs. Pseudobulbs conical to ovoid, sometimes depressed, covered with scarious sheaths. Leaves 1–2, lanceolate, plicate, sheathing at base. Inflorescence lateral, erect; peduncle tall and slender, racemose, many-flowered above. Flowers pedicellate, showy, resupinate; sepals and petals similar, free, erect to spreading; petals broader than sepals; lip strongly 3-lobed, side lobes narrow, oblong, curving upwards, mid-lobe spathulate, often emarginate, with a very narrow claw at the base of which are 2 small ovoid, often pubescent calli and 2 small laterally spreading teeth; column erect, incurved, slender, dilated above, without a foot; pollinia 8, in 2 groups of 4.

About 40 species distributed from tropical Asia to Australia and the Pacific islands.

S.aurea *Lindl.* in J. Hort. Soc. London, 5:34 (1850). Type: Peninsular Malaysia, Malacca, *Veitch & Co.* (holo. K-LINDL).

>   HABITAT: undergrowth in secondary forest.
>   ALTITUDINAL RANGE: 900 – 1200m.
>   DISTRIBUTION: Borneo (K, SA). Also Peninsular Malaysia, Sumatra, Java.

S.confusa *J.J.Sm.* in Bull. Jard. Bot. Buitenzorg, ser. 3, 12:122 (1932). Type: Kalimantan, Sintang, Goenoeng Kelam, *Hallier* 406a (holo. L, iso. BO).

>   HABITAT: rocky places; heath woodland; riverbanks.
>   ALTITUDINAL RANGE: 100 – 1800m.
>   DISTRIBUTION: Borneo (K, SA, SR). Endemic.

S.gracilis *Rolfe ex Hook.f.* in Bot. Mag. 120, t.7366 (1894). Type: Borneo, *Foerstermann,* ex *Sander & Co.* (holo. K).

>   HABITAT: hill and lower montane forest; among rocks under bamboo; mossy forest with *Gymnostoma; Tristania* forest; on ultramafic substrate.
>   ALTITUDINAL RANGE: 800 – 1700m
>   DISTRIBUTION: Borneo (SA, SR). Also Peninsular Malaysia.

S.kimballiana *Hook.f.* in Bot. Mag. 121, t.7443 (1895). Type: Borneo, ex *Sander & Co.* (holo. K).

>   HABITAT: among stones beside rivers; rocky places in hill and lower montane forest on ultramafic substrate.
>   ALTITUDINAL RANGE: sea level – 1500m.
>   DISTRIBUTION: Borneo (SA, SR). Endemic.

S.microchilina *Kraenzl.* in Bot. Jahrb. Syst. 17: 484 (1893). Type: Sumatra, Padang, *Micholitz* s.n. (holo. B).

HABITAT: among moss and scrub; rocky banks; roadsides; mossy rocks beside rivers; secondary vegetation; hill and lower montane forest.
ALTITUDINAL RANGE: 900 – 1700m; one record from 50m.
DISTRIBUTION: Borneo (B, K, SA, SR). Also Peninsular Malaysia, Sumatra.

S.plicata *Blume,* Bijdr.: 401, fig.76 (1825). Type: Java, *Blume* s.n. (holo. ?L).

HABITAT: open grassy places; roadsides; secondary vegetation; clearings; hill forest.
ALTITUDINAL RANGE: sea level – 1200m.
DISTRIBUTION: Borneo (K, SA, SR). Widespread from India to the Pacific islands.

TAINIA Blume

Bijdr.:354 (1825).

Turner, H. (1992). A revision of the orchid genera *Ania* Lindley, *Hancockia* Rolfe, *Mischobulbum* Schltr. and *Tainia* Blume. Orch. Mon. 6:43–100.

Terrestrial herbs. Shoots arising from basal part of last pseudobulb; sterile shoots with 1 terminal leaf. Pseudobulb consisting of only 1 internode, erect, cylindrical to slightly ovoid. Leaf petiolate, elliptic to ovate, acute to acuminate. Inflorescence terminal, rarely lateral, erect, arising from basal part of terminal internode, rarely from subterminal internode of previous shoot. Flowers resupinate, glabrous; sepals and petals free, triangular to ovate, or elliptic to obovate; lateral sepals slightly decurrent along column-foot; lip without a spur, rarely obscurely saccate; column alate; pollinia 8, in 4 pairs, rarely 6.

Fourteen species distributed in Sri Lanka and India north to China, Hong Kong and Japan, south from Burma to New Guinea, Australia and the Pacific islands.

T.maingayi *Hook.f.,* Fl. Brit. Ind. 5:822 (1890). Type: Peninsular Malaysia, Pinang, *Maingay* 1668 (holo. K).

HABITAT: mossy forest; low open forest.
ALTITUDINAL RANGE: unknown.
DISTRIBUTION: Borneo (unspecified). Also Peninsular Malaysia, Sumatra, Java.

T.ovalifolia *(Ames & C.Schweinf.) Garay & W.Kittr.* in Bot. Mus. Leafl. 30 (3): 193 (1986). Type: Sabah, Mt. Kinabalu, Kiau, *Clemens* 93 (holo. AMES).
Eulophia ovalifolia Ames & C.Schweinf., Orch. 6:208 (1920).

HABITAT: hill forest.
ALTITUDINAL RANGE: 900m.
DISTRIBUTION: Borneo (SA). Endemic.

NOTE: Excluded from *Tainia* by Turner (1992) but not assigned elsewhere.

T.paucifolia *(Breda) J.J.Sm.* in Bull. Jard. Bot. Buitenzorg, ser. 2, 8:5 (1912). Type: Java, *Icon.v.Breda.*
Octomeria paucifolia Breda, Gen. Sp. Orch. t.11 (1829).
Tainia latilingua Hook.f., Fl. Brit. Ind. 5: 822 (1890). Type: Peninsular Malaysia, Perak, *Scortechini* 759b (holo. K).

139

Tainia plicata (Blume) Lindl. in J. Linn. Soc., Bot. 31:285 (1896). Type: West Java, *Blume* s.n. (holo. ? L).

HABITAT: mixed lowland dipterocarp forest; hill forest.
ALTITUDINAL RANGE: 200 – 800m.
DISTRIBUTION: Borneo (B, SA). Also Java.

T.purpureifolia *Carr* in Gard. Bull. Straits Settlem. 8: 199 (1935). Type: Sabah, Mt. Kinabalu, Gurulau Spur, *Carr* 3150 (SFN 26597) (holo. SING, iso. AMES, K).

HABITAT: in moss or humus in lower montane forest.
ALTITUDINAL RANGE: 1200 – 1400m.
DISTRIBUTION: Borneo (SA). Endemic.

T.speciosa *Blume*, Bijdr.:354 (1825). Type: Java, Salak, *Blume* s.n. (holo. ?L).

HABITAT: primary vegetation on tertiary granodiorite.
ALTITUDINAL RANGE: 800m.
DISTRIBUTION: Borneo (SA, SR). Also Peninsular Malaysia, Thailand, Java.

T.vegetissima *Ridl.* in J. Linn. Soc., Bot. 38:328 (1908). Type: Peninsular Malaysia, Gunung Tahan, *Robinson* 5314 (holo. BM).

HABITAT: unknown.
ALTITUDINAL RANGE: 1200 – 1300m.
DISTRIBUTION: Borneo (SA, SR). Also Peninsular Malaysia.

SUBTRIBE ARUNDINAE

ARUNDINA Blume

Bijdr.:401 (1825).

Terrestrial herbs. Stems slender, erect, leafy, often swollen at base, pseudobulbs absent. Leaves grass-like, with imbricate sheaths. Inflorescence terminal, sometimes branched, producing a succession of flowers one or two at a time. Flowers resupinate, large and showy, purple to white; sepals and petals free; sepals narrow, dorsal erect, laterals close together behind lip; petals broader than sepals, spreading; lip trumpet-shaped, embracing column, subentire, apex emarginate, disc with 3 thin longitudinal keels; column without a foot; pollinia 8.

Between 2 and 5 species distributed from India and Sri Lanka eastwards to Sulawesi, north to China. *A.graminifolia* is widely naturalised in the Pacific islands.

A.graminifolia *(D.Don) Hochr.* in Bull. New York Bot. Gard. 6: 270 (1910). Type: Nepal, "Hab. ad Suembu Nepaliae superioris, *Hamilton*" (holo. BM).

var. **graminifolia**
Bletia graminifolia D.Don, Prodr. Fl. Nepal: 29 (1825).
Arundina speciosa Blume, Bijdr.:401, Pl.73 (1825). Type: Java, Buitenzorg, Province, Tjanjor, Krawang, etc., *Blume* s.n. (holo. L).
A.philippii Rchb.f. in Linnaea 25:227 (1852). Types: China, locality unknown, *Philippi* s.n. (syn. W); *Fortune* s.n. (syn. W).

HABITAT: secondary vegetation; open grassy areas; roadside banks; rocky places in and beside rivers.
ALTITUDINAL RANGE: sea level – 1600m.
DISTRIBUTION: Borneo (K, SA, SR). Widespread in S. and S.E. Asia from China and India through Indonesia east to the Pacific islands.

var. **revoluta** *(Hook.f.)* A.Lamb in Orch. Borneo 1:66 (1994). Types: Peninsular Malaysia, Perak, *Scortechini* 1504, *Wray* 1979 and *King's collector* (syn. K).
A.revoluta Hook.f., Fl. Brit. Ind. 5:858 (1890).

HABITAT: sandy shingle riverside banks; riverside rocks in lowland rain forest.
ALTITUDINAL RANGE: 100 – 200m.
DISTRIBUTION: Borneo (SR). Also Peninsular Malaysia.

DILOCHIA Lindl.

Gen. Sp. Orch. Pl.:38 (1830).

Terrestrial or epiphytic herbs. Stems tall, stout, erect, leafy, without pseudobulbs. Leaves elliptic, stiff. Inflorescences terminal, usually branched, straight or decurved, many-flowered; floral bracts large, often concave, deciduous. Flowers resupinate, small to medium-sized; sepals and petals similar, free, spreading or not opening widely; lip 3-lobed, side lobes erect, mid-lobe often bilobed, disc with 5 keels; column long, slender, curved; pollinia 8. Fruit globose.

About 6 or 7 species distributed from Burma, Thailand and Peninsular Malaysia, through Indonesia to the Philippines and New Guinea.

D.cantleyi *(Hook.f.) Ridl.* in J. Linn. Soc., Bot. 32: 332 (1896). Types: Peninsular Malaysia, Gunung Batu, Perak, *Cantley* s.n., *Wray* s.n. (syn. K).
   Arundina cantleyi Hook.f., Fl. Brit. Ind. 5:858 (1890).

HABITAT: lower and upper montane forest; oak-laurel forest; landslides; mostly on ultramafic substrate.
ALTITUDINAL RANGE: 1600 – 2200m.
DISTRIBUTION: Borneo (SA). Also Peninsular Malaysia, Sumatra.

D.longilabris *J.J.Sm.* in Bull. Jard. Bot. Buitenzorg, ser. 3, 11:111 (1931). Type: Kalimantan, G.Kemoel, *Endert* 3597 (holo. L).
   Arundina longilabris (J.J.Sm.) Masam., Enum. Phan. Born.:120 (1942).

HABITAT: lower montane ridge forest.
ALTITUDINAL RANGE: 1200m.
DISTRIBUTION: Borneo (K). Also Sulawesi.

D.parviflora *J.J.Sm.* in Bull. Jard. Bot. Buitenzorg, ser. 3, 11: 112 (1931). Type: Kalimantan, G.Kemoel, *Endert* 4262 (holo. L, iso. BO).
   Arundina parviflora (J.J.Sm.) Masam., Enum. Phan. Born.:120 (1942).

HABITAT: among limestone boulders in hill and lower montane forest; padang vegetation in shallow peat overlaying fine sand.
ALTITUDINAL RANGE: 800 – 1800m.
DISTRIBUTION: Borneo (K, SA). Endemic.

D.rigida *(Ridl.) J.J.Wood* in Wood, Beaman & Beaman, Plants of Mt. Kinabalu 2, Orchids: 206 (1993). Type: Sabah, Mt. Kinabalu, *Haviland* 1251 (holo. K).
   Bromheadia rigida Ridl. in Stapf in Trans. Linn. Soc. London, Bot.4: 239(1894).
   Arundina gracilis Ames & C.Schweinf., Orch. 6:96 (1920). Type: Sabah, Mt. Kinabalu, Marai Parai Spur, *Clemens* 370 (holo. AMES, iso. BM).
   Dilochia gracilis (Ames & C.Schweinf.) Carr in Gard. Bull. Straits Settlem. 8:91(1935).

HABITAT: lower and upper montane forest; ridge vegetation, often in mossy, exposed places; montane heath forest of *Dacrydium, Ericaceae* and *Schima* with field layer of *Dipteris*, etc.; oak-laurel forest; mostly on ultramafic substrate.
ALTITUDINAL RANGE: 1300 – 2400m.
DISTRIBUTION: Borneo (SA, SR). Endemic.

D.wallichii *Lindl.*, Gen. Sp. Orch. Pl.: 38 (1830). Type: Singapore, *Wallich* s.n. (holo. K-WALL).
   Arundina wallichii (Lindl.) Rchb.f., Xenia Orch. 2: 13, t.105 (1862).

HABITAT: podsol forest with understory of *Pandanus* and rattans; rocky banks; recorded growing with *Nepenthes fusca, Lycopodium, Rhododendron* and ferns in full sun; hill forest; ridge scrub.
ALTITUDINAL RANGE: 400 – 1500m.
DISTRIBUTION: Borneo (K, SA, SR). Also Peninsular Malaysia, Thailand, Sumatra, Bangka, Java, New Guinea

TRIBE GLOMEREAE

SUBTRIBE GLOMERINAE

AGROSTOPHYLLUM Blume

Bijdr.:368, t.53 (1825).

Epiphytic herbs. Stems clustered, without pseudobulbs, erect or pendent, bilaterally flattened, of many internodes, leafy. Leaves distichous, narrow, rather thin, usually twisted at base to lie in one plane, with black or brown edged imbricate sheaths. Inflorescences terminal, usually globose heads on an elongate axis, or in a panicle, surrounded by bracts. Flowers small, resupinate, numerous, white or yellow, often self-pollinating; sepals and petals similar, free, petals narrower; lateral sepals forming a mentum which contains the saccate lip base; lip entire or 3-lobed, saccate base divided from the blade by a transverse partition; column short or rather long, foot rudimentary; pollinia 8, attached to a solitary viscidium.

Between 40 and 50 species distributed in the Old World tropics from the Seychelles and tropical Asia east to the Pacific islands, with the centre of distribution in New Guinea.

A.arundinaceum *Ridl.* in Sarawak Mus. J.1 (2):36 (1912). Type: Sarawak, Mt. Poe, *Brookes* 5 (holo. SING).

HABITAT: lower montane forest; rocky places; recorded from ultramafic substrate.
ALTITUDINAL RANGE: 600 – 1700m.
DISTRIBUTION: Borneo (SA, SR). Endemic.

A.bicuspidatum *J.J.Sm.* in Icon. Bogor. 2:55 (1903). Type: Java, Bantam, *Blume* s.n. (holo. BO).
   A.callosum auct. non Rchb.f.

HABITAT: hill dipterocarp forest; podsol forest.
ALTITUDINAL RANGE: 200 – 700m.
DISTRIBUTION: Borneo (B, K, SA, SR). Also Burma, Peninsular Malaysia, Thailand, Sumatra, Java, Mentawai, Krakatau; distribution further east uncertain.

A.cyathiforme *J.J.Sm.*, Orch. Java: 291 (1905). Type: Java, Gede, between Salabintana and Tjibeureum, *Smith* s.n. (holo. L).

HABITAT: lower montane forest.
ALTITUDINAL RANGE: 1200m.
DISTRIBUTION: Borneo (SA). Peninsular Malaysia, Sumatra, Java.

A.elongatum *(Ridl.) Schuit.* in Blumea 35 (1):165 (1990). Type: Peninsular Malaysia, Tahan River, *Ridley* s.n. (holo. SING).
   Appendicula elongata Ridl. in Trans. Linn. Soc. London, Bot., ser. 2, 3:375 (1893).
   A.hasseltii Blume, Bijdr.: 304 (1825). Types: Java, Bantam, *Blume* s.n.; Buitenzorg, *Blume* s.n. (syn. BO).
   Agrostophyllum hasseltii (Blume) J.J.Sm. in Icon. Bogor. 2:55 (1903).

HABITAT: lowland forest.
ALTITUDINAL RANGE: 100 – 200m.
DISTRIBUTION: Borneo (K). Also Peninsular Malaysia, Sumatra, Java,

Philippines, New Guinea, Solomon Islands, Caroline Islands (Palau).

A.globigerum *Ames & C.Schweinf.*, Orch. 6:138 (1920). Type: Sabah, Mt. Kinabalu, Marai Parai, *Clemens* 241 (holo. AMES).

> HABITAT: lower montane forest on ultramafic substrate.
> ALTITUDINAL RANGE: 1500m.
> DISTRIBUTION: Borneo (SA). Endemic.

A.glumaceum *Hook.f.*, Fl. Brit. Ind. 5: 821 (1890). Types: Peninsular Malaysia, Perak, *Scortechini* 1810; Perak, Ulu Bubong, *King's collector* 10876 (syn. K).
> ? *A.khasiyanum* sensu Ridl. in J. Linn. Soc., Bot. 31:286 (1896), non Griff.

> HABITAT: lowland and hill forest; often on limestone; mixed forest on sandy soil with scattered limestone rocks; riverine forest.
> ALTITUDINAL RANGE: sea level – 900m.
> DISTRIBUTION; Borneo (B, K, SA, SR). Also Peninsular Malaysia, Sumatra.

A.indifferens *J.J.Sm.* in Mitt. Inst. Allg. Bot. Hamburg 7:44, t.7., fig.34 (1927). Type: Kalimantan, Bukit Mehipit, *Winkler* 725 (holo. HBG, iso. BO).

> HABITAT: primary forest.
> ALTITUDINAL RANGE: 500m.
> DISTRIBUTION: Borneo (K). Endemic.

A.javanicum *Blume*, Bijdr.: 369, t.53 (1825). Type: Java, Buitenzorg, and Bantam Provinces, *Blume* s.n.(holo. BO).

> HABITAT: lower montane forest; moss forest; podsol forest.
> ALTITUDINAL RANGE: 1400 – 2000m.
> DISTRIBUTION: Borneo (SA). Also Sumatra, Java.

A.laterale *J.J.Sm.* in Bull. Jard. Bot. Buitenzorg, ser. 2,13: 50 (1914). Type: Kalimantan, Gunung Labang, *Amdjah* 154 (holo. L).

> HABITAT: unknown.
> ALTITUDINAL RANGE: lowlands.
> DISTRIBUTION.: Borneo (K). Endemic.

A.laxum *J.J.Sm.* in Bull. Jard. Bot Buitenzorg, ser. 3,3:279 (1921). Types: Java, many collections by *Backer, Bakhuisen van den Brink*, etc. (syn. L)

> HABITAT: hill forest.
> ALTITUDINAL RANGE: 700m.
> DISTRIBUTION: Borneo (SA). *A.laxum* occurs in Sumatra and Java.

> NOTE: The specimens *Dewol et al.* SAN 108802 (K, SAN) and *Yii* S. 48488 (SAR) match *A.laxum* vegetatively, but lack of flowers prevents confirmation of identity.

A.longifolium *(Blume) Rchb.f.* in Bonplandia 5:41 (1857). Type: Java, Salak, Pangurango, etc., *Blume* s.n. (holo. BO)
> *Appendicula longifolia* Blume, Bijdr.: 304 (1825).

> HABITAT: unknown.
> ALTITUDINAL RANGE; unknown
> DISTRIBUTION: Borneo (K). Also Peninsular Malaysia, Thailand, Sumatra, Java.

A.majus *Hook.f.*, Fl. Brit. Ind. 5:824 (1890). Types: Peninsular Malaysia, Perak, *Scortechini* 436, *King's collector* 1127 (syn. K).

HABITAT: peat swamp forest; hill forest on sandstone; riverine forest; kerangas forest with *Eugenia, Ficus and Tristania,* etc; lower montane forest; epiphytic on stilt roots.
ALTITUDINAL RANGE: sea level – 1500m.
DISTRIBUTION: Borneo (K, SA, SR). Also Peninsular Malaysia, Singapore, Sumatra, Papua New Guinea (Bougainville), Solomon Islands, Vanuatu.

A.mearnsii *Ames* in Philipp. J. Sci., 8:420 (1913). Type: Philippines, Mindanao, Mt. Malindang, *Mearns & Hutchinson* 4607 (holo. AMES, iso. K).

HABITAT: steep rocky roadside bank with *Arundina,* ferns, etc.
ALTITUDINAL RANGE: 1500 – 1600m.
DISTRIBUTION: Borneo (SA). Also Philippines.

A.saccatum *Ridl.* in J. Linn. Soc., Bot. 31: 286 (1896). Type: Sarawak, Kuching, *Haviland* 2329 (holo. SING, iso. K).

HABITAT: lowland and hill forest, sometimes on ultramafic substrate.
ALTITUDINAL RANGE: sea level – 400m.
DISTRIBUTION: Borneo (SA, SR). Endemic.

A.stipulatum *(Griff.) Schltr.* in Bot. Jahrb. Syst. 45, Beibl. 104: 22 (1911). Type: Peninsular Malaysia, Pinang, *Griffith* s.n. (holo. not located).

Appendicula stipulata Griff., Not. 3:358 (1851).
Agrostophyllum confusum J.J.Sm. in Bull. Jard. Bot. Buitenzorg, ser. 3, 2:37 (1920). Types: Sumatra, Deli, Medan, *Heldt* s.n., cult. Bogor (syn. BO); Lampong, Menggala, *Gusdorf* s.n., cult. Bogor (syn. BO).

HABITAT: unknown.
ALTITUDINAL RANGE: 1400m.
DISTRIBUTION: Borneo (B, K, SA, SR). Also Peninsular Malaysia, Singapore, Thailand, Sumatra, Lingga Archipelago, Riau, Bangka.

A.sumatranum *Schltr. & J.J.Sm.* in Bull. Dép. Agric. Indes Néerl. 15:7 (1908).

var. **borneense** *J.J.Sm.* in Bull. Jard. Bot. Buitenzorg, ser. 3, 2: 37 (1920). Type: Kalimantan, Liang Gagang, *Hallier* 1894 (holo. L).

HABITAT: unknown.
ALTITUDINAL RANGE: unknown.
DISTRIBUTION: Borneo (K). Endemic.

A.tenue *J.J.Sm.* in Bull. Jard. Bot. Buitenzorg, ser. 2, 26:33 (1918). Types: Java, Priangan, Tjibodas, Gede, *Hallier* 229 (syn. L); Pasir Kaboejoetan, Lemboer Tjimaloha, *Kds* n. 26547B (syn. L); Tjadas Malang, Tjibeber, *Winckel* s.n. (syn. L).

HABITAT: unknown.
ALTITUDINAL RANGE: unknown.
DISTRIBUTION: Borneo (SA). Also Peninsular Malaysia, Sumatra, Java.

A.trifidum *Schltr.* in Bot. Jahrb. Syst. 45, Beibl. 104: 22 (1911). Type: Sumatra, Padang - Pandjang, *Schlechter* 16029 (holo. B, destroyed).

HABITAT: primary forest.
ALTITUDINAL RANGE: 100 – 200m.
DISTRIBUTION: Borneo (K). Also Sumatra.

SUBTRIBE POLYSTACHYINAE

POLYSTACHYA Hook.

Exot. Fl. 2:t.103 (1824).

Epiphytic or lithophytic herbs. Stems erect, often pseudobulbous or fusiform, simple or superposed, leafy. Leaves conduplicate, solitary to several. Inflorescences terminal; few to many flowered, racemose or paniculate; peduncle often enclosed in scarious sheaths. Flowers non-resupinate, mostly small; sepals connivent or spreading, free; lateral sepals adnate to column-foot; petals smaller, usually linear; lip uppermost, entire or 3-lobed, articulate with column-foot, disc often farinaceous; column short, with a distinct foot; pollinia 4.

About 200 species centred in Africa, with a couple of species in tropical America. One pantropical species is native to tropical Asia from Sri Lanka and India eastwards to Sulawesi and the Philippines.

P.concreta *(Jacq.) Garay & H.R.Sweet* in Orquideologia 9, 3:206 (1974). Type: Lesser Antilles, Martinique, locality unknown, *Jacquin* s.n. (holo. not located).
Epidendrum concretum Jacq., Enum. Syst. Pl.:30 (1760).
Polystachya flavescens (Blume) J.J.Sm., Orch. Java 6:284 (1905). Type: Java, Salak & Seribu, *Blume* s.n. (holo. BO).
P.pleistantha Kraenzl. in Gard. Chron., ser. 3, 21:118 (1897). Type: Borneo, cult. *Lauche* (holo. B, destroyed).

HABITAT: riverine and hill forest on ultramafic substrate.
ALTITUDINAL RANGE: 200 – 900m.
DISTRIBUTION: Borneo (K, SA). Pantropical; in Asia from Sri Lanka, India and Nicobar Islands to Laos, Vietnam, Thailand, Malaysia and Indonesia to the Philippines.

TRIBE COELOGYNEAE

SUBTRIBE COELOGYNINAE

CHELONISTELE Pfitzer

in Pfitzer & Kraenzl., Pflanzenr. IV. 50. II. B. 7: 136 (1907).

de Vogel, E.F. (1986). Revisions in Coelogyninae (Orchidaceae) II. The genera *Bracisepalum, Chelonistele, Entomophobia, Geesinkorchis* and *Nabaluia.* Orch. Mon. 1:23 – 40.

Epiphytic or lithophytic herbs. Rhizome creeping, rather short. Pseudobulbs ovoid, conical or cylindrical, all turned to one side of rhizome, 1- or 2-leaved. Leaves elliptic to linear, usually tough and coriaceous, petiole deeply sulcate. Inflorescence proteranthous to synanthous, racemose, erect to spreading, few- to many-flowered; rachis fractiflex; floral bracts caducous. Flowers resupinate, small to medium-sized, tender to rather fleshy. Sepals free, concave, spreading. Petals narrower, often rolled backwards. Lip 3-lobed, base narrow, saccate, with more or less parallel sides; side lobes narrow, arising above the base at right angles to the blade and almost in the same plane; mid-lobe lingulate to broadly spathulate or almost orbicular, acute or deeply retuse, disc with 2, rarely 4, keels. Column long, with an entire or toothed apical wing. Pollinia 4, each attached to a small caudicle.

Eleven species were formerly known, all, except *C.sulphurea* var. *sulphurea,* endemic to Borneo. An unidentified species has recently been photographed on Mt. Kinabalu in Sabah. De Vogel (pers. comm.) reports a further two to five new species from Brunei and Sarawak.

C.amplissima *(Ames & C.Schweinf.) Carr* in Gard. Bull. Straits Settlem. 8: 218 (1935). Type: Sabah, Mt. Kinabalu, Kiau, *Clemens* 80 (holo. AMES, iso. L).
  Coelogyne amplissima Ames & C.Schweinf., Orch. 6:21 (1920).
  C. amplissima Ames & C.Schweinf. var. *schweinfurthiana* J.J.Sm. in Bull. Jard. Bot. Buitenzorg, ser. 3, 11:101 (1931). Types: Kalimantan, Gunung Kemoel, *Endert* 4252 (lecto. L); *Endert* 3970 (syn. L).

HABITAT: heath and moss forest; ridge top forest with *Agathis alba,* small rattans etc.; oak-laurel forest; lower montane mossy ericaceous forest; frequently on ultramafic substrate.
ALTITUDINAL RANGE: 800 – 2100m.
DISTRIBUTION: Borneo (K, SA, SR). Endemic.

C.brevilamellata *(J.J.Sm.) Carr* in Gard. Bull. Straits Settlem. 8:216, 217 (1935). Types: Kalimantan, Gunung Kemoel, *Endert* 4427 (lecto. L); *Endert* 4281 (syn. L).
  Coelogyne brevilamellata J.J.Sm. in Bull. Jard. Bot. Buitenzorg, ser. 3, 11: 102 (1931).

HABITAT: dense lower montane forest with an abundance of lianas, rattans, moss and orchids.
ALTITUDINAL RANGE: 1800m.
DISTRIBUTION: Borneo (K). Endemic.

C.dentifera *de Vogel* in Blumea 30:203, Pl. 3a–d (1984). Type: Sarawak, Bukit Mersing, Anap, *Sibat ak Luang* S.21954 (holo. L, iso. K).

HABITAT: mixed dipterocarp forest; lower montane forest.
ALTITUDINAL RANGE: 500 – 1100m.
DISTRIBUTION: Borneo (SR). Endemic.

C.ingloria *(J.J.Sm.) Carr* in Gard. Bull. Straits Settlem. 8:216, 217 (1935). Type: Kalimantan, Gunung Kemoel, *Endert* 3972 (holo. L).
  Coelogyne ingloria J.J.Sm. in Bull. Jard. Bot. Buitenzorg, ser. 3, 11:95 (1931).

HABITAT: lower montane riverine forest; ridge forest.
ALTITUDINAL RANGE: 1200 – 1500m.
DISTRIBUTION: Borneo (K, SA, SR). Endemic.

C.kinabaluensis *(Rolfe) de Vogel* in Blumea 30:203 (1984). Type: Sabah, Mt. Kinabalu, below Pakapaka, *Gibbs* 4260 (holo. BM).
  Sigmatochilus kinabaluensis Rolfe in Gibbs in J. Linn. Soc., Bot. 42:155, Pl.3 (1914).
  Pholidota sigmatochilus J.J.Sm. in Blumea 5:299 (1943).

HABITAT: upper montane oak-laurel forest; in thick moss cushions on branches of shrubs, or terrestial; often on ultramafic substrate.
ALTITUDINAL RANGE: 1900 – 3500m.
DISTRIBUTION: Borneo (SA). Endemic.

C.lamellulifera *Carr* in Gard. Bull. Straits Settlem. 8:78 (1935). Type: Sarawak, Dulit ridge, *Synge* S.531 (holo. K).
  Coelogyne lamellulifera (Carr) Masam., Enum. Phan. Born.: 142 (1942).

HABITAT: hill, lower and upper montane forest.
ALTITUDINAL RANGE: 900 – 2300m.
DISTRIBUTION: Borneo (B, SR). Endemic.

C.lurida *(L.Linden & Cogn.) Pfitzer* in Pflanzenr. 4, 50:138 (1907). Type: *Cult. Jard. Bot. Bruxelles* s.n., Herb. Cogniaux (holo. BR).
  Coelogyne lurida L.Linden & Cogn. in Lindenia 11:32 (1895).
  C. sarawakensis Schltr. in Notizbl. Bot. Gart. Berlin-Dahlem 8: 15 (1921). Type: Sarawak, *Beccari* 3963 (holo. B, destroyed, iso. FI; K, drawing of type).

var. **lurida**

HABITAT: hill, lower and upper montane forest; moss forest; oak-laurel forest; secondary forest; on trunks and moss covered roots.
ALTITUDINAL RANGE: 800 – 2700m.
DISTRIBUTION: Borneo (K, SA, SR). Endemic.

var. **grandiflora** *de Vogel* in Blumea 30:205, Pl. 3e,f (1984). Type: Sarawak, Gunung Mulu, *Lewis* 336 (holo. K).

HABITAT: lower and upper montane moss forest.
ALTITUDINAL RANGE: 1500 – 2000m.
DISTRIBUTION: Borneo (SR). Endemic.

C.ramentacea *J.J.Wood* in Kew Bull. 39:80, fig.6 (1983). Type: Sarawak, Gunung Mulu, *Burtt & Woods* 2192 (holo. E).

HABITAT: unknown.
ALTITUDINAL RANGE: 1330m.
DISTRIBUTION: Borneo (SR). Endemic.

C.richardsii *Carr* in Gard. Bull. Straits Settlem. 8: 79 (1935). Type: Sarawak, Dulit ridge, *Richards* S.59 (holo. K).
  Coelogyne richardsii (Carr) Masam., Enum. Phan. Born.: 144 (1942).

HABITAT: 'elfin forest'.

ALTITUDINAL RANGE: unknown.
DISTRIBUTION: Borneo (B, SR). Endemic.

C.sulphurea *(Blume) Pfitzer* in Pflanzenr. 4, 50:137 (1907). Type: Java, circa flumen Tjapus in monte Salak, *Blume* s.n. H.L.B 902, 322–1084.
    Chelonanthera sulphurea Blume, Bijdr.: 383 (1825).
    Coelogyne cuneata J.J.Sm. in Bull. Jard. Bot. Buitenzorg, ser. 3, 11:97 (1931). Type: Kalimantan, Gunung Kemoel, *Endert* 3897 (holo. L).
    C. pinniloba J.J.Sm. in Bull. Jard. Bot. Buitenzorg, ser. 3, 11:98 (1931). Type: Kalimantan, Gunung Kemoel, *Endert* 4524 (holo. L).
    C. kutaiensis J.J.Sm. in Bull. Jard. Bot. Buitenzorg, ser. 3, 11:99 (1931). Type: Kalimantan, Gunung Kemoel, *Endert* 4380 (holo. L, iso. BO).
    Chelonistele kutaiensis (J.J.Sm.) Carr in Gard. Bull. Straits Settlem. 8:77 (1935).
    C. cuneata (J.J.Sm.) Carr in Gard. Bull. Straits Settlem. 8:217 (1935).
    C. pinniloba (J.J.Sm.) Carr in Gard. Bull. Straits Settlem. 8:218 (1935).

var. **sulphurea**

HABITAT: hill forest; lower montane moss forest; kerangas forest on sandy soil; dipterocarp/Fagaceae/*Casuarina*/*Agathis* forest on ultramafic substrate; 'subalpine shrubbery'; ridge oak-chestnut forest.
ALTITUDINAL RANGE: 600 – 2400m.
DISTRIBUTION: Borneo (B, K, SA, SR). Also Peninsular Malaysia, Sumatra, Java, Philippines.

var.**crassifolia** *(Carr) de Vogel* in Blumea 3:205 (1984). Type: Sabah, Mt. Kinabalu, main spur below Kamborangah, *Carr* C.3565 (SFN 28027) (holo. SING).
    Chelonistele crassifolia Carr in Gard. Bull. Straits Settlem. 8:218 (1935).
    Coelogyne crassifolia (Carr) Masam., Enum. Phan. Born.:140 (1942).

HABITAT: lower and upper montane forest on ultramafic substrate; rocks and boulders, often in rather exposed sites.
ALTITUDINAL RANGE: 1500 – 2400m.
DISTRIBUTION: Borneo (SA). Endemic.

C.unguiculata *Carr* in Gard. Bull. Straits Settlem. 8:77 (1935). Type: Sarawak, Dulit ridge, *Synge* S.399 (holo. K, iso. L).
    Coelogyne unguiculata (Carr) Masam., Enum. Phan. Born.:146 (1942).

HABITAT: moss forest.
ALTITUDINAL RANGE: 1100 – 1400m.
DISTRIBUTION: Borneo (B, SR). Endemic.

COELOGYNE Lindl.

Coll. Bot., sub t.33 (1821?), corr. Lindl. op. cit. sub t.37 (1826).

Epiphytic, rarely lithophytic or terrestrial herbs. Pseudobulbs ovoid, conical or cylindrical, crowded or remote on rhizome, 1- or 2-leaved at apex. Leaves plicate, broad or narrow. Inflorescences erect or pendulous, hysteranthous, synanthous, protantherous or heteranthous, 1- to many-flowered. Flowers resupinate, small to large and showy, opening simultaneously or one at a time. Sepals free, often strongly concave and carinate. Petals free, usually narrower than sepals and often rolled backwards at apex. Lip 3-lobed, rather broad and concave at base, rarely saccate; side

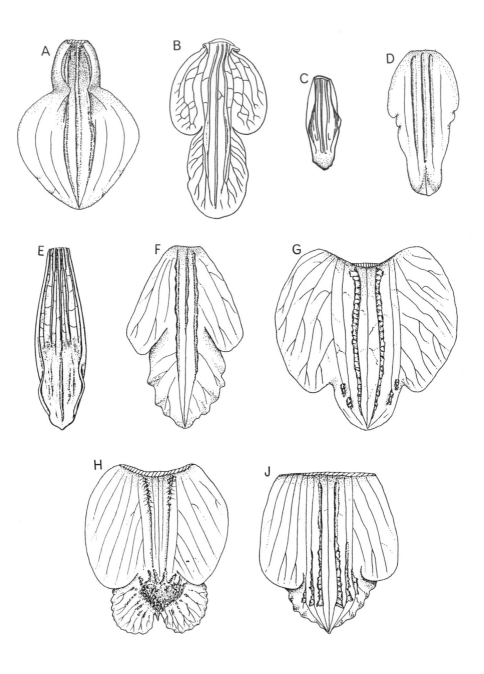

FIG. 11. *Coelogyne* lips. **A**, *C. clemensii* var. *clemensii*, x 3; **B**, *C. compressicaulis*, x 3; **C**, *C. plicatissima*, x 3; **D**, *C. subintegra*, x 3; **E**, *C. craticulaelabris*, x 3; **F**, *C. tenompokensis*, x 3; **G**, *C. hirtella*, x 2; **H**, *C. radioferens*, x 2; **J**, *C. rhabdobulbon*, x 2. Drawn by Sarah Thomas.

lobes erect, broad and widening gradually from base; mid-lobe spreading; disc keeled, keels often extending onto mid-lobe. Column long, with an apical wing. Pollinia 4.

About 300 species distributed from tropical Asia to the Pacific islands, particularly well represented in mainland Asia, Malaysia and Indonesia.

C.albobrunnea *J.J.Sm.* in Bull. Jard. Bot. Buitenzorg, ser. 3, 11:103 (1931). Type: Kalimantan, Gunung Kemoel, *Endert* 3868 (holo. BO, iso. L).

HABITAT: primary forest.
ALTITUDINAL RANGE: 1100m.
DISTRIBUTION: Borneo (K, SR). Endemic.

C.asperata *Lindl.* in J. Hort. Soc. London 4:221 (1849). Type: Borneo, ex cult. *Twisden Hodges* (holo. K-LINDL).

HABITAT: dipterocarp forest; swamp forest; podsol forest; cliffs.
ALTITUDINAL RANGE: sea level – 600m.
DISTRIBUTION: Borneo (K, SA, SR). Widespread from Peninsular Malaysia to the Solomon Islands.

C.balfouriana *Sander,* Cat. Orch.: 6 (1896). Type: ex cult. *Sander* (holo. not located).

HABITAT: unknown.
ALTITUDINAL RANGE: unknown.
DISTRIBUTION: Borneo (SA). Endemic.

C.borneensis *Rolfe* in Kew Bull.: 62 (1893). Type: Borneo, cult. *Linden* 22 (holo. K).

HABITAT: unknown.
ALTITUDINAL RANGE: unknown.
DISTRIBUTION: Borneo (unspecified). Endemic.

C.bruneiensis *de Vogel* in Orch. Mon. 6:11, fig.2, Pl.1a (1992). Type: Brunei, Sungai Ingei, surrounding Batu Melintang, *de Vogel* 27697 (holo. BRUN, iso. K,L).

HABITAT: forest near waterfall.
ALTITUDINAL RANGE: 100 – 200m.
DISTRIBUTION: Borneo (B). Endemic.

C.clemensii *Ames & C.Schweinf.* Orch. 6:23 (1920). Type: Sabah, Mt. Kinabalu, Marai Parai Spur, *Clemens* 227 (holo. AMES, iso. BO, K).

var. **clemensii**

HABITAT: hill, lower and upper montane forest, sometimes on ultramafic substrate.
ALTITUDINAL RANGE: 900 – 2200m.
DISTRIBUTION: Borneo (SA). Endemic.

var. **angustifolia** *Carr* in Gard. Bull. Straits Settlem. 8:212 (1935). Type: Sabah, Mt. Kinabalu, Penibukan ridge, *Carr* C.3091 (SFN 26453) (holo. SING, iso. K).

HABITAT: lower and upper montane forest; mossy forest with *Gymnostoma* & understorey of climbing bamboo; on ultramafic substrate.
ALTITUDINAL RANGE: 800 – 2400m.
DISTRIBUTION: Borneo (SA). Endemic.

var. **longiscapa** *Ames & C.Schweinf.*, Orch. 6: 25 (1920). Type: Sabah, Mt. Kinabalu, *Haslam* s.n. (holo. AMES, iso. BO).

HABITAT: unknown.
ALTITUDINAL RANGE: unknown.
DISTRIBUTION: Borneo (SA). Endemic.

C.compressicaulis *Ames & C.Schweinf.*, Orch. 6:25 (1920). Type: Sabah, Mt. Kinabalu, *Clemens* s.n. (holo. AMES).

HABITAT: low mossy and xerophyllous scrub forest on ultramafic substrate; upper montane forest.
ALTITUDINAL RANGE: 1200 – 2300m.
DISTRIBUTION: Borneo (SA). Endemic.

C.crassiloba *J.J.Sm.* in Mitt. Inst. Allg. Bot. Hamburg 7:28, t.4, fig.18 (1927). Type: Kalimantan, Bukit Mulu, *Winkler* 481 (holo. HBG, drawing at BO).

HABITAT: primary forest.
ALTITUDINAL RANGE: 600 – 700m.
DISTRIBUTION: Borneo (K). Endemic.

C.craticulaelabris *Carr* in Gard. Bull. Straits Settlem. 8:214 (1935). Type: Sabah, Mt. Kinabalu, below Lumu Lumu, *Carr* C. 2665 (SFN 27965) (holo. SING).

HABITAT: lower and upper montane forest; montane ericaceous scrub; ridge forest with *Dacrydium, Podocarpus, Tristania, Phyllocladus, Leptospermum*, etc.; recorded from sandstone.
ALTITUDINAL RANGE: 900 – 2400m.
DISTRIBUTION: Borneo (B, K, SA, SR). Endemic.

C.cumingii *Lindl.* in Bot. Reg. 26, misc. 187:76 (1840). Type: Singapore (holo. K-LINDL).

HABITAT: hill forest on ultramafic substrate.
ALTITUDINAL RANGE: 900 – 1000m.
DISTRIBUTION: Borneo (K, SA). Also Peninsular Malaysia, Singapore, Thailand, Laos, Sumatra, Riau.

C.cuprea *H.Wendl. & Kraenzl.* in Gard. Chron. ser. 3, 11:619 (1892). Type: Origin unspecified, ex *Sander & Co.* (holo. B).

var. **cuprea**

HABITAT: lower montane forest on ultramafic substrate.
ALTITUDINAL RANGE: 1500m.
DISTRIBUTION: Borneo (SA). Also Sumatra.

var. **planiscapa** *J.J.Wood & C.L.Chan* in Lindleyana 5, 2:84, fig.3 (1990). Type: Sabah, Gunung Alab, *Wood* 784 (holo. K).

HABITAT: lower montane ridge forest of *Dacrydium, Podocarpus, Tristania, Phyllocladus, Leptospermum*, etc.; oak-laurel forest.
ALTITUDINAL RANGE: 1200 – 2000m.
DISTRIBUTION: Borneo (SA, SR). Endemic.

C.dayana *Rchb.f.* in Gard. Chron. ser. 2, 21:826 (1884). Type: Borneo, *Veitch* s.n. (holo. W).

Coelogyne pulverula Teijsm. & Binn. in Natuurk. Tijdschr. Ned.-Indië 24:306 (1862). Type: Sumatra, Singalang, *Teijsmann* s.n. (holo. BO).

HABITAT: lowland dipterocarp forest; hill forest; riverine forest; on *Eugenia*; cliffs on ultramafic substrate.
ALTITUDINAL RANGE: 100 – 1100m.
DISTRIBUTION: Borneo (B, SA, SR). Also Peninsular Malaysia, Sumatra, Java, possibly Thailand.

C.distans *J.J.Sm.* in Bull. Dép. Agric. Indes Néerl. 15:2 (1908). Type: Kalimantan, Pontianak, *cult. Bogor* (holo. BO).

HABITAT: riverine forest.
ALTITUDINAL RANGE: lowlands.
DISTRIBUTION: Borneo (K, SA). Endemic.

C.dulitensis *Carr* in Gard. Bull. Straits Settlem. 8:73 (1935). Type: Sarawak, Mt.Dulit, *Synge* S.342 (holo. SING, iso. K).

HABITAT: unknown.
ALTITUDINAL RANGE: 600m.
DISTRIBUTION: Borneo (SA, SR). Endemic.

C.echinolabium *de Vogel* in Orch. Mon. 6:16, fig.6 (1992). Type: Sarawak, Gat, Upper Rejang River, *Clemens* 21639 (holo. L, iso. K, NY).

HABITAT: unknown.
ALTITUDINAL RANGE: unknown.
DISTRIBUTION: Borneo (B, SR). Endemic.

C.endertii *J.J.Sm.* in Bull. Jard. Bot. Buitenzorg, ser. 3, 11:94 (1931). Type: Kalimantan, Long Petak, *Endert* 3200 (holo. L, iso. BO).

HABITAT: primary forest.
ALTITUDINAL RANGE: 800m.
DISTRIBUTION: Borneo (K). Endemic.

C.exalata *Ridl.* in J. Straits Branch Roy. Asiat. Soc. 49:29 (1908). Type: Sarawak, Serapi, Matang, *Ridley* 124/70 (holo. SING, iso. K).

HABITAT: mixed hill dipterocarp forest; lower and upper montane forest; oak-laurel forest.
ALTITUDINAL RANGE: 900 – 2700m.
DISTRIBUTION: Borneo (B, SA, SR). Endemic.

C.foerstermannii *Rchb.f.* in Gard. Chron. n.s.: 262 (1886). Type: 'Sondaic Archipelago', *Foerstermann* s.n. (holo. W).

HABITAT: lowland and hill dipterocarp forest.
ALTITUDINAL RANGE: sea level – 500m.
DISTRIBUTION: Borneo (K, SA, SR). Also Peninsular Malaysia, Sumatra.

C.genuflexa *Ames & C.Schweinf.*, Orch. 6:28 (1920). Type: Sabah, Mt. Kinabalu, Marai Parai Spur, *Clemens* 251 (holo. AMES).

HABITAT: upper montane mossy forest on ultramafic substrate.
ALTITUDINAL RANGE: unknown.
DISTRIBUTION: Borneo (SA). Endemic.

C.gibbifera *J.J.Sm.* in Bull. Jard. Bot. Buitenzorg, ser. 2:53 (1912). Type: Sarawak, Batu Lawai, Ulu Limbang, *Moulton* 12 (holo. BO).

> HABITAT: unknown.
> ALTITUDINAL RANGE: unknown.
> DISTRIBUTION: Borneo (SR). Endemic.

C.harana *J.J.Sm.* in Mitt. Inst. Allg. Bot. Hamburg 7:27, t.4, fig.17 (1927). Type: Kalimantan, Lebang Hara, *Winkler* 346 (holo. HBG, drawing at BO).

> HABITAT: primary forest.
> ALTITUDINAL RANGE: 100 – 200m.
> DISTRIBUTION: Borneo (K). Endemic.

C.hirtella *J.J.Sm.* in Bull. Jard. Bot. Buitenzorg, ser. 3,11: 105 (1931). Type: Kalimantan, Gunung Kemoel, *Endert* 3976 (holo. L, iso. BO).
> *C.radioferens* J.J.Sm. in Mitt. Inst. Allg. Bot. Hamburg 7:33 (1927), non Ames & C.Schweinf.
> *C.radiosa* J.J.Sm. in Bull. Jard. Bot. Buitenzorg, ser. 3, 11: 105 (1931). Type: Kalimantan, Bukit Raja, *Winkler* 954 (lecto. L).

> HABITAT: oak-laurel forest; podsol forest; lower montane forest; secondary kerangas forest; montane heath forest; rocky places.
> ALTITUDINAL RANGE: 1000 – 2000m.
> DISTRIBUTION: Borneo (B, K, SA, SR). Endemic.

C.imbricans *J.J.Sm.* in Bull. Jard. Bot. Buitenzorg, ser. 3, 2: 26 (1920). Type: Kalimantan, Landak, Soengei Menjeeke, *Gravenhorst* 1917, cult.Bogor (holo. L, iso. BO).

> HABITAT: unknown.
> ALTITUDINAL RANGE: unknown.
> DISTRIBUTION: Borneo (K).

C.incrassata *(Blume) Lindl.*, Gen. Sp. Orch. Pl.: 40 (1830). Type: Java, Pantjar, *Blume* s.n. (holo. L).
> *Chelonanthera incrassata* Blume, Bijdr.: 384 (1825).

var. **incrassata**

> HABITAT: unknown.
> ALTITUDINAL RANGE: unknown.
> DISTRIBUTION: Borneo (B, K). Also Sumatra, Java.

var. **valida** *J.J.Sm.* in Bull. Jard. Bot. Buitenzorg, ser. 3, 11: 92 (1930). Type: Kalimantan, Liang Gagang, *Hallier* 3054 (holo. L, iso. BO).

> HABITAT: riverine forest.
> ALTITUDINAL RANGE: 400m.
> DISTRIBUTION: Borneo (K). Endemic.

FIG. 12. *Coelogyne* lips. **A**, *C. monilirachis*, x 1¼; **B**, *C. distans*, x 1¼; **C**, *C. kinabaluensis*, x 1¼ **D**, *C. zurowetzii*, x 1¼; **E**, *C. planiscapa* var. *grandis*, x 1¼; **F**, *C. prasina*, x 3; **G**, *C. exalata*, x 2; **H**, *C. macroloba*, x 2; **J**, *C. rochussenii*, x 2; **K**, *C. papillosa*, x 2; **L**, *C. dulitensis*, x 2; **M**, *C. moultonii*, x 2; **N**, *C. foerstermannii*, x 2½. Drawn by Sarah Thomas.

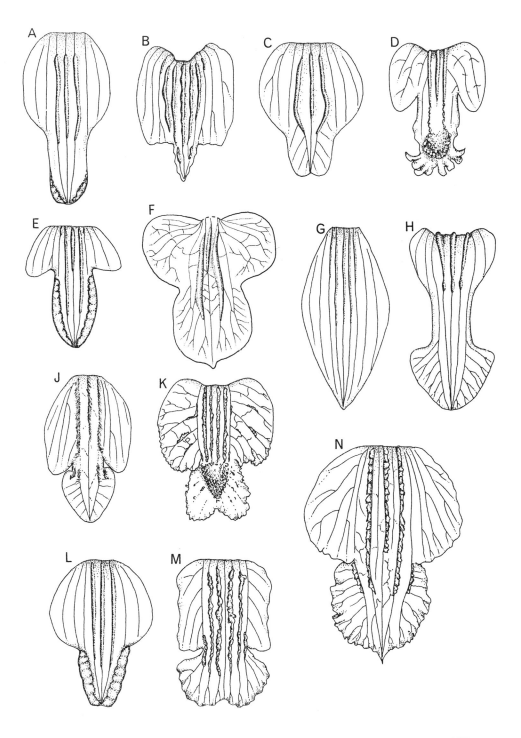

C.kelamensis *J.J.Sm.* in Icon. Bogor. 4:5, t.302 (1910). Type: Kalimantan, Kelam, *Hallier* 2489 (holo. BO).

HABITAT: unknown.
ALTITUDINAL RANGE: unknown.
DISTRIBUTION: Borneo (K). Endemic.

C.kinabaluensis *Ames & C.Schweinf.*, Orch. 6:30 (1920). Type: Sabah, Mt. Kinabalu, Marai Parai Spur, *Clemens* 229 (holo. AMES).

HABITAT: lower montane oak-laurel forest; sometimes on ultramafic substrate.
ALTITUDINAL RANGE: 1200 – 1800m.
DISTRIBUTION: Borneo (SA). Endemic.

C.latiloba *de Vogel* in Orch. Mon. 6: 20, fig.10 (1992). Type: Sabah, Bahagian Pantai Barat, *Collenette* A38 (holo. BM).

HABITAT: lower montane *Leptospermum* scrub.
ALTITUDINAL RANGE: 1600m.
DISTRIBUTION: Borneo (SA). Endemic.

C.longibulbosa *Ames & C.Schweinf.*, Orch. 6: 33 (1920). Type: Sabah, Mt. Kinabalu, Kiau, *Clemens* 79 (holo. BM, iso. AMES, BO, K).

HABITAT: river bed boulders; hill and lower montane forest, sometimes on ultramafic substrate.
ALTITUDINAL RANGE: 400 – 1500m.
DISTRIBUTION: Borneo (SA, SR). Endemic.

C.longpasiaensis *J.J.Wood & C.L.Chan* in Lindleyana 5, 2:87, fig.4 (1990). Type: Sabah, Sipitang District, 4 km south of Long Pa Sia, *Vermeulen & Duistermaat* 946 (holo. K).

HABITAT: low, very open and dry forest on steep sandstone ridges, with open places thickly overgrown with ferns; lower montane forest.
ALTITUDINAL RANGE: 1200 – 1300m.
DISTRIBUTION: Borneo (K, SA). Endemic.

C.macroloba *J.J.Sm.* in Mitt. Inst. Allg. Bot. Hamburg 7:30, t.4, fig.20 (1927). Type: Kalimantan, Bukit Raja, *Winkler* 860 (holo. HBG, drawing at BO).

HABITAT: hill forest; lower montane forest with *Lithocarpus* and *Castanopsis*, etc., and field layer of gingers and small rattans.
ALTITUDINAL RANGE: 600 – 1400m.
DISTRIBUTION: Borneo (B, K, SA, SR). Endemic.

C.membranifolia *Carr* in Gard. Bull. Straits Settlem. 7:2 (1932). Type: Peninsular Malaysia, Pahang, Tembeling, *Carr* s.n. (holo. SING).

HABITAT: on riverside trees; lower montane forest; epiphytic on shrubs and trees 1–3 metres above ground level.
ALTITUDINAL RANGE: 1200m.
DISTRIBUTION: Borneo (SA, SR).

C.meyeriana *Rchb.f.* in Gard. Chron. n.s. 8:134 (1877). Type: Origin not specified, ex cult. *Mayer* (holo. W).

HABITAT: lowland heath forest.
ALTITUDINAL RANGE: 100m.

DISTRIBUTION: Borneo (SA, SR). Also Peninsular Malaysia, Sumatra, Java.

C.monilirachis *Carr* in Gard. Bull. Straits Settlem. 6:206 (1935). Types: Sabah, Mt. Kinabalu, Tenompok, *Carr* C.3366 (SFN 27230) (syn. SING); Tenompok, *Clemens* 28294, 26127, 28316, 28454, 27166 (syn. SING).

HABITAT: lower montane forest.
ALTITUDINAL RANGE: 1200 – 1600m.
DISTRIBUTION: Borneo (SA, SR). Endemic.

C.moultonii *J.J.Sm.* in Bull. Jard. Bot. Buitenzorg, ser. 2, 3:54 (1912). Type: Sarawak, Bukit Labeng Barian, Ulu Limbang, *Moulton* 17 (holo. BO).

HABITAT: rocky areas; landslips; lower montane oak-laurel forest; recorded from ultramafic substrate.
ALTITUDINAL RANGE: 1100 – 2400m.
DISTRIBUTION: Borneo (SA, SR). Endemic.

C.muluensis *J.J.Wood* in Kew Bull. 39(1):76, fig.2 (1984). Type: Sarawak, near Bukit Berar, *Collenette* 2348 (holo. K).

HABITAT: kerangas forest; mixed dipterocarp forest with a kerangas element; alluvial forest.
ALTITUDINAL RANGE: lowlands.
DISTRIBUTION: Borneo (SR). Endemic.

C.naja *J.J.Sm.* in Bull. Jard. Bot. Buitenzorg, ser. 3, 11:93 (1931). Type: Kalimantan, Gunung Kemoel, *Endert* 3708 (holo. BO, iso. L).

HABITAT: lower montane forest.
ALTITUDINAL RANGE: 1200m.
DISTRIBUTION: Borneo (K). Endemic.

C.obtusifolia *Carr* in Gard. Bull. Straits Settlem. 8:205 (1935). Type: Sabah, Mt. Kinabalu, below Bundu Tuhan, *Carr* C.3149 (SFN 27897) (holo. SING, iso. K).

HABITAT: hill and lower montane forest.
ALTITUDINAL RANGE: 600 – 1500m.
DISTRIBUTION: Borneo (SA). Endemic.

C.odoardi *Schltr.* in Notizbl. Bot. Gart. Berlin-Dahlem 8:14 (1921). Type: Sarawak, Mt.Mattan, *Beccari* 1678 (holo. FI).

HABITAT: unknown.
ALTITUDINAL RANGE: 800 – 1900m.
DISTRIBUTION: Borneo (B, SA, SR). Endemic.

C.pallens *Ridl.* in J. Straits Branch Roy. Asiat. Soc. 39:81 (1903). Types: Peninsular Malaysia, Perak, Taiping Hills, *Curtis* s.n., Bujong Malacca, *Ridley* s.n. (syn. SING).

HABITAT: hill dipterocarp forest with *Tristania* on ridges.
ALTITUDINAL RANGE: 1000m.
DISTRIBUTION: Borneo (SA). Also Burma, Thailand, Peninsular Malaysia.

C.pandurata *Lindl.* in Gard. Chron.: 791(1853). Type: Borneo, *Low* s.n. (holo. K-LINDL).

HABITAT: lowland forest; hill forest with *Gymnostoma*; recorded from ultramafic substrate.

ALTITUDINAL RANGE: sea level – 1000m.
DISTRIBUTION: Borneo (SA, SR). Endemic.

C.papillosa *Ridl.* in Trans. Linn. Soc. London, Bot., ser. 2,4: 238 (1894). Type: Sabah, Mt. Kinabalu, *Haviland* 1098 (holo. BM, iso. K).

HABITAT: crevices in granite rocks; upper montane *Leptospermum-Dacrydium* scrub; on either granitic or ultramafic substrate.
ALTITUDINAL RANGE: 2400 – 3600m.
DISTRIBUTION: Borneo (SA). Endemic.

C.peltastes *Rchb.f.* in Gard. Chron. 2:296 (1880). Type: Borneo, *Veitch* s.n. (holo. W).

var. **peltastes**

HABITAT: heath forest; mixed dipterocarp forest; recorded from limestone.
ALTITUDINAL RANGE: sea level – 900m.
DISTRIBUTION: Borneo (K,SA,SR). Endemic.

var. **unguiculata** *J.J.Sm.* in Mitt. Inst. Allg. Bot. Hamburg 7:33, t.5, fig.23 (1927). Type: Kalimantan, Lebang Hara, *Winkler* 347 (holo. HBG, iso. BO).

HABITAT: lowland forest.
ALTITUDINAL RANGE: 100 – 200m
DISTRIBUTION: Borneo (K). Endemic.

C.pholidotoides *J.J.Sm.* in Icon. Bogor. 24, t.106B (1903). Types: Kalimantan, Damus, *Hallier* s.n. (syn. L); Bukit Raja, *Molengraaf* s.n. (syn. L).

HABITAT: mixed dipterocarp forest; sandstone and shale outcrops partially covered with grass and small bushes.
ALTITUDINAL RANGE: 800 – 1500m.
DISTRIBUTION: Borneo (K, SA, SR). Endemic.

C.planiscapa *Carr* in Gard. Bull. Straits Settlem. 8:74 (1935). Type: Sarawak, Dulit ridge, *Synge* S.419 (holo. SING, iso. K).

var. **planiscapa**

HABITAT: moss forest; low and rather open, wet, somewhat podsolic forest with dense undergrowth of *Pandanus* and rattan palms, on sandstone and shale outcrops.
ALTITUDINAL RANGE: 1200 – 1500m.
DISTRIBUTION: Borneo (SA, SR). Endemic.

var. **grandis** *Carr* in Gard. Bull. Straits Settlem. 8:202 (1935). Types: Sabah, Mt. Kinabalu, Penibukan ridge, *Carr* C. 3120 (SFN 27464) (syn. SING, iso. K); above Lumu Lumu, *Clemens* s.n. (syn. SING).

HABITAT: lower and upper montane forest; oak-laurel forest, sometimes on ultramafic substrate.
ALTITUDINAL RANGE: 1300 – 2000m.
DISTRIBUTION: Borneo (SA). Endemic.

C.plicatissima *Ames & C.Schweinf.*, Orch. 6:35 (1920). Type: Sabah, Mt. Kinabalu, Pakka, *Clemens* 204 (holo. AMES, iso. BO, K).

HABITAT: lower and upper montane forest; oak-laurel forest; montane ericaceous scrub; frequently on ultramafic substrate.

ALTITUDINAL RANGE: 1500 – 3000m.
DISTRIBUTION: Borneo (SA). Endemic.

C.prasina *Ridl.* in J. Linn. Soc., Bot. 32:326 (1896). Type: Peninsular Malaysia, Kedah Peak, *Ridley* 5131 (holo. SING).
HABITAT: hill forest.
ALTITUDINAL RANGE: 1200 – 1500m.
DISTRIBUTION: Borneo (SA). Also Peninsular Malaysia, Sulawesi, Maluku (Ternate).

C. radicosa *Ridl.* in J. Fed. Malay States Mus. 6:57 (1915). Type: Peninsular Malaysia, Gunung Kerbau, *Dyak collector* s.n. (holo. SING, iso. K).

HABITAT: podsol forest; lower montane dipterocarp forest.
ALTITUDINAL RANGE: 1000 – 1300m.
DISTRIBUTION: Borneo (SA). Endemic.

NOTE: The specimens *Beaman* 8844 (K) and *Wood* 666 (K) match *C.radicosa* except that the pseudobulbs are broader towards the base and the peduncle is terete.

C.radioferens *Ames & C.Schweinf.*, Orch. 6:38 (1920). Type: Sabah, Mt. Kinabalu, Pakka, *Clemens* 200 (holo. AMES).

HABITAT: lower and upper montane forest; oak-*Ericaceae* moss forest; oak-laurel forest; frequently on ultramafic substrate.
ALTITUDINAL RANGE: 1300 – 2700m.
DISTRIBUTION: Borneo (B, K, SA, SR). Endemic.

C.reflexa *J.J.Wood & C.L.Chan* in Lindleyana 5 (2):87, fig.5 (1990). Type: Sabah, Gunung Alab, *de Vogel* 8666 (holo. K, iso. L).

HABITAT: lower montane *Leptospermum/Dacrydium* forest on sandstone and shale soils; mossy low forest; shrubs on landslips; sometimes on ultramafic substrate.
ALTITUDINAL RANGE: 1200 – 1900m.
DISTRIBUTION: Borneo (SA). Endemic.

C.rhabdobulbon *Schltr.* in Notizbl. Bot. Gart. Berlin-Dahlem 8:15 (1921). Type: Sarawak, *Beccari* 1868 (holo. B, destroyed, iso. FI, lecto. Carr drawing in K).
C.pulverula auct. non Teijsm.& Binn.

HABITAT: heath forest; hill forest; lower montane oak-laurel forest.
ALTITUDINAL RANGE: 800 – 1500m.
DISTRIBUTION: Borneo (SA, SR). Endemic.

C.rigidiformis *Ames & C.Schweinf.*, Orch. 6:40 (1920). Type: Sabah, Mt. Kinabalu, Kiau, *Clemens* 71 (holo. AMES).

HABITAT: hill and lower montane forest.
ALTITUDINAL RANGE: 900 – 1600m.
DISTRIBUTION: Borneo (SA). Endemic.

C.rochussenii *de Vries*, Ill. Orch. Ind. Or. Néerl. t.2, t.11, fig.6 (1854). Type: Java, Gunung Salak, *cult.Bogor* (probably not collected; lecto. de Vriese, Ill. Orch. Livr. 2, t.11, fig.6 (1854).
C.stellaris Rchb.f. in Gard.Chron., ser. 2, 25:8 (1886). Type: Borneo, *Lobb* s.n. (holo. W).

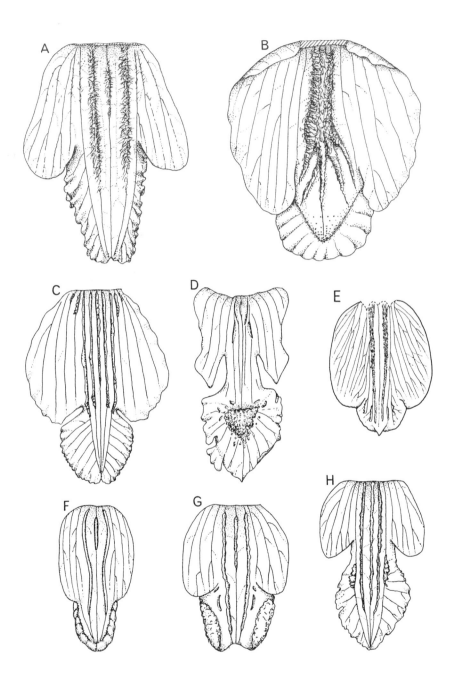

FIG.13. *Coelogyne* lips. **A**, *C. speciosa* ; **B**, *C. membranifolia* ; **C**, *C. sanderiana* ; **D**, *C. pandurata* ; **E**, *C. bruneiensis* ; **F**, *C. cuprea* var. *planiscapa* ; **G**, *C. planiscapa* var. *planiscapa*; **H**, *C. cumingii*. All x 1¼. Drawn by Sarah Thomas.

HABITAT: podsolic dipterocarp/*Dacrydium* forest on very wet sandy soil; riverine secondary forest; hill forest.
ALTITUDINAL RANGE: sea level – 1500m.
DISTRIBUTION: Borneo (B, K, SA, SR). Also Peninsular Malaysia, Thailand, Sumatra, Java, Sulawesi, Maluku, Philippines.

C.rupicola *Carr* in Gard. Bull. Straits Settlem. 8:210 (1935). Type: Sabah, Mt. Kinabalu, main spur above Kamborangah, *Carr* C. 3552 (SFN 27793) (holo. SING, iso. K).

HABITAT: lower montane forest; upper montane scrub; rocky places on ultramafic substrate.
ALTITUDINAL RANGE: 1500 – 2700m.
DISTRIBUTION: Borneo (SA). Endemic.

C.sanderiana *Rchb.f.* in Gard. Chron. ser. 3, 1:764 (1887). Type: Origin not specified, *Foerstermann* s.n. (holo. W).

HABITAT: secondary hill forest; moss forest on sandstone ridges.
ALTITUDINAL RANGE: 800m.
DISTRIBUTION: Borneo (B, K, SA, SR). Also Sumatra.

C.septemcostata *J.J.Sm.* in Icon. Bogor. 2:23, t.106A (1903). Type: 'Sumatra; Borneo, *Nieuwenhuis* s.n.' (holo. BO).

HABITAT: lowland forest; riverine forest; often on ultramafic substrate.
ALTITUDINAL RANGE: sea level – 900m.
DISTRIBUTION: Borneo (SA, SR). Also Sumatra.

C.speciosa *(Blume) Lindl.*, Gen. Sp. Orch. Pl.:39 (1830). Type: Java, Salak, *Blume* s.n. (holo. BO).
Chelonanthera speciosa Blume, Bijdr.:384, t.52 (1825).

HABITAT: unknown.
ALTITUDINAL RANGE: unknown.
DISTRIBUTION: Borneo (K). Also Sumatra, Java.

C.squamulosa *J.J.Sm.* in Bull. Dép. Agric. Indes Néerl. 15:3 (1908). Type: Kalimantan, Semedoem, *Hallier* s.n. (holo. BO).

HABITAT: unknown.
ALTITUDINAL RANGE: unknown.
DISTRIBUTION: Borneo (K). Endemic.

C.subintegra *J.J.Sm.* in Bull. Dép. Agric. Indes Néerl. 22:12 (1909). Type: Kalimantan, Gunung Kenepai, *Hallier* 1721 (holo. BO).

HABITAT: unknown.
ALTITUDINAL RANGE: unknown.
DISTRIBUTION: Borneo (K, SA). Endemic.

C.swaniana *Rolfe* in Kew Bull.:183 (1894). Type: 'Philippines', cult. *Sander & Co.* (holo. K).
C.quadrangularis Ridl. in J. Linn. Soc., Bot. 32:323 (1896). Type: Peninsular Malaysia, Gunung Hijan, *Ridley* s.n. (holo. SING).

HABITAT: hill dipterocarp forest; kerangas forest; secondary forest; sandstone and shale outcrops partially covered with grass and small bushes; also recorded from ultramafic substrate.

ALTITUDINAL RANGE: 600 – 1000m.
DISTRIBUTION: Borneo (SA, SR). Also Peninsular Malaysia, Sumatra. The type was probably incorrectly recorded from the Philippines.

C.tenompokensis *Carr* in Gard. Bull. Straits Settlem. 8:203 (1935). Types: Sabah, Mt. Kinabalu, main spur above Tenompok, *Carr* C.3270 (SFN 27501) (syn. SING, isosyn. AMES,K); Tenompok, *Clemens* 27191 & 29126 (syn. SING).

HABITAT: moss forest; lower montane forest; oak-laurel forest.
ALTITUDINAL RANGE: 1200 – 2100m.
DISTRIBUTION: Borneo (SA, SR). Endemic.

C.tenuis *Rolfe* in Kew Bull.:171 (1893). Type: Borneo, *Linden* s.n. (holo. K).
Coelogyne bihamata J.J.Sm. in Mitt. Inst. Allg. Bot. Hamburg 7:29, t.4, fig.19 (1927). Type: Kalimantan, Bukit Mulu, *Winkler* 517 (holo. HBG).

HABITAT: lower montane forest.
ALTITUDINAL RANGE: 500 – 1000m.
DISTRIBUTION: Borneo (K). Endemic.

C.testacea *Lindl.* in Bot. Reg.:28, misc. 34 (1842). Type: Singapore, *Loddiges* s.n. (holo. K-LINDL).

HABITAT: podsol forest
ALTITUDINAL RANGE: 400 – 500m.
DISTRIBUTION: Borneo (K, SA). Also Peninsular Malaysia, Singapore, Sumatra.

C.venusta *Rolfe* in Gard. Chron. ser. 3, 35:259 (1904). Type: 'China, Yunnan', cult. *Glasnevin* (holo. K).

HABITAT: sandstone and shale outcrops with grass and small bushes; hill forest; lower montane ridge forest; low and open podsol forest; frequently on ultramafic substrate.
ALTITUDINAL RANGE: 900 – 2100m.
DISTRIBUTION: Borneo (B, SA, SR). Endemic. The Chinese origin of the type collection is false.

C.vermicularis *J.J.Sm.* in Icon. Bogor. 3:9, t.204 (1906). Type: Borneo, *Nieuwenhuis* s.n. (holo. BO).

HABITAT: unknown.
ALTITUDINAL RANGE: unknown.
DISTRIBUTION: Borneo (K, SA, SR). Endemic.

C.zurowetzii *Carr* in Orchid Rev. 42:44 (1934). Type: Kalimantan, Sambas, *Zurowetz* s.n. (holo. SING).

HABITAT: high kerangas forest with dominant *Dacrydium* and *Tristania;* podsolic dipterocarp - *Dacrydium* forest; rather open,swampy forest on ultramafic substrate.
ALTITUDINAL RANGE: 400 – 700m.
DISTRIBUTION: Borneo (K, SA). Endemic.

DENDROCHILUM Blume

Bijdr.:398 (1825).

Epiphytic, lithophytic, rarely terrestrial herbs. Pseudobulbs tufted, narrow, fusiform or ovoid, 1-leaved. Leaves flat, linear to narrowly elliptic, coriaceous. Inflorescence lateral, slender, suberect to pendulous, spicate or racemose, usually densely many-flowered, synanthous (in section *Platyclinis*) or almost always heteranthous (in section *Dendrochilum*). Flowers small, usually resupinate, thin-textured. Sepals subequal, spreading. Lateral sepals adnate to base of column. Petals smaller than sepals, often erose. Lip 3-lobed or entire, usually with small side lobes and a large mid-lobe, disc 2- or 3-keeled. Column usually short, curved, with or without a foot, with narrow lateral arms (stelidia) and an often toothed apical wing or hood around anther. Pollinia 4.

Between 100 and 150 species distributed from mainland Asia east to the Philippines and New Guinea, particularly well represented in the montane areas of Sumatra and Borneo.

Full detailed line drawings of all the new taxa described below will appear in a revision of the Bornean species currently in preparation.

D.acuiferum *Carr* in Gard. Bull. Straits Settlem. 8:227 (1935). Type: Sabah, Mt. Kinabalu, near Paka Paka, *Carr* C.3550 (SFN 27645) (holo. SING, iso. AMES, K).

HABITAT: upper montane *Leptospermum* scrub, on ultramafic substrate; open granite slopes.
ALTITUDINAL RANGE: 3000m.
DISTRIBUTION Borneo (SA). Endemic.

D.alatum *Ames*, Orch. 6:45, Pl.82, fig.3 (1920). Type: Sabah, Mt. Kinabalu, Marai Parai Spur, *Clemens* 383 (holo. AMES, iso. K, SING).

HABITAT: lower and upper montane scrub; rocky places; moss forest; oak-laurel forest; on ultramafic and granitic substrates.
ALTITUDINAL RANGE: 1700 – 3200m.
DISTRIBUTION: Borneo (SA). Endemic.

D.alpinum *Carr* in Gard. Bull. Straits Settlem. 8:235 (1935). Type: Sabah, Mt. Kinabalu, below Sayat Sayat, *Carr* C.3545 (SFN 27624) (holo. SING, iso. AMES, K).

HABITAT: granite rocks in *Leptospermum* scrub.
ALTITUDINAL RANGE: 2400 – 3700m.
DISTRIBUTION: Borneo (SA). Endemic.

D.angustilobum *Carr* in Gard. Bull. Straits Settlem. 8:222 (1935). Type: Sabah, Mt. Kinabalu, Tenompok, *Carr* C.3233 (SFN 26874) (holo. SING, iso. AMES, K).

HABITAT: lower montane oak-laurel forest.
ALTITUDINAL RANGE: 1400 – 1600m.
DISTRIBUTION: Borneo (SA). Endemic.

D.angustipetalum *Ames*, Orch. 6:47, Pl.83b (1920). Type: Sabah, Mt. Kinabalu, Marai Parai Spur, *Clemens* 270 (holo. AMES).

HABITAT: lower montane forest on ultramafic substrate.
ALTITUDINAL RANGE: 1200 – 2000m
DISTRIBUTION: Borneo (SA). Endemic.

163

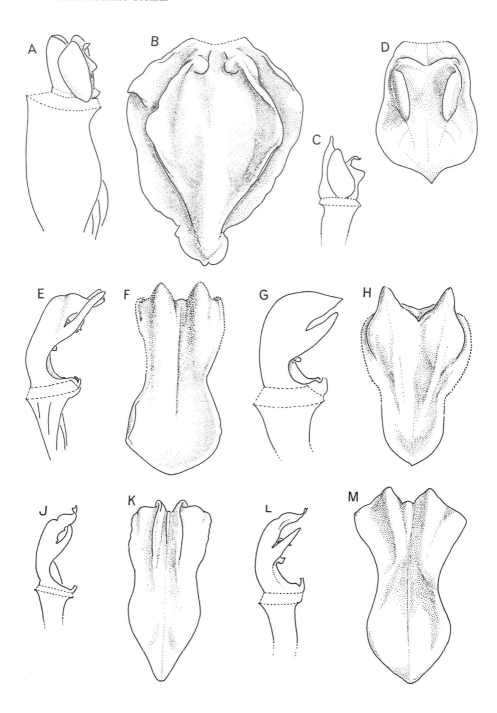

FIG. 14. *Dendrochilum* columns and lips. **A** & **B**, *D. alpinum*, x 9; **C** & **D**, *D. alatum*, x 8; **E** & **F**, *D. brevilabratum* var. *brevilabratum*, x 10 & x 27; **G** & **H**, *D. crassum*, x 14 & x 20; **J** & **K**, *D. intermedium*, x 7 & x 14; **L** & **M**, *D. pallideflavens*, x 16 & x 14. Drawn by Mutsuko Nakajima.

D.anomalum *Carr* in Gard. Bull. Straits Settlem. 8:87 (1935). Type: Sarawak, Mt.Dulit, Ulu Koyan, *Richards* 2497 (holo. SING, iso. K).

HABITAT: heath forest; hill forest on ultramafic substrate.
ALTITUDINAL RANGE: 600 – 1000m.
DISTRIBUTION: Borneo (SA, SR). Endemic.

D.auriculilobum *J.J.Wood* **sp.nov.** sectionis *Platyclinidis D.hologynaiki* Carr affine, sed pseudobulbis multo brevioribus crassioribus, inflorescentiis laxioribus, floribus majoribus, labio lobis lateralibus auriculatis iugoque basali transversali carnoso instructo distinguitur. Typus: East Malaysia, Sabah, Sipitang District, Sungai Rurun headwaters, c.1700m, December 1986, *Vermeulen & Duistermaat* 1057 (holotypus L, herbarium material only, isotypus K, spirit material only).

Terrestrial herb. Pseudobulbs 2.8 – 3 x 0.6 – 0.8 cm, cylindrical, 2 – 3 cm apart on rhizome. Basal sheaths 3.5 – 8 cm long, pale fawn speckled pale brown, enclosing young pseudobulbs and basal part of peduncle, becoming fibrous. Leaves with petiole 5 – 6 cm long, blade 20 – 25 x 1.2 – 1.3 cm, linear-lanceolate, acute, thin-textured. Inflorescence many flowered; peduncle 20 – 25 cm long, filiform, wiry; rachis 18 cm long, curving, with about 6 imbricate sterile bracts up to 1 mm long at base; floral bracts 3 – 3.2 x 4 – 4.5 mm, broadly ovate, obtuse, margin hyaline, prominently veined, ramentaceous, involute, entirely enclosing pedicel with ovary. Flowers arranged in two ranks, each flower 2 – 2.5 mm apart, very pale greenish, lip green, somewhat fragrant. Pedicel with ovary 2.3 – 2.4 mm long, straight. Sepals and petals spreading. Dorsal sepal 4 x 1.7 – 1.8 mm, ovate-elliptic, acute. Lateral sepals 4 x 2 – 2.1 mm, ovate, acute. Petals 3.9 – 4 x 1.7 – 1.8 mm, elliptic, acute. Lip 2.5 mm long, 3 mm wide across side lobes, concave, cup-like; mid-lobe 1.5 – 1.6 x 1.5 mm, oblong-ovate, obtuse, with a small obtuse apical mucro, 3-nerved, with a fleshy, transverse basal ridge; side lobes 1.1 x 0.2 – 0.3 mm, auriculate, acute. Column 0.8 x 1 mm , oblong, arms (stelidia), foot and apical hood absent. Anther cap c. 0.6 – 0.7 x 0.6 – 0.7 mm, cucullate, apiculate. Fig.22, H & J.

D.auriculilobum is related to *D.hologyne* Carr, described from Sarawak and also recorded from Sipitang District in S.W. Sabah. *D.auriculilobum* is distinguished by its much shorter, thicker pseudobulbs, laxer inflorescences, larger flowers and lip with auriculate side lobes and a fleshy transverse basal ridge.

This interesting plant is known only from the type which was collected in low and open mossy ridge forest with a dense undergrowth of bamboo and rattan palms.

The specific epithet is derived from the Latin *auriculatus*, with ear-like appendages, and *lobus*, lobe, referring to the side lobes of the lip.

D.brevilabratum *(Rendle) Pfitzer* in Engler, Pflanzenr. Orch. Monand. 4, 50:89 (1907). Type: Sarawak, Baram, *Hose* 52 (holo. ?BM).
Platyclinis brevilabrata Rendle in J. Bot. 39:173 (1901).
Acoridium brevilabratum (Rendle) Rolfe in Orchid Rev. 12:220 (1904).

var. **brevilabratum**

HABITAT: lowland forest; ridge forest on sandy loam soil; trees in paddyfield clearings.
ALTITUDINAL RANGE: sea level – 300m.
DISTRIBUTION: Borneo (K, SR).

var.**petiolatum** *J.J.Wood* in Kew Bull. 39 (1):78, fig.3 (1984). Type: Sarawak, Gunung Mulu National Park, between Sungei Berar and Sungei Mentawai, *Nielsen* 675 (holo. AAU, iso. K).

HABITAT: mixed lowland dipterocarp forest.
ALTITUDINAL RANGE: 100 – 300m.
DISTRIBUTION: Borneo (SR). Endemic.

D.bulbophylloides *Schltr.* in Notizbl. Bot. Gart. Berlin-Dahlem 8:16 (1921). Type:
Sarawak, Gunung Mattan, *Beccari* 3036 (holo. FI, iso. L, drawing by Carr at K).

HABITAT: unknown.
ALTITUDINAL RANGE: unknown.
DISTRIBUTION: Borneo (SR). Endemic.

D.conopseum *Ridl.* in Stapf. in Trans. Linn. Soc. London, Bot., ser. 2, 4:236 (1894).
Type: Sabah, Mt. Kinabalu, Tampassuk, Koung, *Haviland* 1381 (holo. K, iso. SAR).

HABITAT: mixed hill dipterocarp forest; lower montane forest; sometimes on
ultramafic substrate.
ALTITUDINAL RANGE: 400 – 1100m.
DISTRIBUTION: Borneo (SA, SR). Endemic.

D.cornutum *Blume*, Bijdr.: 939 (1825). Type: Java, Buitenzorg, and Tjanjor Province,
Blume s.n. (holo. ?L).

HABITAT: unknown.
ALTITUDINAL RANGE: unknown.
DISTRIBUTION: Borneo (unspecified). Also Sumatra, Java.

D.corrugatum *(Ridl.) J.J.Sm.* in Recueil Trav. Bot. Néerl. 1:65 (1904). Type: Sabah,
Mt. Kinabalu, Marai Parai, *Haviland* s.n. (holo. not located).
Platyclinis corrugata Ridl. in Stapf in Trans. Linn. Soc. London, Bot. 4:233 (1894).

HABITAT: lower montane forest on ultramafic substrate.
ALTITUDINAL RANGE: 1700m.
DISTRIBUTION: Borneo (SA). Endemic.

D.crassifolium *Ames*, Orch. 6:49, Pl.84a (1920). Type: Sabah, Mt. Kinabalu, *Haslam*
s.n. (holo. AMES).

var. **crassifolium**

HABITAT: lower montane forest; low, rather open, somewhat podsolic forest
with a dense field layer of *Pandanus* and rattan.
ALTITUDINAL RANGE: 1300 – 1800m.
DISTRIBUTION: Borneo (SA, SR). Endemic.

var. **murudense** *J.J.Wood* **var. nov.** sectionis *Platyclinidis* a varietate typica foliis
ellipticis 4.4 – 8 x 1 – 2 cm metientibus petiolis 0.8 – 1.5 cm longis ad instar forma
eorum *D.lewisii* J.J.Wood, floribus aliquantum majoribus sepalis petalis labelloque
longioribus differt. Typus: East Malaysia, Sarawak, Fourth Division, Kelabit
Highlands, Mt. Murud, 2400 m, 7 April 1970, *Nooteboom & Chai* 01995 (holotypus K,
herbarium and spirit material; material labelled *Nooteboom & Chai* 01995 at L is
D.dewindtianum W.W.Sm. var. *dewindtianum*).

Tufted epiphyte. Pseudobulbs 0.6 – 2 x 0.8 cm, conical, greenish-yellow, crowded
together on rhizome, enclosed in membranous sheaths when young. Leaves with a
thick leathery texture, petiole 0.8 – 1.5 cm, deeply sulcate, blade 4.4 – 8 x 1 – 2 cm,
elliptic or oblong-elliptic, obtuse and mucronate. Inflorescence erect, densely many-
flowered; peduncle 3.5 – 6.5 cm long; rachis 16–20 cm long, quadrangular; floral
bracts 4 – 7 mm long, ovate, mucronate, membranous, with prominent nerves.

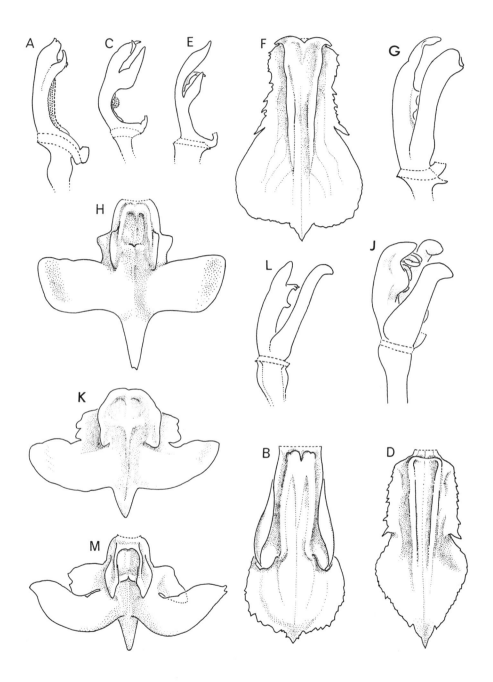

FIG. 15. *Dendrochilum* columns and lips. **A** & **B**, *D. anomalum*, x 7 & x 9; **C** & **D**, *D. dewindtianum* var. *dewindtianum*, x 7 & x 10; **E** & **F**, *D. dewindtianum* var. *sarawakense*, x 7 & x 9; **G** & **H**, *D. exasperatum*, x 6; **J** & **K**, *D. gibbsiae* (variant), x 6; **L** & **M**, *D. gibbsiae* (variant), x 7. Drawn by Mutsuko Nakajima.

Flowers 'greenish', each borne 4.5 – 6 mm apart. Pedicel with ovary 2.5–3 mm long, cylindrical, slightly curved. Dorsal sepal 0.9 – 1 x 2.1 – 2.8 mm, oblong-lanceolate, acuminate. Lateral sepals 9 – 9.2 x 2.2 – 3 mm, ovate-lanceolate, acute to acuminate. Petals 8 – 8.1 x 0.9 – 2.1 mm, lanceolate, acuminate, minutely erose. Lip 6.6 mm in total length, 2 – 2.1 mm wide at base, stipitate to column foot by a tiny claw; mid–lobe 4 x 2.8 – 3 mm, broadly elliptic, acuminate, margins minutely erose; side lobes 3 x c. 0.8 mm, triangular, acute to acuminate, outer margin erose to lacinulate; disc with two fleshy raised keels linked by a basal swelling, the keels terminating at or just beyond junction of mid- and side lobes, central nerve prominent. Column 4 – 4.5 mm long, curved, broader above, foot 0.5 – 1 mm long, apical hood 0.8 mm wide, ovate, entire, arms (stelidia) 2 – 2.5 mm long, ligulate, obtuse, fleshy, basal. Anther cap cucullate, with a dorsal tooth. Fig.22, A & B.

Other material examined: Sarawak, Fourth Division, Mt. Murud, second sandstone summit, moss forest, 2250 m, 12 September 1982, *Yii Puan Ching* S. 44430A (K, L, SAR, mixed collection with S.44430, *D.dewindtianum* W.W.Sm. var. *sarawakense* Carr).

This differs from var. *crassifolium* in having elliptic leaves measuring 4.4 – 8 x 1 – 2 cm with petioles 0.8 – 1.5 cm long and resembling in shape those of *D. lewisii* J.J.Wood. The flowers are somewhat larger, with longer sepals, petals and lip.

The varietal epithet refers to the type locality, Mt. Murud (2422 m), the highest mountain in Sarawak.

D.crassilabium *J.J. Wood* **sp.nov.** sectionis *Platyclinidis D.pachyphyllo* J.J.Wood arcte affinis sed foliis plerumque longioribus atque coriaceis vice carnosorum crassorumque, inflorescentia quam foliis plerumque longiore, sepalis petalisque obtusis plane crassis carnosisque, labello parum longiore lobo mediano spathulato non profunde retuso lobis lateralibus laceratis, brachiis columnae quam cucullo apicali brevioribus distinguitur. Typus: Indonesia, Kalimantan Timur, Apo Kayan, Gunung Sungai Pendan, East Ridge, east of Long Nawan, 1300m, 13 October 1991, *de Vogel* & *Cribb* s.n., Leiden cult. no.913205A (holotypus L, isotypus K).

Tufted epiphyte. Pseudobulbs 1 – 1.6 x 0.3 – 0.4 cm , ovoid-elliptic or narrowly fusiform, clothed in sheaths when young. Leaves narrowly linear-ligulate, acute, coriaceous, petiole 2 – 5 mm long, sulcate, blade 5 – 8.5 x 0.45 – 0.5 cm. Inflorescences 2 to 3(–5)-flowered, pendulous, usually longer than the leaves; peduncle 5.5 – 12 cm long, filiform; rachis 0.7 – 1.8 cm, thicker than peduncle, fractiflex; floral bracts 3 – 5 mm long, ovate-acuminate, brown. Flowers with dull brownish-salmon, translucent sepals and petals, lip paler brownish-salmon, column brownish olive-green, anther cap cream. Pedicel with ovary 2 mm long, clavate. Dorsal sepal 7 x 2.1 mm, oblong-elliptic, obtuse, curved. Lateral sepals 7.5 x c.2.8 mm, obliquely ovate-elliptic, obtuse. Petals 7 x 2mm, oblong, obtuse, erose. Lip 7 – 7.1 mm long, 3-nerved, minutely papillose; mid-lobe 5.5 x 3.5 mm, spathulate, shallowly retuse, thick and fleshy, gently curved at base; side lobes 1.5 – 2 mm long, lacerate; disc with 2 low, fleshy basal ridges forming a U shape. Column 3 mm long, 1 mm wide across apical hood, apical hood truncate, often irregularly toothed, arms (stelidia) 2 mm long, shorter than apical hood, ligulate or oblong-ligulate, obtuse, sometimes falcate, with a short foot. Anther cap 0.8   0.9 x 1 mm, cucullate. Fig.22, C & D.

Other material examined: Sarawak, Fourth Division, route from Ulu Sungai Limbang to Bukit Buli, 1540m, October 1987, *Awa* & *Lee* S.50774 (K, SAR).
D.crassilabium is identical in habit to *D.pachyphyllum* J.J.Wood and has similarly coloured flowers. The leaves of *D.crassilabium,* however are coriaceous, never thick and fleshy, and normally proportionately longer. The inflorescence is usually longer

than the leaves. The sepals and petals are obtuse and the thick, fleshy lip is slightly longer and has a shallowly retuse, spathulate mid-lobe and lacerate side lobes. The column arms are shorter than the apical column hood. The thin-textured, coriaceous leaves and thick, fleshy, spathulate, retuse lip of *D.crassilabium* are distinctive. However, intermediates between it and *D.pachyphyllum* may well exist.

The material of *Awa & Lee* S.50774 at K appears to be a mixed collection of both species. The plant mounted at the top of the sheet has long leaves and more closely matches *D.crassilabium*, but the inflorescence is shorter. The flowers are rather badly preserved.

The specific epithet is derived from the Latin *crassus*, thick and *labiatus*, lipped, referring to the fleshy lip.

D.crassum *Ridl.* in J. Linn. Soc., Bot. 32:288 (1896). Type: Peninsular Malaysia, Perak, Hermitage Hill, *Ridley* s.n. (holo. ?BM).

HABITAT: hill forest; steep rocky roadside banks with *Melastoma, Nepenthes, Lycopodium, Gahnia*, ferns,etc., on sandstone and shale outcrops.
ALTITUDINAL RANGE: 1200 – 1500m.
DISTRIBUTION: Borneo (SA). Also Peninsular Malaysia.

D.cruciforme *J.J.Wood* **sp.nov.** sectionis *Platyclinidis D.sublobato* Carr, speciei Sarawakensi arcte affinis sed floribus aliquantum majoribus labello cruciformi manifestius trilobato carinis basalibus magis valde perspicuis atque columna cucullo acuto acuminato vel bifido instructa pede destituta distinguitur; habitu structuraque florali *D.dolichobrachio* (Schltr.) Merr. etiam similis, sed labello carinis brevioribus lobis lateralibus latioribus triangularibus vel triangulari-ovatis lobo mediano multo minore anguste triangulari-acuminato cuspidato instructa; a *D.devogelii* J.J.Wood pedicello cum ovario breviore, labello cruciato atque columnae brachiis brevioribus etiam distingui potest; a *D.gibbsiae* Rolfe statura multo pumiliore, foliis brevioribus angustioribus graminiformibus, inflorescentia breviore floribus parum minoribus, labello trilobato vice quinquelobati atque columna longiore differt. Typus: East Malaysia, Sabah, Mt. Kinabalu, Penibukan, 1200 – 1500m, January 1933, *J. & M.S.Clemens* s.n. (holotypus K, herbarium and spirit material, isotypus AMES, E, HBG, herbarium material only).

var. **cruciforme**

Clump-forming epiphyte. Pseudobulbs 0.5 – 2.5 x 0.2 – 0.4 cm, cylindrical or narrowly fusiform, enclosed in sheaths when young. Leaves narrowly linear, subacute to acute, narrowed into a slender petiole; petiole 0.5 – 2 cm long, blade 5 – 11.5 x 0.1 – 0.3(0.4) cm. Inflorescence gently curving, densely many-flowered, each flower borne 1 – 2 mm apart; peduncle 2.5 – 6.5 cm long; rachis 3.5 – 5(–6.5) cm long, quadrangular; floral bracts 1 – 1.5 x 1 mm, ovate acute. Flowers variously described as 'pinkish-cream with purple spots', 'cream-green', 'cream-yellow', 'cream' or 'pure white', lip usually with a dark purple-brown blotch at base of side lobes and on apex of keels, rarely pale greenish-white with a greenish lip, with a 'musty scent' according to Bacon. Pedicel with ovary 0.5 – 1(–2)mm long, clavate. Dorsal sepal 2.2 – 2.6(–3) x 0.6 – 0.9 mm, oblong-elliptic, acute. Lateral sepals 2.1 – 2.5(–2.8) x 0.8 – 1 mm, ovate-elliptic, acute. Petals 2 – 2.1 x 0.4 – 0.6 mm, narrowly elliptic, acute. Lip 2 mm long, 1.8 – 2 mm wide across side lobes, cruciform, 3-nerved, with 2 rounded basal keels joined at the base by a transverse ridge; mid-lobe narrowly triangular-acuminate, cuspidate; side lobes triangular or triangular-acute, acute or subacute, often somewhat falcate. Column 1 – 1.2 x 0.2 – 0.3 mm, hood acute, acuminate or bifid, arms (stelidia) basal, linear-ligulate, obtuse, 0.9 – 1.1 mm long, foot absent. Anther cap cucullate. Fig.24, A & B.

Other material examined: Sabah, Kota Kinabalu to Sinsuron road, Mile 27, from roadside stall, 1971, *Bacon* 187 (E); Mt. Kinabalu, Penibukan, 1200 – 1500m, 20 December 1932, *J.& M.S.Clemens* 30471 (E) and 10 January 1933, *J.& M.S.Clemens* 30826 (AMES, BM ,E, K); Penibukan, 1200m, 18 March 1933, *J.& M.S.Clemens* 32220 (BM, E); Mt. Kinabalu, Penataran Basin, 900 – 1200m, 27 June 1933, *J.& M.S.Clemens* 34329 (AMES, BM, E, HBG, L); Penataran Basin, 900m, 31 August 1933, *J.& M.S.Clemens* 40134 (AMES, BM, E); Mt. Kinabalu, Tinekuk Falls, 1800m, 1 November 1933, *J.& M.S.Clemens* 50278 (AMES, BM, K); Penibukan 1500m, 11 November 1933, *J.& M.S.Clemens* 50322 (AMES, BM, K); Penibukan, 1200 – 1500m, 12 January 1933, *J.& M.S.Clemens* s.n. (BM, E); Marai Parai, 1500 m, 22 November 1931, *Collector unknown* SFN 36563 (SING); Mt.Alab, south ridge, 2000m, 31 October 1986, *de Vogel* 8661 (K, L).

D.cruciforme is closely related to *D.sublobatum* Carr, from Sarawak, but is distinguished by its slightly larger flowers with a more distinctive three-lobed cruciform lip, with more pronounced basal keels, and column with an acute, acuminate or bifid hood and no foot. The habit and floral structure also resembles *D.dolichobrachium* (Schltr.)Merr., but the lip of *D.cruciforme* has shorter keels, broader triangular or triangular-ovate side lobes and a much smaller narrowly triangular-acuminate, cuspidate mid-lobe. It may also be distinguished from *D.devogelii* J.J.Wood by the shorter pedicel with ovary, cruciform lip and shorter column arms. It differs from *D.gibbsiae* Rolfe by its much dwarfer stature, shorter, narrower grass-like leaves, shorter inflorescence with slightly smaller flowers, a three, rather than five-lobed lip and longer column.

Clemens 40134 is a more robust specimen with pseudobulbs up to 2.5 x 0.4 cm, leaf blades up to 11.5 x 0.4 cm and pure white flowers lacking the characteristic dark blotch on the lip. *De Vogel* 8661 is a variant with a longer filiform pedicel, greenish-white flowers, again without a dark blotch on the lip, less pronounced side lobes and a deeply bifid apical column hood. It more closely resembles *D.sublobatum* Carr.

The degree of lobing of the lip in *D.cruciforme* appears quite variable, some specimens having more pronounced side lobes than others. This is also true of the widespread *D.gibbsiae*. The shape of the apical column hood also ranges from acute to acuminate and entire to deeply bifid, but is nearly always longer than the arms. The full range of variation is difficult to ascertain given the limited material available. However, one distinctive variant which seems to warrant recognition, is described below.

Habitat details are not provided by the Clemenses. De Vogel, however, records the habitat on Mt.Alab as *Leptospermum/Dacrydium* forest about 8 metres high growing on a steep east facing slope on weathered sandstone and shale.

The specific epithet is derived from the Latin *cruciformis*, cross-shaped, in reference to the lip.

var. **longicuspum** *J.J.Wood* **var.nov.** varietas nova a varietate *cruciformi* lobis lateralibus labelli aliquantum angustis atque falcatis, lobo mediano longiore anguste triangulari acuminato, columnae brachiis columnae cucullo apicali tridentato longioribus differt. Typus: East Malaysia, Sabah, Mount Kinabalu, Kadamaian River, 2000m, August 1933, *Carr* 3675, SFN 28004 (holotypus K, isotypus SING).

Floral bracts c.3.5 x 2 mm. Flowers pale yellow, base of lip greenish-yellow. Dorsal sepal 3 x 0.9 mm. Lateral sepals 2.9 – 3 x 0.9 – 1 mm. Petals 2.5 – 2.6 x c. 0.7 mm. Lip 2.5 mm long, 1.5 mm wide across side lobes; mid-lobe 1 mm long, narrowly triangular, acuminate; side lobes narrow, somewhat falcate. Column 1 x 0.2 – 0.3 mm, hood tridentate, shorter than arms, arms 1 – 2 mm long, ligulate, obtuse, foot absent.

This differs from var. *cruciforme* in having a lip with a longer, acuminate mid-lobe and narrower, somewhat falcate side lobes and column arms longer than the apical tridentate hood. It was incorrectly determined as *D.corrugatum* J.J.Sm. by Carr.

The varietal epithet is derived from the Latin *longus*, long and *cuspis*, a sharp, rigid point, in reference to the mid-lobe of the lip.

D.cupulatum *J.J.Wood* **sp.nov.** sectionis *Eurybrachii D.fimbriato* Ames affinis sed pseudobulbis seorsum dispositis, foliis lineari-ligulatis atque floribus paullum minoribus sepalo dorsali breviore sepalis lateralibus non falcatis petalis integris labio entegro valde concavo cupulato callis centralibus minutis rotundatis duobus instructo et columnam breviorem habenti distinguitur; praeterea a *D.alato* Ames pseudobulbis seorsum dispositis, inflorescentia densiore floribus paullum minoribus petalis integris labio breviore valde concavo cupulato apice apiculato callis minutis atque columna breviore truncata instructo eam etiam distinguere potest. Typus: East Malaysia, Sarawak, Mt.Mulu National Park, above Camp 4, c.1700m, 21 March 1981, *Lamb* MAL 12 (holotypus K, herbarium material only).

Creeping, clump-forming epiphyte. Rhizome tough, branching profusely, producing numerous wiry roots. Pseudobulbs 0.9 – 2.3 x 0.4 – 0.8 cm, cylindrical or elliptic, borne 0.3 – 2.4 cm apart on rhizome. Basal sheaths pale brown, unspotted, concealing young pseudobulbs at first, but soon becoming fibrous. Leaves erect, petiole 0.1 – 0.5 cm long, blade 4 – 10 x 0.4 – 0.5 (– 0.8)cm, linear-ligulate, conduplicate at base, apex slightly carinate, apiculate. Inflorescence erect, densely many-flowered; peduncle 1.5 – 3 cm long, greenish-yellow; rachis 6 – 14 cm long, quadrangular, greenish-yellow; floral bracts 2 – 2.5 x 1.1 – 1.2 mm, ovate, apiculate, carinate. Flowers 3 mm across, borne in 2 ranks, each flower 1 mm apart, greenish-yellow, very pale greenish with a brighter green lip or white with a greenish-white lip, with a slightly spicy scent. Pedicel with ovary 1.8 mm long, slightly curved. Sepals and petals spreading. Dorsal sepal 2 x 1 mm, ovate-elliptic, acute. Lateral sepals 2 x 1mm, ovate-elliptic, acute. Petals c. 1.6 – 1.7 x 0.8 – 0.9 mm, ovate-elliptic, acute. Lip 1.1 x 1.5 – 1.6 mm, broadly ovate, concave, cupulate, apex apiculate, with a basal ridge and 2 small rounded central calli. Column 0.2 x 0.2 – 0.3 mm, cuneate, truncate, arms (stelidia) 0.5 mm long, basal, oblong, somewhat truncate, foot absent. Anther cap minute, cucullate. Fig. 23, J – L.

Other material examined: Sabah, Crocker Range, Kimanis to Keningau Road, 1400 m, 6 April 1985, *Kitayama* 893 (UKMS); same locality, 1500 – 1600 m, December 1986, *Vermeulen & Duistermaat* 667 (K, L). Sarawak, Mt.Mulu National Park, southern summit, c. 2100 m, 28 April 1978, *Argent & Coppins* 1126 (E, K); Mt. Mulu National Park, Camp 4, 1700m, 18 March 1978, *Hansen* 498 (C, K); Mt. Mulu National Park, west ridge near Camp 4, c. 1880m, 22 March 1978, *Nielsen* 806 (AAU, K).

D.cupulatum is related to *D.fimbriatum* Ames from Sabah but is distinguished by its well-spaced pseudobulbs, linear-ligulate leaves and slightly smaller flowers with a shorter dorsal sepal, non-falcate lateral sepals, entire petals, an entire, strongly concave, cupulate lip with two tiny rounded central calli and a shorter column. It may also be distinguished from *D.alatum* Ames, also from Sabah, by the well-spaced pseudobulbs, denser inflorescence with slightly smaller flowers with entire petals, shorter, strongly concave, cupulate lip with an apiculate apex and tiny calli, and a shorter, truncate column. The spikes of tiny crowded flowers look remarkably like those of *D.sublobatum* Carr, a species belonging to section *Platyclinis* so far only recorded from Sarawak. This, however, has crowded pseudobulbs and longer leaves. On closer inspection, the flowers of *D.sublobatum* have a decurved, slightly lobed

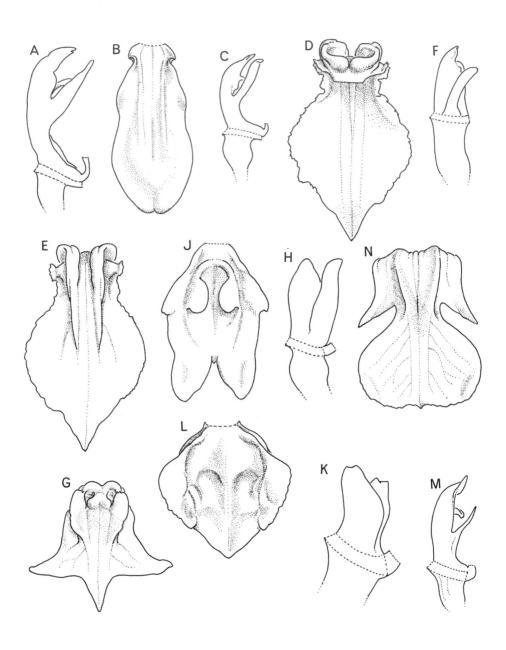

FIG. 16. *Dendrochilum* columns and lips. **A** & **B**, *D. globigerum*, x 6 & x 5; **C**, *D. gracilipes*, column, x 6; **D**, *D. gracilipes*, lip (back view), x 7; **E**, *D. gracilipes*, lip (front view), x 7; **F** & **G**, *D. grandiflorum*, x 5; **H** & **J**, *D. haslamii*, x 5; **K** & **L**, *D. hologyne*, x 12 & x 9; **M** & **N**, *D. imbricatum*, x 5 & x 7. Drawn by Mutsuko Nakajima.

ovate-acuminate lip with a large horseshoe-shaped callus and a column with a conspicuous hood and long linear arms.

The specific epithet is derived from the Latin *cupula*, cup, in reference to the concave, cup-like lip.

D.devogelii *J.J. Wood* **sp.nov.** sectionis *Platyclinidis D.cruciformi* J.J.Wood affinis sed labello pandurato parte basali haud lobata pro portione longiore, brachiis columnae cucullo apicali longioribis apice decurvatis, pedicello longiore distinguitur. Typus: East Malaysia, Sabah, Sipitang District, ridge between Maga River and Pa Sia River, 1450m, 18 October 1986, *de Vogel* 8376 (holotypus L, isotypus K).

Tufted epiphyte. Pseudobulbs 1 – 1.5 x 0.4 – 0.5 cm, cylindrical. Leaves 9.5 x 0.3 cm, linear-ligulate, acute. Inflorescence laxly many-flowered; peduncle 2.2 cm long; rachis 6.5 cm long, with a 3.5 mm long sterile bract at base; floral bracts 1 – 2 mm long, ovate, acute. Flowers pale green. Pedicel with ovary 3 mm long, very narrowly clavate. Dorsal sepal 2.5 x 0.8 mm, oblong, acute. Lateral sepals 2 – 2.1 x 0.8 – 0.9 mm, oblong, slightly falcate, acute. Petals 2 x 0.5 – 0.6 mm, linear, acute, margins a little uneven. Lip 1 – 1.1 x 0.9 – 1 mm, pandurate, with a fleshy, u-shaped keeled basal callus; mid-lobe acute, apiculate, mucronate; side lobes rounded to subacute. Column 2 mm long, arms (stelidia) basal, 2 – 2.1 mm long, ligulate, obtuse, tips decurved, apical hood entire, shorter than arms. Anther cap cucullate.

D.devogelii is only known from the type collection which consists of only two pseudobulbs, one leaf and one inflorescence. It is related to *D.cruciforme* but is distinguished by the pandurate lip with a proportionately longer basal unlobed portion, column arms longer than the apical hood and decurved at the apex, and the longer pedicel.

The habitat is described as 'rather dense primary forest up to 30 metres high dominated by *Agathis* and *Lithocarpus* on poor sandy soil, bedrock sandstone, much leaf litter, little undergrowth'.

The specific epithet honours Dr. E.F. de Vogel of the Rijksherbarium, Leiden, The Netherlands, who collected the type.

D.dewindtianum *W.W.Sm.* in Notes Roy. Bot. Gard. Edinburgh 8:321 (1915). Types: Sabah, Mt. Kinabalu, *native collector* 68 (syn. E); *native collector* 99 (syn. E, isosyn. K, SAR).
   D.lobongense Ames, Orch. 6:59 (1920). Type: Sabah, Mt. Kinabalu, Lobong, *Clemens* 116 (holo. AMES), **syn.nov.**
   D.perspicabile Ames, Orch. 6:62, Pl.82, fig.4 (1920). Type: Sabah, Mt. Kinabalu, *Clemens* 202 (holo. AMES, iso. BO, K, SING).

var. **dewindtianum**

HABITAT: lower montane forest; upper montane scrub; rocky places; mostly on ultramafic substrate
ALTITUDINAL RANGE: 1500 – 3000m.
DISTRIBUTION: Borneo (SA, SR). Endemic.

var. **sarawakense** *Carr* in Gard. Bull. Straits Settlem. 8:83 (1935). Types: Sarawak, Dulit, *Shackleton* S.186 (syn. SING, isosyn. K); Dulit Ridge, *native collector* in *Synge* S.558 (syn. SING, isosyn. K).

HABITAT: moss forest on sandstone.
ALTITUDINAL RANGE: 700 – 2300m.
DISTRIBUTION: Borneo (SR). Endemic.

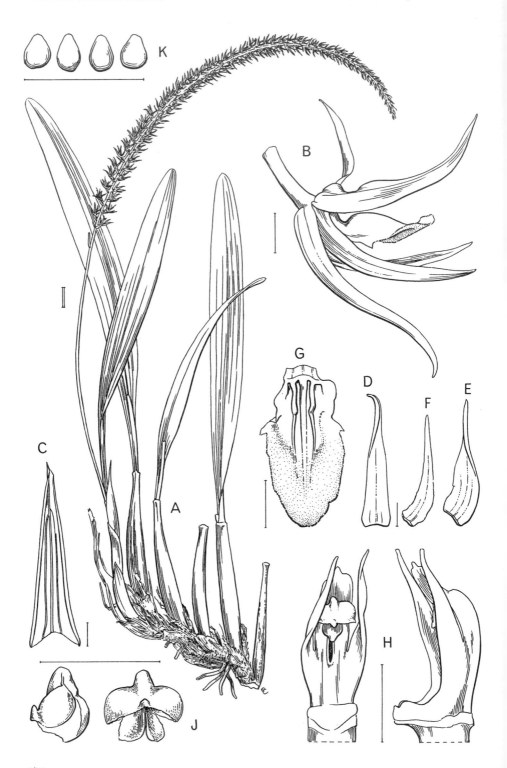

D.dolichobrachium *(Schltr.) Merr.*, Bibl. Enum. Born. Pl.:147 (1921). Type: Kalimantan, Long Dett, *Schlechter* 13557 (holo. B, destroyed, sketch by Carr in K). *Platyclinis dolichobrachia* Schltr. in Bull. Herb. Boissier, ser. 2, 6:301 (1906).

HABITAT: lower montane ridge forest.
ALTITUDINAL RANGE: 1500 – 1600m.
DISTRIBUTION: Borneo (K, SR). Endemic.

D.dulitense *Carr* in Gard. Bull. Straits Settlem. 8:84 (1935). Type: Sarawak, Dulit ridge, *Synge* S.435 (holo. SING, iso. K).

HABITAT: moss forest.
ALTITUDINAL RANGE: 1200 – 1400m.
DISTRIBUTION: Borneo (B, SA, SR). Endemic.

D.exasperatum *Ames*, Orch. 6:50 (1920). Type: Sabah, Mt. Kinabalu, Marai Parai Spur, *Clemens* 396 (holo. AMES).

HABITAT: hill forest; lower montane forest.
ALTITUDINAL RANGE: 900 – 1600m.
DISTRIBUTION: Borneo (SA, SR). Endemic.

D.fimbriatum *Ames*, Orch. 6:51 (1920). Type: Sabah, Mt. Kinabalu, Marai Parai Spur, *Clemens* 248 (holo. AMES).

HABITAT: lower montane forest; *Leptospermum* scrub on ridges; on ultramafic substrate.
ALTITUDINAL RANGE: 1600 – 1700m.
DISTRIBUTION: Borneo (SA). Endemic.

D.geesinkii *J.J. Wood* **sp.nov.** sectionis *Platyclinidis D.crassifolio* Ames affinis, sed pseudobulbis multo longioribus, bracteis floralibus paulo longioribus, floribus aliquanto minoribus petalisque angustioribus, lobis labelli lateralibus auriculatis indistinctis, brachiis columnae cucullum apicalem eiusdem superantibus distinguitur. Typus: Indonesia, Kalimantan Timur, base of Gunung Tapa Sia, between Long Bawan and Panado, 1400 m, 22 July 1981, *Geesink* 9180 (holotypus L, herbarium material only, isotypus K, spirit material only, BO, KYO, herbarium material only).

Epiphyte with a creeping rhizome enclosed in fibrous sheaths. Pseudobulbs 6 – 8 x 0.6 – 0.8 cm, each about 0.5 – 0.6 cm apart, cylindrical, enclosed by three 1.5 – 4 cm long, acute to acuminate sheaths when immature. Leaves 12.6 – 23.5 x 1.4 – 2.2 cm, oblong-elliptic, obtuse and mucronate to subacute, narrowed and conduplicate at base, coriaceous. Inflorescences densely many-flowered; peduncle 13 – 17 cm long, naked; rachis up to 23 cm long; floral bracts 5 – 7 mm long, 1 – 1.1 mm wide at base, subulate, acuminate. Flowers pale green. Pedicel with ovary 2 mm long. Dorsal sepal 6 mm long, 1 mm wide (across base), linear-lanceolate, acuminate. Lateral sepals 5.5 mm long, 1 mm wide (across base), lanceolate, slightly falcate, acute. Petals 5mm long, c. 0.8 mm wide (across base), linear-lanceolate, slightly falcate, acute, minutely erose. Lip 3.8 – 4 x 2 mm, oblong-elliptic, obtuse, with tiny auriculate side lobes, margin minutely erose-papillose, disc with 3 fleshy keels near base, grading into prominent nerves which terminate towards apex. Column 2 mm

FIG. 17. *Dendrochilum geesinkii*. **A**, habit; **B**, flower (side view); **C**, floral habit; **D**, dorsal sepal; **E**, lateral sepal; **F**, petal; **G**, lip (front view); **H**, column (front and side views); **J**, anther cap (side and back views); **K**, pollinia. A-K from *Geesink* 9180. Scale: single bar = 1 mm; double bar = 1 cm. Drawn by Eleanor Catherine.

long, with a foot, arms (stelidia) borne from near base, 2.1 – 2.2 mm long, acute, slightly exceeding hood, hood c. 0.2 mm wide, slightly irregularly toothed. Anther cap cucullate. Fig.17.

Other material examined: Kalimantan Timur, between Long Bawan and Panado, 1400m, 9 July 1981, *Geesink* 8989 (L). Sarawak, Fourth Division, route to Batu Lawi, Bario, 1290 m, 3 August 1985, *Awa & Lee* S.50586 (AAU, K, KEP, L, MEL, SING).

This species is related to *D.crassifolium* Ames but may be distinguished by the much longer pseudobulbs, slightly longer floral bracts, and slightly smaller flowers with narrower petals, obscure, auriculate lip side lobes and column arms exceeding the apical column hood.

D.geesinkii was collected by the late Rob Geesink of the Rijksherbarium, Leiden during a joint Indonesian-Japanese expedition sponsored by LIPI, Jakarta and the Ministry of Education, Tokyo. It was found growing in hill forest on sandstone. The collection from Sarawak was found on the top of a ridge in mixed hill dipterocarp forest.

D.gibbsiae *Rolfe* in J. Linn. Soc., Bot. 42:147 (1914). Type: Sabah, Mt. Kinabalu, Marai Parai Spur, *Gibbs* 4087 (holo. K).
  D.kinabaluense Rolfe, loc. cit.:148 (1914). Type: Sabah, Mt. Kinabalu, above Lobang to Pakapaka, *Gibbs* 4252 (holo. K).
  D.quinquelobum Ames, Orch. 6:63, Pl.82, ii (1920). Type: Sabah, Mt. Kinabalu, Kiau, *Clemens* 361 (holo. AMES, iso. BO, K).

HABITAT: on moss covered trees in hill, lower and upper montane forest; in moss cushions on ground in wet *Dacrydium/Podocarpus* forest; oak-laurel forest; low mossy and xerophyllous scrub forest on ultramafic substrate; steep rocky banks; sandstone and shale outcrops fully exposed to sun; moss forest with *Gymnostoma* and undergrowth of climbing bamboo; on limestone boulders in primary forest.
ALTITUDINAL RANGE: 900 – 2400m.
DISTRIBUTION: Borneo (B, K, SA, SR). Endemic.

D.globigerum *(Ridl.) J.J.Sm.* in Recueil Trav. Bot. Néerl. 1:64 (1904). Type: Sarawak, Serapi, *Haviland* 169 (holo. K).
  Platyclinis globigera Ridl. in J. Linn. Soc., Bot. 31:266 (1896).
  P.minima Ridl. in J.Straits Branch Roy. Asiat. Soc. 49:30 (1908), **syn.nov**. Type: Sarawak, Tiang Layu, *Hewitt* 14 (holo. SING, iso. K).
  P.minor Ridl. in J.Straits Branch Roy. Asiat. Soc. 49:30 (1908), **syn.nov**. Type: Sarawak, Gunung Santubong, *Hewitt* s.n. (holo. SING, iso. K).
  Dendrochilum minimum (Ridl.) Ames in Merr., Bibl. Enum. Born. Pl.:148 (1921), **syn.nov**.
  D.minus (Ridl.) Ames in Merr., Bibl. Enum. Born. Pl.:148 (1921), **syn.nov**.

HABITAT: hill and lower montane forest.
ALTITUDINAL RANGE: 700 – 900m.
DISTRIBUTION: Borneo (SA, SR). Endemic.

D.gracile *(Hook.f.) J.J.Sm.* in Recueil Trav. Bot. Néerl. 1:69 (1904). Type: Peninsular Malaysia, Perak, Larut, *King's collector* 3280 (holo. K).
  Platyclinis gracilis Hook.f., Fl. Brit. Ind. 5:708 (1890).
  Dendrochilum lyriforme J.J.Sm. in Brittonia 1:107 (1931). Type: Sarawak, Mt. Poi, *Clemens* 20313 (holo. L, iso. K).

HABITAT: lower montane ridge forest; low stature forest on ultramafic substrate.
ALTITUDINAL RANGE: 1300 – 1800m.
DISTRIBUTION: Borneo (B, SA, SR). Also Peninsular Malaysia, Sumatra, Java.

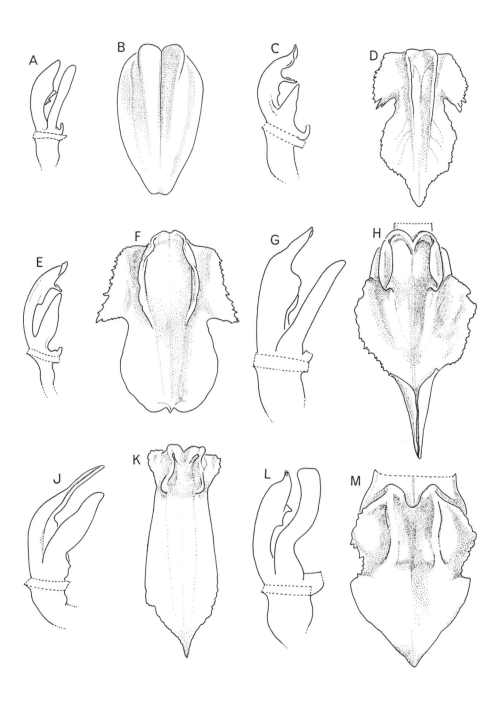

FIG. 18. *Dendrochilum* columns and lips. **A** & **B**, *D. integrilabium*, x 7 & x 13; **C** & **D**, *D. kamborangense*, x 9; **E** & **F**, *D. lacteum*, x 7 & x 9; **G** & **H**, *D. lancilabium*, x 7; **J** & **K**, *D. lewisii*, x 8; **L** & **M**, *D. longipes*, x 8 & x 12. Drawn by Mutsuko Nakajima.

177

D.gracilipes *Carr* in Gard. Bull. Straits Settlem. 8:81 (1935). Type: Sarawak, Dulit ridge, *Synge* S.418 (holo. SING, iso. K).

HABITAT: moss forest; 'sand forest'.
ALTITUDINAL RANGE: 900 – 1600m.
DISTRIBUTION: Borneo (SA, SR). Endemic.

D.graminoides *Carr* in Gard. Bull. Straits Settlem. 8:229 (1935). Type: Sabah, Mt. Kinabalu, near Pinansak, *Carr* C.3680 (SFN 28006) (holo. SING, iso. AMES,K).

HABITAT: hill and lower montane forest, sometimes on ultramafic substrate.
ALTITUDINAL RANGE: 900 – 2700m.
DISTRIBUTION: Borneo (B, SA). Endemic.

D.grandiflorum *(Ridl.) J.J.Sm.* in Recueil Trav. Bot. Néerl. 1:66 (1904). Type: Sabah, Mt. Kinabalu, *Haviland* 1142 (holo. K).
Platyclinis grandiflora Ridl. in Stapf in Trans. Linn. Soc. London, Bot. 4:233 (1894).

HABITAT: upper montane forest; *Leptospermum recurvum* scrub; rocky streamsides in shade; on ultramafic substrate.
ALTITUDINAL RANGE: 1500 – 3800m.
DISTRIBUTION: Borneo (SA). Endemic.

D.gravenhorstii *J.J.Sm.* in Bull. Jard. Bot. Buitenzorg, ser. 3, 11:28 (1920). Type: Kalimantan, Upper Kapoeas, Sungai Talaj, *Gravenhorst* s.n. (holo. BO, iso. L).

HABITAT: kerangas forest.
ALTITUDINAL RANGE: 400 – 1000m.
DISTRIBUTION: Borneo (K, SA, SR). Endemic.

NOTE: Listed as *'Dendrobium gravenhorstii'* by Masamune (1942).

D.hamatum *Schltr.* in Feddes Repert. 9:340 (1911). Type: Sarawak, Undup, *Beccari* 476 (holo. B, destroyed, iso. FI).

HABITAT: unknown.
ALTITUDINAL RANGE: unknown.
DISTRIBUTION: Borneo (SR). Endemic.

D.haslamii *Ames,* Orch. 6:53, Pl.85 (1920). Type: Sabah, Mt. Kinabalu, *Haslam* s.n. (holo. AMES).

HABITAT: upper montane ridge forest, frequently on ultramafic substrate.
ALTITUDINAL RANGE: 2400 – 3100m.
DISTRIBUTION: Borneo (SA, SR). Endemic.

D.havilandii *Pfitzer* in Engler, Pflanzen. Orch. Monand. 4, 50:107 (1907). Type: Sarawak, *Haviland* 2346 (holo. K, iso. SING).
D.hewittii J.J.Sm. in Bull. Dép. Agric. Indes Néerl. 22:14 (1909), **syn.nov**. Type: Sarawak, Quop, *Hewitt* s.n. (holo. BO).

HABITAT: limestone rocks and cliffs in light shade.
ALTITUDINAL RANGE: sea level – 400m.
DISTRIBUTION: Borneo (SR). Endemic.

D.hologyne *Carr* in Gard. Bull. Straits Settlem. 8:89 (1935). Type: Sarawak, Dulit ridge, *Synge* S.513 (holo. SING, iso. K).

HABITAT: moss forest; podsol forest; ridge top forest with *Agathis*, small rattans, etc.

ALTITUDINAL RANGE: 100 – 1600m.

DISTRIBUTION: Borneo (SA, SR). Endemic.

D.hosei *J.J.Wood* **sp.nov.** sectionis *Platyclinidis* forma labelli structuraque columnae *D.gibbsiae* Rolfe persimilis, sed foliis ellipticis duris coriaceis 5.3 – 5.7 cm latis inflorescentiaque aliquantum rigida pedunculo rhachidique crassiore non eis *D.longipedis* J.J.Sm. dissimilibus statim distinguitur; carinae duae in disco labelli perarcte parallele dispositae sed non cum porca basali transversali sunt conjunctae. Typus: East Malaysia, Sarawak, Bahagian Kapit, northern Hose Mountains, base of ridge leading to Bukit Batu, 1200m, 1 December 1991, *de Vogel* 1244 (holotypus L, isotypus K, both spirit material only).

Epiphyte. Pseudobulbs up to 5 x 2 cm, elliptic, slightly wrinkled, enclosed in brown sheaths up to 8 cm long when young. Leaves tough, coriaceous, petiole 5 – 6 cm long, deeply sulcate, blade 22 – 25 x 5.3 – 5.7 cm, elliptic, acute, cuneate at base. Inflorescence many-flowered, rather rigid, flowers borne 5 – 6 mm apart; peduncle 22 cm long, 2 – 2.8 mm wide, tough; rachis 14 cm long, 2 – 3 mm wide, fleshy; floral bracts 3 – 5 x 2 mm, triangular-ovate, acute. Flower colour not recorded. Pedicel with ovary c.1.8 mm long, clavate, curved. Dorsal sepal 4 x 1.9 mm, oblong-elliptic to ovate-elliptic, apiculate. Lateral sepals 4 x 1.9 mm, ovate-elliptic, apiculate. Petals 3.5 – 3.6 x 1.6 – 1.7 mm, elliptic, acute. Lip 2.9 – 3 mm long, 4 mm wide across side lobes, pandurate, margins slightly erose towards base; mid-lobe with erect oblong-ovate, obtuse, slightly falcate lateral lobules and a tooth-like apiculate, mucronate central lobule; side lobes obscure, oblong, margins rather uneven to erose; disc with a basal fleshy callus and two separate fleshy crest-like keels either side which almost touch the basal ridge at their base. Column 2.1 mm long, slightly carinate dorsally, foot absent, arms (stelidia) basal, 2 mm long, ligulate, obtuse, decurved at apex, apical hood ovate, entire. Anther cap cucullate. Fig.23, D & E.

The lip shape and column structure of this rather unattractive species bear a striking resemblance to the widespread *D.gibbsiae* Rolfe. The broad leaves of *D.hosei* are quite different however, and similar to those of *D.longifolium* Rchb.f., while the rather rigid inflorescence is not unlike *D.longipes* J.J.Sm. The two keels on the disc are aligned very close to, but are not united with the transverse basal ridge. In many species, including *D.gibbsiae*, these are united to form a u- or m-shaped structure.

The specific epithet refers to the Hose Mountains in Sarawak which were named after the Rev. George F. Hose (1838 – 1922), Bishop of Singapore, Labuan and Sarawak, from 1881 until 1908.

D.imbricatum *Ames*, Orch. 6:54, Pl.82, fig.1 (1920). Type: Sabah, Mt. Kinabalu, Kiau, *Clemens* 179 (holo. AMES, iso. BO, K, SING).

HABITAT: hill and lower montane forest; kerangas vegetation.

ALTITUDINAL RANGE: 600 – 1500m.

DISTRIBUTION: Borneo (K, SA, SR). Endemic.

D.imitator *J.J.Wood* **sp.nov.** sectionis *Platyclinidis D.lacteo* Carr arcte affinis sed vaginis rhizomatis maculatis, bracteis floralibus quam pedicello-cum-ovario brevioribus, floribus minoribus, pedicello-cum-ovario longiore, sepalis petalisque brevioribus paullo angustioribus acutis vice acuminatis, sepalis lateralibus reflexis, labello breviore carina mediana multo magis manifesta, brachiis columnae bifidis distinguitur. Typus: East Malaysia, Sabah, Sipitang District, trail from Long Pa Sia to Long Samado, c.4 km S.S.W. of Long Pa Sia, 1300m, 22 October 1986, *de Vogel* 845 (holotypus L, herbarium material, isotypus K, spirit material only).

Pendulous epiphyte. Rhizome branching, up to 18 cm long, rooting at nodes. Pseudobulbs 1.2 – 3.5(–4.5) x 0.2 – 0.3 cm, narrowly cylindrical or narrowly fusiform, enclosed in several brown-spotted acute sheaths 1.5 – 3.5 cm long. Leaves thin-textured, petiole 0.3 – 0.6 cm long, sulcate, blade 4 – 10 x 0.8 – 1.4 cm, elliptic, acute, with numerous small transverse nerves. Inflorescence 6- to many-flowered, lax; peduncle 4.5 – 9 cm long, filiform, minutely furfuraceous at base; rachis 2.5 – 10.5 cm long, quadrangular, minutely furfuraceous, with 1 or 2 ovate, acute, basal sterile bracts; floral bracts 2 – 2.5 mm long, ovate, acute. Flowers borne 4 – 5 mm apart, pale green or dull pale greenish-brown, lip with a large central brownish blotch, median keel pale green, column cream. Pedicel with ovary 4 mm long, clavate, slightly curved. Dorsal sepal 4.5 – 5 x 1 – 1.2 mm, lanceolate, acute, concave. Lateral sepals 4 – 4.5 x 1.5 – 1.9 mm, obliquely ovate-elliptic, acute, reflexed. Petals 4 – 5 x 0.9 – 1 mm, ligulate, subacute, slightly falcate. Lip 3.2 – 4 mm long, 1.6 – 2.1 mm wide across mid-lobe, stipitate to column-foot by a narrowly oblong claw; mid-lobe obovate, obtuse, sometimes with a small apical mucro; side lobes triangular, acute, irregularly erose-denticulate; disc with 3 fleshy papillose keels, the outer contiguous at base and extending to at or just below base of mid-lobe, median keel broad at base, narrowing at the middle and usually becoming pronounced again at the apex, extending almost to apex of mid-lobe. Column 3.5 mm long , arcuate, foot 0.5 – 0.8 mm long, apical hood ovate, cucullate, arms (stelidia) 1.5 mm long, arising just above base, ligulate, bifid at apex. Anther cap cucullate, with a dorsal tooth. Fig.22G.

Other material examined: Sabah, Sipitang District, trail from Long Pa Sia to Long Samado, near crossing with Sungai Malabid, 1300m, December 1986, *Vermeulen & Duistermaat* 912 (K, L); Sipitang District, Maga River, confluence with Pa Sia River, 1350m, 19 October 1986, *de Vogel* 8426 (L).

D.imitator is distinguished from the closely related *D.lacteum* Carr by the spotted rhizome sheaths, floral bracts shorter than the pedicel with ovary, smaller flowers with a longer pedicel with ovary, shorter and a little narrower, acute rather than acuminate sepals and petals, reflexed lateral sepals, shorter lip with a much more pronounced median keel, and bifid arms on the column.

De Vogel 2079 (K, L, spirit material only) collected at 1300m on Gunung Pagon in Brunei is similar to *D.imitator* but has minute hairs at the base of broader sepals and petals, non reflexed lateral sepals as in *D.lacteum*, and a broader lip. The pronounced median keel on the lip and the bifid column arms are, however, typical of *D.imitator*. Further collections are necessary before its status can be clarified.

D.imitator is recorded as a trunk epiphyte at the type locality which is dry, rather low and open ridge forest on soil probably derived from sandstone and shales. This forest is interspersed with small open patches of grass and *Gleichenia* ferns. It also grows as a branch epiphyte overhanging water in dense primary riverine forest up to 30 metres high on soil derived from sandstone. *Vermeulen & Duistermaat* 912 was collected in open podsol forest.

The specific epithet is derived from the Greek *imitator*, to imitate, in reference to the close resemblance of this species to *D.lacteum*.

D.integrilabium *Carr* in Gard. Bull. Straits Settlem. 8:85 (1935). Type: Sarawak, Mt.Dulit, Ulu Koyan, *Richards* S.484 (holo. SING, iso. K).

HABITAT: podsol forest; 'sandy forest'; rocky banks in moss forest.
ALTITUDINAL RANGE: 800 – 1300m.
DISTRIBUTION: Borneo (SA, SR). Endemic.

D.intermedium *Ridl.* in J. Straits Branch Roy. Asiat. Soc. 50:135 (1908). Type: Sarawak, Mt.Matang, *Hewitt* s.n. (holo. SING).
Dendrobium ridleyi Ames in Merr., Bibl. Enum. Born. Pl.:164 (1921), nom. superfl.

HABITAT: moss forest; kerangas forest.
ALTITUDINAL RANGE: unknown.
DISTRIBUTION: Borneo (B, SR). Endemic.

NOTE: Ames (1921) mistakenly proposed the name *Dendrobium ridleyi* in the belief that '*Dendrobium intermedium* Ridley' was a later homonym of *D.intermedium* Teijsm.& Binn. (1853). The fact is that Ridley (1908) was describing *Dendrochilum intermedium* and Ames' name is consequently superfluous.

D.joclemensii *Ames*, Orch. 6:55, Pl.83 (1920). Type: Sabah, Mt. Kinabalu, Marai Parai Spur, *Clemens* 247 (holo. AMES).

HABITAT: lower montane forest, sometimes on ultramafic substrate.
ALTITUDINAL RANGE: 1800 – 2000m.
DISTRIBUTION: Borneo (SA). Endemic.

D.johannis-winkleri *J.J.Sm.* in Mitt. Inst. Allg. Bot. Hamburg 7:36, t.5, fig.26 (1927). Type: Kalimantan, Bukit Tilung, *Winkler* 1495 (holo. HBG, ? iso. BO).

HABITAT: unknown.
ALTITUDINAL RANGE: 800m.
DISTRIBUTION: Borneo (K). Endemic.

D.kamborangense *Ames*, Orch. 6:57, Pl.84 (1920). Type: Sabah, Mt. Kinabalu, Kamborangah, *Clemens* 205 (holo. AMES, iso. BO, K, SING).

HABITAT: upper montane mossy forest, often in exposed places on ultramafic substrate.
ALTITUDINAL RANGE: 1500 – 2900m.
DISTRIBUTION: Borneo (SA). Endemic.

D.kingii *(Hook.f)* *J.J.Sm.* in Recueil Trav. Bot. Néerl. 1:76 (1904). Types: Peninsular Malaysia, Perak, *Scortechini* s.n. and *King's collector* s.n. (syn. K).
Platyclinis kingii Hook.f., Fl. Brit. Ind. 5:708 (1890).
Platyclinis sarawakensis Ridl. in J. Linn. Soc. Bot. 31:267 (1896). Type: Sarawak, *Biggs* s.n., cult. Penang (holo. SING).
Dendrochilum sarawakense (Ridl.) J.J.Sm. in Recueil Trav. Bot. Néerl. 1:66 (1904).
D.palawanense Ames, Orch. 2:103 (1908), **syn. nov.** Type: Philippines, Palawan, Mt. Pulgar, 1250m, *Foxworthy* 553 (holo. AMES).
D.bicallosum J.J.Sm. in Bull. Dép. Agric. Indes Néerl. 22:17 (1909), non Ames, **syn.nov.**
D.bigibbosum J.J.Sm. in Bull. Dép. Agric. Indes Néerl. 45:13 (1911), **syn.nov.** Types: Kalimantan, Sungai Keriboeng, *Hallier* 1312 (syn. BO, isosyn. K, L); Sarawak, Mt. Bengkaum, *Brookes* s.n. (syn. BO, isosyn. K).

HABITAT: podsol forest on sandy soils; on *Eugenia* beside stream.
ALTITUDINAL RANGE: 300 – 900m.
DISTRIBUTION: Borneo (K, SA, SR). Also Peninsular Malaysia, Philippines (Palawan).

D.lacinilobum *J.J.Wood* et *A.Lamb* **sp.nov.** sectionis *Platyclinidis* habitu *D.gramineo* (Ridl.)Holttum, species malayana, atque *D.kamborangensi* Ames, borneensis, similis. Ab ambabus speciebus sepalis petalisque albis, lobis lateralibus labelli distinctis irregulariter laciniatis, carinis basalibus duabus lateruliformibus (unaquaque 0.4 mm lata) atque brachiis apicalibus columnae instructa differt. Praeterea a *D.gramineo* petalis latioribus labelloque parum latiore, et a *D.kamborangensi* floribus minoribus, petalis labelloque angustioribus distinguitur. Typus: East Malaysia, Sabah, Crocker

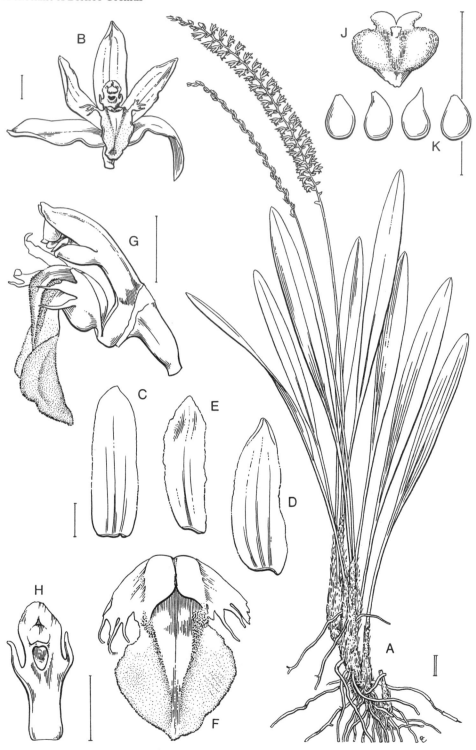

Range, Keningau and Tambunan District border, Ulu Apin Apin, 1500m, 5 November 1991, *Lamb & Surat* in *Lamb* AL 1390/91 (holotypus K, herbarium and spirit material).

Tufted epiphyte forming large clumps. Pseudobulbs 1.5–4 x 0.3–0.5 cm, narrowly fusiform, covered in pale brown, darker brown mottled fibrous sheaths. Leaves narrowly elliptic or ligulate-elliptic, gradually tapering into petiole, apex obtuse and mucronate, thin textured, petiole (2–)4–5 cm long, blade 8.5–11 (–17) x 1–1.8 cm. Inflorescences dense, with over 50 flowers in two ranks; peduncle (9–)15–16 cm long, terete, naked, pale green; rachis 10–13 cm long, 4-angled, yellowish; floral bracts 2–3 mm long, 2 mm wide when flattened, ovate-oblong, obtuse, concave, involute, prominently nerved, brown. Flowers 0.9 cm across, sweetly scented. Pedicel with ovary 1 mm long. Sepals and petals creamy-white. Dorsal sepal 4.5 x 1.5 mm, oblong, acute. Lateral sepals 4.5 x 1.5–1.6 mm, oblong-elliptic, acute. Petals 4 x 1.1–1.2 mm, oblong-elliptic, acute, minutely erose-papillose. Lip 3 mm long when flattened, 3-lobed, creamy-white or very pale green with yellowish-cinnamon or ochre keels; mid-lobe strongly recurved in natural position, 2 mm long, 1 mm wide at base, expanding into a 1.5 mm wide ovate-elliptic, rounded or subacute blade, minutely toothed at base, surface minutely papillose, with 2 low, raised keel-like and minutely papillose ridges extending from base and terminating near apex, with a furrow between; side lobes spreading, 0.8 – 1 mm long, apex irregularly laciniate; disc with 2 flange-like basal keels, each about 0.4 mm wide, which curve toward the middle and meet, tapering off and terminating at base of mid-lobe. Column 2 mm long, pink, with a foot, arms (stelidia) subacute, borne near apex, either side of stigmatic cavity, extending as high as rostellum, apical hood ovate, entire. Anther cap ovate, cucullate, 0.2 x 0.2 mm. Fig.19.

Other material examined: Sabah, top of Kimanis road, c.1220m, 14 April 1983, *Bailes & Cribb* 632 (K); Tenom District, Sapong, Gunung Anginon, 900m, April 1965, *H.F.Comber* 4032 (K); Keningau to Kimanis road, 1500 – 1600m, Dec.1986, *Vermeulen & Duistermaat* 663 (K, L); Tenom District, above Kallang Waterfall, 1000m, 11 October 1986, *de Vogel* 8172 (L) and 8173 (L).

D.lacinilobum is similar in habit to *D.gramineum* (Ridl.) Holttum from Peninsular Malaysia and *D.kamborangense* Ames from Borneo. It differs from both in having white sepals and petals, distinct irregularly laciniate lip side lobes, two flange-like basal keels each about 0.4 mm wide and apical column arms. It is further distinguished from *D.gramineum* by its wider petals and slightly broader lip, and from *D.kamborangense* by its smaller flowers with narrower petals and lip.

It grows as an epiphyte in thick moss on the trunks and branches of trees in hill and lower montane forest on sandstone and shale ridges.

The specific epithet is derived from the Latin *laciniatus*, slashed into narrow divisions with tapering pointed incisions, and *lobus*, lobe, referring to the distinctive side lobes of the lip.

D.lacteum *Carr* in Gard. Bull. Straits Settlem. 8:223 (1935). Type: Sabah, Mt. Kinabalu, above Lumu Lumu, *Carr* C.3608 (SFN 27892) (holo. SING, iso. AMES, K).

HABITAT: lower montane forest.
ALTITUDINAL RANGE: 1400 – 1800m.
DISTRIBUTION: Borneo (SA, SR). Endemic.

FIG. 19. *Dendrochilum lacinilobum.* **A**, habit; **B**, flower (front view); **C**, dorsal sepal; **D**, lateral sepal; **E**, petal; **F**, lip (front view); **G**, pedicel with ovary, column and lip (side view); **H**, column (front view); **J**, anther cap (back view); **K**, pollinia. A-K from *Lamb & Surat* in *Lamb* AL 1390/91. Scale: single bar = 1mm; double bar = 1 cm. Drawn by Eleanor Catherine.

D.lancilabium *Ames*, Orch. 6:58, Pl.83 (1920). Type: Sabah, Mt. Kinabalu, Marai Parai Spur, *Clemens* 280 (holo. AMES, iso. K, SING).

HABITAT: upper montane ridge top forest; low mossy and xerophyllous scrub forest on ultramafic substrate.
ALTITUDINAL RANGE: 1200 – 2400m.
DISTRIBUTION: Borneo (SA, SR). Endemic.

D.lewisii *J.J.Wood* in Kew Bull. 39 (1):78, fig.4 (1984). Type: Sarawak, Gunung Mulu, *Lewis* 366 (holo. K).

HABITAT: ridge top upper montane forest.
ALTITUDINAL RANGE: 2200 – 2300m.
DISTRIBUTION: Borneo (SR). Endemic.

D.linearifolium *Hook.f.*, Fl. Brit. Ind. 5:782 (1890). Types: Peninsular Malaysia, Perak, *Scortechini* s.n. (syn. K); Batang Padong, *Wray* 1512 (syn. K).

HABITAT: ridge top upper montane forest.
ALTITUDINAL RANGE: 1500 – 1600m.
DISTRIBUTION: Borneo (SA, SR). Also Peninsular Malaysia, Sumatra.

D.longifolium *Rchb.f.* in Bonplandia 4:329 (1856). Type: Origin unknown, *Consul Schiller*, cult. *Stange* (holo. W).

HABITAT: lower montane forest on limestone; roadside cuttings on sandstone and shale outcrops.
ALTITUDINAL RANGE: 600 – 1500m.
DISTRIBUTION: Borneo (SA, SR). Widespread from Peninsular Malaysia and Sumatra to New Guinea.

D.longipes *J.J.Sm.* in Bull. Jard. Bot. Buitenzorg, ser. 2, 3:55 (1912). Type: Sarawak, Batu Lawi, Ulu Limbang, *Moulton* 15 (holo. L, iso. SAR).
D.mantis J.J.Sm. in Bull. Jard. Bot. Buitenzorg, ser. 3, 11:108 (1931), **syn.nov**. Type: Kalimantan, Gunung Kemoel, *Endert* 3991 (holo. L).

HABITAT: montane ericaceous scrub; moss forest; ridge top montane forest; among boulders.
ALTITUDINAL RANGE: 1700 – 2400m.
DISTRIBUTION: Borneo (K, SR). Endemic.

D.longirachis *Ames*, Orch. 6:60 (1920). Type: Sabah, Mt. Kinabalu, Kiau, *Clemens* 332 (holo. AMES, iso. BO).

HABITAT: 'sand forest'; riverine lower montane forest with *Rhododendron* understorey; sometimes on ultramafic substrate.
ALTITUDINAL RANGE: 1200 – 1500m.
DISTRIBUTION: Borneo (SA, SR). Endemic.

D.lumakuense *J.J.Wood* **sp.nov**. sectionis *Platyclinidis* formas quasdam *Dendrochili dewindtiani* varietatis *dewindtiani* habitu accedit, sed floribus viridibus labio oblongo-pandurato lobis lateralibus deficientibus distinguitur. Typus: East Malaysia, Sabah, Sipitang District, Mt.Lumaku, 1800m, December 1963, *J.B. Comber* 108 (holotypus K, herbarium and spirit material).

Clump-forming epiphyte. Pseudobulbs 0.8 – 1.5 x 0.6 – 1 cm, ovoid, yellowish, wrinkled in dried material, enclosed in greyish-fawn black-spotted sheaths when young, caespitose. Leaves erect, petiole 0.5 – 1 cm long, sulcate, blade 3 – 5 x 0.7 –

0.8 cm, ligulate, subacute, minutely apiculate, conduplicate at base. Inflorescence erect to gently curving, densely many-flowered; peduncle 3.5 – 4 cm long, naked; rachis 8 – 18 cm long, both brown; floral bracts 3 mm long, subulate, acute. Flowers open simultaneously, green, unscented, borne 2 – 2.5 mm apart. Pedicel with ovary 3 mm long, narrow. Dorsal sepal 7 – 8 x 1 – 1.2 mm, linear-lanceolate, acute. Lateral sepals 6.5 – 7 x 1 – 1.2 mm, linear-lanceolate, acute, slightly falcate. Petals 6 – 6.8 x 1.8 – 2 mm, spathulate, acute or subacute. Lip 3 – 3.1 x 1.1 – 1.2 mm, oblong-pandurate, obtuse, with 2 papillose basal keels, curved in the basal portion. Column 2 mm long, gently curved, with a foot, hood irregularly toothed, arms (stelidia) borne between stigma and hood, subulate, acute, equalling apex of hood. Anther cap cucullate.

D.lumakuense resembles certain forms of *D.dewindtianum* var.*dewindtianum* in habit, but can be distinguished by the green flowers with an oblong-pandurate lip lacking side lobes. The flower structure is remarkably similar to many of the species in section *Dendrochilum*, but the synanthous inflorescence is typical of section *Platyclinis*.

It is, so far, only known from the type collection which was described as growing in 80% sun at the top of a 15 metre high tree. Further habitat details were not provided.

The specific epithet records the type locality, Mt. Lumaku in S.W. Sabah.

D.meijeri *J.J.Wood* **sp.nov.** sectionis *Platyclinidis D.dolichobrachio* (Schltr.) Merr. arcte affinis, sed floribus minoribus sepalis petalisque modo 2 – 2.1 mm longis, labello 1.6 mm longo lobo medio 2 mm lato, columnae brachiis quam cucullo apicali longioribus distinguitur. Typus: Indonesia, Kalimantan Timur, Kutai, Mount Beratus, near Balikpapan, Terrace Sulau Mandau, 1000m, 18 July 1952, *Meijer* 882 (holotypus K, herbarium and spirit material, isotypi A, BO, L).

Tufted epiphyte. Rhizome 5 – 6 cm long, slender, creeping, rooting profusely. Pseudobulbs 0.6 – 1.5 x 0.2 – 0.4 cm, narrowly cylindrical or fusiform, enclosed in sheaths when young. Leaves linear-ligulate, attenuated below, apex obtuse and minutely mucronate, petiole 0.5 – 1 cm long, blade 3.5 – 7.5 x 0.2 – 0.6 cm. Inflorescence many-flowered, lax; peduncle 4 – 7 cm long; rachis 3.5 – 8 cm long; floral bracts 1 – 2 mm long, ovate, acute. Flowers 'yellow-green, stalk light brown-red'. Pedicel with ovary 2 mm long, clavate. Dorsal sepal 2 – 2.1 x 0.9 mm, elliptic, acute. Lateral sepals 2 x 0.9 mm, ovate-elliptic, acute. Petals 2 – 2.1 x 0.6 mm, lanceolate, acuminate. Lip 1.6 mm long, 2 mm wide across mid-lobe, pandurate; mid-lobe transversely oblong, somewhat concave at centre, lobules obtuse, with a triangular-acute, tooth-like central lobe; side lobes keel-like, slightly falcate, obtuse; disc with two prominent raised keels extending from near base to at or just beyond junction of mid- and side lobes. Column 1.4 – 1.5 mm long, foot absent, hood ovate, obtuse, arms (stelidia) 1.6 – 1.7 mm long, ligulate, obtuse. Anther cap cucullate. Fig.23, F – H.

Other material examined: Kalimantan Timur, Kutai, Mount Beratus, near Balikpapan, Terrace Sulau Mandau, 1000m, 7 July 1952, *Meijer* 599 (BO, L); Mount Beratus, Terrace Berikan bulu, 800 – 850m, 18 July 1952, *Meijer* 843 (BO, K, L); Mount Beratus, Terrace Sulau Mandau, 1000, 18 July 1952, *Meijer* 867 (BO, L) and *Meijer* 874 (BO, L).

This tiny flowered species is closely allied to *D.dolichobrachium* (Schltr.) Merr., also first described from Kalimantan. It is distinguished by its smaller flowers with sepals and petals only 2 – 2.1 mm long, lip 1.6 mm long and 2 mm wide across the mid-lobe, and column arms longer than the apical hood. It grows in mossy forest where it has been recorded as an epiphyte on a variety of trees including *Dacrydium*

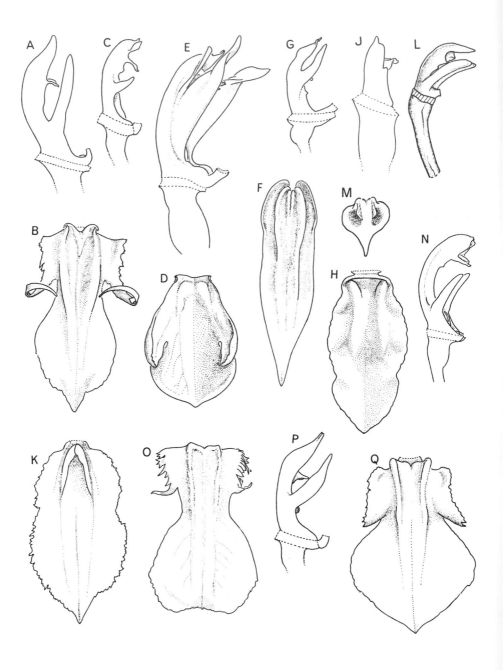

and *Eugenia* as well as members of the *Euphorbiaceae, Lauraceae* and *Meliaceae* families.

The specific epithet honours W.Meijer who collected the type.

D.micranthum *Schltr.* in Bull. Herb. Boissier ser. 2, 6:303 (1906). Type: Kalimantan, Long Gombeng, *Schlechter* 13561 (holo. B, destroyed).

HABITAT: forest on limestone.
ALTITUDINAL RANGE: unknown.
DISTRIBUTION: Borneo (K). Endemic.

D.microscopicum *J.J.Sm.* in Bull. Jard. Bot. Buitenzorg, ser. 3, 11:110 (1931). Type: Kalimantan, Gunong Kemoel, *Endert* 4103 (holo. BO, iso. K, L).

HABITAT: lower montane ridge forest.
ALTITUDINAL RANGE: 1500m.
DISTRIBUTION: Borneo (K). Endemic.

D.minimiflorum *Carr* in Gard. Bull. Straits Settlem. 8:90 (1935). Type: Sarawak, Dulit ridge, *Richards* S.476 (holo. SING, iso. K, L).

HABITAT: moss forest; on small trees in 'sand forest'.
ALTITUDINAL RANGE: 1000–1300m.
DISTRIBUTION: Borneo (SR). Endemic.

D.mucronatum *J.J.Sm.* in Brittonia 1:106 (1931). Type: Sarawak, Mt. Poi, *Clemens* 22587 (holo. L, iso. BO, K).

HABITAT: moss forest.
ALTITUDINAL RANGE: 1200–1800m.
DISTRIBUTION: Borneo (SA, SR). Endemic.

D.muluense *J.J.Wood* in Kew Bull. 39(1): 80, fig.5 (1984). Type: Sarawak, Gunung Mulu National Park,*Nielsen* 143 (holo. AAU, iso. K).

HABITAT: ridge top upper montane forest; oak-laurel forest; moss forest.
ALTITUDINAL RANGE: 1700–2200m.
DISTRIBUTION: Borneo (B, SA, SR). Endemic.

D.ochrolabium *J.J.Wood* **sp. nov.** sectionis *Platyclinidis D.dulitensi* Carr affinis sed foliis latioribus anguste ellipticis minute praesertim in pagina inferiore nigro-punctatis vel-ramentaceis, floribus pallide viridis labello silaceo pariete exteriore pedicelli-cum-ovario pellucido sepalis petalisque paullo minoribus papillis basalibus destitutis labello manifeste minore et angustiore sine nervo mediano elevato stelidiis columnae ligulatis acutis differt. Typus: East Malaysia, Sabah, Sipitang District, ridge east of Maga River, c.1.5 km south of confluence with Pa Sia River, 1450 m, 17 October 1986, *de Vogel* 8351 (holotypus L, herbarium material only, isotypus K, herbarium and spirit material).

Tufted epiphytic herb. Pseudobulbs 0.5–1.2 x 0.2–0.3 cm, ovoid to elliptic, or oblong-cylindrical. Leaves very minutely black punctate-ramentaceous, particularly on lower surface, petiole 0.1–0.3 cm long, blade 1.2–2.6(–3) x 0.3–0.4 cm, narrowly

FIG. 20. *Dendrochilum* columns and lips. **A** & **B**, *D. longirachis*, x 7; **C** & **D**, *D. muluense*, x 10; **E** & **F**, *D. planiscapum*, x 7 & x 9; **G** & **H**, *D. simplex*, x 7 & x 15; **J** & **K**, *D. stachyodes*, x 13 & x 9; **L** & **M**, *D. sublobatum*, x 15; **N** & **O**, *D. subulibrachium*, x 13; **P** & **Q**, *D. tenompokense*, x 7. A–K & N–Q drawn by Mutsuko Nakajima; **L** & **M** drawn by Jeffrey Wood.

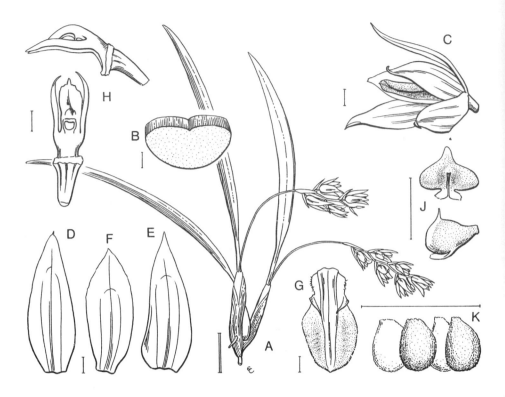

FIG. 21. *Dendrochilum pachyphyllum.* **A**, habit; **B**, transverse section through leaf; **C**, flower (side view); **D**, dorsal sepal; **E**, lateral sepal; **F**, petal; **G**, lip (front view); **H**, pedicel with ovary and column (front and side views); **J**, anther-cap (back and side views); **K**, pollinia. **A-K** from *Lamb* AL 674/86. Scale: single bar = 1 mm; double bar = 1 cm. Drawn by Eleanor Catherine.

elliptic, obtuse and usually minutely mucronate, margin minutely papillose, appearing raised and thickened in dried material. Inflorescence 5–8 flowered, flowers borne 2–3 mm apart; peduncle 1.2–2 cm long, filiform; rachis 1.5–2.5 cm long, quadrangular, sparsely ramentaceous; floral bracts 2–2.5 x 1.1–1.2 mm, oblong-elliptic, acute, concave, clasping pedicel with ovary, sparsely ramentaceous. Flowers with 'pale green' or 'very pale greenish' sepals and petals, lip 'ochre, middle part brownish, back green with two longitudinal brown stripes' or 'orange with pale green lateral lobes', column 'pale green', slightly scented. Pedicel with ovary 1.9–2 mm long, outer wall fleshy and translucent. Dorsal sepal 3.9–4 x 1.1 mm, narrowly ovate-elliptic, acute. Lateral sepals 3.9–4 x 1.2–1.3 mm, ovate-elliptic, acute. Petals 3 x 1.1 mm, ovate-elliptic, acute. Lip 2.8 mm long, 1.2–1.3 mm wide across side lobes, 1.4–1.5 mm wide across mid-lobe, stipitate to column-foot by a tiny claw, minutely papillose; mid-lobe obovate, obtuse, convex, minutely papillose ; side lobes auriculate, irregularly toothed; disc with two somewhat papillose raised keels joined at the base to a narrow transverse flange, each terminating a little above base of mid-lobe. Column c. 1.8 mm long, with a foot, curved, arms (stelidia) c. 0.8 mm long, ligulate, acute, shorter than apical hood, arising at middle, apical hood ovate, very obscurely 3-lobed. Anther cap minute, cucullate, extinctoriform. Fig.24, J & K.

188

Other material examined: Sabah, Sipitang District, Long Pa Sia to Long Samado trail, near crossing with Sungai Malabid, c. 1300 m, December 1986, *Vermeulen & Duistermaat* 967 (L).

D.ochrolabium is related to *D.dulitense* Carr from Sarawak but differs in having slightly broader, narrowly elliptic, minutely black punctate-ramentaceous leaves, pale green flowers with an ochre lip, pedicel with ovary with a translucent outer wall, slightly smaller sepals and petals lacking basal papillae, a distinctly smaller, narrower lip without an elevated median nerve, and ligulate, acute stelidia.

The type locality is in open, low, dry stunted forest 5 to 10 m high on a sandstone ridge, with a dense undergrowth of terrestrial orchids and other herbs. It is also recorded from very low and open podsol forest.

The specific epithet is derived from the Latin *ochraceus*, ochre-yellow or yellowish-brown, and *labium*, lip, referring to the colour of the lip.

D.oxylobum *Schltr.* in Feddes Repert. 9:431 (1911). Type: Sarawak, Kuching, *Beccari* 1125 (holo. FI).
    D.viridifuscum J.J.Sm. in Bull. Jard. Bot. Buitenzorg, ser. 2:11 (1917). Type: Kalimantan, Kota Waringin, *van Nouhuys* s.n., cult. Bogor no. 16 (holo. BO).

HABITAT: podsol forest; lower montane forest on sandstone.
ALTITUDINAL RANGE: 400 – 900m.
DISTRIBUTION: Borneo (SA,SR). Endemic.

D.pachyphyllum *J.J.Wood et A.Lamb* **sp.nov.** ab omnibus speciebus ceteris foliis brevibus acutis crassis carnosisque, pedunculis pendentibus filiformibus, floribus brunneo-salmoneus distinguitur. Typus: East Malaysia, Sabah, Crocker Range, Gunung Alab, Sinsuron Road, 1650m, 7 November 1986, *Lamb* AL 674/86 (holotypus K, spirit material only).

Tufted clump-forming epiphytic or lithophytic herb. Pseudobulbs 0.7–1.2 x 0.3–0.5 to 1.8 x 0.2–0.3 cm, oblong-elliptic to narrowly fusiform, clothed in acute, brown sheaths when young. Leaves narrowly linear-elliptic, acute, thick and fleshy, sulcate above, curved, petiole 0.6 – 1.3 cm long, narrow, sulcate, blade 2 – 7 x 0.3 – 0.6 cm. Inflorescences 3–5- flowered, pendulous, shorter than leaves; peduncle 1 – 7 cm long, filiform, yellowish-green; rachis 0.6 – 1.8 cm long, thicker than peduncle, fractiflex, salmon-pink; floral bracts 4–7 mm long, ovate-acuminate, brown. Flowers c. 1 cm across. Pedicel with ovary 1–2 mm long, brownish-pink. Sepals acute, somewhat carinate, translucent pink to salmon-pink or pale brownish, mid vein darker red to brownish-pink. Dorsal sepal 7 – 9 x 2 – 3 mm, ovate-elliptic, curved. Lateral sepals 7 – 9 x 2.5 – 3 mm, ovate-elliptic. Petals 6 – 8 x 2 – 2.8 mm, elliptic, acute, minutely erose towards apex, translucent pink to salmon-pink. Lip 5.5–7 mm long, 2 mm wide at base, 3–3.5 mm wide across mid-lobe, 3-lobed, spathulate, 3-nerved, green at base; mid-lobe 4.5–5 mm long, oblong-elliptic, obtuse, rather fleshy at centre, shallowly concave, margin thin, erose, white or dull yellow, central area red to brownish-red; side lobes minutely toothed, free portion 1–1.5 mm long, linear-triangular, acute, white; disc with a raised keel 1 mm wide extending from base of lip to upper portion of mid-lobe, its base retuse, its margins raised. Column 4 mm long, with a foot, pink, its apex irregularly 3–4-toothed, the central tooth longer than the outer, arms (stelidia) 4 x 0.5–0.6 mm, linear-ligulate, subacute, borne from near column base, exceeding apex. Anther cap ovate-cordate, minutely papillose, white. Fig.21.

Other material examined: Sabah, Sinsuron Road, 1470 m, September 1979, *Collenette* 49/79 (E); Kota Kinabalu to Tambunan road, km 56, 1700 m, July 1986, *Vermeulen & Chan* 413 (L); Sipitang District, Long Pa Sia to Long Samado trail, near crossing with Sungai Malabid, 1300 m, December 1986, *Vermeulen & Duistermaat* 965 (L); Gunung Alab, south ridge, 2000 m, 31 October 1986, *de Vogel* 8646 (L). Sarawak, Fourth Division, route from Ulu Sungai Limbang to Bukit Buli, 1540 m, *Awa & Lee* S.50774 (K,SAR, material at K is a mixed collection with *D.crassilabium*).

D.pachyphyllum belongs to section *Platyclinis* and is distinguished from all other species by its short, acute, thick and fleshy leaves, pendulous filiform peduncles and brownish salmon-pink flowers. It grows as an epiphyte in mixed hill dipterocarp forest, in lower montane mossy forest, on mossy sandstone rocks and shale banks along roadside cuttings and in very low and open podsol forest.

The specific epithet is derived from the Greek *pachy*, thick or stout, and *phyllum*, leaf, referring to the fleshy leaves.

D.pallideflavens *Blume*, Bijdr.:399, fig.52 (1825). Type: Java, Pantjar, *Blume* s.n. (holo. BO).

> HABITAT: podsol forest.
> ALTITUDINAL RANGE: 400m.
> DISTRIBUTION: Borneo (B, SA). Also Burma, Peninsular Malaysia, Thailand, Java.

Dendrochilum pandurichilum *J.J.Wood* **sp.nov.** sectionis *Platyclinidis*, aliquantum intra hanc sectionem sejuncta, scilicet ab omnibus speciebus aliis statura pumila, labio valde pandurato carinis duobus prominentibus oblongis rotundatis aliformibus instructo atque columna arcuata brachiis lateralibus carentibus distingui potest. Typus: East Malaysia, Sarawak, Sipitang District, Mt.Lumaku, 1500 m, December 1963, *J.B. Comber* 102 (holotypus K, herbarium and spirit material).

Clump-forming epiphyte. Pseudobulbs 5 – 7 x 2 – 3 mm, ovoid-elliptic, yellowish, enclosed in pale brown sheaths when young. Leaves 0.8 – 3.5 x 0.2 – 0.4 cm, petiole 1 – 3 mm long, blade narrowly ligulate, obtuse and mucronate. Inflorescence laxly 3 to 8 – flowered, opening from top down; peduncle 1 – 3.5 cm long, filiform, naked; rachis 0.6 – 1.8 cm long; floral bracts 2 – 3 mm long, subulate, acute. Flowers pale orange. Pedicel with ovary 1.8 mm long, clavate. Dorsal sepal 3 x 1 – 1.1 mm, ovate-elliptic, subacute, concave, cucullate, median nerve prominent and dorsally carinate. Lateral sepals 2.9 x 1.5 – 1.6 mm, triangular-ovate, subacute, somewhat concave. Petals 2.9 – 3 x c.0.8 – 0.9 mm, linear, acute, slightly falcate. Lip 2.1 – 2.8 mm long, 2.4 – 2.5 mm wide across keels, pandurate, side lobes absent, mid-lobe trilobulate, 2 – 2.1 mm wide when flattened, outer lobules obtuse, median lobule tooth-like, subacute; disc with prominent oblong, rounded wing-like keels each 0.8 x 1 mm, looking, at first sight, like side lobes. Column 3 mm long (when straightened), narrow, arcuate, hood entire, with a foot, arms (stelidia) absent. Anther cap cucullate. Fig.23, A – C.

Other material examined: Brunei, Belait District, Badas Forest Reserve, 10 m, 17 January 1992, *de Vogel* s.n., cult Leiden no. 911260A (K, L). Sabah, Sipitang District, N.W. of Long Pa Sia to Long Semado trail, on Sarawak border, 1900 m, 24 October 1986, *de Vogel* 8569 (L). Sarawak, Fourth Division, Sungai Pa Mario, Ulu Limbang, route to Mt.Batu Lawi, 1530 m, October 1987, *Awa & Lee* S.50731 (K, L, SAR).

This species is rather isolated within the section and may be distinguished from all others by its dwarf stature, strongly pandurate lip with prominent wing-like keels and arcuate column lacking the characteristic arms.

It is recorded between 1500 and 1900 m in Sabah and Sarawak. The habitat of the type locality on Mount Lumaku in S.W. Sabah is lower montane forest, while in Sarawak it is recorded from riparian forest. It is curious, therefore, that it has also been collected in lowland kerangas forest at only 10 metres above sea level in

Brunei. Here the forest grows on pure white sand, is about 30 metres in height, contains numerous large *Agathis* and has an undergrowth of slender pole trees. De Vogel 8569, although sterile, is almost certainly this species. The data label records the floral buds, since lost, as pink. It was collected in primary ridge forest up to 20 metres high on sandstone with an understorey of climbing bamboos, rattan palms and an abundance of small trees.

The specific epithet is derived from the Latin *panduratus*, fiddle-shaped, and the Greek *chilus*, lipped, referring to the distinctive and elegant lip shape.

D.papillilabium *J.J.Wood* **sp. nov.** sectionis *Platyclinidis* formas latifolias *D.linearifolii* Hook.f. (*Platyclinis pulchellae* Ridl.), species Peninsulae Malaysianae simulans, sed floribus minoribus, labello obscure lobato papilloso-piloso, brachiis columnae in medio columnae sitis cucullo aequantibus vel eo subaequantibus distinguitur. Typus: East Malaysia, Sabah, Crocker Range, Keningau to Kimanis road, montane ridge forest on sandstone, 1500–1600 m, December 1986, *Vermeulen & Duistermaat 666* (holotypus L, herbarium material only).

Tufted epiphyte producing many long, branching roots. Pseudobulbs 0.5–1 x 0.4–0.5 cm, ovoid to elliptic, enclosed by several acute to acuminate greyish brown sheaths up to 2 cm long when young. Leaves coriaceous, petiole 0.2–1 cm long, sulcate, blade 1.5–5.5 x 0.6–1 cm, oblong-elliptic, obtuse and mucronate to acute. Inflorescence many flowered, upper flowers opening first, each borne 2–5 mm apart; peduncle 1.7–4 cm long; rachis 5–7 cm long; floral bracts 3–4(–5) mm long, ovate-elliptic, acute. Flowers 'light green', 'pale greenish, lip brighter green'. Pedicel with ovary 2.2 mm long, clavate. Sepals and petals with a few brown ramentaceous scales at the base. Dorsal sepal 6 x 2 mm, ovate-elliptic, acute, concave, slightly carinate. Lateral sepals 5.5 x 2–2.1 mm, slightly obliquely ovate-elliptic, acute, slightly carinate. Petals 5 x 2.2 mm, oblong to oblong-elliptic, acute, slightly concave. Lip 3.5–3.8 mm long, 1.5 mm wide across side lobes, 1.1–1.2 mm wide across mid-lobe, decurved; mid-lobe elliptic, obtuse, minutely papillose; side lobes shallowly rounded, margin irregularly toothed, upper surface papillose-hairy except along margins; disc with two papillose-hairy fleshy ridges joined near base of lip and terminating on lower part of mid-lobe. Column 3.5 mm long, foot 0.6–0.7 mm long, curved, arms (stelidia) 2 mm long, falcate, acute, almost equal or subequal to apical hood, arising at the middle, apical hood irregularly toothed, rostellum prominent, lower margin of stigmatic cavity developed into a distinct bilobed flange. Anther cap 0.9 x 0.9 mm, cucullate, with a triangular-acute 'tail'. Fig.24, L & M.

Other material examined: Sabah, Beluran Districat, Bukit Monkobo, mossy forest, 1500 m, 15 March 1982, *Aban* SAN 95230 (K, L, SAN).

D.papillilabium is distinguished from all other Bornean species by the character combination of an inflorescence which opens from the top down, papillose-hairy lip and bilobed flange-like lower margin of the stigmatic cavity. These characters also distinguish it from broad-leaved forms of *D.linearifolium* Hook.f. from Peninsular Malaysia, named *Platyclinis pulchella* Ridl., which it superficially resembles.

The specific epithet is derived from the Latin *papillatus*, having papillae, and *labium*, lip, referring to the papillose lip.

D.planiscapum *Carr* in Gard. Bull. Straits Settlem. 8:228 (1935). Type: Sabah, Mt. Kinabalu, Tenompok, *Carr* 3663 (SFN 28020) (holo. SING, iso. AMES, K).

HABITAT: lower and upper montane forest; moss forest; epiphytic on *Castanopsis*.
ALTITUDINAL RANGE: 1300 – 2400m.
DISTRIBUTION: Borneo (SA). Endemic.

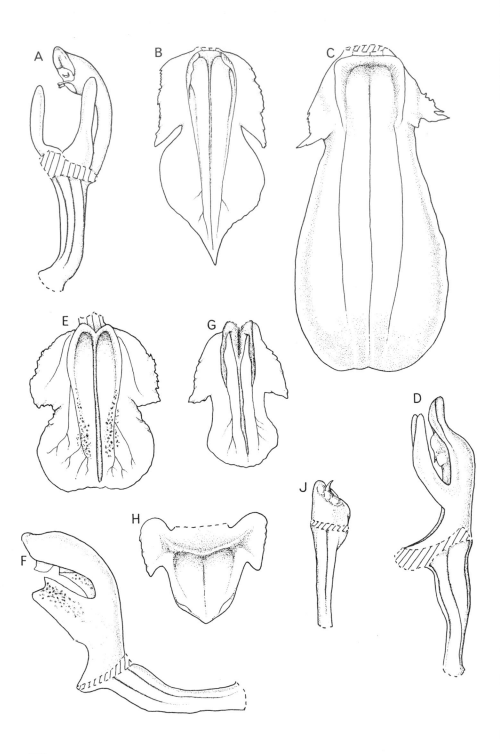

D.pterogyne *Carr* in Gard. Bull. Straits Settlem. 8:236 (1935). Type: Sabah, Mt. Kinabalu, Pakka-Pakka, *Carr* 3541 (SFN 27597) (holo. SING, iso. AMES, K).

HABITAT: upper montane forest; *Leptospermum* scrub; streamside forest; sometimes on ultramafic substrate.
ALTITUDINAL RANGE: (900 –)2600 – 3800m.
DISTRIBUTION: Borneo (SA). Endemic.

D.pubescens *L.O.Williams* in Bot. Mus. Leafl. 6:58 (1938). Type: Sarawak, Gunung Temabok, Upper Barami Valley, *Moulton* 6763 (holo. AMES, iso. SING).

HABITAT: unknown.
ALTITUDINAL RANGE: 900m.
DISTRIBUTION: Borneo (B, SR). Endemic.

D.remotum *J.J.Sm.* in Bull. Jard. Bot. Buitenzorg, ser. 2, 3:57 (1912). Type: Sarawak, Batu Lawi, Ulu Limbang, *Moulton* 10 (holo. BO, iso. SAR).

HABITAT: unknown.
ALTITUDINAL RANGE: unknown.
DISTRIBUTION: Borneo (SR). Endemic.

D.scriptum *Carr* in Gard. Bull. Straits Settlem. 8:234 (1935). Type: Sabah, Mt. Kinabalu, above Kamborangah, *Carr* 3597 (holo. SING).

HABITAT: upper montane forest on ultramafic substrate.
ALTITUDINAL RANGE: 2600m.
DISTRIBUTION: Borneo (SA). Endemic.

D.simile *Blume*, Bijdr.: 400 (1825). Type: Java, Buitenzorg, & Tjanjor provinces, *Blume* s.n. (holo. BO).

HABITAT: unknown.
ALTITUDINAL RANGE: unknown.
DISTRIBUTION: Borneo (SR). Also Peninsular Malaysia, Sumatra, Java.

D.simplex *J.J.Sm.* in Bull. Dép. Agric. Indes Néerl. 22:13 (1909). Type: Kalimantan, Liang Gagang, *Hallier* 2646 (holo. BO, iso. K,L).

HABITAT: podsolic dipterocarp - *Dacrydium* forest on very wet sandy soil; lower montane forest, mostly on ultramafic substrate.
ALTITUDINAL RANGE: 400 – 1500m.
DISTRIBUTION: Borneo (K, SA). Endemic.

D.stachyodes *(Ridl.) J.J.Sm.* in Recueil Trav. Bot. Néerl. 1:77 (1904). Type: Sabah, Mt. Kinabalu, *Haviland* 1097 (holo. BM, iso. K, SING).
Platyclinis stachyodes Ridl. in Trans. Linn. Soc. London, Bot.: 4:234 (1894).

HABITAT: in granite rock crevices; among rocks in *Leptospermum* scrub; in the open or among scrub; also recorded from ultramafic substrate.
ALTITUDINAL RANGE: 2400 – 3700m.
DISTRIBUTION: Borneo (SA). Endemic.

FIG. 22. Lips and columns of newly described *Dendrochilum.* **A & B,** *D. crassifolium* var. *murudense,* x 9; **C & D,** *D. crassilabium,* x 12; **E & F,** *D. sp. aff. imitator,* x 12; **G,** *D. imitator,* x 12; **H & J,** *D. auriculilobum,* x 12. Drawn by Jeffrey Wood.

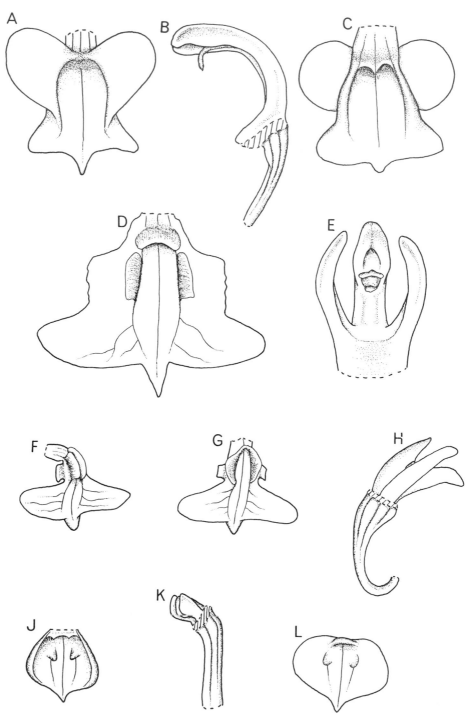

D.subintegrum *Ames* in Orch. 6:65 (1920). Type: Sabah, Mt. Kinabalu, Lobong cave, *Clemens* 285 (holo. AMES).

HABITAT: low open podsol forest, often with *Pandanus* and rattan understorey; lower montane forest.
ALTITUDINAL RANGE: 1200 – 1500m.
DISTRIBUTION: Borneo (SA, SR). Endemic.

D.sublobatum *Carr* in Gard. Bull. Straits Settlem. 8:86 (1935). Type: Sarawak, Dulit ridge, *Synge* S.406 (holo. SING, iso. K).

HABITAT: on small trees in moss forest.
ALTITUDINAL RANGE: 1200 – 1300m.
DISTRIBUTION: Borneo (SR). Endemic.

D.subulibrachium *J.J.Sm.* in Bull. Jard. Bot. Buitenzorg, ser. 3, 11:109 (1931). Type: Kalimantan, Long Petak, *Endert* 3221 (holo. L, iso. BO).

HABITAT: primary hill forest; lower montane ridge forest.
ALTITUDINAL RANGE: 800 – 1730m.
DISTRIBUTION: Borneo (K,SR). Endemic.

D.suratii *J.J.Wood* in Lindleyana 7 (2):77, fig.3 (1991). Type: Sabah, Tambunan District, Mt. Trus Madi, *Surat* in *Wood* 905 (holo. K).

HABITAT: upper montane ericaceous forest, growing on exposed branches and twigs.
ALTITUDINAL RANGE: 2400 – 2500m.
DISTRIBUTION: Borneo (SA). Endemic.

D.tardum *J.J.Sm.* in Bull. Dép. Agric. Indes Néerl. 15:6 (1908). Type: Borneo, locality and collector unknown (holo. BO).

HABITAT: unknown.
ALTITUDINAL RANGE: unknown.
DISTRIBUTION: Borneo (unspecified). Endemic.

D.tenompokense *Carr* in Gard. Bull. Straits Settlem. 8:225 (1935). Type: Sabah, Mt. Kinabalu, Tenompok, *Carr* 3623 (SFN 27891) (holo. SING, iso. AMES, K).

HABITAT: lower montane forest; moss forest; oak-laurel forest.
ALTITUDINAL RANGE: 1500 – 1800m.
DISTRIBUTION: Borneo (SA, SR). Endemic.

D.tenuitepalum *J.J.Wood* **sp.nov.** sectionis *Platyclinidis* a *D.mucronato* J.J.Sm. arcte cognato sepalis petalisque longioribus parum angustoribusque, petalis necnon non erosis atque labello integro anguste ovato-elliptico minute papilloso distinguitur. Typus: East Malaysia, Sabah, Sipitang District, ridge between Sungai Maga headwaters and Sungai Malabid headwaters, 1600m, December 1986, *Vermeulen* & *Duistermaat* 1008 (holotypus L, isotypus K).

FIG. 23. Lips and columns of newly described *Dendrochilum*. **A**, *D. pandurichilum*, lip (front view), x 12; **B**, *D. pandurichilum*, column (side view), x 12; **C**, *D. pandurichilum*, lip (back view), x 12; **D** & **E**, *D. hosei*, x 15 & x 20; **F-H**, *D. meijeri*, x 15; **J**, *D. cupulatum*, lip, natural position, x 15; **K**, *D. cupulatum*, column (side view), x 50; **L**, *D. cupulatum*, lip, flattened, x 15. Drawn by Jeffrey Wood.

Creeping, clump-forming epiphyte. Rhizome tough, producing numerous wiry roots. Pseudobulbs 1 – 1.8 x 0.4 – 0.6 cm, ovate-elliptic, gently curved, borne 0.5 cm apart on rhizome, enclosed in sheaths when young. Leaves linear-ligulate, obtuse and mucronate or apiculate, conduplicate at base, petiole 2.5 – 5 cm long, sulcate, blade 10.5 – 21.5 x 0.4 – 0.6 cm. Inflorescence many-flowered; peduncle 6 – 9 cm long, lower half enclosed in 2 – 3 acute to acuminate sheaths 1 – 4 cm long; rachis 14 – 18 cm long, gently curving; floral bracts 2 – 5 mm long, ovate-elliptic, acuminate. Flowers borne 3 – 4 mm apart, very pale greenish, unscented. Pedicel with ovary 2.5 – 2.6 mm long, narrowly clavate. Sepals and petals narrowly lanceolate, acuminate. Dorsal sepal 6 – 6.5 mm long, 0.8 – 0.9 mm wide at base. Lateral sepals 7 – 7.1 mm long, 1 – 1.1 mm wide at base. Petals 7 – 7.1 mm long, 1 mm wide at base. Lip 2.8 mm long, 1 mm wide near base, narrowly ovate-elliptic, acute, entire, minutely papillose, margin irregular in places, erect at base, disc with two low papillose keels extending from base and terminating at or just beyond middle of lip, not linked by a transverse basal ridge. Column 2 mm long, with a foot, hood minutely serrulate, arms (stelidia) c. 1.8 mm long, a little longer than hood, ligulate, obtuse to subacute, minutely serrulate at apex, inserted a little above base. Anther cap cucullate, extinctoriform. Fig.24, N & O.

D.tenuitepalum is distinguished from the closely related *D.mucronatum* J.J.Sm. from Sarawak by its longer and slightly narrower sepals and petals, non-erose petals and entire, narrowly ovate-elliptic, minutely papillose lip. It was collected in open, low mossy forest and is known only from the type.

The specific epithet is derived from the Latin *tenui*, slender, thin and *tepalum*, a division of the perianth, either sepal or petal, referring to the long, narrow sepals and petals.

D.transversum *Carr* in Gard. Bull. Straits Settlem. 8:233 (1935). Type: Sabah, Mt. Kinabalu, Marai Parai, *Carr* 3477 (SFN 27431) (holo. SING, iso. K).

HABITAT: upper montane forest on ultramafic substrate.
ALTITUDINAL RANGE: 2100 – 3000m.
DISTRIBUTION: Borneo (SA). Endemic.

D.trilobum *(Ridl.) Ames* in Merr., Bibl. Enum. Born. Pl.:149 (1921). Type: Sarawak, Mt.Penrissen, *Moulton* s.n. (holo. not located).
Platyclinis trilobus Ridl. in Sarawak Mus. J. 1:35 (1912).

HABITAT: unknown.
ALTITUDINAL RANGE: 1100 – 1200m.
DISTRIBUTION: Borneo (SR). Endemic.

D.trusmadiense *J.J. Wood* in Lindleyana 5 (2):93, fig.8 (1990). Type: Sabah, Tambunan District, Mt. Trus Madi, *Wood* 886 (holo. K, iso. L, UKMS).

HABITAT: upper montane mossy ericaceous forest.
ALTITUDINAL RANGE: 1900 – 2000m.
DISTRIBUTION: Borneo (SA). Endemic.

ENTOMOPHOBIA de Vogel

Blumea 30:199 (1984);

de Vogel, E.F. (1986). Revisions in Coelogyninae (Orchidaceae) II. The genera *Bracisepalum, Chelonistele, Entomophobia, Geesinkorchis* and *Nabaluia.* Orch. Mon. 1:23 – 40.

Epiphytic or lithophytic herbs. Rhizome short, creeping. Pseudobulbs 2 – 5cm long, all turned to one side of rhizome, 2-leaved. Leaves 16 – 65 x 1 – 1.9cm, linear, acute, petiolate. Inflorescence proteranthous to synanthous, racemose, densely 15 – 42-flowered, erect; floral bracts persistent, ovate, acuminate to truncate, folded along midrib. Flowers resupinate, all turned to one side, not opening widely, glabrous; sepals 7 – 10mm long, ovate-oblong, concave; petals spathulate; lip 5 – 8mm long, with lateral margins adnate to basal half of column, base deeply saccate, separated from front part by a transverse, high, slightly bent, fleshy callus which more or less fits into the stigmatic cavity, apical margins undulate; column 4 – to 5mm long, laterally flattened, foot absent, apex 3-lobed; pollinia 4.

A monotypic genus endemic to Borneo.

E.kinabaluensis *(Ames) de Vogel* in Blumea 3:199 (1984). Type: Sabah, Mt. Kinabalu, Marai Parai Spur, *Clemens* 279 (holo. AMES).
Pholidota kinabaluensis Ames, Orch. 6:68 (1920).

HABITAT: very low and open podsol forest; low scrub on limestone; lower montane ridge forest on ultramafic substrate.
ALTITUDINAL RANGE: 900 – 2100m.
DISTRIBUTION: Borneo (B, K, SA, SR). Endemic

GEESINKORCHIS de Vogel

Blumea 30:199 (1984);

de Vogel, E.F. (1986). Revisions in Coelogyninae (Orchidaceae) II. The genera *Bracisepalum, Chelonistele, Entomophobia, Geesinkorchis* and *Nabaluia.* Orch. Mon. 1:23 – 40.

Epiphytic or terrestrial herbs. Rhizome creeping, short. Pseudobulbs close together, flattened, 2-leaved. Leaves linear-lanceolate, acute, petiolate. Inflorescences erect, rigid, many-flowered, at first proteranthous, continuing to produce flowers successively long after pseudobulbs and leaves are mature; peduncle flattened; rachis fractiflex; floral bracts imbricate, distichous, conduplicate. Flowers resupinate; sepals and petals free; sepals deeply concave; petals linear, curved or rolled backwards; lip pandurate when flattened, hypochile somewhat saccate, convex in front, with erect or decurved side lobes, with 2 wing-like keels, with or without a central callus, epichile 2-lobed, broadly spathulate when flattened, usually hairy or finely laciniate at base. Column short, deeply concave. Pollinia 4.

Four species endemic to Borneo, two of which remain undescribed.

G.alaticallosa *de Vogel* in Blumea 30:201, Pl.1 d-i (1984). Type: Kalimantan, *Geesink* 8965 (holo. L, iso. BO).

HABITAT: podsol forest; summit ridge scrub vegetation.
ALTITUDINAL RANGE: 400 – 900m.
DISTRIBUTION: Borneo (K, SA, SR). Endemic.

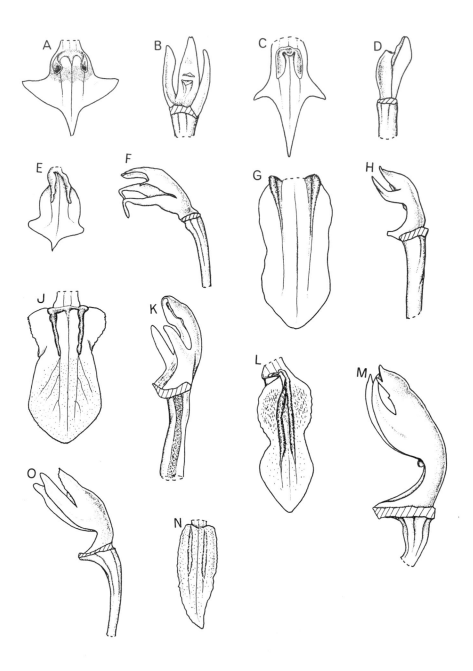

G.phaiostele *(Ridl.) de Vogel* in Blumea 30:201, Pl.1c (1984). Types: Sarawak, Mt. Poa (?Poi), *Lewis* s.n. (lecto. K); Sarawak, Mt. Santubong, *Haviland* s.n. (syn. K).

Coelogyne phaiostele Ridl. in J. Straits Branch Roy. Asiat. Soc. 54:51 (1910).

C.ridleyana Schltr. in Feddes Repert. 8:561 (1910). Types: Sarawak, Mt. Santubong, *Brooks* s.n. (lecto. BM, isolecto. K); *Hewitt* s.n. (syn. ?).

Pholidota triloba J.J.Sm. in Brittonia 1:105 (1931). Type: Sarawak, Mt. Poi, *Clemens* 20398 (holo. L, iso. NY).

HABITAT: hill and mountain tops and ridges.
ALTITUDINAL RANGE: 800 – 2000m.
DISTRIBUTION: Borneo (K, SA, SR). Endemic.

NABALUIA Ames

Orch.6:70, Pl.87 (1920).

de Vogel, E.F. (1986). Revisions in Coelogyninae (Orchidaceae) II. The genera *Bracisepalum, Chelonistele, Entomophobia, Geesinkorchis* and *Nabaluia*. Orch. Mon. 1:23 – 40.

Epiphytic or lithophytic herbs. Rhizome short, creeping. Pseudobulbs all turned to one side of rhizome, more or less flattened, smooth, 2-leaved. Leaves ovate-oblong, to linear, petiole sulcate. Inflorescences synanthous, racemose, erect, rigid; rachis fractiflex or more or less straight; floral bracts distichous, caducous, folded along midrib. Flowers resupinate, alternating in 2 rows, many open together; sepals ovate to oblong, margins more or less recurved; petals linear, usually rolled back; lip divided into a short, saccate hypochile and a long, more or less flat, spathulate, retuse epichile, hypochile 2- or 4-lobed, 2- to 5-keeled, disc with a horseshoe-shaped callus between hypochile and epichile; column with a 3-lobed hood, foot absent; pollinia 4, rather flattened.

Three species endemic to Borneo.

N.angustifolia *de Vogel* in Blumea 30:202, Pl.2 d & e (1984). Type: Sabah, Mt. Kinabalu, *Clemens* 27159 (holo. AMES, iso. BM, BO, K, L).

HABITAT: lower and upper montane forest; oak-laurel forest; sometimes on ultramafic substrate.
ALTITUDINAL RANGE: 1400 – 2700m.
DISTRIBUTION: Borneo (SA). Endemic.

N.clemensii *Ames*, Orch. 6:71, Pl.87 (1920). Type: Sabah, Mt. Kinabalu, Kamborangah, *Clemens* 210 (holo. AMES, iso. BM, BO, K, SING).

Chelonistele keithiana W.W.Sm. in Notes Roy. Bot. Gard. Edinburgh 13:188 (1921). Type: Sabah, Mt. Kinabalu, *Moulton* 103 (holo. E, iso. K).

C.clemensii (Ames) Carr in Gard. Bull. Straits Settlem. 8:220 (1935).

HABITAT: oak-laurel forest; lower and upper montane forest, sometimes on ultramafic substrate.
ALTITUDINAL RANGE: 1200 – 2900m.
DISTRIBUTION: Borneo (SA). Endemic.

FIG. 24. Lips and columns of newly described *Dendrochilum*. **A & B**, *D. cruciforme* var. *cruciforme*, x 15 & x 20; **C & D**, *D. cruciforme* var. *longicuspum*, x 15 & x 20; **E & F**, *D. devogelii*, x 25 & x 15; **G & H**, *D. lumakuense*, x 15 & x 12; **J & K**, *D. ochrolabium*, x 15; **L & M**, *D. papillilabium*, x 15; **N & O**, *D. tenuitepalum*, x 12 & x 15. Drawn by Jeffrey Wood.

N.exaltata *de Vogel* in Blumea 30:202, Pl.2 a-c (1984). Type: Sarawak, *Burtt & Martin* B5259 (holo. E, iso. SAR).

HABITAT: on boulders; upper montane mossy ericaceous forest.
ALTITUDINAL RANGE: 2000 – 2500m.
DISTRIBUTION: Borneo (SA, SR). Endemic.

PHOLIDOTA Lindl. ex Hook.

Exot. Fl.2:t.138 (Jan. 1825).

de Vogel, E.F.(1988). Revisions in Coelogyninae (Orchidaceae) III. The genus *Pholidota*. Orch. Mon 3:1–118.

Pendulous or erect epiphytic or lithophytic herbs. Pseudobulbs closely - or well-spaced, slender or swollen, 1- or 2-leaved. Leaves narrowly elliptic, ovate or oblong, coriaceous. Inflorescence terminal, emerging from apex of pseudobulb, slender, racemose, laxly or densely many flowered, distichous, pendent; floral bracts large, concave, persistent. Flowers small, fleshy, resupinate, white. Dorsal sepal concave or convex, broadly ovate to elliptic; lateral sepals concave or convex, ovate to ovate-oblong, often carinate. Petals ovate to linear. Lip sessile, with a saccate basal hypochile and subentire or 3- to 4-lobed, deflexed epichile. Column short, with a broad hooded apex. Pollinia 4.

Twenty nine species distributed in mainland and S.E. Asia, New Guinea, Australia and the Pacific islands.

P.articulata *Lindl.*, Gen. Sp. Orch. Pl.:38 (1830). Type: India, *Wallich* 1992 (lecto. K-WALL).

HABITAT: unknown.
ALTITUDINAL RANGE: near sea level.
DISTRIBUTION: Borneo (K). Also India, Nepal, Bhutan, China (Yunnan), Burma, Peninsular Malaysia, Thailand, Laos, Cambodia, Vietnam, Sumatra, Java, Sulawesi.

P.carnea *(Blume) Lindl.*, Gen. Sp. Orch. Pl.:37 (1830). Type: Java, Salak & Gede, *Blume* 271 (lecto.: L, iso. BO, L).
Crinonia carnea Blume, Bijdr.:339 (1825).

var. **carnea**

HABITAT: lower montane forest; rarely terrestrial in loam in rock crevices along newly opened roadside banks; podsol forest; oak-laurel forest; moss forest.
ALTITUDINAL RANGE: 900 – 2100m.
DISTRIBUTION: Borneo (SA,SR). Also Peninsular Malaysia, Thailand, Sumatra, Java, Bali, Lombok, Flores, Sulawesi, Philippines, New Guinea.

P.clemensii *Ames*, Orch. 6:66 (1920). Type: Sabah, Mt. Kinabalu, Marai Parai, *Clemens* 390 (holo. AMES).
?*P.dentiloba* J.J.Sm. in Bull. Jard. Bot. Buitenzorg, ser. 3, 11:107 (1931). Type: Kalimantan, Bukit Raja, *Winkler* 994 (holo. HBG, probably lost).

HABITAT: lower montane oak-laurel forest, with *Agathis*; moss forest on limestone; open kerangas forest.
ALTITUDINAL RANGE: 1200 – 3400m.
DISTRIBUTION: Borneo (K, SA, SR). Endemic.

P.gibbosa *(Blume) de Vriese*, Ill. Orch., Pl.5, fig.1 (1854). Type: Java, *Blume* s.n. (lecto. L).

Chelonanthera gibbosa Blume, Bijdr.: 383 (1825).

Pholidota clypeata Lindl. in J. Hort. Soc. London 5:37 (1850). Type: Borneo, cult. *Kenrick*, bought from *Messrs. Low & Co.* in 1847 (holo. K-LINDL).

Coelogyne clypeata (Lindl.) Rchb.f. in Ann. Bot. Syst. 6:237 (1861).

Chelonanthera clypeata (Lindl.) Pfitzer in Engl., Pflanzenr. 32:142 (1907).

Pholidota caduca Ridl. in J. Linn. Soc., Bot. 31: 288 (1896). Type: Sarawak, *Haviland* s.n. (holo. SING).

HABITAT: hill and lower montane forest; kerangas forest, sometimes with *Agathis;* on rocks; rarely in loam in rock crevices along newly constructed roads.

ALTITUDINAL RANGE: lowlands – 1900m.

DISTRIBUTION: Borneo (B, K, SA, SR). Also Peninsular Malaysia, Sumatra, Java, Bali, Sulawesi, Solomon Islands (Guadalcanal).

P.imbricata *Hook.*, Exot. Fl.2, t.138 (1825). Type: Nepal, *Wallich* s.n. (holo. not preserved, lecto. drawing and description in Exotic Fl., t.138).

P.beccarii Schltr. in Notizbl. Bot. Gart. Berlin-Dahlem 8:17 (1921). Type: Sumatra, Singalang, Padang, *Beccari* s.n. (holo. FI).

HABITAT: lowland, hill and lower montane forest.

ALTITUDINAL RANGE: sea level – 1100m.

DISTRIBUTION: Borneo (K, SA, SR). Also China (Yunnan), Nepal, India (Sikkim), East Pakistan, Burma, Nicobar Islands, Andaman Islands, Sri Lanka, Peninsular Malaysia, Thailand, Laos, Vietnam, Sumatra, Java, Lesser Sunda Islands, Sulawesi, Maluku, Philippines, New Guinea, Australia (Queensland), Solomon Islands, Santa Cruz Islands, Vanuatu and other Pacific islands.

P.mediocris *de Vogel* in Orch. Mon. 3:12, fig.2 (1988). Type: Kalimantan, *Geesink* 9108 (holo. L).

HABITAT: unknown.

ALTITUDINAL RANGE: 1300 – 2000m.

DISTRIBUTION: Borneo (K, SR). Endemic.

P.pectinata *Ames*, Orch. 6:69, Pl.86 (1920). Type: Sabah, Mt. Kinabalu, Marai Parai, *Clemens* 273 (holo. AMES, iso. BM, E, K, S, SING).

HABITAT: lower montane oak-laurel ridge forest; low open mossy kerangas forest.

ALTITUDINAL RANGE: 1600 – 2400m.

DISTRIBUTION: Borneo (K, SA, SR). Endemic.

P.schweinfurthiana *L.O.Williams.* in Bot. Mus. Leafl. 6:60 (1938). Type: Sarawak, Gunung Temabok, Upper Barami Valley, *Moulton* 6678 (holo. AMES).

HABITAT: ridge top forest with *Agathis alba*, small rattans, etc.; low open kerangas forest.

ALTITUDINAL RANGE: 1300 – 1700m.

DISTRIBUTION: Borneo (B, SA, SR). Endemic.

P.sulcata *J.J.Sm.* in Icon. Bogor. 2:27, t.107A (1903). Type: cult. hort. Bogor, *Hallier* s.n. (holo. BO, later dupl. BO, L).

HABITAT: very low open kerangas forest with dense bushes and patches of open higher forest with *Eugenia, Ficus, Tristania*, etc.

ALTITUDINAL RANGE: 1100 – 1500m.

DISTRIBUTION: Borneo (K, SA, SR). Endemic.

P.ventricosa *(Blume) Rchb.f.* in Bonplandia 5:43 (1857). Type: Java, Tjapus River, Salak, *Blume* s.n. (holo.:L).
Chelonanthera ventricosa Blume, Bijdr.:383 (1825).

HABITAT: lower montane forest; open kerangas forest, sometimes with patches of open higher forest with *Eugenia, Ficus, Tristania*,etc., field layer of *Cyperus*, creeping *Araceae* and *Bromheadia*.
ALTITUDINAL RANGE: 500 – 1700m.
DISTRIBUTION: Borneo (K, SA, SR). Also in Peninsular Malaysia, Sumatra, Java, Sulawesi, Philippines, New Guinea.

TRIBE PODOCHILEAE

SUBTRIBE ERIINAE

ASCIDIERIA Seidenf.

Nordic J. Bot. 4 (1):44 (1984).

Epiphytic, rarely terrestrial, herbs. Stems up to 25cm long, rather slender, covered with thin sheaths below, 2-leaved at apex. Leaves up to 25 x 1.2cm, linear to linear-elliptic, acute to acuminate. Inflorescences erect, 1 or 2 emerging from near stem apex, densely pubescent, up to 16cm long, with many flowers arranged in whorls about 5mm apart, each whorl containing up to 10 flowers. Flowers non-resupinate, very small, 4 x 3mm, white; sepals densely pubescent; dorsal sepal free, ovate-elliptic; lateral sepals united at base and attached directly to base of column, triangular-ovate, acute; petals smaller than sepals; lip cymbiform, deeply concave; column short, foot absent; pollinia 8.

A monotypic genus distributed in Thailand, Peninsular Malaysia, Sumatra and Borneo.

A.longifolia *(Hook.f.) Seidenf.* in Nordic J. Bot. 4 (1):44 (1984). Type: Peninsular Malaysia, Perak, Ulu Batang Padang, *Wray* 1541 (holo. K).
Eria longifolia Hook.f., Fl. Brit. Ind. 5:790 (1890).
E.verticillaris Kraenzl. in Bot. Jahrb. Syst. 44, Beibl. 101:29 (1910). Type: Sarawak, *Beccari* 2453 (holo. B, destroyed, iso. FI).

HABITAT: 'sand forest'; lower montane forest; mossy rocks; oak-laurel forest; sometimes on ultramafic substrate.
ALTITUDINAL RANGE: 900 – 1900m.
DISTRIBUTION: Borneo (B, SA, SR). Also Peninsular Malaysia, Thailand, Sumatra.

CERATOSTYLIS Blume

Bijdr.:304, t.56 (1825).

Epiphytic herbs. Roots fibrous. Stems simple or branched, tufted, 1-leaved, sometimes terete and rush-like, with thin, brown, often reticulate basal sheaths, pseudobulbs absent. Leaves narrow, coriaceous, fleshy or subterete, rarely thin-textured. Flowers resupinate, small, rarely larger and showy, solitary or a few in a small cluster of bracts. Sepals erect, connivent; lateral sepals forming a saccate or spur-like mentum with column-foot. Petals narrower than sepals. Lip adnate to column-foot by a long incumbent claw, entire, usually with longitudinal calli. Column short, dilated above, 2-lobed or with 2 spathulate erect arms; foot long. Pollinia 8, sessile.

About 100 species distributed from tropical Asia to the Pacific islands.

C.alata *Carr* in Gard. Bull. Straits Settlem. 8:93 (1935). Type: Sarawak, Mt. Dulit, near Long Kapa, *Synge* S.137 (holo. SING, iso. K).

HABITAT: unknown
ALTITUDINAL RANGE: 500m.
DISTRIBUTION: Borneo (SR). Endemic.

C.ampullacea *Kraenzl.* in Bot. Jahrb. Syst. 17:487 (1893). Type: ?Sumatra, *collector unknown* (holo. B,destroyed).

HABITAT: low mossy and xerophyllous scrub forest on extreme ultramafic substrate; lower and upper montane forest.
ALTITUDINAL RANGE: 1200 – 3200m.
DISTRIBUTION: Borneo (SA). Also Peninsular Malaysia, Thailand, Sumatra.

[**C.beccariana** *Kraenzl.* in Bot. Jahrb. Syst. 44, Beibl. 101:20 (1910). Type: Sarawak, *Beccari* 1341 (holo. B, destroyed, iso. FI).

HABITAT: unknown
ALTITUDINAL RANGE: unknown.
DISTRIBUTION: Borneo (SR). Endemic.]

NOTE: Examination of a photograph of the Beccari type at FI shows it to be an *Eria* belonging to section Cylindrolobus, probably *E.nutans* Lindl.

C.borneensis *J.J.Sm.* in Mitt. Inst. Allg. Bot. Hamburg 7:45, t.7, fig.35 (1927). Type: Kalimantan, Bukit Tilung, *Winkler* 1496 (holo. HBG, drawing at BO).

HABITAT: primary forest.
ALTITUDINAL RANGE: 800m.
DISTRIBUTION: Borneo (K, SA). Endemic.

C.crassilingua *Ames & C.Schweinf.*, Orch. 6:135 (1920). Type: Sabah, Mt. Kinabalu, Marai Parai, *Clemens* 260 (holo. AMES, iso. K).

HABITAT: lower and upper montane forest, usually on ultramafic substrate; in open areas or in low xerophyllous scrub.
ALTITUDINAL RANGE: 1200 – 2300m.
DISTRIBUTION: Borneo (SA). Endemic.

C.gracilis *Blume* Bijdr.:306 (1825). Type: Java, Seribu and Pantjar, *Blume* s.n.(holo. BO).

HABITAT: lower montane forest.
ALTITUDINAL RANGE: 1100 – 1200m.
DISTRIBUTION: Borneo (SR). Also Peninsular Malaysia, Sumatra, Java.

C.longisegmenta *Ames & C.Schweinf.*, Orch. 6:136 (1920). Type: Sabah, Mt. Kinabalu, Kiau, *Clemens* 93 (holo. AMES).

HABITAT: hill forest.
ALTITUDINAL RANGE: 900m.
DISTRIBUTION: Borneo (K, SA). Endemic.

C.pendula *Hook.f.*, Fl. Brit. Ind. 5.826 (1890). Type: Peninsular Malaysia, Perak, Larut, *King's collector* 3847 (holo. K).

HABITAT: unknown.
ALTITUDINAL RANGE: 1200m.
DISTRIBUTION: Borneo (K, SA, SR). Also Peninsular Malaysia, Sulawesi.

C.radiata *J.J.Sm.*, Orch. Java: 295, fig.225 (1905). Type: Java, Soekaboemi, Garoet and south Preangen, *Raciborski* s.n. (holo. BO).

HABITAT: lower montane forest; podsol forest; oak-laurel forest.
ALTITUDINAL RANGE: 1100 – 1900m.
DISTRIBUTION: Borneo (SA). Also Peninsular Malaysia, Thailand, Sumatra, Java.

C.subulata *Blume*, Bijdr.:306 (1825). Type: Java, Salak, Pantjar, *Blume* s.n. (holo. BO).

HABITAT: lowland forest; hill dipterocarp forest; lower montane oak-laurel forest.
ALTITUDINAL RANGE: 400 – 1700m.
DISTRIBUTION: Borneo (K, SA). Also Nicobar Islands, Peninsular Malaysia, Thailand, ?Cambodia, Sumatra, Java, Seram, Philippines, New Guinea, Solomon Islands, Vanuatu.

C.tjihana *J.J.Sm.* in Bull. Dép. Agric. Indes Néerl. 45:13 (1911). Type: Kalimantan, Sungai Tjihan, *Nieuwenhuis* 1849 (holo. L).

HABITAT: unknown.
ALTITUDINAL RANGE: unknown.
DISTRIBUTION: Borneo (K). Endemic.

ERIA Lindl.

Bot. Reg. 11: t.904 (1825).

Epiphytic or, rarely, terrestrial polymorphic herbs. Stems either pseudobulbous (short or long, slender or thick), or unthickened and cane-like, 2- to many-leaved, rarely 1-leaved. Leaves flat or terete, thin-textured to coriaceous. Inflorescence terminal or axillary, racemose or rarely 1-flowered, rachis often hirsute or woolly. Flowers mostly small to medium-sized, rarely showy. Sepals subequal, free, rarely connate, glabrous or hirsute. Lateral sepals adnate to column-foot forming a short to long and spur-like or saccate mentum. Petals similar to dorsal sepal. Lip simple or 3-lobed, adnate to apex of column-foot, incumbent, rarely versatile, disc with or without keels. Column short, broad, more or less 2-winged, with a prominent foot. Pollinia 8, attached in fours by narrow bases to viscidium.

About 370 species widespread in tropical Asia, extending east to New Guinea, Australia and the Pacific islands.

E.angustifolia *Ridl.* in Trans. Linn. Soc. London, Bot. 4:237 (1894). Type: Sabah, Mt. Kinabalu, *Haviland* s.n. (holo. not located).

HABITAT: lower montane forest.
ALTITUDINAL RANGE: 1800m.
DISTRIBUTION: Borneo (SA). Endemic.

E.atrovinosa *Ridl.* in Gard. Bull. Straits Settlem. 5:14, Pl.7B (1929). Type: Peninsular Malaysia, Pahang, Mentakab, *Carr* s.n. (holo. SING).

HABITAT: podsol forest.
ALTITUDINAL RANGE: 400 – 500m.
DISTRIBUTION: Borneo (SA). Also Peninsular Malaysia.

E.aurantia *J.J.Sm.* in Bull. Jard. Bot. Buitenzorg, ser. 2, 3:10 (1912). Type: Kalimantan, *Nieuwenhuis* 1901, cult. Bogor no.2148 (holo. BO).
    E.aurantiaca J.J.Sm. in Bull. Dép. Agric. Indes Néerl. 45:19 (1911), non Ridl.
    Trichotosia aurantiaca (J.J.Sm.) Kraenzl. in Engler, Pflanzenr. Orch. Monand. IV. II. B. 21:173 (1911).
    Eria smithii Ames in Merr., Bibl. Enum. Born. Pl.:174 (1921).

HABITAT: podsol forest including *Eugenia, Podocarpus, Leptospermum.* etc. with understorey of *Pandanus.*
ALTITUDINAL RANGE: 1200 - 1300m.
DISTRIBUTION: Borneo (K, SA). Endemic.

E.aurantiaca *Ridl.* in J. Straits Branch Roy. Asiat. Soc. 54:50 (1910). Types: Sarawak, Kuching, *Haviland* s.n., *Moulton* s.n. (syn. SING).

HABITAT: unknown.
ALTITUDINAL RANGE: sea level.
DISTRIBUTION: Borneo (SR). Endemic.

E.bancana *J.J.Sm.* in Bull. Jard. Bot. Buitenzorg, ser. 3, 2:53 (1920). Type: Bangka, *Bünnemeijer* s.n., cult. Bogor (holo. L).

HABITAT: lower montane forest.
ALTITUDINAL RANGE: 1600m.
DISTRIBUTION: Borneo (SA). Also Bangka.

E.berringtoniana *Rchb.f.* in Gard. Chron.:666 (1872). Type: Borneo, cult. *Berrington* (holo. W).

HABITAT: unknown.
ALTITUDINAL RANGE: unknown.
DISTRIBUTION: Borneo (unspecified). Endemic.

E.bifalcis *Lindl.* in J. Linn. Soc., Bot. 3:56 (1859). Type: Borneo, *Lobb* s.n. (holo. K-LINDL).

HABITAT: unknown.
ALTITUDINAL RANGE: unknown.
DISTRIBUTION: Borneo (unspecified). Endemic.

E.biflora *Griff.*, Not. 3:302 (1851). Type: Burma, Peenma, Mergue Herb., *Griffith* 830 (holo. K-LINDL).

HABITAT: lower montane forest.
ALTITUDINAL RANGE: 1400 – 1700m.
DISTRIBUTION: Borneo (SA). Also India (Sikkim), Burma, Laos, Vietnam, Thailand, Peninsular Malaysia, Sumatra, Java, Bali.

E.bigibba *(Benth.ex Kraenzl.) Rchb.f.* in Gard. Chron., ser. 2, 22:680 (1884). Type: Borneo, *Linden* s.n. (holo. W).
    Tainia bigibba Benth. ex Kraenzl. in Engl., Pflanzenr., Orch. Mon. Dendr., 2:165 (1911).
    T.bigibba (Rchb.f.) P. F. Hunt in Kew Bull. 26 (1):181 (1971), comb. inval.

HABITAT: unknown.
ALTITUDINAL RANGE: unknown.
DISTRIBUTIONB: Borneo (unspecified). Endemic.

E.biglandulosa *J.J.Sm.* in Bull. Jard. Bot. Buitenzorg, ser. 2, 13:25 (1914). Type: Kalimantan, Kota Waringin, *van Nouhuys* 20, cult. Bogor (holo. L).

HABITAT: unknown.
ALTITUDINAL RANGE: unknown.
DISTRIBUTION: Borneo (K). Endemic.

E.borneensis *Rolfe* in J. Linn. Soc., Bot., 42:150 (1914). Type: Sabah, Mt. Kinabalu, Kiau, *Gibbs* 3955 (holo. K).

HABITAT: hill forest.
ALTITUDINAL RANGE: 900m.
DISTRIBUTION: Borneo (SA). Endemic.

E.bractescens *Lindl.* in Bot. Reg. 27: misc. 18, no.46 (1841). Type: Singapore, *Loddiges* 214 (holo. K-LINDL).

HABITAT: unknown.
ALTITUDINAL RANGE: unknown.
DISTRIBUTION: Borneo (K). Also India, Nepal, Burma, Laos, Cambodia, Vietnam, Thailand, Peninsular Malaysia, Singapore, Indonesia (excluding Java), Philippines and, doubtfully, New Guinea.

E.brookesii *Ridl.* in J. Straits Branch Roy. Asiat. Soc. 50: 136 (1908). Type: Sarawak, Bidi, *Ridley* s.n. (holo. SING, iso. K).

HABITAT: 'heath forest'; moss forest on sandstone; lower montane forest; ridge forest with epiphytic rhododendrons; sometimes on ultramafic substrate.
ALTITUDINAL RANGE: 900 – 1900m.
DISTRIBUTION: Borneo (SA, SR). Endemic.

E.caricifolia *J.J.Wood* in Lindleyana 5, 2:93, fig.9 (1990). Type: Sarawak, Gunung Temabok, Upper Baram valley, *Moulton* 6673 (holo. K, iso. SING).

var. **caricifolia**

HABITAT: unknown.
ALTITUDINAL RANGE: 1200m.
DISTRIBUTION: Borneo (SR). Endemic.

var. **glabra** *J.J.Wood* in Lindleyana 5, 2:95 (1990). Type: Sabah, Sipitang District, slope north of Sungai Malabid, west of Long Pa Sia to Long Semado trail, *Vermeulen & Duistermaat* 1000 (holo. K).

HABITAT: lower montane *Agathis-Lithocarpus* forest; podsol forest.
ALTITUDINAL RANGE: 1300 – 1500m.
DISTRIBUTION: Borneo (SA). Endemic.

E.carnea *J.J.Sm.* in Bull. Dép. Agric. Indes Néerl. 5:12 (1907). Type: Kalimantan, Liang Gagang, *Hallier* s.n. (holo. L).
Trichotosia carnea (J.J.Sm.) Kraenzl. in Engler, Pflanzenr. Orch. Monand. IV. II. B. 21:153 (1911).

HABITAT: unknown.
ALTITUDINAL RANGE: unknown.
DISTRIBUTION: Borneo (K). Endemic.

E.carnosissima *Ames & C.Schweinf.*, Orch. 6:120 (1920). Type: Sabah, Mt. Kinabalu, Kiau, *Clemens* 314 (holo. AMES).

HABITAT: lower montane forest.
ALTITUDINAL RANGE: 900 – 1400m.
DISTRIBUTION: Borneo (SA). Endemic.

E.cepifolia *Ridl.* in J. Linn. Soc., Bot. 31:282 (1896). Type: Sarawak, *Hose* s.n. (holo. SING).

HABITAT: lowland forest; lower montane forest.
ALTITUDINAL RANGE: sea level – 1400m.
DISTRIBUTION: Borneo (SA, SR). Also Peninsular Malaysia, Thailand.

E.citrina *Ridl.* in Kew Bull.:204 (1924). Type: Kalimantan, Pontianak, received from orchid dealer in Singapore, comm. *Ridley* (holo. K).

HABITAT: lowland, hill and lower montane forest.
ALTITUDINAL RANGE: sea level – 1200m.
DISTRIBUTION: Borneo (K, SA). Also Peninsular Malaysia.

E.consanguinea *J.J.Sm.* in Bull. Jard. Bot. Buitenzorg, ser. 3, 11:131 (1931). Type: Kalimantan, Gunung Kemoel, *Endert* 4315 (holo. L, iso. BO).

HABITAT: lower montane forest.
ALTITUDINAL RANGE: 1600m.
DISTRIBUTION: Borneo (K). Endemic.

E.convallariopsis *Kraenzl.* in Bot. Jahrb. Syst. 44:28 (1910). Type: Sarawak, *Beccari* 3639 (holo. B, destroyed).

HABITAT: unknown.
ALTITUDINAL RANGE: unknown.
DISTRIBUTION: Borneo (SR). Endemic.

E.cordifera *Schltr.* in Feddes Repert. Beih. 1:664 (1912).

subsp. **borneensis** *J.J.Wood* in Kew Bull. 39 (1):84, fig.8 (1984). Type: Sarawak, 'Rob Anderson Camp', near Brunei border, *Nielsen* 652 (holo. AAU, iso. K).

HABITAT: mixed forest on sandy soil with scattered limestone rocks; mixed dipterocarp forest beside rivers.
ALTITUDINAL RANGE: lowlands.
DISTRIBUTION: Borneo (B, SR). Endemic.

NOTE: *E.cordifera* subsp. *cordifera* is found in Papua New Guinea.

E.crassipes *Ridl.* in J. Linn. Soc., Bot. 38:327 (1908). Type: Peninsular Malaysia, Gunung Tahan, *Robinson & Wray* 5336 (holo. BM).

HABITAT: upper montane ericaceous forest; moss forest; low open scrub on ridges.
ALTITUDINAL RANGE: 1900 – 3000m.
DISTRIBUTION: Borneo (SA, SR). Also Peninsular Malaysia, Sumatra.

E.crucigera *Ridl.* in J. Linn. Soc., Bot. 31:280 (1896). Type: Sarawak, Trusan River, *Haviland* s.n. (holo. SING, iso. SAR).

HABITAT: lowland forest on limestone.
ALTITUDINAL RANGE: 200m.
DISTRIBUTION: Borneo (SR). Endemic.

E.curtisii *Rchb.f.* in Gard. Chron., ser. 2, 14:685 (1880). Type: Borneo, *Curtis* s.n. (holo. W).

HABITAT: unknown.
ALTITUDINAL RANGE: unknown.
DISTRIBUTION: Borneo (unspecified). Endemic.

E.cymbidifolia *Ridl.* in J. Bot. 36:212 (1898). Types: Kalimantan, Pontianak, cult. Singapore Bot. Gard., *Ridley* s.n. (lecto. SING, chosen here); Sumatra, Deli Baros, cult.1910 (syn. SING).

var. **cymbidifolia**

HABITAT: lower montane forest; oak-chestnut forest; oak-laurel forest; recorded from ultramafic substrate.
ALTITUDINAL RANGE: 900 – 2400m.
DISTRIBUTION: Borneo (K, SA). Also Sumatra.

var. **pandanifolia** *J.J.Wood* in Lindleyana 5, 2:95 (1990). Type: Sarawak, Mt. Dulit (Ulu Tinjar), near Long Kapa, *Synge* S.154 (holo. K).

HABITAT: hill forest; lower montane forest; upper montane forest; mostly on ultramafic substrate.
ALTITUDINAL RANGE: 300 – 2700m.
DISTRIBUTION: Borneo (SA, SR). Endemic.

E.densa *Ridl.* in J. Linn. Soc., Bot. 31:281 (1896). Types: Sarawak, *Haviland* s.n.; Peninsular Malaysia, Perak, Larut and Hermitage hills, ?*Ridley* (syn. SING).

HABITAT: unknown.
ALTITUDINAL RANGE: lowlands.
DISTRIBUTION: Borneo (K, SR). Also Peninsular Malaysia, Thailand, Sumatra, Mentawai Islands.

E.discolor *Lindl.* in J. Linn. Soc., Bot. 3:51 (1859). Type: India, Sikkim, *J.D.Hooker* 168 (holo. K-LINDL).

HABITAT: epiphyte on *Dacrydium* in 'heath forest'; hill dipterocarp forest; lower montane forest.
ALTITUDINAL RANGE: 100 – 1600m.
DISTRIBUTION: Borneo (SA, SR). Also India (Sikkim), Burma, Laos, Vietnam, Thailand, Peninsular Malaysia, Sumatra, Java.

E.dulitensis *Carr* in Gard. Bull. Straits Settlem. 8:97 (1935). Type: Sarawak, Dulit ridge, *Synge* S. 463 (holo. K).

HABITAT: montane ridge forest.
ALTITUDINAL RANGE: 1200 – 1300m.
DISTRIBUTION: Borneo (SR). Endemic.

E.eurostachys *Ridl.* in J. Straits Branch Roy. Asiat. Soc. 50:136 (1908). Type: Sarawak, Matang, *Hewitt* s.n. (holo. SING, iso. K).

HABITAT: lowland forest.
ALTITUDINAL RANGE: lowlands.
DISTRIBUTION: Borneo (SA). Endemic.

E.farinosa *Ames & C.Schweinf.*, Orch. 6:122 (1920). Type: Sabah, Mt. Kinabalu, Marai Parai, *Clemens* 283 (holo. AMES).

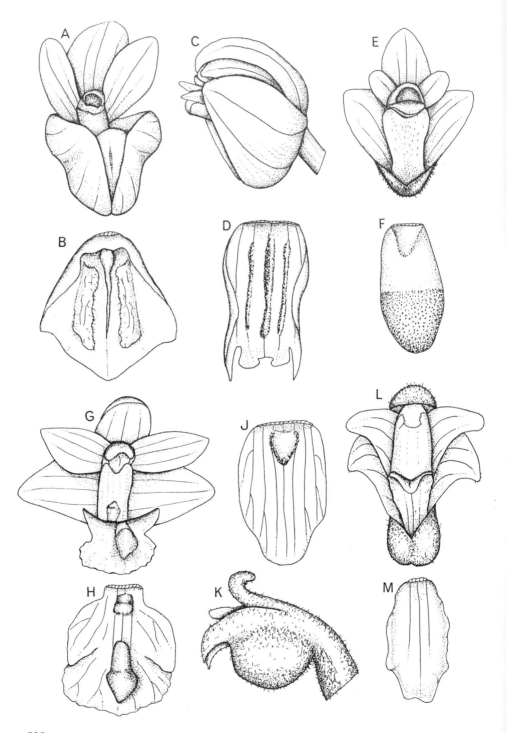

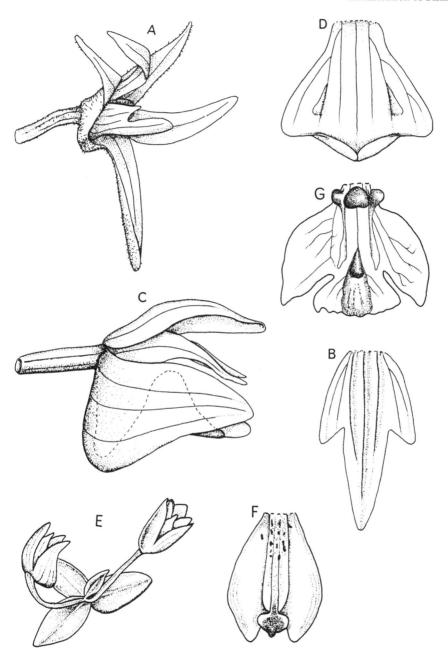

FIG. 25. *Eria* flowers and lips. **A** & **B**, *E. cymbidifolia* var. *cymbidifolia*, x 5; **C** & **D**, *E. aurantia*, x 2½; **E** & **F**, *E. pannea*, x 3; **G** & **H**, *E. paniculata*, x 4; **J** & **K**, *E. pulchella*, x 4; **L** & **M**, *E. robusta*, x 6. Drawn by Sarah Thomas.

FIG. 26. *Eria* flowers and lips. **A** & **B**, *E. javanica*, x 3; **C** & **D**, *E. punctata*, x 3; **E** & **F**, *E. jenseniana*, x 1 & x 3; **G**, *E. magnicallosa*, x 12. Drawn by Sarah Thomas.

HABITAT: oak-laurel forest; lower montane forest; sometimes on ultramafic substrate.
ALTITUDINAL RANGE: 1200 – 2000m.
DISTRIBUTION: Borneo (K, SA). Endemic.

E.floribunda *Lindl.* in Bot. Reg. 29, misc. 43 (1843). Type: Singapore, *Prince* s.n., comm. to *Wallich* no.7408 (holo. K-LINDL).

HABITAT: lowland and hill forest; lower montane oak-laurel forest.
ALTITUDINAL RANGE: lowlands – 2400m
DISTRIBUTION: Borneo (B, K, SA, SR). Also Burma, Vietnam, Thailand, Peninsular Malaysia, Sumatra, Mentawai and Riau Islands, Philippines.

E.gibbsiae *Rolfe* in Gibbs in J. Linn. Soc., Bot. 42:151 (1914). Type: Sabah, Mt. Kinabalu, Kiau, *Gibbs* 3960 (holo. K).

HABITAT: secondary hill forest.
ALTITUDINAL RANGE: 900m.
DISTRIBUTION: Borneo (SA, ?SR). Endemic.

E.graminea *Ridl.* in J. Fed. Malay States Mus. 8, 4:103 (1917). Type: Sumatra, Korinchi Peak, Barong Bharu, west side of Barisan Range, *Robinson & Boden Kloss* s.n. (holo. SING).

HABITAT: lower montane ridge forest.
ALTITUDINAL RANGE: 700m.
DISTRIBUTION: Borneo (K). Also Sumatra.

NOTE: The specimen *de Vogel* 1077 (K, L) is similar to *E.graminea* but has shorter, broader leaves.

E.grandis *Ridl.* in Stapf. in Trans. Linn. Soc. London, Bot. 4:237 (1894). Type: Sabah, Mt. Kinabalu, *Haviland* 1157 (holo. SING, iso. K, SAR).

HABITAT: terrestial among granite rocks and below upper montane scrub; epiphytic in moss forest; under *Leptospermum*; low forest with oaks, *Dacrydium* and *Tristania* on ultramafic substrate.
ALTITUDINAL RANGE: 2100 – 3700m.
DISTRIBUTION: Borneo (SA). Endemic.

E.hallieri *J.J.Sm.* in Icon. Bogor. 3:35, t.214 (1906). Type: Kalimantan, Pontianak, Sungai Kenepai, *Hallier* s.n. (holo. K).
Trichotosia hallieri (J.J.Sm.) Kraenzl. in Engler, Pflanzenr. Orch. Monand. IV. 50. II. B. 21:160 (1911).

HABITAT: unknown.
ALTITUDINAL RANGE: unknown.
DISTRIBUTION: Borneo (K). Endemic.

E.hosei *Rendle* in J. Bot. 39:174 (1901). Type: Sarawak, Baram District, *Hose* 476 (holo. K).

HABITAT: unknown.
ALTITUDINAL RANGE: unknown.
DISTRIBUTION: Borneo (K, SR). Endemic.

E.hyacinthoides *(Blume) Lindl.*, Gen. Sp. Orch. Pl.:66 (1830). Type: Java, Salak, *Blume* s.n. (holo. BO).

Dendrolirium hyacinthoides Blume, Bijdr.:346 (1825).

HABITAT: unknown.
ALTITUDINAL RANGE: unknown.
DISTRIBUTION: Borneo (unspecified). Also Peninsular Malaysia, Sumatra, Java, Bali.

E.ignea *Rchb.f.* in Gard. Chron. ser. 2, 15:782 (1881). Type: Borneo, imported by *Veitch* (holo. W).
Eria cinnabarina Rolfe in Kew Bull.:183 (1894). Type: Borneo, introduced by *Linden* (holo. K).

HABITAT: podsol forest with *Dacrydium*.
ALTITUDINAL RANGE: 200 – 500m.
DISTRIBUTION: Borneo (SA). Endemic.

E.iridifolia *Hook.f.*, Fl. Brit. Ind. 5:790 (1890). Type: Peninsular Malaysia, Gunung Batu Pateh, *Wray* s.n. (holo. K).
Eria longispica Rolfe in Bot. Mag. 133: t.8171 (1907). Type: Borneo, introduced by *Linden* (holo. K).

HABITAT: hill and lower montane forest; low stature forest on ultramafic substrate; rocky places.
ALTITUDINAL RANGE: 800 – 1600m.
DISTRIBUTION: Borneo (SA). Also Peninsular Malaysia, Sumatra, Java.

E.javanica *(Sw.) Blume*, Rumphia 2:23 (1836). Type: East Java, *Thunberg* s.n.(holo. S?).
Dendrobium javanicum Sw. in Kgl. Sv. Vet. Akad. nya Handl. 21: 247 (1800).
Eria striolata Rchb.f. in Gard. Chron., ser. 3, 3:554 (1888). Type: Papua New Guinea, imported by *Messrs. Linden* (holo. W).

HABITAT: dipterocarp forest; limestone screes in lowland forest; peat swamp forest; on mangrove trees in swamp forest; lower montane oak-laurel forest.
ALTITUDINAL RANGE: sea level – 2400m.
DISTRIBUTION: Borneo (K, SA, SR and probably also B). Widespread in S. and S.E. Asia from India to New Guinea.

E.jenseniana *J.J.Sm.* in Bull. Jard. Bot. Buitenzorg, ser. 3, 2:50 (1920). Types: Sumatra, ?Deli, cult. *Jensen* (syn. BO); Ophir districts, Taloe, *Groeneveldt* s.n., cult. under no. 396 (syn. BO).

HABITAT: hill forest on ultramafic substrate.
ALTITUDINAL RANGE: 600m.
DISTRIBUTION: Borneo (SA). Also Sumatra.

E.kinabaluensis *Rolfe* in Gibbs in J. Linn. Soc., Bot. 42:151 (1914). Type: Sabah, Mt. Kinabalu, below Pakapaka, *Gibbs* 4227 (holo. K).

HABITAT: lower and upper montane forest; sometimes on ultramafic substrate.
ALTITUDINAL RANGE: 1500 – 3000m.
DISTRIBUTION: Borneo (SA). Endemic.

E.lanuginosa *J.J.Wood* in Lindleyana 5(2):95, fig.10 (1990). Type: Sabah, Sipitang District, banks of Pa Sia River near Long Pa Sia, *Wood* 725 (holo. K).

HABITAT: on riverside trees.
ALTITUDINAL RANGE: 900 – 1000m.
DISTRIBUTION: Borneo (SA, SR). Endemic.

E.latibracteata *Ridl.* in J. Linn. Soc., Bot. 32:293 (1896). Type: Peninsular Malaysia, Sungai Ujong, *Braddon* s.n. (holo. SING).

HABITAT: hill forest on ultramafic substrate.
ALTITUDINAL RANGE: 500 – 600m.
DISTRIBUTION: Borneo (K, SA). Also Peninsular Malaysia.

E.latiuscula *Ames & C.Schweinf.*, Orch. 6:125 (1920). Type: Sabah, Mt. Kinabalu, *Clemens* s.n. (holo. AMES).

HABITAT: lower and upper montane forest; oak-laurel forest; sometimes on ultramafic substrate.
ALTITUDINAL RANGE: 1100 – 2100m.
DISTRIBUTION: Borneo (SA). Endemic.

E.leiophylla *Lindl.* in J. Linn. Soc., Bot.:3:57 (1859). Type: Sarawak, *Lobb* s.n. (holo. K-LINDL).

HABITAT: oak-laurel forest; podsol forest with *Dacrydium* and dipterocarps on very wet sandy soil; lower montane forest with *Gymnostoma sumatrana*, small rattans etc. on ultramafic substrate; in crowns of *Gonystylus bancanus* growing in peat swamp forest.
ALTITUDINAL RANGE: sea level – 2000m.
DISTRIBUTION: Borneo (K, SA, SR). Also Peninsular Malaysia, Sumatra, Sulawesi, Seram.

E.leptocarpa *Hook.f.*, Fl. Brit. Ind. 5:805 (1890). Types: Peninsular Malaysia, Perak, *Scortechini* s.n. and *King's collector* 10339 (syn. K).

HABITAT: unknown.
ALTITUDINAL RANGE: lowlands.
DISTRIBUTION: Borneo (K, SR). Also Peninsular Malaysia, Sumatra.

E.linearifolia *Ames* in Merr., Bibl. Enum. Born. Pl.:172 (1921). Type: Sarawak, *Hose* s.n. (holo. SING).
Eria elongata Ridl. in J. Linn. Soc., Bot. 31:284 (1896), non Blume.

HABITAT: riverine forest; hill forest; on ultramafic substrate.
ALTITUDINAL RANGE: 800 – 1000m.
DISTRIBUTION: Borneo (K, SA, SR). Endemic.

E.lobata *(Blume) Rchb.f.* in Bonplandia 5:55 (1857). Type: Java, Salak, Pantjar, Seribu etc., *Blume* s.n. (holo. BO).
Mycaranthes lobata Blume, Bijdr.:352 (1825).

HABITAT: riverine and ridge forest.
ALTITUDINAL RANGE: 200 – 1200m.
DISTRIBUTION: Borneo (K). Also Sumatra, Java.

E.longerepens *Ridl.* in J. Linn. Soc., Bot. 31:282 (1896). Types: Sarawak, *Haviland* s.n.; Singapore, Kranji and Sungai Morai, *Ridley* s.n. (syn. SING).
Sarcopodium beccarianum Kraenzl. in Engler, Pflanzenr. Orch. Monand. IV. 50. II. B. 21:322 (1910). Type: Sarawak, *Beccari* 3655 (holo. FI).
Dendrobium beccarianum (Kraenzl.) Masam., Enum. Phan. Born. :152 (1942).
Katherinea beccariana (Kraenzl.) A.D. Hawkes in Lloydia 19:95 (1956).

HABITAT: lowland forest; mangrove; *Agathis* dominated heath forest on podsol.
ALTITUDINAL RANGE: sea level – ?200m.
DISTRIBUTION: Borneo (B, K, SA, SR). Also Singapore, Peninsular Malaysia.

E.macrophylla *Ames & C.Schweinf.*, Orch. 6:127 (1920). Type: Sabah, Mt. Kinabalu, *Haslam* s.n. (holo. AMES).

HABITAT: unknown.
ALTITUDINAL RANGE: unknown.
DISTRIBUTION: Borneo (SA). Endemic.

E.magnibracteata *J.J.Wood*, **nom. nov.**
E.latibracteata Rolfe in Kew Bull.:194 (1898) and in Bot. Mag. t.7605 (1898), non Ridl. Type: Borneo, imported by *Sander & Co.* (holo. K).

HABITAT: unknown.
ALTITUDINAL RANGE: unknown.
DISTRIBUTION: Borneo (unspecified). Endemic.

E.magnicallosa *Ames & C.Schweinf.*, Orch. 6:129, Pl.92 (1920). Type: Sabah, Mt. Kinabalu, Kiau, *Clemens* 32 (holo. AMES).

HABITAT: hill forest; oak-laurel forest; *Tristania* forest with bamboo understorey; 'pole forest'; lower montane forest with *Drimys*, rattans, etc.
ALTITUDINAL RANGE: 900 – 2400m.
DISTRIBUTION: Borneo (B, SA). Endemic.

E.major *Ridl.ex Stapf* in Trans. Linn. Soc. London, Bot. ser. 2, 4:237 (1894). Type: Sabah, Mt. Kinabalu, 1800m, *Haviland* 1250 (holo. K).
E.scortechinii Stapf in Trans. Linn. Soc. London, Bot. ser. 2, 4:237 (1894) nomen delendum, non Hook.f.
E.villosissima Rolfe in Gibbs in J. Linn. Soc., Bot. 42:150 (1914). Type: Sabah, Mt. Kinabalu, Marai Parai, *Gibbs* 4090 (holo. K).

HABITAT: lower and upper montane forest; oak-laurel forest; low mossy and xerophyllous scrub forest on ultramafic substrate; damp rocky places.
ALTITUDINAL RANGE: 1100 – 2900m.
DISTRIBUTION: Borneo (SA). Also Maluku (Ternate).

E.megalopha *Ridl.* in J. Straits Branch Roy. Asiat. Soc. 33:23 (1900). Type: Sarawak, *Shelford* s.n. (holo. SING).

HABITAT: mixed dipterocarp forest; lower montane forest.
ALTITUDINAL RANGE: lowlands – 1200m.
DISTRIBUTION: Borneo (B, SR). Endemic.

E.melaleuca *Ridl.* in Sarawak Mus.J. 1(2):33 (1912). Type: Sarawak, Maropok Mountains, Ulu Lawas, *Moulton* 1 (holo. SING, iso. K).

HABITAT: riverine forest; hill dipterocarp forest; 'heath forest' including *Gymnostoma sumatrana* on podsolic soil.
ALTITUDINAL RANGE: sea level – 1000m.
DISTRIBUTION: Borneo (B, SA, SR). Endemic.

E.monophylla *Schltr.* in Bull. Herb. Boissier ser. 2, 6:461 (1906). Type:Kalimantan, Long Sele, *Schlechter* 13485 (holo. B,destroyed).

HABITAT: hill forest on ultramafic substrate.
ALTITUDINAL RANGE: 800m.
DISTRIBUTION: Borneo (K, SA). Endemic.

E.moultonii *Ridl.* in Sarawak Mus. J. 1:33 (1912). Type: Sarawak, Ulu Lawas, *Moulton* s.n. (holo. SING).

HABITAT: unknown.
ALTITUDINAL RANGE: unknown.
DISTRIBUTION: Borneo (SR). Endemic.

E.neglecta *Ridl.* in J. Linn. Soc., Bot. 31:283 (1896). Types: Sarawak, *Haviland* s.n.; Singapore, Kranji; Penang, *Curtis* s.n. (syn. SING).

HABITAT: moss forest on limestone; mossy limestone rocks.
ALTITUDINAL RANGE: lowlands – 1400m.
DISTRIBUTION: Borneo (SR). Also Peninsular Malaysia, Singapore, Lingga Archipelago, Thailand.

E.nieuwenhuisii *J.J.Sm.* in Bull. Jard. Bot. Buitenzorg, ser. 3, 2:51 (1920). Type: Kalimantan, Sungai Dengei, Nieuwenhuis expedition, *Jaheri* 792 (holo. L).

HABITAT: unknown.
ALTITUDINAL RANGE: unknown.
DISTRIBUTION: Borneo (K). Endemic.

E.nutans *Lindl.* in Bot. Reg. 16, misc. 83 (1840). Type: Singapore, *Cuming* 129 (holo. K-LINDL).

HABITAT: kerangas forest with dominant *Dacrydium* and *Tristania* on podsolic soil; rather open swampy forest on serpentine; 'sand forest'; mossy limestone rocks; mangrove swamp; hill forest; lower montane forest.
ALTITUDINAL RANGE: sea level – 1500m.
DISTRIBUTION: Borneo (K, SA, SR). Also Peninsular Malaysia, Singapore, Thailand, Riau and Lingga Archipelagos, Natuna.

E.obliqua *(Lindl.) Lindl.* in J. Linn. Soc., Bot. 3:55 (1859). Type: Singapore, *Cuming* s.n., cult.*Loddiges* no 121 (holo. K-LINDL).
Mycaranthes obliqua Lindl. in Bot. Reg. 26, misc. 184 (1840).

HABITAT: in crowns of *Gonystylus bancanus*; hill dipterocarp forest; lower montane forest.
ALTITUDINAL RANGE: sea level – 1500m.
DISTRIBUTION: Borneo (K, SA, SR). Also Peninsular Malaysia, Singapore, Sumatra, Bangka.

E.oblitterata *(Blume) Rchb.f.* in Bonplandia 5:55 (1857). Type: Java, Salak, Gede, etc., *Blume* s.n. (holo. BO).
Mycaranthes oblitterata Blume, Bijdr.:353 (1825).
Eria kingii Hook. f., Fl. Brit. Ind. 5:790 (1890), non F. Muell. Types: Peninsular Malaysia, Perak, Larut, *King's collector* 3311 & 6491 (syn. K); Perak, *King's collector* 8153 (syn. K); Perak, *Scortechini* s.n. (syn. K).
E.major sensu Ridl. in J. Linn. Soc., Bot. 32:288 (1896), non Ridl. (1894).
E.ridleyi Rolfe in Gibbs in J. Linn. Soc., Bot. 42:150 (1914). Type: based on same types as *E.kingii*.

HABITAT: hill and lower montane forest; rocky places.
ALTITUDINAL RANGE: 200 – 1600m.
DISTRIBUTION: Borneo (K, SA, SR). Also Peninsular Malaysia, Thailand, Cambodia, Vietnam, Sumatra, Java, Bali.

E.ornata *(Blume) Lindl.*, Gen. Sp. Orch. Pl.:66 (1830). Type: Java, Buitenzorg and Bantam Provinces, *Blume* s.n. (holo. BO).
Dendrolirium ornatum Blume, Bijdr.:345 (1825).

HABITAT: hill and lower montane forest on limestone.
ALTITUDINAL RANGE: 300 – 1500m.
DISTRIBUTION: Borneo (K, SA, SR). Also Peninsular Malaysia, Thailand, Sumatra, Java, Sulawesi, Philippines.

E.ovilis *J.J.Sm.* in Bull. Jard. Bot. Buitenzorg, ser. 3:63 (1912). Type: Sarawak, Gunung Batu Lawi, Ulu Limbang, *Moulton* 6 (holo. BO, iso. SAR).

HABITAT: unknown.
ALTITUDINAL RANGE: unknown.
DISTRIBUTION: Borneo (SR). Endemic.

E.pachycephala *Kraenzl.* in Bot. Jahrb. Syst. 44, Beibl. 101:26 (1910). Type: Sarawak, *Beccari* 3597 (holo. FI).

HABITAT: unknown.
ALTITUDINAL RANGE: unknown.
DISTRIBUTION: Borneo (SR). Endemic.

E.paniculata *Lindl.* in Wall., Pl. As. Rar. 1:32, t.36 (1830). Types: India and Bangladesh, Pundua, Sylhet, *Wallich* Cat. 1971 (holo. K-LINDL).

HABITAT: unknown.
ALTITUDINAL RANGE: unknown.
DISTRIBUTION: Borneo (unspecified). Also Bangladesh, India, Nepal, Burma, Thailand, Laos, Cambodia, Vietnam.

E.pannea *Lindl.* in Bot. Reg. 28: misc. 64, no.79 (1842). Type: Singapore, *Loddiges* 252 (holo. K-LINDL).

HABITAT: hill dipterocarp forest, on sandstone; peat swamp forest.
ALTITUDINAL RANGE: 800 – 1200m.
DISTRIBUTION: Borneo (K, SA, SR). Also China (Yunnan), India, Bhutan, Burma, Thailand, Laos, Cambodia, Vietnam, Peninsular Malaysia, Singapore, Sumatra, Riau Archipelago, Bangka.

E.pellipes *Hook.f.*, Fl. Brit. Ind. 5:802 (1890). Type: Peninsular Malaysia, Johor, Mt. Ophir, *Maingay* 2676 (lecto. K).
E.teretifolia Griff., Not. 3:298 (1851), non Griff. (1848). Type: Peninsular Malaysia, Mt. Ophir, *Griffith* s.n. (holo. W).

HABITAT: lower montane ridge forest; low stature forest on ultramafic substrate.
ALTITUDINAL RANGE: 900 – 1500m.
DISTRIBUTION: Borneo (SA, SR). Also Peninsular Malaysia, Thailand.

E.pseudocymbiformis *J.J.Wood* in Kew Bull. 39(1):84, Fig.9 (1984). Type: Sarawak, Gunung Mulu National Park, Gunung Api, below Pinnacle Camp, *Nielsen* 486 (holo. AAU, iso. K).
E.cymbidifolia Ridl. var. *longipes* Carr in Gard. Bull. Straits Settlem. 8:99 (1935). Type: Sarawak, Dulit Ridge, *Synge* S. 416 (holo. SING, iso. K).

var. **pseudocymbiformis**

HABITAT: lower montane forest on limestone; oak-laurel forest, moss forest; sometimes on ultramafic substrate.

ALTITUDINAL RANGE: 800 – 2100m.
DISTRIBUTION: Borneo (SA, SR). Endemic.

var. **hirsuta** *J.J.Wood* in Lindleyana 5(2): 97 (1990). Type: Sabah, Mt. Kinabalu, head of Columbon River, *Clemens* 33938 (holo. BM, iso. K).

HABITAT: lower and upper montane mossy ridge forest; oak-laurel forest with *Agathis*; low mossy and xerophyllous scrub forest on ultramafic substrate; damp rocky areas.
ALTITUDINAL RANGE: 900 – 2900m.
DISTRIBUTION: Borneo (SA). Endemic.

E.pseudoleiophylla *J.J.Wood* in Orchid Rev. 89 (1053):209 (1981). Type: Sabah, Mt. Kinabalu, *Puasa* 154 (holo. K, iso. SAN).

HABITAT: hill and lower montane forest.
ALTITUDINAL RANGE: 800 – 1700m.
DISTRIBUTION: Borneo (SA, SR). Also Sulawesi, Papua New Guinea.

E.pudica *Ridl.* in J. Linn. Soc., Bot. 32:294 (1896). Types: Singapore, Changi, *Ridley* s.n. (syn. SING, isosyn. K); Peninsular Malaysia, Johor, Batu Pahat, Kwala Kahang, *Ridley* s.n. (syn. SING).

HABITAT: lowland forest.
ALTITUDINAL RANGE: sea level – 200m.
DISTRIBUTION: Borneo (SA). Also Peninsular Malaysia, Singapore.

E.pulchella *Lindl.* in Bot. Reg. 27: misc. 52, no. 106 (1841). Type: Singapore, *Cuming* s.n. (holo. K-LINDL).

HABITAT: rocky beaches; on *Shorea*; peat swamp forest; hill dipterocarp forest; lower montane forest; podsol forest.
ALTITUDINAL RANGE: sea level – 800m.
DISTRIBUTION: Borneo (K, SA, SR). Also Peninsular Malaysia, Singapore, Thailand, Sumatra, Java.

E.punctata *J.J.Sm.* in Bull. Dép. Agric. Indes Néerl. 13:38 (1907). Type: West Java, cult. Bogor (holo. BO).

HABITAT: 'transition forest'.
ALTITUDINAL RANGE: 1000m.
DISTRIBUTION: Borneo (SR). Also Peninsular Malaysia, Sumatra, Java.

E.rigida *Blume* in Mus. Bot. Lugd. Bat. 2:183 (1856). Type: Kalimantan, Gunung Pamathon, *Korthals* s.n. (holo. BO).
   E.elongata Blume, Mus. Bot. Lugd. Bat. 2:183 (1856). Type: 'in Sumatra, Borneo, etc.', *Blume* s.n. (syn. ?BO).
   E.pendula Ridl. in J. Straits Branch Roy. Asiat. Soc. 39:78 (1903). Type: Peninsular Malaysia, Selangor, Kuala Lumpur Caves, *Kelsall* s.n. (holo. SING).

HABITAT: lower montane forest.
ALTITUDINAL RANGE: 1500m.
DISTRIBUTION: Borneo (K, SA, SR). Also Peninsular Malaysia, Thailand, Sumatra, New Guinea.

E.robusta *(Blume) Lindl.*, Gen. Sp. Orch. Pl.:69 (1830). Type: Java, Buitenzorg and Tjanjor provinces, *Blume* s.n. (holo. BO).
   Dendrolirium robustum Blume, Bijdr.:347 (1825).

Eria aeridostachya Reichb.f. ex Lindl. in J. Linn. Soc., Bot. 3:48 (1859). Type: Fiji, *Seemann* 609 (holo. W).

Eria purpureocentrum J.J.Sm. in Bull. Jard. Bot. Buitenzorg, ser. 3,5:75 (1922). Types: Sumatra, Agam, Sungai Djanih, *Jacobson* s.n., cult. under numbers 998 & 999 (syn. BO); locality unknown, *Groeneveldt* s.n., cult. under number 1956 (syn. BO); Laras Talang, Bukit Niroe, *Bünnemeijer* 5347 (syn. BO).

HABITAT: oak-laurel forest; hill, lower and upper montane forest; low mossy and xerophyllous scrub forest on ultramafic substrate.
ALTITUDINAL RANGE: 400 – 3000m.
DISTRIBUTION: Borneo (B, K, SA, SR). Also in Peninsular Malaysia, Thailand, Sumatra, Java, Solomon Islands, New Caledonia, Fiji, Samoa.

E.saccifera *Hook.f.*, Fl. Brit. Ind. 5:797 (1890). Type: Peninsular Malaysia, Gunung Batu Pateh, *Wray* 1215 (holo. K).

HABITAT: riverine forest; lower montane forest.
ALTITUDINAL RANGE: 200 – 1500m.
DISTRIBUTION: Borneo (K). Also Peninsular Malaysia.

E.sarrasinorum *Kraenzl.* in Bot. Jahrb. Syst. 44, Beibl. 101:29 (1910). Types: Sulawesi, Tomohon, *Sarrasin* 426, 433a and 439; Sarawak, *Beccari* 2488 (syn. B, destroyed).

HABITAT: unknown.
ALTITUDINAL RANGE: unknown.
DISTRIBUTION: Borneo (SR). Also Sulawesi.

E.sonkaris *Rchb.f.* in Bonplandia 5:55 (1857). Type: Sumbawa, Gunung Sonkar, *Zollinger* 1176 (holo. K-LINDL).

HABITAT: unknown.
ALTITUDINAL RANGE: unknown.
DISTRIBUTION: Borneo (unspecified). Also Sumbawa.

E.tenuiflora *Ridl.* in J. Linn. Soc., Bot. 32:291 (1896). Types: Peninsular Malaysia, Johor, Batu Pahat, *Ridley* s.n.; Perak, Hermitage Hill, *Curtis* s.n.; Pahang, Pekan, *Ridley* s.n.; Singapore, Sungai Morai, Toas, *Ridley* s.n. (syn. SING).

HABITAT: 'heath forest' on raised podsolic coastal plateaux; podsol forest.
ALTITUDINAL RANGE: sea level – 500m.
DISTRIBUTION: Borneo (SA, SR). Also Burma, Peninsular Malaysia, Singapore, Thailand, Vietnam, Sumatra, Java.

E.triloba *Ridl.* in J. Straits Branch Roy. Asiat. Soc. 49:31 (1908). Type: Sarawak, Santubong, *Hewitt* 9 (holo. SING, iso. K).

HABITAT: mangrove swamp; lowland forest.
ALTITUDINAL RANGE: sea level.
DISTRIBUTION: Borneo (K, SR). Endemic.

E.versicolor *J.J.Sm.* in Bull. Dép. Agric. Indes Néerl. 22:33 (1909). Type: Kalimantan, *Jaheri* 1385 (holo. BO).

HABITAT: riverine forest.
ALTITUDINAL RANGE: 300m.
DISTRIBUTION: Borneo (K). Endemic.

E.xanthocheila *Ridl.* in Mat. Fl. Mal. Pen. 1:102 (1907). Type: Peninsular Malaysia, Selangor, near Klang, *Ridley* 10272 (holo. SING).

HABITAT: podsol forest; low stature forest on ultramafic substrate; hill and lower montane forest.
ALTITUDINAL RANGE: 500 – 1600m.
DISTRIBUTION: Borneo (K, SA, SR). Also Burma, Peninsular Malaysia, Thailand, Sumatra, Mentawai, Java.

PORPAX Lindl.

Bot. Reg. 31: misc. 62, no.66 (1845).

Tiny clump-forming epiphytic or lithophytic herbs. Pseudobulbs crowded together, flattened, covered by a sheath which disintegrates into a fine fibrous network or in radiating fibres. Leaves 2 – 3, often appearing after the flowers, oblong to elliptic, obtuse or acute, sometimes minutely hairy along margins. Inflorescence 1-flowered, borne either from base of pseudobulb, breaking through the sheath, or from apex of a developed pseudobulb; peduncle and pedicel very short, the flower appearing sessile at edge or centre of pseudobulb; floral bract conspicuous, enclosing lower part of flower. Flowers orange-red to deep dull red, sometimes flushed greenish-yellow; dorsal sepal connate with lateral sepals at least at its base; lateral sepals connate nearly, or completely, to apex, forming a tube and forming a mentum with column-foot; petals narrow, shorter than sepals; lip minute, much shorter than petals, usually recurved, obscurely 3-lobed, disc with a basal callus; column minute, with a large broad rostellum which more or less covers the stigma entrance; pollinia 8.

Eleven species distributed on mainland Asia from India through Thailand and Indochina to Peninsular Malaysia, and one outlying species in Borneo.

P.borneensis *J.J.Wood & A.Lamb* in Wood, Beaman & Beaman, Plants of Mt. Kinabalu 2, Orchids:300, fig.49, Pl.74A (1993). Type: Sabah, Sandakan Zone, Gunung Tawai (Tavai), S. of Telupid, *Lamb* AL 1164/89 (holo. K).

HABITAT: on rocks and buttress roots in hill forest on ultramafic substrate.
ALTITUDINAL RANGE: 600 – 1000m.
DISTRIBUTION: Borneo (SA). Endemic.

SARCOSTOMA Blume

Bijdr.: 339, t.45 (1825).

Small sympodial epiphytes. Rhizomes short, branched. Stems very short, caespitose, pseudobulbs absent. Leaves 1-3 per stem, linear, thick and fleshy. Inflorescences terminal, short, bearing one flower at a time. Flowers resupinate, c.8mm across, lasting one day only. Sepals spreading. Lateral sepals adnate to column-foot to form a mentum. Petals narrower than sepals. Lip 3-lobed, narrow at base, apex thickened, disc with 2 hairy ridges. Column short, broad, apex entire; stigma large, simple. Anther cap 2-chambered. Pollinia 4.

Four species distributed in Peninsular Malaysia, Sumatra, Java, Borneo and Sulawesi.

S.borneense *Schltr.* in Notizbl. Bot. Gart. Berlin-Dahlem 8:18 (1921). Type: Sarawak, Kutein, *Beccari* 257 (holo. B, destroyed, iso. FI).

HABITAT: unknown.
ALTITUDINAL RANGE: unknown.
DISTRIBUTION: Borneo (SR). Endemic.

TRICHOTOSIA Blume

Bijdr.:342 (1825).

Epiphytic, rarely terrestrial herbs. Stems long or short, leafy throughout except at base, usually covered throughout with reddish-brown, rarely white, hispid hairs, sometimes hairs restricted to leaf-sheaths and inflorescences. Inflorescences lateral, from any node, piercing the leaf-sheath, short and few-flowered, or long, pendulous and many-flowered; floral bracts hairy, at right angles to the rachis, large, concave. Flowers small to medium-sized, not opening widely, resupinate; sepals red-hairy on outside, laterals adnate to column-foot forming a mentum; lip entire to obscurely 3-lobed, disc with or without keels, sometimes papillose; column with a foot; pollinia 8.

About 50 species distributed from mainland Asia, through S.E. Asia to New Guinea and the Pacific islands.

T.annulata *Blume,* Bijdr.:343 (1825). Type: Java, Salak, *Blume* s.n. (holo. L).

HABITAT: lower montane forest.
ALTITUDINAL RANGE: 1700m.
DISTRIBUTION: Borneo (SA, SR). *T.annulata* occurs in Sumatra and Java.

NOTE: Several collections are close to *T.annulata* but have narrower leaves and smaller flowers.

T.aporina *(Hook.f.) Kraenzl.* in Engler, Pflanzenr. Orch. Monand. IV.50. II. B. 21:150 (1911). Types: Peninsular Malaysia, Perak, *King's collector, Scortechini* s.n. (syn. K).
Eria aporina Hook.f., Fl. Brit. Ind. 5:808 (1890).

HABITAT: lower montane forest; oak-laurel forest.
ALTITUDINAL RANGE: 300 – 1700m.
DISTRIBUTION: Borneo (SA, SR). Also Peninsular Malaysia.

T.aurea *(Ridl.) Carr* in Gard. Bull. Straits Settlem. 8:99 (1935). Type: Sarawak, Kuching, *Hewitt* s.n. (holo. SING, iso. K).
Eria aurea Ridl. in J. Straits Branch Roy. Asiat. Soc. 49:31 (1908).
E.rhombilabris J.J.Sm. in Mitt. Inst. Allg. Bot. Hamburg 7:50, t.8, fig.40 (1927). Type: Kalimantan, Bukit Mulu, *Winkler* 524 (holo. HBG).

HABITAT: lower montane forest; 'sand forest'; mixed dipterocarp forest; oak-laurel forest; very low and open podsol forest; low, open mossy ridge forest with bamboos and rattans.
ALTITUDINAL RANGE: 600 – 2000m.
DISTRIBUTION: Borneo (B, K, SA, SR). Endemic.

T.brevipedunculata *(Ames & C.Schweinf.) J.J.Wood* in Wood, Beaman & Beaman, Plants of Mt. Kinabalu 2, Orchids: 333 (1993). Type: Sabah, Mt. Kinabalu, Marai Parai, *Clemens* 255 (holo. AMES).

Eria brevipedunculata Ames & C. Schweinf., Orch. 6:118 (1920).

HABITAT: lower montane forest; oak-laurel forest, with *Agathis;* sometimes on ultramafic substrate.
ALTITUDINAL RANGE: 1200 – 2100m.
DISTRIBUTION: Borneo (SA). Endemic.

T.brevirachis *(J.J.Sm.) J.J.Wood,* **comb.nov.**
Eria brevirachis J.J.Sm. in Bull. Jard. Bot. Buitenzorg, ser. 3, 11:134 (1931). Type: Kalimantan, Gunung Kemoel, *Endert* 3971 (holo. L).

HABITAT: lower montane ridge forest.
ALTITUDINAL RANGE: 1500m.
DISTRIBUTION: Borneo (K). Endemic.

T.canaliculata *(Blume) Kraenzl.* in Engler, Pflanzenr. Orch. Monand. IV. 50. II. B. 21:155 (1911). Type: Kalimantan, Martapura, *Korthals* s.n. (holo. BO).
Eria canaliculata Blume, Mus. Bot. Lugd. Bat. 2:184 (1856).

HABITAT: unknown.
ALTITUDINAL RANGE: unknown.
DISTRIBUTION: Borneo (K, SR). Also Sulawesi.

T.conifera *(J.J.Sm.) J.J.Wood,* **comb.nov.**
Eria conifera J.J.Sm. in Bull. Dép. Agric . Indes Néerl. 22:27 (1909). Type: Kalimantan, Bukit Mili, *Nieuwenhuis* 1899 (holo. BO).

HABITAT: unknown.
ALTITUDINAL RANGE: unknown.
DISTRIBUTION: Borneo (K). Endemic.

T.ferox *Blume,* Bijdr.:342 (1825). Type: Java, Salak, *Blume* s.n. (holo. BO).
Eria ferox (Blume) Blume, Mus. Bot. Lugd. Bat. 2:184 (1856).

HABITAT: mossy submontane forest and upper montane ericaceous shrubbery; moss forest.
ALTITUDINAL RANGE: 1500 – 3000m.
DISTRIBUTION: Borneo (K, SA, SR). Also Peninsular Malaysia, Thailand, Sumatra, Java, Lombok.

T.fusca *(Blume) Kraenzl.* in Pflanzenr. Orch. Monand. IV. 50. II. B. 21:145 (1911). Type: Java, "in sylvis montanis provinciarum occidentalium Javae", *Blume* s.n. (holo. ?L).
Eria fusca Blume, Mus. Bot. Lugd. Bat. 2:183 (1856).

HABITAT: unknown.
ALTITUDINAL RANGE: unknown.
DISTRIBUTION: Borneo (SR). Also Java.

T.gracilis *(Hook.f.) Kraenzl.* in Engler, Pflanzenr. Orch. Monand. IV. 50. II. B. 21:143 (1911). Type: Peninsular Malaysia, Perak, *Scortechini* s.n. (holo. K).
Eria gracilis Hook.f., Fl. Brit. Ind. 5:806 (1890).
E. oligantha Hook.f., loc. cit.:807 (1890). Type: Peninsular Malaysia, Penang, *Maingay* 1629 (holo. K).

HABITAT: lowland dipterocarp forest.
ALTITUDINAL RANGE: sea level – 200m.
DISTRIBUTION: Borneo (K, SA, SR). Also Peninsular Malaysia, Thailand.

T.hispidissima *(Ridl.) Kraenzl.* in Engler, Pflanzenr. Orch. Monand. IV. 50. II. B. 21:136 (1911). Type: Peninsular Malaysia, Selangor, Ulu Langat, *Goodenough* s.n. (holo. SING).
Eria hispidissima Ridl., in J. Bot. 36:213 (1898).

HABITAT: unknown.
ALTITUDINAL RANGE: unknown.
DISTRIBUTION: Borneo (unspecified). Also Peninsular Malaysia.

T.jejuna *(J.J.Sm.) J.J.Wood,* **comb.nov.**
Eria jejuna J.J.Sm. in Bull. Jard. Bot. Buitenzorg, ser. 3, 11:133 (1931). Type: Kalimantan, Gunung Kemoel, *Endert* 3603 (holo. L).

HABITAT: lower montane forest.
ALTITUDINAL RANGE: 1200m.
DISTRIBUTION: Borneo (K). Endemic.

T.lacinulata *(J.J.Sm.) Carr* in Gard. Bull. Straits Settlem. 8:101 (1935). Type: Kalimantan, Bukit Raja, *Winkler* 996 (holo. HBG).
Eria lacinulata J.J.Sm. in Mitt. Inst. Allg. Bot. Hamburg 7:49, t.7, fig.39 (1927).

HABITAT: unknown.
ALTITUDINAL RANGE: 1400m.
DISTRIBUTION: Borneo (K). Endemic.

T.lawiensis *(J.J.Sm.) J.J.Wood,* **comb.nov.**
Eria lawiensis J.J.Sm. in Bull. Jard. Bot. Buitenzorg, ser. 2, 3:62 (1912). Type: Sarawak, Gunung Batu Lawi, Ulu Limbang, *Moulton* 14 (holo. BO, iso. SAR).

HABITAT: lower montane forest.
ALTITUDINAL RANGE: 700 – 800m.
DISTRIBUTION: Borneo (SR). Endemic.

NOTE: This species is incorrectly cited as *'Dendrochilum lawiense'* by Ames (1921) & Masamune (1942).

T.microphylla *Blume,* Bijdr.:343 (1825). Type: Java, Tjitjalobak, Salak, *Blume* s.n. (holo. BO).
Eria microphylla (Blume) Blume, Mus. Bot. Lugd. Bat. 2:184 (1856).

HABITAT: swamp forest; oak-laurel forest; lower montane forest.
ALTITUDINAL RANGE: sea level – 1500m.
DISTRIBUTION: Borneo (K, SA, SR). Also Peninsular Malaysia, Thailand, Vietnam, Sumatra, Java.

T.mollicaulis *(Ames & C.Schweinf.) J.J.Wood* in Wood, Beaman & Beaman, Plants of Mt. Kinabalu 2, Orchids: 335 (1993). Type: Sabah, Mt. Kinabalu, Kiau, *Clemens* 66 (holo. AMES, iso. SING).
Eria mollicaulis Ames & C.Schweinf., Orch. 6:131 (1920).

HABITAT: hill and lower montane forest; lower montane Rhododendron scrub with field layer of *Dipteris conjugata.*
ALTITUDINAL RANGE: 900 – 1700m.
DISTRIBUTION: Borneo (K, SA). Endemic.

T.odoardi *Kraenzl.* in Bot. Jahrb. Syst. 44, Beibl. 101:21 (1910). Type: Sarawak, *Beccari* 1873 (holo. FI).
Eria odoardi (Kraenzl.) Ames in Merrill, Bibl. Enum. Born. Pl.:172 (1921).

A checklist of Borneo Orchids

HABITAT: unknown.
ALTITUDINAL RANGE: unknown.
DISTRIBUTION: Borneo (SR). Endemic.

T.pauciflora *Blume,* Bijdr.:343, fig.11 (1825). Type: Java, Pantjar, *Blume* s.n. (holo. BO).
Eria pauciflora (Blume) Blume, Mus. Bot. Lugd. Bat. 2:183 (1856).
E.tuberosa Hook.f., Fl. Brit. Ind. 5:807 (1890). Type: Peninsular Malaysia, Perak, *Scortechini* 1118 (holo. K).

HABITAT: hill forest on limestone.
ALTITUDINAL RANGE: 300 – 500m.
DISTRIBUTION: Borneo (SA, SR). Also Peninsular Malaysia, Sumatra, Java, Bali.

T.pilosissima *(Rolfe) J.J.Wood* in Wood, Beaman & Beaman, Plants of Mt. Kinabalu 2, Orchids: 336 (1993). Type: Sabah, Mt. Kinabalu, above Lobang, *Gibbs* 4117 (holo. K, iso. BM).
Eria pilosissima Rolfe in J. Linn. Soc., Bot.42:152 (1914).

HABITAT: lower montane forest; oak-laurel forest; sometimes on ultramafic substrate.
ALTITUDINAL RANGE: 1200 – 2400m.
DISTRIBUTION: Borneo (B, SA, SR). Endemic.

T.rubiginosa *(Blume) Kraenzl.* in Engler, Pflanzenr. Orch. Monand. IV. 50. II. B. 21:155 (1911). Type: Kalimantan, Sakumbang, *Korthals* s.n. (holo. BO).
Eria rubiginosa Blume, Mus. Bot. Lugd. Bat. 2:184 (1856).

HABITAT: hill dipterocarp forest; lower montane ridge forest.
ALTITUDINAL RANGE: 100 – 1700m.
DISTRIBUTION: Borneo (K, SA, SR). Endemic.

T.sarawakensis *Carr* in Gard. Bull. Straits Settlem. 8:100 (1935). Type: Sarawak, Dulit ridge, *Synge* S.402 (holo. SING, iso. K).
Eria sarawakensis (Carr) Masam., Enum. Phan. Born.:184 (1942).

HABITAT: moss forest.
ALTITUDINAL RANGE: 1200m.
DISTRIBUTION: Borneo (SR). Endemic.

T.spathulata *(J.J.Sm.) Kraenzl.* in Engl., Pflanzenr. Orch. Monand. IV. 50. II. B. 21:140 (1911). Type: Kalimantan, Kapoeas, *Teysmann* s.n. (holo. BO).
Eria spathulata J.J.Sm. in Bull. Dép. Agric. Indes Néerl. 22:28 (1909).

HABITAT: hill forest; on limestone.
ALTITUDINAL RANGE: 800m.
DISTRIBUTION: Borneo (K). Endemic.

T.teysmannii *(J.J.Sm.) Kraenzl.* in Engl., Pflanzenr. Orch. Monand. IV. 50. II. B. 21:145 (1911). Types: Kalimantan, Salimbouw, Kapoeas, *Teysmann* s.n.,'Sabalouw' = Sabalomo, *Teysmann* s.n. (syn. BO); Sungei Sambas, *Hallier* 1134 (syn. BO, isosyn. K).
Eria teysmannii J.J.Sm. in Bull. Dép. Agric. Indes Néerl. 22:29 (1909).
Trichotosia dajakorum Kraenzl. in Bot. Jahrb. Syst. 44, Beibl. 101:22 (1910). Type: Kalimantan, Sebalomo, *Teysmann* 10894 (holo. K).
Eria dajakorum (Kraenzl.) Ames in Merr., Bibl. Enum. Born. Pl.:169 (1921).

HABITAT: lowland forest; riverine forest.
ALTITUDINAL RANGE: lowlands – 500m.

DISTRIBUTION: Borneo (K, SR). Also Siberut Island off Sumatra.

T.unguiculata *(J.J.Sm.) Kraenzl.* in Engl., Pflanzenr. Orch. Monand. IV. 50. II. B. 21:147 (1911). Type: Kalimantan, Bukit Batoe Ajoh, *Jaheri* 1653 (holo. BO).
  Eria unguiculata J.J.Sm. in Bull. Dép. Agric. Indes Néerl. 22:30 (1909).

HABITAT: shrubbery on exposed limestone.
ALTITUDINAL RANGE: 1700 – 1800m.
DISTRIBUTION: Borneo (K, SR). Endemic.

T.velutina *(Lodd.ex Lindl.) Kraenzl.* in Engl., Pflanzenr. Orch. Monand. IV. 50. II. B. 21:140 (1911). Type: Singapore, *Cuming* s.n. (holo. K-LINDL).
  Eria velutina Lodd.ex Lindl. in Bot. Reg. 26, misc. 86 no. 209 (1840).

HABITAT: lowland riverine forest; secondary forest; swamp forest.
ALTITUDINAL RANGE: lowlands.
DISTRIBUTION: Borneo (B, K, SA, SR). Also Burma, Peninsular Malaysia, Singapore, Thailand, Vietnam, Sumatra, Java, Lingga Archipelago, Bangka, Papua New Guinea(doubtful).

T.vestita *(Lindl.) Kraenzl.* in Engl., Pflanzenr. Orch. Monand. IV. 50. II. B. 21:151, fig.32 (1911). Type: Singapore, *Wallich* 2005 (holo. K-LINDL).
  Eria vestita Lindl. in Bot. Reg. 30, misc. 76, no. 79 (1844).

HABITAT: riverine forest; podsol forest; hill forest.
ALTITUDINAL RANGE: sea level – 500m.
DISTRIBUTION: Borneo (K, SA, SR). Also Peninsular Malaysia, Sumatra.

SUBTRIBE PODOCHILINAE

APPENDICULA Blume

Bijdr.:297 (1825).

Epiphytic, lithophytic, rarely terrestrial herbs. Stems erect or pendulous, simple or branched, pseudobulbs absent. Leaves distichous, flat, twisted at the base so that the blades all lie in one plane. Inflorescence terminal, lateral or both, short or long. Flowers small, resupinate, white or greenish; lateral sepals connate at base to column-foot to form a mentum; lip with a round or concave basal appendage, sometimes lengthened into small keels, mid-lobe often with a median keel or callus; column with a foot; pollinia 6, on a solitary forked caudicle or 2 separate ones.

About 60 species distributed from tropical Asia to the Pacific islands.

A.anceps *Blume*, Bijdr.:299 (1825). Type: West Java, Blume s.n. (holo. BO).

HABITAT: lowland, hill and lower montane forest; moss forest; alluvial forest; mixed dipterocarp forest.
ALTITUDINAL RANGE: sea level – 1600m.
DISTRIBUTION: Borneo (B, K, SA, SR). Also Peninsular Malaysia, Thailand, Sumatra, Riau Archipelago, Bangka, Sulawesi, Natuna, Philippines.

A.babiensis *J.J.Sm.* in Bot. Jahrb. Syst. 48:102 (1912). Type: Kalimantan, Batu babi, *Winkler* 2805 (holo. BO).

HABITAT: riverine forest.
ALTITUDINAL RANGE: 400m.
DISTRIBUTION: Borneo (K). Endemic.

A.buxifolia *Blume*, Bijdr.:300(1825). Type: Java, Pantjar, *Blume* s.n. (holo. BO, iso. K).
Dendrobium lycopodioides Lindl. in J. Linn. Soc., Bot. 3:13 (1859). Type: Sarawak, *Lobb* s.n. (holo. K-LINDL).
D.tmesipteris Lindl. in J. Linn. Soc., Bot. 3:13 (1859). Type: Sarawak, *Lobb* s.n. (holo. K-LINDL).
Appendicula frutex Ridl. in J. Linn. Soc., Bot. 31:302 (1896). Type: Sarawak, Pengkula Ampat, *Haviland* 2337 (holo. SING, iso. K).
Podochilus buxifolius (Blume) Schltr. in Mém. Herb. Boissier 21:52 (1900).

HABITAT: lowland forest; lower montane forest.
ALTITUDINAL RANGE: 300 – 1500m., rarely around sea-level.
DISTRIBUTION: Borneo (K, SA, SR). Also Sumatra, Java, Philippines.

A.calcarata *Ridl.* in J. Linn. Soc., Bot. 31:302(1896). Type: Sarawak, *Haviland* 2334 & 3145 (syn. SING, isosyn. K).
Podochilus calcaratus (Ridl.) Schltr. in Mém. Herb. Boissier 21:41(1900).

HABITAT: hill and lower montane forest.
ALTITUDINAL RANGE: 700 – 900m.
DISTRIBUTION: Borneo (K, SA, SR). Endemic.

A.congesta *Ridl.* in Stapf in Trans. Linn. Soc. London, Bot. 4:239 (1894). Type: Sabah, Mt. Kinabalu, Penokok, *Haviland* 1302 (holo. SING, iso. K).
Podochilus congestus (Ridl.) Schltr. in Mém. Herb. Boissier 21:59 (1900).
Chilopogon kinabaluensis Ames & C. Schweinf. Orch. 6:141, Pl.93 (1920). Type:

Sabah, Mt. Kinabalu, Marai Parai, *Clemens* 230 (holo. AMES, iso. BO, K).
Appendicula kinabaluensis (Ames & C.Schweinf.) J.J.Sm. in Bull. Jard. Bot. Buitenzorg, ser. 3, 5:65 (1922).

HABITAT: hill and lower montane forest; oak-laurel forest.
ALTITUDINAL RANGE: 700 – 2100m.
DISTRIBUTION: Borneo (SA, SR). Endemic.

A.cornuta *Blume*, Bijdr.:302 (1825). Type: Java, Seribu and Pantjar, *Blume* s.n.(holo. BO).

HABITAT: hill dipterocarp forest; lower montane forest; open podsol forest; *Agathis/Lithocarpus* forest; low scrub on limestone.
ALTITUDINAL RANGE: 500 – 1400m.
DISTRIBUTION: Borneo (B, K, SA, SR). Widespread in S.E. Asia, west to India, north to Hong Kong and east to Sulawesi.

A.crispa *J.J.Sm.* in Bull. Jard. Bot. Buitenzorg, ser. 3, 11:127 (1931). Type: Kalimantan, Gunung Kemoel, *Endert* 3804 (holo. L).

HABITAT: primary forest.
ALTITUDINAL RANGE: 1100m.
DISTRIBUTION: Borneo (K). Endemic.

A.cristata *Blume*, Bijdr.:298 (1825). Type: Java, Pantjar, *Blume* s.n. (holo. BO).
Appendicula divaricata Ames & C.Schweinf., Orch. 6:143 (1920). Type: Sabah, Mt. Kinabalu, Kiau, *Clemens* 137 (holo. AMES).

HABITAT: hill and lower montane forest; podsol forest with undergrowth of *Pandanus* and rattan; *Gymnostoma sumatrana* forest on ultramafic substrate; mixed dipterocarp forest.
ALTITUDINAL RANGE: 600 – 1500m.
DISTRIBUTION: Borneo (B, K, SA). Also Sumatra, Java, Sulawesi.

A.dajakorum *J.J.Sm.* in Bull. Jard. Bot. Buitenzorg, ser. 3, 8:44 (1926). Types: Kalimantan, Amai Ambit, *Hallier* 3212 & 3197; Liang Gagang, *Hallier* 2808; Damoes, *Hallier* 622 (syn. L).

HABITAT: unknown.
ALTITUDINAL RANGE: unknown.
DISTRIBUTION: Borneo (K). Endemic.

A.damusensis *J.J.Sm.* in Bull. Jard. Bot. Buitenzorg, ser. 2, 8:44 (1912). Type: Kalimantan, Damoes, *Hallier* 505 (holo. L).

HABITAT: unknown.
ALTITUDINAL RANGE: unknown.
DISTRIBUTION: Borneo (K). Endemic.

A.floribunda *(Schltr.) Schltr.* in Feddes Repert. Beih. 1:355 (1912). Type: Kalimantan, locality unknown, *Korthals* s.n.(holo. B, destroyed).
Podochilus floribundus Schltr. in Mém. Herb. Boissier, 21:58 (1900).

HABITAT: hill forest on ultramafic substrate.
ALTITUDINAL RANGE: 600m.
DISTRIBUTION: Borneo (SA). Also Sumatra.

A.foliosa *Ames & C.Schweinf.*, Orch. 6:145 (1920). Type: Sabah, Mt. Kinabalu, Kiau, *Clemens* 361 (holo. AMES, iso. BO, K, SING).

HABITAT: hill and lower montane forest; dipterocarp forest; oak-laurel forest; sometimes on ultramafic substrate.
ALTITUDINAL RANGE: 100 – 2400m.
DISTRIBUTION: Borneo (SA, SR). Endemic.

A.fractiflexa *J.J.Wood* in Wood, Beaman & Beaman, Plants of Mt. Kinabalu, 2, Orchids: 89, fig.7 (1993). Type: Sabah, Mt. Kinabalu, Pinosuk Plateau, between East Mesilau River & Menteki River, *Wood* 827 (holo. K).

HABITAT: lower montane forest.
ALTITUDINAL RANGE: 1200 – 1700m.
DISTRIBUTION: Borneo (SA). Endemic.

A.linearifolia *Ames & C.Schweinf.*, Orch. 6:147 (1920). Type: Sabah, Mt. Kinabalu, Marai Parai, *Clemens* 286 (holo. AMES, iso. BO, K, SING).

HABITAT: lower montane forest on ultramafic substrate.
ALTITUDINAL RANGE: 1400 – 2100m.
DISTRIBUTION: Borneo (SA). Endemic.

A.longirostrata *Ames & C.Schweinf.*, Orch. 6:149 (1920). Type: Sabah, Mt. Kinabalu, Marai Parai, *Clemens* 387 (holo. AMES, iso. BO, K, SING.).

HABITAT: lower montane forest; oak-laurel forest; secondary forest; rocky roadside banks with *Arundina*, ferns, etc.; sometimes on ultramafic substrate.
ALTITUDINAL RANGE: 1100 – 2700m.
DISTRIBUTION: Borneo (B, SA, SR). Endemic.

A.lucida *Ridl.* in J. Linn. Soc., Bot. 32:392 (1896). Types: Singapore, Kranji, *Ridley* s.n.(syn. SING, isosyn. K); Chan Chu Kang, *Ridley* s.n.(syn. SING); Peninsular Malaysia, Selangor, Seppan, *Ridley* s.n.(syn. SING).

HABITAT: mangrove; riverine forest; lowland dipterocarp forest; low stature forest on ultramafic substrate; hill dipterocarp, *Fagaceae, Gymnostoma* and *Agathis* forest.
ALTITUDINAL RANGE: sea level – 1000m.
DISTRIBUTION: Borneo (K, SA, SR). Also Peninsular Malaysia, Singapore, Riau Archipelago.

A.magnibracteata *Ames & C.Schweinf.*, Orch. 6:151 (1920). Type: Sabah, Mt. Kinabalu, Marai Parai, *Clemens* 282 (holo. AMES).

HABITAT: hill and lower montane forest, sometimes on ultramafic substrate.
ALTITUDINAL RANGE: 900 – 1400m.
DISTRIBUTION: Borneo (SA). Endemic.

A.merrillii *Ames* in Philipp. J. Sci. 8:418 (1913). Type: Philippines, Mindanao, Zamboanga District, *Merrill* 8135 (holo. AMES, iso. K).
Chilopogon merrillii (Ames) Ames, Orch. 5:91 (1915).

HABITAT: unknown.
ALTITUDINAL RANGE: unknown.
DISTRIBUTION: Borneo (SA). Also Philippines.

A.minutiflora *Ames & C.Schweinf.*, Orch. 6:153 (1920). Type: Sabah, Mt. Kinabalu, Kiau, *Clemens* 333 (holo. AMES).

HABITAT: hill forest on ultramafic substrate.
ALTITUDINAL RANGE: 400 – 500m.
DISTRIBUTION: Borneo (SA). Endemic.

A.pauciflora *Blume*, Bijdr.:300 (1825). Type: Java, Gede, *Blume* s.n. (holo. BO).

HABITAT: mixed dipterocarp forest on shale.
ALTITUDINAL RANGE: lowlands.
DISTRIBUTION: Borneo (B). *A.pauciflora* occurs in Sumatra and Java.

NOTE: The specimen *Boyce* 371 (K) is similar in habit to *A.pauciflora* but has larger flowers.

A.pendula *Blume*, Bijdr.:298 (1825). Type: Java, Buitenzorg, Bantam and Tjanjor Provinces, *Blume* s.n. (holo. BO, iso. K).
  A.latibracteata J.J.Sm. in Bull. Jard. Bot. Buitenzorg, ser. 2, 13:37 (1914). Type: Kalimantan, locality unknown, *Nieuwenhuis* 1896, cult. Bogor. (holo. BO).
  A.latibracteata J.J.Sm. var. *rajana* J.J.Sm. in Mitt. Inst. Allg. Bot. Hamburg, 7:48, t.7, fig.38 (1927). Type: Kalimantan, Bukit Raja, *Winkler* 925 (holo. HBG).

HABITAT: lowland and hill dipterocarp forest; alluvial forest; lower montane forest; wet mossy rocks; trees overhanging rivers; tree fern trunks; peat swamp forest.
ALTITUDINAL RANGE: sea level – 1800m.
DISTRIBUTION: Borneo (B, K, SA, SR). Also Peninsular Malaysia, Thailand, Sumatra, Java, Natuna.

A.pilosa *J.J.Sm.* in Icon. Bogor. 2:53, t.110A (1903). Type: Kalimantan, Sungai Tjehan, *Nieuwenhuis* s.n.(holo. BO).
  A.niahensis Carr in Gard. Bull. Straits Settlem. 8:95 (1935). Type: Sarawak, Niah, *Synge* S.601(holo. SING, iso. K).

HABITAT: lowland forest on steep sandstone slopes and scree; dipterocarp forest.
ALTITUDINAL RANGE: sea level – 700m.
DISTRIBUTION: Borneo (K, SR). Endemic.

A.polita *J.J.Sm.* in Bull. Dép. Agric. Indes Néerl. 22:41 (1909). Type: Kalimantan, Sungai Kenepai and Sungai Sekedouw, *Hallier* 2115 (holo. BO, iso. K).

HABITAT: unknown.
ALTITUDINAL RANGE: unknown.
DISTRIBUTION: Borneo (K). Endemic.

A.purpureifolia *J.J.Sm.* in Bull. Jard. Bot. Buitenzorg, ser. 3, 8:43 (1926). Type: Kalimantan, Bukit Tiloeng, *Dakkus* s.n., cult. Bogor no. 340 (holo. L).

HABITAT: unknown.
ALTITUDINAL RANGE: lowlands.
DISTRIBUTION: Borneo (K). Endemic.

A.recondita *J.J.Sm.* in Bull. Jard. Bot. Buitenzorg, ser. 3, 11: 128 (1931). Type: Kalimantan, West Koetai, Long Hoet, *Endert* 2532 (holo. L).

HABITAT: lowland forest.
ALTITUDINAL RANGE: 100 – 600m.
DISTRIBUTION: Borneo (K, SA). Endemic.

A.reflexa *Blume*, Bijdr.:301 (1825). Type: Java, Tjapus, Pantjar, *Blume* s.n. (holo. BO).

HABITAT: lowland dipterocarp forest.
ALTITUDINAL RANGE: sea level – 300m.
DISTRIBUTION: Borneo (B, K, SA, SR). Widespread all over S.E. Asia, eastwards to New Guinea and the Pacific islands.

A.rostellata *J.J.Sm.* in Icon. Bogor. 4:13, t.305 (1910). Type: Kalimantan, Liang Gagang, *Hallier* s.n.(holo. L).

HABITAT: hill forest on sandstone.
ALTITUDINAL RANGE: 300 – 900m.
DISTRIBUTION: Borneo (K, SA). Endemic.

A.rupicola *(Ridl.) Rolfe* in Gibbs in J. Linn. Soc., Bot. 42:159 (1914). Types: Sarawak, Bidi, *Ridley* 11792; Batu, *Hewitt* s.n. (syn. SING).
Podochilus rupicolus Ridl. in J. Straits Branch Roy. Asiat. Soc. 50:142 (1908).

HABITAT: hill and secondary forest.
ALTITUDINAL RANGE: 300m.
DISTRIBUTION: Borneo (SA, SR). Endemic.

A.spathilabris *J.J.Sm.* in Bull. Jard. Bot. Buitenzorg, ser. 3, 11:129 (1931). Type: Kalimantan, West Koetai, Long Hoet, *Endert* 2621 (holo. L).

HABITAT: riverine forest.
ALTITUDINAL RANGE: 100 – 200m.
DISTRIBUTION: Borneo (K). Endemic.

A.tenuifolia *J.J.Wood* in Kew Bull. 39(1):90, fig.11 (1984). Type: Sarawak, Gunung Mulu National Park, *Hansen* 166 (holo. C, iso. K).

var. **tenuifolia**

HABITAT: lower montane forest on steep mossy limestone slopes; riparian forest.
ALTITUDINAL RANGE: 600 – 900m.
DISTRIBUTION: Borneo (SR). Endemic.

var. **filiformis** *J.J.Wood* in Kew Bull. 39(1):90, fig.12 (1984). Type: Sarawak, Gunung Mulu National Park, Sungai Berar, *Lamb* MAL 11 (holo. K).

HABITAT: podsol forest; mixed dipterocarp forest with a kerangas element on sandstone and shales.
ALTITUDINAL RANGE: 100 – 200m.
DISTRIBUTION: Borneo (SR).

A.torta *Blume*, Bijdr.:303 (1825). Type: Java, Seribu and Pantjar, *Blume* s.n. (holo. BO, iso. K).

HABITAT: hill forest; lower montane ridge forest; dipterocarp, Fagaceae, *Gymnostoma, Agathis* forest; oak-laurel forest; steep roadside banks with sandstone and shale outcrops; very low and open podsol forest; mostly on ultramafic substrate.
ALTITUDINAL RANGE: 300 – 1600m.
DISTRIBUTION: Borneo (K, SA, SR). Also Peninsular Malaysia, Sumatra, Java.

A.uncata *Ridl.* in J. Linn. Soc., Bot. 32:390 (1896)

subsp. **sarawakensis** *J.J.Wood* in Kew Bull. 39(1):92, fig.13 (1984). Type: Sarawak, Gunung Mulu National Park, plateau N.W. of Melinau Gorge, *Nielsen* 387(holo. AAU).

HABITAT: kerangas forest; on root pneumatophores in peat swamp forest; lower montane forest; sclerophyllous ridge vegetation.
ALTITUDINAL RANGE: sea level – 700m.
DISTRIBUTION: Borneo (SA, SR). Endemic.

NOTE: *A.uncata* subsp. *uncata* is found in Peninsular Malaysia.

A.undulata *Blume*, Bijdr.:301 (1825). Type: Java, Pantjar, *Blume* s.n. (holo. BO).

var. **undulata**

HABITAT: riverine forest.
ALTITUDINAL RANGE: 200 – 500m.
DISTRIBUTION: Borneo (K, SA, SR). Also Peninsular Malaysia, Sumatra, Java.

var. **longicalcarata** *(Rolfe) Ames* in Merr., Bibl. Enum. Born. Pl.:179 (1921). Type: Sarawak, *Lobb* s.n., introduced by Messrs. Linden (holo. K).
Podochilus longicalcaratus Rolfe in Kew Bull.:186 (1894).
P.undulatus (Blume) Schltr. var. *longicalcaratus* (Rolfe) Schltr. in Mém. Herb. Boissier, 21:43 (1900).

HABITAT: lowland forest; mixed hill dipterocarp forest; forest on dolomitic limestone; riparian forest.
ALTITUDINAL RANGE: sea level – 300m.
DISTRIBUTION: Borneo (K, SA, SR). Also Sumatra, Java, Philippines.

A.xytriophora *Rchb.f.* in Seem., Fl. Vit.:299 (1868). Type: ?Philippines, *Cuming* 2149 (holo. W).

HABITAT: lower montane forest on limestone; moss forest on limestone.
ALTITUDINAL RANGE: 900 – 1200m.
DISTRIBUTION: Borneo (SR). Doubtfully occurring in Philippines.

POAEPHYLLUM Ridl.

Mat. Fl. Malay Pen. 1:108 (1907).

Epiphytic, rarely terrestrial herbs. Stems simple, leafy, covered in sheathing leaf-bases and resembling *Appendicula*. Leaves narrow, distichous. Inflorescence lateral, axillary, usually short. Flowers minute to medium-sized, resupinate or non-resupinate; sepals acute, laterals adnate to column-foot forming a short mentum; petals often smaller and narrower than sepals; lip entire to 3-lobed, lacking a basal appendage and always adnate to the sides of the column-foot to form a sac; column with a short foot; pollinia 8.

Ten species distributed from the Nicobar Islands to Peninsular Malaysia, Sumatra, Java, Borneo, Sulawesi, the Philippines and New Guinea.

P.hansenii *J.J.Wood* in Kew Bull. 39(1):88, fig.10 (1984). Type: Sarawak, Gunung Mulu National Park, confluence of Sungai Lansat and Sungai Tutuh, *Hansen* 40 (holo. C, iso. K).

HABITAT: steep river banks in alluvial forest; riverine forest.
ALTITUDINAL RANGE: lowlands – 600m.
DISTRIBUTION: Borneo (K, SR). Endemic.

P.pauciflorum *(Hook.f.) Ridl.*, Mat. Fl. Mal. Pen. 1:109 (1907). Type: Peninsular Malaysia, Perak, Hulu Kwantu, *Scortechini* 929 (holo. K).
   Agrostophyllum pauciflorum Hook.f., Fl. Brit. Ind., 5:824 (1890).

HABITAT: lowland and hill forest.
ALTITUDINAL RANGE: sea level – 600m.
DISTRIBUTION: Borneo (SA, SR). Also Peninsular Malaysia, Sumatra, Java, Philippines (Sulu Archipelago).

P.podochiloides *(Schltr.) Ridl.* in Trans. Linn. Soc. London, Bot. 9:192 (1916). Type: Papua New Guinea, Torricelli Mountains, *Schlechter* 14367 (holo. B,destroyed, iso. K).
   Eria podochiloides Schltr. in K. Schum. & Lauterb., Nachtr. Fl. Schutzgeb. Südsee:182 (1905).

HABITAT: lower montane forest on limestone.
ALTITUDINAL RANGE: 1000m.
DISTRIBUTION: Borneo (K). Also New Guinea.

P.uniflorum *J.J.Wood* in Lindleyana 5(2):97, fig.11 (1990). Type: Sabah, Crocker Range, Kimanis Road, *Clements* 3228 (holo. K).

HABITAT: roadside cuttings on shale, with *Melastoma, Nepenthes, Lycopodium, Gahnia,* ferns, etc.
ALTITUDINAL RANGE: 1000 – 1200m.
DISTRIBUTION: Borneo (SA). Endemic.

PODOCHILUS Blume

Bijdr.:295, t.12 (1825).

Small, rather delicate epiphytes or lithophytes, often forming dense mats. Stems slender, tufted, leafy, erect to spreading. Leaves short, distichous, lying in one plane by the twisting of the sheathing bases, not jointed on the sheaths. Inflorescences terminal, lateral or both, rarely exceeding 2cm long, few- to many-flowered. Flowers minute or small, resupinate, white or green, often with purple markings; sepals and petals adnate at base, or free; lateral sepals adnate to column-foot forming an often spur-like mentum; petals smaller than sepals; lip narrow, entire or obscurely 3-lobed, with a simple or bilobed basal appendage; column short, foot long and often curved upwards; pollinia 4.

About 60 species distributed from India and Sri Lanka to China, south and east through Indonesia and New Guinea to the Pacific islands.

P.auriculigerus *Schltr.* in Notizbl. Bot. Gart. Berlin-Dahlem, 8:18 (1921). Type: Sarawak, Gunung Mattan, *Beccari* 1881 (holo. B,destroyed, iso. FI).

HABITAT: unknown.
ALTITUDINAL RANGE: unknown.
DISTRIBUTION: Borneo (SR). Endemic.

P.bilabiatus *J.J.Sm.* in Bull. Jard. Bot. Buitenzorg, ser. 3, 2:33 (1920). Type: Sumatra, Benkoelen, Bukit Barisan, Rimbo Pengadang, *Ajoeb* 87 (holo. L).

HABITAT: primary forest.
ALTITUDINAL RANGE: 500m.

DISTRIBUTION: Borneo (K). Also Sumatra.

P.cucullatus *J.J.Sm.* in Bull. Dép. Agric. Indes Néerl., 22:40 (1909). Type: Sarawak, Mt. Poe, *Hewitt* s.n. (holo. BO).

HABITAT: unknown.
ALTITUDINAL RANGE: unknown.
DISTRIBUTION: Borneo (K, SR). Endemic.

P.densiflorus *Blume*, Rumphia 4:44, t.192, fig.5, t.200B (1848). Types: New Guinea, *Zippelius* s.n., *Latour* s.n. (syn. BO).

HABITAT: unknown.
ALTITUDINAL RANGE: unknown.
DISTRIBUTION: Borneo (unspecified). Also New Guinea.

P.gracilis *(Blume) Lindl.*, Gen Sp. Orch. Pl.235 (1833). Type: Java, Pantjar and Seribu, *Blume* s.n.(holo. BO).
Platysma gracilis Blume, Bijdr.:296, t.43 (1825).

HABITAT: lowland forest.
ALTITUDINAL RANGE: 100 – 200m.
DISTRIBUTION: Borneo (K). Also Sumatra, Java, Bali.

P.lucescens *Blume*, Bijdr.:295, Pl.12 (1825). Type: Java, Salak, Pantjar, Meggamedung, etc., *Blume* s.n.(holo. BO).

HABITAT: lowland and hill mixed dipterocarp forest; lower montane forest; mossy forest on ultramafic substrate, with *Gymnostoma* and undergrowth of climbing bamboo; mossy rocks, near waterfalls; recorded among moss on buttresses of *Shorea*.
ALTITUDINAL RANGE: sea level – 1500m.
DISTRIBUTION: Borneo (B, K, SA, SR). Also Burma, Thailand, Peninsular Malaysia, Sumatra, Java, Lingga Archipelago, Anambas Islands, Natuna, Sulawesi, Philippines.

P.malabaricus *Wight*, Icon. Pl. Ind. Or. 5(1):20, Pl.1748, fig.2 (1851). Type: India, Malabar, *Jerdon* s.n. (holo. not located).

HABITAT: unknown.
ALTITUDINAL RANGE: unknown.
DISTRIBUTION: Borneo (SA). Also India, Sri Lanka.

P.microphyllus *Lindl.*, Gen. Sp. Orch. Pl.234 (1833). Type: Peninsular Malaysia, Pinang, *Porter* s.n., Wallich Cat. No.7335a (holo. K-LINDL, K-WALL).

HABITAT: lowland, hill and lower montane forest; damp mossy rocks; streamsides.
ALTITUDINAL RANGE: 200 – 1400m.
DISTRIBUTION: Borneo (B, K, SA, SR). Also Burma, Cambodia, Vietnam, Thailand, Peninsular Malaysia, Sumatra, Java, Lingga Archipelago.

P.obovatipetalus *J.J.Sm.* in Mitt. Inst. Allg. Bot. Hamburg, 7:47, t.7, fig.37 (1927). Type: Kalimantan, Bukit Mehipit, *Winkler* 622a (holo. HBG, iso. BO).

HABITAT: lower montane forest.
ALTITUDINAL RANGE: 500 – 1100m.
DISTRIBUTION: Borneo (K). Endemic.

P.oxyphyllus *Schltr.* in Bull. Herb. Boissier, ser. 2, 6:308 (1906). Type: Kalimantan, Koetai, Long Dett, *Schlechter* 13558 (holo. B, destroyed).

HABITAT: unknown.
ALTITUDINAL RANGE: sea level.
DISTRIBUTION: Borneo (SR). Endemic.

P.sciuroides *Rchb.f.* in Bonplandia 5:41 (1857). Type: Sumatra, Lampong, *Zollinger* 3090 (holo. W).

HABITAT: hill forest; on rocks in lower montane forest; low stature forest on ultramafic substrate; montane Fagaceae, Dipterocarpaceae, *Agathis* forest.
ALTITUDINAL RANGE: 100 – 1700m.
DISTRIBUTION: Borneo (K, SA, SR). Also Sumatra, Riau Archipelago, Sumbawa.

P.serpyllifolius *(Blume) Lindl.* in J. Linn. Soc., Bot. 3:37 (1859). Type: Java, Seribu, *Blume* s.n. (holo. BO).
Cryptoglottis serpyllifolia Blume, Bijdr.:297 (1825).

HABITAT: mixed hill dipterocarp forest including *Tristania*; mossy rocks, saplings, etc. in lower montane forest.
ALTITUDINAL RANGE: 700 – 1800m.
DISTRIBUTION: Borneo (B, SA, SR). Also Sumatra, Java.

P.similis *Blume* in Rumphia 4:44, t.200A (1848). Type: Borneo, *Blume* s.n. (holo. BO).

HABITAT: mangrove forest.
ALTITUDINAL RANGE: sea level.
DISTRIBUTION: Borneo (K, SR).

P.spathulatus *J.J.Sm.* in Bull. Dép. Agric. Indes Néerl. 22:40 (1909). Type: Kalimantan, *Jaheri* s.n. (holo. BO).

HABITAT: unknown.
ALTITUDINAL RANGE: unknown.
DISTRIBUTION: Borneo (K). Endemic.

P.tenuis *(Blume) Lindl.*, Gen. Sp. Orch. Pl.235 (1833). Type: Java, Buitenzorg and Bantam Provinces, *Blume* s.n. (holo. BO).
Apista tenuis Blume, Bijdr.:296 (1825).

HABITAT: riparian forest; lower montane forest; low stature forest on ultramafic substrate; oak-laurel forest.
ALTITUDINAL RANGE: lowlands – 1900m.
DISTRIBUTION: Borneo (K, SA, SR). Also Peninsular Malaysia, Sumatra, Java.

SUBTRIBE THELASIINAE

OCTARRHENA Thwaites

Enum. Pl. Zeyl.:305 (1861).

Epiphytic herbs. Stems elongate, leafy towards apex, branching from the rooting base. Leaves distichous, terete or laterally compressed, jointed on their equitant sheaths. Inflorescences lateral, short, spiciform, racemose, secund, densely many-flowered. Flowers minute, resupinate, greenish-yellow; sepals and petals free, spreading; petals smaller than sepals; lip entire, sessile at base of column, concave, sometimes uncinnate; column very short, without a foot; pollinia 8.

About 20 species distributed from Sri Lanka to the Pacific islands, the majority of which are endemic to New Guinea.

O.condensata *(Ridl.) Holttum* in Gard. Bull. Singapore 11:285 (1947). Type: Peninsular Malaysia, Gunung Tahan, *Robinson & Wray* 5487 (holo. BM).
 Oberonia condensata Ridl. in J. Linn. Soc., Bot. 38:322 (1908).

HABITAT: hill, lower and upper montane forest.
ALTITUDINAL RANGE: 900 – 1900m.
DISTRIBUTION: Borneo (K, SA, SR). Also Peninsular Malaysia, Sumatra, Java, New Guinea, Solomon Islands, Vanuatu.

O.parvula *Thwaites*, Enum. Pl. Zeyl.:305 (1861). Type: Sri Lanka, Central Province, *Thwaites* C.P.3072 (holo. K).

HABITAT: hill forest; oak-laurel forest; lower montane Fagaceae, *Agathis*, Dipterocarpaceae forest; open kerangas with *Eugenia*, *Ficus* and *Tristania*; very open dry forest on sandstone ridges.
ALTITUDINAL RANGE: 900 – 1800m.
DISTRIBUTION: Borneo (SA). Also Sri Lanka, Peninsular Malaysia, Sumatra, Java, Philippines.

PHREATIA Lindl.

Gen. Sp. Orch. Pl.:63 (1830).

Epiphytic herbs. Stems pseudobulbous or caulescent, the former 1- to 3-leaved, the latter with up to 12 leaves, very short or elongate. Leaves distichous or arranged in a fan, erect to spreading, terminal, sometimes fleshy, jointed on equitant sheaths. Inflorescences lateral or arising from base of pseudobulb, racemose, densely many-flowered. Flowers minute, resupinate, pale green or white, often self-pollinating; sepals and petals similar, spreading; lateral sepals adnate to column-foot to form a mentum; lip entire or obscurely 3-lobed, usually without a spur, concave; column short, with a distinct foot; pollinia 8.

About 150 species distributed from mainland Asia, through S.E. Asia east to Australia, New Guinea and the Pacific islands, most species occuring in Indonesia and New Guinea.

P.amesii *Kraenzl.* in Engl., Pflanzenr. Orch. Monand. IV. 50. II. B. 23:16 (1911). Types: Philippines, Luzon, Bataan Province, Mt. Mariveles, Lamao River, *Whitford* 1361 (syn. AMES); *Leiberg* 6061 and *Merrill* 3854 (syn. AMES).
   P. myosurus Ames, Orch. 2:203 (1908), non (Forst.) Ames.

HABITAT: lower montane forest.
ALTITUDINAL RANGE: 1200 – 1500m.
DISTRIBUTION: Borneo (SA). Also Philippines.

P.aristulifera *Ames*, Orch. 2:199 (1908). Type: Philippines, Luzon, Benguet Province, Mt.Santo Tomas, *Williams* 1955 (holo. NY).

HABITAT: lower montane forest.
ALTITUDINAL RANGE: 1200 – 1300m.
DISTRIBUTION: Borneo (SA). Also Philippines.

P.densiflora *(Blume) Lindl.*, Gen. Sp. Orch. Pl.:64 (1830). Type: Java, Tjapus, Salak, *Blume* s.n. (holo. BO).
   Dendrolirium densiflorum Blume, Bijdr.:350 (1825).

HABITAT: lower and upper montane forest, sometimes on ultramafic substrate.
ALTITUDINAL RANGE: 1200 – 2100m.
DISTRIBUTION: Borneo (K, SA, SR). Also Peninsular Malaysia, Thailand, Sumatra, Java, Maluku (Ambon, Aru Islands), Philippines.

P.listrophora *Ridl.* in J. Linn. Soc., Bot. 32:307 (1896). Types: Peninsular Malaysia, Perak, Larut Hills, *Ridley* s.n. (syn. SING); Kedah, Gunung Raya, *Curtis* s.n. (syn. SING).

HABITAT: 'sand forest'; lower montane oak-chestnut forest with *Agathis alba*, *Tristania* and small rattans; moss forest; riverside rocks.
ALTITUDINAL RANGE: 900 – 1800m.
DISTRIBUTION: Borneo (SA). Also Peninsular Malaysia, Thailand, Sulawesi, Seram.

P.monticola *Rolfe* in J. Linn. Soc., Bot. 42:152 (1914). Type: Sabah, Mt. Kinabalu, Kiau, *Gibbs* 3959 (holo. K).

HABITAT: hill and lower montane forest.
ALTITUDINAL RANGE: 800 – 1500m.
DISTRIBUTION: Borneo (K, SA). Endemic.

P.secunda *(Blume) Lindl.*, Gen. Sp. Orch. Pl.:64 (1930). Type: Java, Pantjar, *Blume* s.n. (holo. BO).
   Dendrolirium secundum Blume, Bijdr.:350 (1825).
   Phreatia minutiflora Lindl. in J. Linn. Soc., Bot. 3:62 (1859). Type: Borneo, *Lobb* s.n. (holo. K-LINDL).
   Eria minutiflora (Lindl.) Rchb.f. in Seemann, Fl. Vit.:301 (1868).

HABITAT: coastal forest; oak-laurel forest; lower montane forest and scrub; kerangas; hill dipterocarp forest; rocky roadside cuttings on shale.
ALTITUDINAL RANGE: sea level – 2000m.
DISTRIBUTION: Borneo (K, SA, SR). Also Vietnam, Thailand, Peninsular Malaysia, Sumatra, Mentawai Islands, Java, Seram, Sulawesi, Philippines.

P.sulcata *(Blume) J.J.Sm.*, Orch. Java: 505 (1905). Type: Java, Gede, *Blume* s.n. (holo. BO).
   Dendrolirium sulcatum Blume, Bijdr.:347 (1825).

HABITAT: lower montane forest.
ALTITUDINAL RANGE: 1100 – 1800m.
DISTRIBUTION: Borneo (K, SA). Also Peninsular Malaysia, Thailand, Sumatra, Java, Bali, Maluku(Ambon), Seram, Philippines.

THELASIS Blume

Bijdr.:385, t.75 (1825).

Epiphytic herbs. Stems either pseudobulbous, 1- or 2-leaved, with sheaths and sometimes additional smaller leaves at base, or short and unthickened with several leaves in 2 close opposite ranks, laterally compressed and overlapping at base. Leaves narrow, rather thin. Inflorescence lateral, many-flowered. Flowers very small, greenish-yellow or white, resupinate; sepals and petals similar, only spreading at the apex; column short, without a foot; pollinia 8, in two groups of 4.

About 20 species distributed from India eastwards to New Guinea and the Pacific islands.

T.borneensis *Schltr.* in Bull. Herb. Boissier, ser. 2, 6:340 (1906). Type: Kalimantan, West Koetei, Long Sele, *Schlechter* 13522 (holo. B, destroyed).

HABITAT: unknown.
ALTITUDINAL RANGE: unknown.
DISTRIBUTION: Borneo (K, SR). Endemic.

T.capitata *Blume*, Bijdr.:386 (1825). Type: Java, Salak, Pantjar, *Blume* s.n. (holo. BO).
T. ochreata Lindl. in J. Linn. Soc., Bot. 3:63 (1859). Type: Sarawak, *Lobb* s.n. (holo. K-LINDL).

HABITAT: hill dipterocarp forest on ultramafic substrate; podsol forest.
ALTITUDINAL RANGE: 600 – 1500m.
DISTRIBUTION: Borneo (K, SA). Also Peninsular Malaysia, Sumatra, Java, Christmas Island (Indian Ocean), Seram, Philippines.

T.carinata *Blume*, Bijdr.:386 (1825). Type: Java, Salak, *Blume* s.n. (holo. BO, iso. K).

HABITAT: coastal forest (on *Lumnitzera littorea*); lowland and hill forest, on limestone.
ALTITUDINAL RANGE: sea level – 900m.
DISTRIBUTION: Borneo (K, SA, SR). Also Peninsular Malaysia, Thailand, Sumatra, Java, Bali, Sulawesi, Buru, Philippines, New Guinea.

T.carnosa *Ames & C.Schweinf.*, Orch. 6:204 (1920). Type: Sabah, Mt. Kinabalu, Kiau, *Clemens* 88 (holo. AMES).

HABITAT: hill & lower montane forest.
ALTITUDINAL RANGE: 900m.
DISTRIBBUTION: Borneo (SA). Endemic.

T.cebolleta *J.J.Sm.* in Bull. Dép. Agric. Indes Néerl. 22:38 (1909). Type: Sarawak, Quop, *Hewitt* s.n. (holo. BO, iso. K).

HABITAT: lowland forest, on limestone.
ALTITUDINAL RANGE: lowlands.
DISTRIBUTION: Borneo (SA, SR). Endemic.

T.macrobulbon *Ridl.* in J. Linn. Soc., Bot. 32:393 (1896). Type: Peninsular Malaysia, Maxwell's Hill, *Ridley* s.n. (holo. SING).

HABITAT: unknown.
ALTITUDINAL RANGE: 200 – 300m.
DISTRIBUTION: Borneo (K, SR). Also Peninsular Malaysia.

T.micrantha *(Brogn.) J.J.Sm.* in Orch. Java: 495 (1905). Type: Irian Jaya, Waigiou, Offack, *Duperrey* s.n. (holo. P).
Oxyanthera micrantha Brogn. in Duperr., Voy. Coquill. Phan.:198, t.37B (1834).

HABITAT: lowland, hill and lower montane forest; mossy forest on limestone and ultramafic substrate.
ALTITUDINAL RANGE: lowlands – 1600m.
DISTRIBUTION: Borneo (B, K, SA, SR). Also Burma, Vietnam, Thailand, Peninsular Malaysia, Sumatra, Java and several islands east to Irian Jaya (Waigeo), north to the Philippines.

T.obtusa *Blume*, Bijdr.:386 (1825). Type: Java, Pangoerangoe, *Blume* s.n. (holo. BO).

HABITAT: hill dipterocarp forest; lower montane forest.
ALTITUDINAL RANGE: 300 – 1600m.
DISTRIBUTION: Borneo (SA, SR). Also Java.

T.pygmaea *(Griff.) Blume*, Fl. Javae, ser. 2, 1, Orch. (Coll. Orch. Arch. Ind.):22 (1858). Type: Nepal, *Lawrence* s.n. (holo. CAL).
Euproboscis pygmaea Griff. in Calcutta J. Nat. Hist. 5:371, t.26 (1845).
Thelasis triptera Rchb.f. in Bonplandia 3:219 (1855). Type: Philippines, *Cuming* 2062 (holo. W).
T.elongata Blume, Bot. Mus. Lugd. Bat. 2:187 (1856). Type: Java, Salak, Megamedung, etc., *Blume* s.n. (syn. BO).

HABITAT: peat swamp forest; mossy limestone rocks in lower montane forest; mixed dipterocarp forest.
ALTITUDINAL RANGE: sea level – 1500m.
DISTRIBUTION: Borneo (K, SA, SR). Also China, Hong Kong, India, Nepal, Andaman Islands, Burma, Vietnam, Thailand, Peninsular Malaysia, Sumatra, Mentawai Islands, Java, Bali, Lingga Archipelago, Natuna, Anambas, Sulawesi, Philippines, Papua New Guinea (Admiralty Islands), Solomon Islands.

T.variabilis *Ames & C.Schweinf.*, Orch. 6:205 (1920). Type: Sabah, Mt. Kinabalu, Kiau, *Clemens* 84 (holo. AMES).

HABITAT: hill and lower montane forest.
ALTITUDINAL RANGE: 800 – 1600m.
DISTRIBUTION: Borneo (SA). Endemic.

TRIBE DENDROBIEAE

SUBTRIBE DENDROBIINAE

DENDROBIUM Sw.

in Nova Acta Regiae Soc. Sci. Upsal., ser. 2, 6:82 (1799).

Epiphytic, lithophytic, rarely terrestrial, polymorphic herbs. Stems either (a) rhizomatous, (b) erect and many-noded, (c) erect and 1-noded or several-noded from a many-noded rhizome, or (d) without a rhizome, the new stems of many nodes arising from base of old ones; 1 or 2 cm to 5m long, tough or fleshy, swollen at base or along the whole length, often pseudobulbous, more or less covered with sheathing leaf-bases and cataphylls. Leaves 1 to many, borne at apex or distichously along stem, linear, lanceolate, oblong or ovate, papery or coriaceous, usually bilobed or emarginate at apex. Inflorescences racemose, 1- to many-flowered, erect, horizontal or pendulous, either lateral or terminal. Flowers often showy, small to large, resupinate or non-resupinate, ephemeral or long-lived. Sepals short to filiform. Lateral sepals adnate to the elongated column-foot forming a mentum, often spur-like at the base, 0.1 – 3cm long. Petals narrower or broader than sepals. Lip entire to 3-lobed, base joined to column-foot, often forming a closed spur with the lateral sepals to which it may be joined laterally for a short distance, disc with 1 – 7 keels, calli rarely present. Column short, foot long, apical stelidia present. Pollinia 4 in adpressed pairs, naked, i.e. without caudicles or stipes.

Approximately 1400 species distributed from India across to Japan, south to Malaysia, Indonesia, east to New Guinea, Australia, New Zealand and the Pacific islands. The second largest orchid genus in Borneo.

D.acerosum *Lindl.* in Bot. Reg. 30, misc. 86 (1841). Type: Singapore, *Cuming* 357 (holo. K-LINDL).

HABITAT: lowland dipterocarp forest.
ALTITUDINAL RANGE: lowlands.
DISTRIBUTION: Borneo (K, SA). Also Burma, Peninsular Malaysia, Thailand, Riau Archipelago, Sumatra, Sulawesi.
GENERIC SECTION: *Strongyle.*

D.aciculare *Lindl.* in Bot. Reg. 26:81, misc. 188 (1840). Type: Singapore, *Cuming* 174 (holo. K-LINDL).
D.koeteianum Schltr. in Bull. Herb. Boissier, ser. 2, 6:456 (1906). Type: Kalimantan, West Koetei, Samarinda, *Schlechter* 13342 (holo. B, destroyed).

HABITAT: unknown.
ALTITUDINAL RANGE: lowlands.
DISTRIBUTION: Borneo (K). Also Singapore, Thailand.
GENERIC SECTION: *Rhopalanthe.*

D.acuminatissimum *(Blume) Lindl.*, Gen. Sp. Orch. Pl.:86 (1830). Type: Java, Salak, *Blume* s.n. (holo. BO).
Grastidium acuminatissimum Blume, Bijdr.:333 (1825).

HABITAT: open ridge vegetation; lower montane forest.
ALTITUDINAL RANGE: 900 – 1500m.

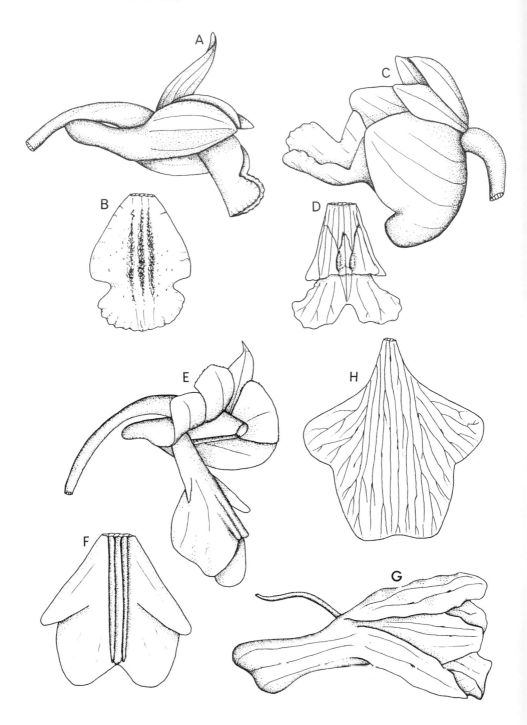

FIG. 27. *Dendrobium* flowers and lips. **A** & **B**, *D. bifarium*, x 5; **C** & **D**, *D. grande*, x 5; **E** & **F**, *D. moquettianum*, x 2½; **G** & **H**, *D. sanguinolentum*, x 2½. Drawn by Sarah Thomas.

DISTRIBUTION: Borneo (SA). Also Sumatra, Java, Philippines.
GENERIC SECTION: *Grastidium.*

D.alabense *J.J.Wood* in Lindleyana 5(2):90, fig.6 (1990). Type: Sabah, Crocker Range, Gunung Alab, summit ridge, *Wood* 777 (holo. K).

HABITAT: lower montane ridge forest with *Dacrydium, Podocarpus, Phyllocladus, Leptospermum, Rhododendron*, bamboo, rattans, *Gahnia*, etc; low mossy and xerophyllous scrub forest on extreme ultramafic substrate; montane oak forest; *Leptospermum, Phyllocladus, Dacrydium* and *Drimys* scrub on steep rocky slopes.
ALTITUDINAL RANGE: 1500 – 2400m.
DISTRIBUTION: Borneo (SA). Endemic.
GENERIC SECTION: *Rhopalanthe.*

D.aloifolium *(Blume) Rchb.f.* in Ann. Bot. Syst. 6:279 (1861). Type: Java, Salak, *Blume* s.n. (holo. BO).
Macrostomium aloefolium Blume, Bijdr.:335 (1825).
Dendrobium lobbii Lindl. in J. Linn. Soc., Bot. 3:3 (1858). Type: Labuan, *Lobb* s.n. (holo. K-LINDL), non Teijsm. & Binn.

HABITAT: lowland and hill forest, often on ultramafic substrate.
ALTITUDINAL RANGE: lowlands – 600m.
DISTRIBUTION: Borneo (K, SA, SR). Also Burma, Peninsular Malaysia, Singapore, Vietnam, Laos, Sumatra, Anambas Island, Riau and Lingga Archipelago, Java, Bangka, Bali.
GENERIC SECTION: *Aporum.*

D.anosmum *Lindl.* in Bot. Reg. 31:32, misc. 41 (1845). Type: Philippines, *Loddiges* s.n. (holo. K-LINDL).

HABITAT: lowland forest; riverine forest;lower montane forest.
ALTITUDINAL RANGE: sea level – 1500m.
DISTRIBUTION: Borneo (K, SA, SR). Widespread from Sri Lanka to Indochina and Malaysia through Indonesia to the Philippines and New Guinea.
GENERIC SECTION: *Dendrobium.*

D.anthrene *Ridl.* in J. Linn. Soc., Bot. 31:272 (1896). Type: Sarawak, *Hose* s.n. (holo. SING, iso. K).

HABITAT: unknown.
ALTITUDINAL RANGE: unknown.
DISTRIBUTION: Borneo (SR). Also Sumatra.
GENERIC SECTION: *Calcarifera.*

D.arcuatum *J.J.Sm.*, Orch. Java: 357 (1905). Types: Java, Simpolan, *Koorders* 20759B & 21781B (syn. BO).

HABITAT: mixed hill forest on limestone.
ALTITUDINAL RANGE: 600m.
DISTRIBUTION: Borneo (SA). Also Java.
GENERIC SECTION: *Calcarifera.*

D.atrorubens *Ridl.* in J. Linn. Soc., Bot. 32:247 (1896). Type: Peninsular Malaysia, Kedah, Gunung Jerai, *Ridley* 5140 (holo. SING).

HABITAT: riverine forest.
ALTITUDINAL RANGE: sea level – 600m.

DISTRIBUTION: Borneo (SA, SR). Also Peninsular Malaysia.
GENERIC SECTION: *Oxystophyllum*.

D.attenuatum *Lindl.* in J. Linn. Soc., Bot. 3:17 (1859). Type: Borneo, *Lobb* s.n. (holo. K-LINDL).

HABITAT: ridge top kerangas forest transitional to mixed dipterocarp forest.
ALTITUDINAL RANGE: 100 – 200m.
DISTRIBUTION: Borneo (unspecified). Endemic.
GENERIC SECTION: *Conostalix*.

D.babiense *J.J.Sm.* in Bot. Jahrb. Syst. 48:98 (1912). Type: Kalimantan, Batu Babi, *Winkler* 2804 (holo. HBG).

HABITAT: 'sand forest'.
ALTITUDINAL RANGE: 900 – 1000m.
DISTRIBUTION: Borneo (K, SR). Endemic.
GENERIC SECTION: *Aporum*.

D.beamanianum *J.J.Wood & A.Lamb* in Wood, Beaman & Beaman, Plants of Mt. Kinabalu 2, Orchids: 163, fig.18 (1993). Type: Sabah, Mt. Kinabalu, Pinosuk Plateau, c.8km.E.S.E. of Desa Dairy, *Beaman* 10722 (holo. K).

HABITAT: lower montane forest on ultramafic substrate.
ALTITUDINAL RANGE: 1200 – 1700m.
DISTRIBUTION: Borneo (SA). Endemic.
GENERIC SECTION: *Conostalix*.

D.bicornutum *Schltr.* in Bull. Herb. Boissier 2, 6:454 (1906). Type: Kalimantan, Moeara Kelindjau, *Schlechter* 13564 (holo. B, destroyed).

HABITAT: unknown.
ALTITUDINAL RANGE: unknown.
DISTRIBUTION: Borneo (K). Endemic.
GENERIC SECTION: *Aporum*.

D.bifarium *Lindl.*, Gen. Sp. Orch. Pl.:81 (1830). Type: Peninsular Malaysia, Penang, *Wallich* 2002 (holo. K-WALL).

HABITAT: lower montane forest; moss forest; oak-laurel forest; Fagaceae, *Dacrydium*, *Leptospermum* forest on sandstone, ultramafic and dioritic substrate; on trees near shingle beaches.
ALTITUDINAL RANGE: 600 – 1800m, one record from near sea level.
DISTRIBUTION: Borneo (K, SA, SR). Also Peninsular Malaysia, Singapore, Thailand, Maluku(Ambon).
GENERIC SECTION: *Distichophyllum*.

D.blumei *Lindl.*, Gen. Sp. Orch. Pl.:88 (1830). Type: Java, Pantjar, *Blume* 1935 (holo. K-LINDL).
   D.tuberiferum Hook.f., Fl. Brit. Ind. 5:728 (1890). Type: Peninsular Malaysia, Perak, *Scortechini* s.n. (holo. K, drawing only).

HABITAT: kerangas forest.
ALTITUDINAL RANGE: sea level – 900m.
DISTRIBUTION: Borneo (K, SR). Also Peninsular Malaysia, Thailand, Java, Philippines.
GENERIC SECTION: *Rhopalanthe*.

D.bostrychodes *Rchb.f.* in Gard. Chron., ser. 2, 14:748 (1880). Type: Borneo, *Boxall* s.n. (holo. W, sketch only).

HABITAT: unknown.
ALTITUDINAL RANGE: unknown.
DISTRIBUTION: Borneo (unspecified). Endemic.
GENERIC SECTION: *Formosae.*

D.calcariferum *Carr* in Gard. Bull. Straits Settlem. 8:107 (1935). Type: Sarawak, Gunung Balapan, Ulu Tinjar, *Richards* 2404 (holo. SING, iso. K).

HABITAT: hill forest; in moss on limestone rocks.
ALTITUDINAL RANGE: 600 – 900m.
DISTRIBUTION: Borneo (SR). Endemic.
GENERIC SECTION: *Calcarifera.*

D.cinereum *J.J.Sm.* in Bull. Jard. Bot. Buitenzorg, ser. 3, 2:78 (1920). Types: Kalimantan, locality unknown, *Nieuwenhuis expedition,* cult. Bogor no. 1386 (syn. L); Sungai Merase, *Sakeran* s.n., cult. Bogor no. 1821 (syn. L).

HABITAT: oak-laurel forest; stunted mossy ridge forest including *Ericaceae;* lower montane ridge forest on sandstone and shale.
ALTITUDINAL RANGE: 900 – 1500m.
DISTRIBUTION: Borneo (K, SA). Endemic.
GENERIC SECTION: *Calcarifera.*

D.cinnabarinum *Rchb.f.* in Gard. Chron., 14:166 (1880). Type: Borneo, *Veitch* s.n. (holo. W).

var. **cinnabarium**

HABITAT: lower montane forest; moss forest; low scrubby forest on peaty accumulations among limestone pinnacles; riverine forest; margin of kerangas forest; lower montane ridge forest with epiphytic *Rhododendron.*
ALTITUDINAL RANGE: 700 – 1900m.
DISTRIBUTION: Borneo (B, SA, SR). Endemic.
GENERIC SECTION: *Rhopalanthe.*

var. **angustitepalum** *Carr* in Gard. Bull. Straits Settlem., 8:103 (1935). Type: Sarawak, Mt. Dulit, Ulu Koyan, *Richards* S.481 (holo. SING).

D.sanguineum Rolfe in Gard. Chron., ser. 3, 18:292 (1895), non Sw. Type: Labuan, *Low & Co.* (holo. K).
D.holttumianum A.D.Hawkes & A.H.Heller in Lloydia 20(2):121 (1957).

HABITAT: 'sand forest'; lower montane forest; ridge tops; mixed *Gymnostoma* and heath forest.
ALTITUDINAL RANGE: 900 – 1400m.
DISTRIBUTION: Borneo (SA, SR). Endemic.
GENERIC SECTION: *Rhopalanthe.*

var. **lamelliferum** *Carr* in Gard. Bull. Straits Settlem. 8:103 (1935). Type: Sarawak, Dulit ridge, *Synge* S.473 (holo. SING, iso. K).

HABITAT: moss forest.
ALTITUDINAL RANGE: 1300m.
DISTRIBUTION: Borneo (SR). Endemic.
GENERIC SECTION: *Rhopalanthe.*

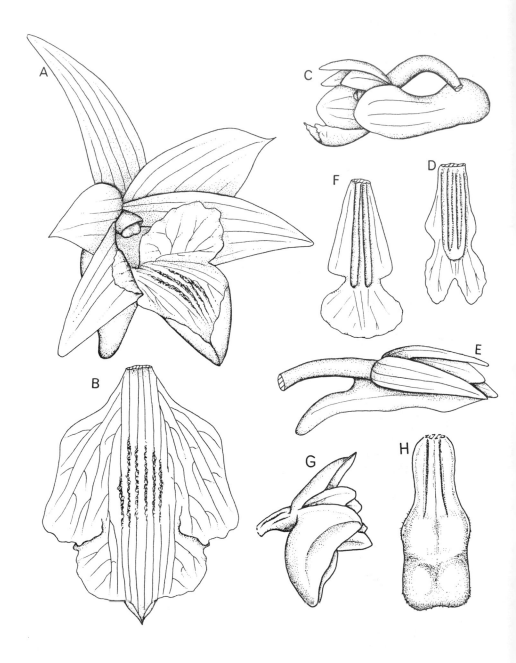

FIG. 28. *Dendrobium* flowers and lips. **A** & **B**, *D. crumenatum*, x 3; **C** & **D**, *D. rosellum*, x 5; **E** & **F**, *D. pandaneti*, x 3; **G** & **H**, *D. helvolum*, x 3 & x 5. Drawn by Sarah Thomas.

PLATE 1

A. **Acanthephippium eburneum.** Sarawak. Photo P. J. Cribb.

B. **Ania ponggolensis.** Sabah. Photo A. Lamb.

C. **Arachnis flosaeris.** Sabah. Photo P. J. Cribb.

D. **Arachnis grandisepala.** Sabah. Photo A. Lamb.

E. **Ascochilopsis lobata.** Sabah. Photo A. Lamb.

F. **Brachypeza zamboangensis.** Sabah. Photo A. Lamb.

PLATE 2

A. **Bulbophyllum acuminatum.** Sabah, cult. R.B.G. Kew. Photo J. R. Oddy.

B. **Bulbophyllum beccarii.** Kalimantan Tengah, cult. Java. Photo J. B. Comber.

C. **Bulbophyllum beccarii.** Kalimantan Tengah, cult. Java. Photo J. B. Comber.

D. **Bulbophyllum dearei.** Sarawak. Photo J. Dransfield.

E. **Bulbophyllum grandilabre.** Sabah. Photo J. B. Comber.

F. **Bulbophyllum habrotinum.** Sabah. Photo A. Lamb.

PLATE 3

A. **Bulbophyllum kemulense.** Sabah. Photo A. Lamb.
B. **Bulbophyllum lemniscatoides.** Kalimantan Timur, cult. R.B.G. Kew. Photo R.B.G. Kew.
C. **Bulbophyllum lobbii.** Sabah. Photo S. Collenette.
D. **Bulbophyllum lohokii.** Sabah. Photo P. J. Cribb.
E. **Bulbophyllum mastersianum.** Sabah. Photo P. J. Cribb.
F. **Bulbophyllum refractilingue.** Sabah. Photo J. B. Comber.

PLATE 4

A. **Bulbophyllum similissimum.** Sabah. Photo J. B. Comber.

B. **Bulbophyllum stormii.** Kalimantan Timur. Photo P. J. Cribb.

C. **Calanthe undulata.** Kalimantan Timur. Photo P. J. Cribb.

D. **Chelonistele amplissima.** Sabah. Photo J. B. Comber.

E. **Chelonistele sulphurea** var. **sulphurea.** Sabah. Photo. A. Lamb.

F. **Chelonistele unguiculata.** Brunei, cult. R.B.G. Kew. Photo R.B.G. Kew.

PLATE 5

A. **Cleisomeria lanatum.** Singapore Botanic Garden. Photo J. B. Comber.
B. **Coelogyne hirtella.** Sabah. Photo J. B. Comber.
C. **Coelogyne longpasiaensis.** Sabah. Photo J. B. Comber.
D. **Coelogyne muluensis.** Sarawak. Photo A. Lamb.
E. **Coelogyne pandurata.** Cult. R.B.G. Kew. Photo P. J. Cribb.
F. **Coelogyne radioferens.** Sabah. Photo R. S. Beaman.

PLATE 6

A. **Coelogyne testacea.** Kalimantan Timur. Photo P. J. Cribb.

B. **Corybas muluensis.** Sabah. Photo J. B. Comber.

C. **Corybas muluensis.** Sabah. Photo J. B. Comber.

D. **Corybas piliferus.** Sabah. Photo J. B. Comber.

E. **Cymbidium bicolor subsp. pubescens.** Sabah, cult. R.B.G. Kew. Photo D. J. DuPuy.

F. **Cymbidium rectum.** Sabah, cult. R.B.G. Kew. Photo D. J. DuPuy.

PLATE 7

A. **Cystorchis salmoneus.** Sabah. Photo J. B. Comber.

B. **Dendrobium cinereum.** Sabah. Photo P. J. Cribb.

C. **Dendrobium cinnabarinum** var. **cinnabarinum.** Sabah. Photo R. S. Beaman.

D. **Dendrobium cinnabarinum** var. **cinnabarinum.** Sabah. Photo J. B. Comber.

E. **Dendrobium cinnabarinum** var. **cinnabarinum.** Sarawak. Photo J. Dransfield.

F. **Dendrobium cinnabarinum** var. **cinnabarinum.** Sarawak. Photo J. Dransfield.

PLATE 8

A. **Dendrobium cinnabarinum** var. **angustitepalum.** Sabah. Photo P. J. Cribb.
B. **Dendrobium cinnabarinum** var. **angustitepalum.** Sabah. Photo J. B. Comber.
C. **Dendrobium cymboglossum.** Sabah. Photo P. J Cribb.
D. **Dendrobium lambii.** Sabah. Photo S. Collenette.
E. **Dendrobium lancilobum.** Sabah. Photo J. B. Comber.
F. **Dendrobium lowii.** Cult. South Africa. Photo L. Vogelpoel.

PLATE 9

A. **Dendrobium ovipostoriferum.** Sabah. Photo S. Collenette.
B. **Dendrobium sanguinolentum.** Sabah. Photo A. Lamb.
C. **Dendrochilum hologyne.** Sabah. Photo J. B. Comber.
D. **Dendrochilum muluense.** Sarawak. Photo G. Lewis.
E. **Dendrochilum pachyphyllum.** Sabah. Photo A. Lamb.
F. **Diplocaulobium vanleeuwenii.** Sabah. Photo J. Dransfield.

PLATE 10

A. **Dyakia hendersoniana.** Sarawak. Photo A. Lamb.

B. **Epigeneium speculum.** Sabah. Photo P. J. Cribb.

C. **Epigeneium zebrinum.** Peninsular Malaysia. Photo J. Dransfield.

D. **Eria aurantia.** Sabah. Photo J. B. Comber.

E. **Eria aurantia.** Sabah. Photo J. B. Comber.

F. **Eria javanica.** Cult. Bogor Botanic Garden. Photo P. J. Cribb.

PLATE 11

A. **Eria lanuginosa.** Sabah. Photo J. B. Comber.
B. **Eria** sp. section **Cylindrolobus.** Sabah. Photo A. Lamb.
C. **Eria** sp. section **Hymeneria.** Sabah, cult. R.B.G. Kew. Photo R.B.G. Kew.
D. **Flickingeria luxurians.** Sabah. Photo J. B. Comber.
E. **Grammatophyllum scriptum.** Philippines, cult. R.B.G. Kew. Photo R.B.G. Kew.
F. **Habenaria marmorophila.** Sarawak. Photo J. Dransfield.

PLATE 12

A. **Kuhlhasseltia rajana.** Kalimantan Timur. Photo P. J. Cribb.
B. **Liparis bicuspidata.** Sabah. Photo A. Lamb.
C. **Liparis grandiflora.** Sarawak. Photo R.B.G. Edinburgh.
D. **Malaxis metallica.** Sabah. Photo R. S. Beaman.
E. **Mischobulbum scapigerum.** Sabah. Photo M. J. S. Sands.
F. **Nephelaphyllum aureum.** Sabah. Photo A. Lamb.

PLATE 13

A. **Paphiopedilum kolopakingii.** Kalimantan, cult. R.B.G. Kew. Photo P. J. Cribb.
B. **Paphiopedilum philippinense.** Sabah. Photo A. Lamb.
C. **Paphiopedilum sanderianum.** Sarawak. Photo A. Lamb.
D. **Paphiopedilum sanderianum.** Sarawak, cult. R.B.G. Kew. Photo P. J. Cribb.
E. **Paphiopedilum stonei.** Sarawak, cult. R.B.G. Kew. Photo P. J. Cribb.
F. **Paphiopedilum supardii.** Kalimantan, cult. Java. Photo P. J. Cribb.

PLATE 14

A. **Paraphalaenopsis laycockii.** Cult. Germany. Photo E. Lückel.

B. **Phaius callosus.** Cult. Java. Photo P. J. Cribb.

C. **Phalaenopsis gigantea.** Sabah. Photo A. Lamb.

D. **Phalaenopsis gigantea.** Sabah, cult. R.B.G. Kew. Photo R.B.G. Kew.

E. **Phalaenopsis violacea.** Sarawak, cult. R.B.G. Kew. Photo P. J. Cribb.

F. **Phalaenopsis sp. nov.** Sabah. Photo P. J. Cribb.

PLATE 15

A. **Plocoglottis hirta.** Sarawak. Photo J. Dransfield.

B. **Pteroceras biserratum.** Sabah. Photo A. Lamb.

C. **Pteroceras erosulum.** Sabah. Photo J. B. Comber.

D. **Rhynchostylis gigantea.** Sabah. Photo A. Lamb.

E. **Robiquetia transversisaccata.** Sabah. Photo A. Lamb.

F. **Spathoglottis confusa.** Sabah. Photo R. S. Beaman.

PLATE 16

A. **Tainia vegetissima.** Peninsular Malaysia. Photo J. Dransfield.
B. **Thrixspermum tortum.** Sabah, cult. R.B.G. Kew. Photo R.B.G. Kew.
C. **Trichoglottis uexkulliana.** Sabah, cult. R.B.G. Kew. Photo R.B.G. Kew.
D. **Trichotosia gracilis.** Kalimantan Timur. Photo J. Dransfield.
E. **Vanda dearei.** Cult. R.B.G. Kew. Photo R.B.G. Kew.
F. **Zeuxine strateumatica.** Hong Kong. Photo P. J. Cribb

D.compressimentum *J.J.Sm.* in Bull. Jard. Bot. Buitenzorg, ser. 3, 10:63 (1928). Types: Sumatra, Gunung Malintang, *Bünnemeijer* 4253 (syn. L); Laras Talang, Bukit Gombak, *Bünnemeijer* 5682 (syn. L); Bukit Niroe, *Bünnemeijer* s.n. (syn. L).

HABITAT: unknown.
ALTITUDINAL RANGE: unknown.
DISTRIBUTION: Borneo (SA, SR). Also Sumatra.
GENERIC SECTION: *Calcarifera.*

D.concinnum *Miq.*, Fl. Ind. Bat. 3:644 (1859). Types: Maluku, 'Amboina'; 'Java'(types not located).
D.carnosum (Blume) Rchb.f. in Ann. Bot. Syst. 6:280 (1861). Type: Java, Salak, Pantjar,etc., *Blume* s.n.(holo. BO).
D.atropurpureum auct., non (Blume) Miq.

HABITAT: lowland, hill and lower montane forest; secondary forest.
ALTITUDINAL RANGE: sea level – 1200m.
DISTRIBUTION: Borneo (SA, SR). Also Burma, Peninsular Malaysia, Thailand, Cambodia, Vietnam, Sumatra, Java, Maluku, Riau Archipelago.
GENERIC SECTION: *Oxystophyllum.*

D.connatum *(Blume) Lindl.*, Gen. Sp. Orch. Pl.:89 (1830). Type: Java, Salak, *Blume* s.n. (holo. BO).
Onychium connatum Blume, Bijdr.:328 (1825).

HABITAT: unknown.
ALTITUDINAL RANGE: unknown.
DISTRIBUTION: Borneo (SA). Also Sumatra, Java.
GENERIC SECTION: *Distichophyllum.*

D.corallorhizon *J.J.Sm.* in Bull. Jard. Bot. Buitenzorg, ser. 3, 11:140 (1931). Type:Kalimantan, West Koetai, Gunung Kemoel, *Endert* 4377 (holo. L).

HABITAT: lower montane forest.
ALTITUDINAL RANGE: 1600m.
DISTRIBUTION: Borneo (K). Endemic.
GENERIC SECTION: *Calcarifera.*

D.crabro *Ridl.* in J. Straits Branch Roy. Asiat. Soc., 50:133 (1908). Type: Sarawak, Matang, *Hewitt* s.n. (holo. SING).

HABITAT: unknown.
ALTITUDINAL RANGE: unknown.
DISTRIBUTION: Borneo (SR). Endemic.
GENERIC SECTION: *Calcarifera.*

D.crucilabre *J.J.Sm.* in Mitt. Inst. Allg. Bot. Hamburg 7:55, t.8, fig.45 (1927). Type: Kalimantan, Bukit Raja, *Winkler* 942 (holo. HBG).

HABITAT: lower montane forest.
ALTITUDINAL RANGE: 1200 – 1300m.
DISTRIBUTION: Borneo (K). Endemic.
GENERIC SECTION: *Aporum.*

D.crumenatum *Sw.* in J. Bot. (Schrader) 2:237 (1799). Types: 'habitat in India orientali, Java'.
D.crumenatum Sw. var. *parviflorum* Ames & C.Schweinf., Orch. 6:100 (1920). Type: Sabah, Mt. Kinabalu, Kiau, *Clemens* 188 (holo. AMES, iso. BO, K, SING).

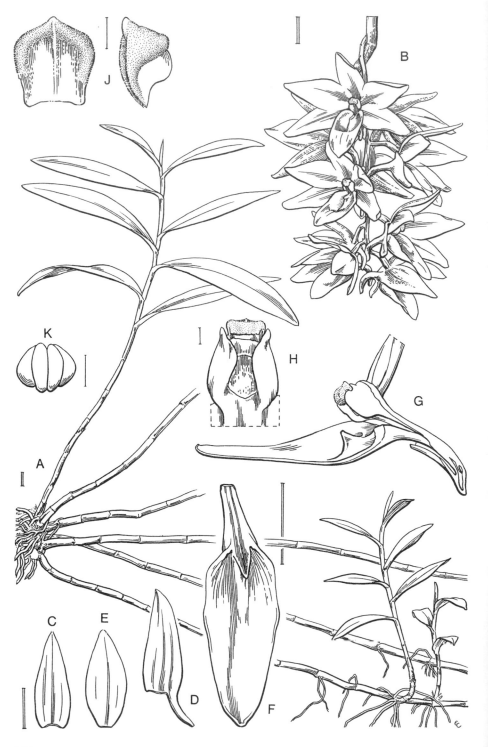

HABITAT: lowland and hill forest; also recorded growing on old coral at sea level.
ALTITUDINAL RANGE: sea level – 900m.
DISTRIBUTION: Borneo (K, SA, SR).Widespread in India, Sri Lanka, China and all over S.E. Asia.
GENERIC SECTION: *Rhopalanthe.*

D.cumulatum *Lindl.* in Gard. Chron.:756 (1855). Type: Origin unknown (holo. K-LINDL).

HABITAT: lower montane forest; peat swamp forest.
ALTITUDINAL RANGE: 1500m.
DISTRIBUTION: Borneo (SA, SR). Also Nepal, Bhutan, India, Burma, Thailand, Laos, Cambodia, Vietnam.
GENERIC SECTION: *Calcarifera.*

D.curtisii *Rchb.f.* in Gard. Chron., ser. 2, 16:102 (1881). Type: Borneo, *Curtis* s.n. (holo. W).

HABITAT: unknown.
ALTITUDINAL RANGE: unknown.
DISTRIBUTION: Borneo (unspecified). Endemic.
GENERIC SECTION: *Calcarifera.*

D.cymboglossum *J.J.Wood et A.Lamb* **sp. nov.** at aliis speciebus sectionis mento angusto valde incurvo unciformi, labio oblong-elliptico valde concavo naviculari callo basali licino spiniformi instructo distinguitur. Typus: East Malaysia, Sabah, Sandakan, 29.4.1895, *Governor Creagh* s.n. (holotypus K, herbarium material only).

Epiphytic herb. Stems 30-100 cm long, to 0.8 cm wide, producing keikis from several nodes. Leaves 11-12 x 1.8-2 cm, narrowly elliptic, acute to acuminate. Inflorescences up to 8-flowered, pendulous; peduncle 1.5-2 cm long; rachis 3.5-5.5 cm long; floral bracts 3-4 mm long, acute to acuminate. Flowers c.3.5 cm across. Pedicel with ovary 1.5-1.7 cm long, narrowly clavate, pale green, streaked purple-red. Sepals cream to yellowish, densely covered in purple-red streaks on the exterior. Dorsal sepal 2.4-2.9 x 0.9-1 cm, narrowly elliptic, apex acute and slightly cucullate. Lateral sepals fused to column-foot, free portion 2.6-3 x 0.9-1.1 cm, elliptic, asymmetrical, apex acute and slightly cucullate. Mentum 1.3-2 cm long, 0.3 cm wide at base, narrow, strongly hamate, acute, purple-red streaked. Petals 2.5-3 x 1-1.2 cm, elliptic, acute, cream to whitish-yellow flushed pink. Lip 2.5 cm long, 0.8-1 cm wide unflattened, blade entire, oblong-elliptic, obtuse, deeply concave, cymbiform, yellow to off white flushed orange on the sides towards the base, with pinkish apical spots; claw 2 mm wide; callus 4 mm long, c. 2.5 mm wide at base, upcurved, fleshy, triangular-acute, thorn-like, yellow with dark orange banding on the disc in front. Column 5 x 4 mm, with short oblong, obtuse wings, cream, foot 2 cm long, hamate. Anther cap 3 x 2 mm, oblong-ovate, apiculate. Fig.29.

FIG. 29. *Dendrobium cymboglossum.* **A**, habit; **B**, inflorescence; **C**, dorsal sepal; **D**, lateral sepal; **E**, petal; **F**, lip (front view); **G**, pedicel with ovary, column with anther cap, and longitudinal section through lip and mentum (side view); **H**, column and anther cap (front view); **J**, anther cap (back and side views); **K**, pollinia. **A** from *Creagh* s.n.; **B** from a photograph by *Lamb;* **C-K** from *Lim Weng Hee* 6.18. Scale: single bar = 1 mm; double bar = 1 cm. Drawn by Eleanor Catherine.

Other material examined: Sabah, Tawao, 1922 or 1923, *Elmer* 21889 (K, herbarium material only); Sepilok Forest Reserve near Sandakan, July 1979, *Lamb* SAN 93402 (K, SAN, herbarium material only); Taman Tawao, comm. 29.3.1990, *Lim Weng Hee* 6.18 (K, spirit material only); cult. Poring Orchid Garden near Ranau, 16.11.1989, *Lohok* in *Cribb* 89/48 (K, spirit material only).

This member of section *Calcarifera* is distinguished by the narrow, strongly incurved hook-shaped mentum and the unusual oblong-elliptic, strongly concave, boat-shaped entire lip with an upturned thorn-like basal callus. The flowers superficially resemble *D.sarawakense* Ames which also has a strongly hooked, though broader, mentum. *D.cymboglossum* can be distinguished at once by the entire, boat-shaped lip.

D.cymbulipes *J.J.Sm.* in Mitt. Inst. Allg. Bot. Hamburg 7:51, t.8, fig.41 (1927). Type: Kalimantan, Bukit Raja, *Winkler* 885 (holo. HBG, drawing at BO).

HABITAT: lower montane forest; moss forest; oak-chestnut ridge forest with *Agathis alba, Tristania*, small rattans etc.; oak-laurel forest; sometimes on ultramafic substrate.
ALTITUDINAL RANGE: 900 – 2400m.
DISTRIBUTION: Borneo (K, SA, SR). Endemic.
GENERIC SECTION: *Rhopalanthe.*

D.distachyon *Lindl.* in J. Linn. Soc., Bot. 3:13 (1859). Type: Borneo, *Lobb* s.n. (holo. K-LINDL).

HABITAT: unknown.
ALTITUDINAL RANGE: unknown.
DISTRIBUTION: Borneo (K, SR). Also Natuna Islands.
GENERIC SECTION: *Distichophyllum.*

D.elephantinum *Finet* in Bull. Soc. Bot. France 50:373, t.11, fig.20-31 (1903). Type: Sarawak, *Beccari* 3656 (holo. P, iso. FI).

HABITAT: unknown.
ALTITUDINAL RANGE: unknown.
DISTRIBUTION: Borneo (SR). Endemic.
GENERIC SECTION: *Distichophyllum.*

D.ellipsophyllum *T.Tang & F.T.Wang* in Acta Phytotax.Sin.1,1:81 (1951). Type: Burma, Moulmein, *Peche* s.n. (holo. K, iso. BM).

HABITAT: unknown.
ALTITUDINAL RANGE: unknown.
DISTRIBUTION: Borneo (K, SA, SR). Also China, Burma, Laos, Cambodia, Vietnam, Thailand.
GENERIC SECTION: *Distichophyllum.*

D.endertii *J.J.Sm.* in Bull. Jard. Bot. Buitenzorg, ser. 3, 11:138 (1931). Type: Kalimantan, West Koetai, Gunung Kemocl, *Endert* 3730 (holo. L).

HABITAT: lower montane forest.
ALTITUDINAL RANGE: 1000 – 1100m.
DISTRIBUTION: Borneo (K). Endemic.
GENERIC SECTION: *Calcarifera.*

D.erosum *(Blume) Lindl.*, Gen. Sp. Orch. Pl.:86(1830). Type: Java, Salak, *Blume* s.n.

(holo. BO, iso. L).

Pedilonum erosum Blume, Bijdr.:323 (1825).

HABITAT: mixed oak-chestnut forest with *Dacrydium* and *Phyllocladus;* mossy forest on sandstone ridges.
ALTITUDINAL RANGE: 1600 – 1700m.
DISTRIBUTION: Borneo (SA). Also Peninsular Malaysia, Thailand, Sumatra, Java, New Guinea, Solomon Islands, Vanuatu.
GENERIC SECTION: *Calyptrochilus.*

D.erythropogon *Rchb.f.* in Gard. Chron., ser. 2, 24:198 (1885). Type: 'ex ins. Sond.', imported by *Low & Co.* (holo. W).

HABITAT: unknown.
ALTITUDINAL RANGE: unknown
DISTRIBUTION: Borneo (unspecified). Endemic.
GENERIC SECTION: *Formosae.*

D.gracile *(Blume) Lindl.,* Gen. Sp. Orch. Pl.:91 (1830). Type: Java, Gede, *Blume* s.n. (holo. BO).

Onychium gracile Blume, Bijdr.:327 (1825).

HABITAT: lower montane forest; moss forest.
ALTITUDINAL RANGE: 1500 – 1700m.
DISTRIBUTION: Borneo (K, SA, SR). Also Sumatra, Java.
GENERIC SECTION: *Rhopalanthe.*

D.gramineum *Ridl.* in J. Straits Branch Roy. Asiat. Soc. 50:132 (1908). Types: Sarawak, Matang, *Ridley* s.n., *Hewitt* s.n. (syn. SING).

HABITAT: unknown.
ALTITUDINAL RANGE: unknown.
DISTRIBUTION: Borneo (SR). Endemic.
GENERIC SECTION: *Conostalix.*

D.grande *Hook.f.,* Fl. Brit. Ind. 5:724 (1890). Type: Peninsular Malaysia, Perak, *Scortechini* s.n. (holo. K).

HABITAT: lowland forest; low stature forest on ultramafic substrate.
ALTITUDINAL RANGE: sea level – 1000m.
DISTRIBUTION: Borneo (K, SA, SR). Also Andaman Islands, Burma, Peninsular Malaysia, Thailand.
GENERIC SECTION: *Aporum.*

D.grootingsii *J.J.Sm.* in Bull. Jard. Bot. Buitenzorg, ser. 2, 25:33 (1917). Types: Kalimantan, Moeara Tewe, *Grootings* s.n. (syn. L); Sungai Sedalir, *Amdjah* 251 and 293 (syn. L); Gunung Labang, *Amdjah* 265 (syn. L).

HABITAT: lowland forest.
ALTITUDINAL RANGE: sea level – 200m.
DISTRIBUTION: Borneo (K). Endemic.
GENERIC SECTION: *Rhopalanthe.*

D.gynoglottis *Carr* in Gard. Bull. Straits Settlem. 8:105 (1935). Type: Sarawak, Mt. Dulit, Ulu Koyan, *Synge* S.505 (holo. SING, iso. K).

HABITAT: 'sandy forest'.
ALTITUDINAL RANGE: 900m.

DISTRIBUTION: Borneo (SR). Endemic.
GENERIC SECTION: *Rhopalanthe.*

D.hallieri *J.J.Sm.* in Bull. Jard. Bot. Buitenzorg, ser. 2, 8:40 (1912). Type: Kalimantan, Kelam, *Hallier* 2313 (holo. BO, iso. K).

HABITAT: unknown.
ALTITUDINAL RANGE: sea level.
DISTRIBUTION: Borneo (K). Endemic.
GENERIC SECTION: *Formosae.*

D.hamaticalcar *J.J.Wood & Dauncey* in Wood, Beaman & Beaman, Plants of Mt. Kinabalu 2, Orchids: 168, fig.20 (1993). Type: Sabah, Tambunan District, foothills of Mt. Trus Madi, near Kaingaran, *Bacon* 142, cult. Edinburgh Botanic Garden no. C6614 & C7336 (holo. E,iso. K).

HABITAT: lowland and hill forest.
ALTITUDINAL RANGE: 400 – 900m.
DISTRIBUTION: Borneo (SA). Endemic.
GENERIC SECTION: *Calcarifera.*

D.helvolum *J.J.Sm.* in Bot. Jahrb. Syst. 48:99 (1912). Types: Kalimantan, Hayup, *Winkler* 2518 (syn. HBG); Kwaru, *Winkler* 3084 (syn. HBG).

HABITAT: mangrove forest.
ALTITUDINAL RANGE: sea level.
DISTRIBUTION: Borneo (K). Endemic.
GENERIC SECTION: *Oxystophyllum.*

D.hendersonii *A.D.Hawkes & A.H.Heller* in Lloydia 20:120 (1957). Type: Sumatra, Indragiri District, Sungai Lalah, *Schlechter* 13297 (holo. B,destroyed).
Dendrobium fugax Schltr. in Bull. Herb. Boissier, ser. 2, 6:455 (1906), non Rchb.f.

HABITAT: riverine forest; hill forest.
ALTITUDINAL RANGE: 200 – 300m.
DISTRIBUTION: Borneo (K, SA). Also Peninsular Malaysia, Thailand, Sumatra.
GENERIC SECTION: *Rhopalanthe.*

D.hepaticum *J.J.Sm.* in Bull. Jard. Bot. Buitenzorg, ser. 2, 25:48(1917). Type: Kalimantan, Gunung Kelam, *Hallier* 1894 (holo. L).

HABITAT: unknown.
ALTITUDINAL RANGE: unknown.
DISTRIBUTION: Borneo (K). Endemic.
GENERIC SECTION: *Distichophyllum.*

D.heterocarpum *Lindl.*, Gen. Sp. Orch. Pl.:78 (1830). Type: Nepal, *Wallich* drawing (holo. K-LINDL).

HABITAT: hill and lower montane forest on sandstone ridges.
ALTITUDINAL RANGE: 900 – 1100m.
DISTRIBUTION: Borneo (SA). Widespread from China(Yunnan), Sri Lanka, India, Nepal and Bhutan to Burma, Thailand, Laos, Vietnam, Peninsular Malaysia, Sumatra, Java, Sulawesi and the Philippines.
GENERIC SECTION: *Dendrobium.*

D.hosei *Ridl.* in Trans. Linn. Soc. London, Bot., ser. 2, 3:363 (1893). Type: Peninsular Malaysia, Tahan River, *Ridley* s.n. (holo. SING).

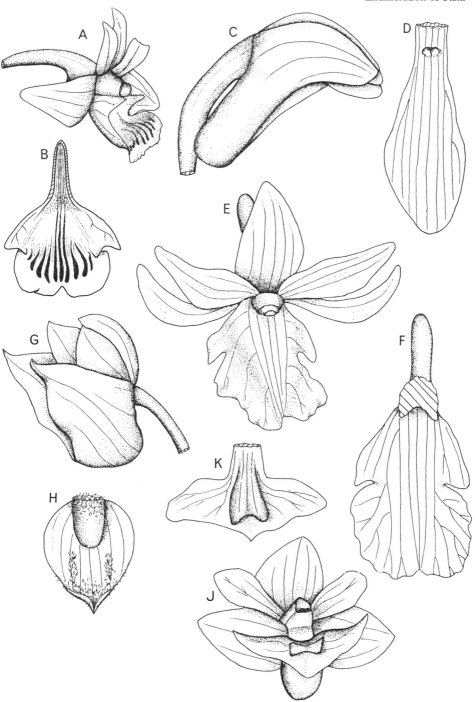

FIG. 30. *Dendrobium* flowers and lips. **A & B**, *D. hosei*, x 2½; **C & D**, *D. microglaphys*, x 6; **E & F**, *D. pachyanthum*, x 3; **G & H**, *D. linguella*, x 2½; **J & K**, *D. bahiense*, x 4. Drawn by Sarah Thomas.

A checklist of Borneo Orchids

D.multicostatum J.J.Sm. in Bull. Dép. Agric. Indes Néerl. 5:11 (1907). Type: Kalimantan, Bukit Lesoeng, *Nieuwenhuis* s.n. (holo. :BO).

HABITAT: lowland forest; riverine forest; podsol forest with *Dacrydium*; mixed dipterocarp forest.
ALTITUDINAL RANGE: sea level – 800m.
DISTRIBUTION: Borneo (B, K, SA, SR). Also Peninsular Malaysia.
GENERIC SECTION: *Distichophyllum.*

D.incurvociliatum *J.J.Sm.* in Bull. Jard. Bot. Buitenzorg, ser. 3, 11:137 (1931). Type: Kalimantan, West Koetai, Gunung Kemoel, *Endert* 4154 (holo. L).

HABITAT: lower montane forest.
ALTITUDINAL RANGE: 1200 – 1500m.
DISTRIBUTION: Borneo (K, SA). Endemic.
GENERIC SECTION. *Rhopalanthe.*

D.indivisum *(Blume) Miq.*, Fl. Ind. Bat. 3:630 (1859). Type: Java, Seribu, *Blume* s.n. (holo. BO).
Aporum indivisum Blume, Bijdr.:334 (1825).
Dendrobium eulophotum Lindl. in J. Linn. Soc., Bot. 3:5 (1859). Type: Burma, Mergui, *Griffith* drawing (holo. K- LINDL).

var. **indivisum**

HABITAT: lowland forest; coffee plantations; mangrove forest.
ALTITUDINAL RANGE: lowlands and hills.
DISTRIBUTION: Borneo (K, SA, SR). Also Burma, Laos, Thailand, Peninsular Malaysia, Singapore, Sumatra, Java, Riau Archipelago, Lingga Archipelago, Maluku, Philippines (Palawan).
GENERIC SECTION: *Aporum.*

var. **pallidum** *Seidenf.* in Opera Bot. 83:216, fig.146 (1985). Type: not designated.

HABITAT: hill forest.
ALTITUDINAL RANGE: 600m.
DISTRIBUTION: Borneo (SA). Also Burma, Peninsular Malaysia, Thailand.
GENERIC SECTION: *Aporum.*

D.indragiriense *Schltr.* in Feddes Repert. 9:164 (1911). Type: Sumatra, Indragiri District, Sungai Lalah, *Schlechter* 13277 (holo. B,destroyed).
D.inconspicuiflorum J.J.Sm. in Bull. Jard. Bot. Buitenzorg, ser. 2, 25:42 (1917). Type: Kalimantan, Gunung Djempanga, *Amdjah* s.n., cult Bogor no. 68 (holo. BO).

HABITAT: riverine forest.
ALTITUDINAL RANGE: 100 – 200m.
DISTRIBUTION: Borneo (K, SA, SR). Also Andaman Islands, Peninsular Malaysia, Thailand, Sumatra.
GENERIC SECTION: *Grastidium.*

D.junceum *Lindl.* in Bot. Reg. 28, misc. 11 (1842). Type: said to be imported from Singapore, but most likely from the Philippines, *Loddiges* catalogue no. 356 (holo. K-LINDL).

HABITAT: unknown.
ALTITUDINAL RANGE: unknown.
DISTRIBUTION: Borneo (SR). Also the Philippines.
GENERIC SECTION: *Rhopalanthe.*

252

D.kenepaiense *J.J.Sm.* in Bull. Jard. Bot. Buitenzorg, ser. 2, 25:46 (1917). Type: Kalimantan, Gunung Kenepai, *Hallier* 2451 (holo. BO).

HABITAT: unknown.
ALTITUDINAL RANGE: unknown.
DISTRIBITION: Borneo (K). Endemic.
GENERIC SECTION: *Distichophyllum.*

D.kentrophyllum *Hook.f.*, Fl. Brit. Ind. 5:725 (1890). Type: Peninsular Malaysia, Perak, *Scortechini* 1247 (holo. K).

HABITAT: peat swamp forest; hill and lower montane forest; ridge top forest with *Agathis alba,* often in open sunny positions.
ALTITUDINAL RANGE: sea level – 1300m.
DISTRIBUTION: Borneo (K, SA, SR). Also Peninsular Malaysia, Thailand, Sumatra.
GENERIC SECTION: *Strongyle.*

D.kiauense *Ames & C.Schweinf.*, Orch. 6:103 (1920). Type: Sabah, Mt. Kinabalu, Kiau, *Clemens* 176 (holo. AMES, iso. BO).
  D.rajanum J.J.Sm. in Mitt. Inst. Allg. Bot. Hamburg 7:56, t.9, fig.46 (1927). Type: Kalimantan, Bukit Raja, *Winkler* 940 (holo. HBG, iso. BO).

HABITAT: hill forest; lower montane oak-chestnut forest; sometimes on ultramafic substrate.
ALTITUDINAL RANGE: 800 – 1500m.
DISTRIBUTION: Borneo (K, SA). Endemic.
GENERIC SECTION: *Aporum.*

D.korthalsii *J.J.Sm.* in Bull. Jard. Bot. Buitenzorg, ser. 2, 25: 40 (1917). Type: Kalimantan, Sungai Njerakat, *Rutten* 480 (holo. L).

HABITAT: mangrove forest.
ALTITUDINAL RANGE: sea level.
DISTRIBUTION: Borneo (K). Endemic.
GENERIC SECTION: *Aporum.*

D.lambii *J.J.Wood* in Kew Bull. 38(1):79, fig.1 (1983). Type: Sabah, Gunung Alab, *Lamb* s.n. (holo. K).

HABITAT: semi-moss forest with *Dacrydium;* lower montane forest.
ALTITUDINAL RANGE: 1600 – 1800m.
DISTRIBUTION: Borneo (SA). Endemic.
GENERIC SECTION: *Distichophyllum.*

D.lamellatum *(Blume) Lindl.*, Gen. Sp. Orch. Pl.:89 (1830). Type: Java, Passir Ipis, *Blume* s.n. (holo. BO).
  Onychium lamellatum Blume, Bijdr.:326, fig.10 (1825).

HABITAT: lowland and hill forest, on ultramafic substrate.
ALTITUDINAL RANGE: lowlands – 600m.
DISTRIBUTION: Borneo (K, SA, SR). Also Burma, Peninsular Malaysia, Thailand, Riau Archipelago, Sumatra, Java, Natuna Islands.
GENERIC SECTION: *Calcarifera.*

D.lamelluliferum *J.J.Sm.* in Mitt. Inst. Allg. Bot. Hamburg, 7:52, t.8, fig.42 (1927). Type: Kalimantan, Serawai, Lebang Hara, *Winkler* 341 (holo. HBG).

HABITAT: hill forest; lower and upper montane forest; ridge top forest with *Agathis alba*, small rattans etc.; moss forest; oak-laurel forest.
ALTITUDINAL RANGE: 150 – 2200m.
DISTRIBUTION: Borneo (K, SA, SR). Endemic.
GENERIC SECTION: *Rhopalanthe.*

D.lampongense *J.J.Sm.* in Bull. Dép. Agric. Indes Néerl., 15:14 (1908). Type: Sumatra, Lampong, *Brautigam* s.n. (holo. BO).

HABITAT: hill forest; secondary forest.
ALTITUDINAL RANGE: 200 – 300m.
DISTRIBUTION: Borneo (K, SA). Also Sumatra.
GENERIC SECTION. *Calcarifera.*

D.lancilobum *J.J.Wood* in Lindleyana 5(2):90, fig.7 (1990). Type: Sabah, Sipitang district, east of Long Pa Sia to Long Miau trail, *Vermeulen & Duistermaat* 1131 (holo. K).

HABITAT: very open kerangas forest; lower montane forest.
ALTITUDINAL RANGE: 900 – 1700m.
DISTRIBUTION: Borneo (K, SA). Endemic.
GENERIC SECTION: *Rhopalanthe.*

D.lawiense *J.J.Sm.* in Bull. Jard. Bot. Buitenzorg, ser. 2, 3:60 (1912). Type: Sarawak, Gunung Batu Lawi, Ulu Limbang, *Moulton* 8 & 9 (holo. BO).

HABITAT: unknown.
ALTITUDINAL RANGE: unknown.
DISTRIBUTION: Borneo (SR). Endemic.
GENERIC SECTION: *Rhopalanthe.*

D.leonis *(Lindl.) Rchb.f.* in Ann. Bot. Syst. 6:280 (1861). Type: Singapore, *Cuming* 2068 (holo. K-LINDL).
Aporum leonis Lindl. in Bot. Reg. 26,59, misc. 126 (1840).

HABITAT: lowland forest; podsol forest with *Dacrydium.*
ALTITUDINAL RANGE: sea level – 500m.
DISTRIBUTION: Borneo (SA, SR). Also Peninsular Malaysia, Laos, Vietnam, Cambodia, Thailand, Sumatra, Bangka, Natuna, Anambas Islands.
GENERIC SECTION: *Aporum.*

D.limii *J.J.Wood* **sp.nov** ab aliis speciebus sectionis conventu distinguitur his characteribus: labium proprium latum, lobis lateralibus magnis falcatis lobo medio parvo bilobulato compositum; petala oblonga, truncata, erosa; folia teretia. Typus: East Malaysia, Sabah, Taman Tawau, 1.12.1989, *Lim Weng Hee* TT 5.4 (holotypus K, herbarium and spirit material).

Epiphytic herb. Stems 20-60 (-90) cm long, slender, 1.8 mm wide, lowermost node swollen to 3 mm wide, internodes 3-5 cm long. Leaves 4.5-12 cm long, to 1.7 mm wide, terete, not sulcate. Flowers 1.5 cm long, 0.5 cm wide, borne in succession from small groups of chaffy bracts at the nodes of the leafless apical part of the stem, ephemeral. Floral bracts 2.5 mm long, oblong-elliptic, obtuse, hyaline. Pedicel with ovary 6 mm long. Sepals translucent white, tip of lateral sepals suffused pink. Dorsal sepal 4.5-5 x 3 mm, oblong, truncate, apical margins somewhat cucullate. Lateral sepals 7 mm wide, triangular-ovate, obtuse, anterior margin 1.2 cm long, fused to column-foot. Mentum 1.5 cm long, conical, curved. Petals 4 x 2.5 mm, oblong, truncate, margins erose above, white, suffused pink at base. Lip 1 x 1.3 cm, white with 2 purple basal spots; mid-lobe small, bilobulate, lobules 2 x 2 mm, oblong-spathulate, margins irregularly crenulate and overlapping side lobes; side lobes large, 1 cm long, c.4.8 mm wide at the middle, 2 mm wide near apex, falcate, obtuse; disc without ridges. Column 1.5 mm long, foot 1.5 mm long, with a rounded apical gland. Anther cap 1 x 1 mm, ovate, obtuse, minutely papillose. Fig.31.

An offshoot or 'keiki' from the parent plant, together with a single flower preserved in alcohol, was sent by Dr. Lim Weng Hee of the Malaysian Agricultural Research and Development Institute (MARDI) for identification in December 1989.

D.limii is distinguished from other members of section *Rhopalanthe*, to which it belongs, by the character combination of a distinctive broad lip with large falcate side lobes, a small bilobulate mid-lobe, oblong, truncate, erose petals and terete leaves.

D.linguella *Rchb.f.* in Gard. Chron. 18:552 (1882). Type: 'Malayan Archipelago', imported by *Veitch & Sons* (holo. W).

HABITAT: riverine forest; hill forest; roadside cuttings on shale with *Rhododendron, Nepenthes fusca, Lycopodium,* ferns,etc.
ALTITUDINAL RANGE: 100 – 1300m.
DISTRIBUTION: Borneo (SA). Also Peninsular Malaysia, Vietnam, Thailand, Sumatra, Anambas Islands.
GENERIC SECTION: *Breviflores.*

D.lobatum *(Blume) Miq.,* Fl. Ind. Bat. 3:631 (1859). Type: Java, Buitenzorg Province, *Blume* s.n. (holo. BO).
Aporum lobatum Blume, Bijdr.:334 (1825).

HABITAT: kerangas forest; lowland forest.
ALTITUDINAL RANGE: lowlands.
DISTRIBUTION: Borneo (K, SR). Also Peninsular Malaysia, Sumatra, Java.
GENERIC SECTION: *Aporum.*

D.lobbii *Teijsm. & Binn.* in Natuurk. Tijdschr. Ned.-Indië 5:491 (1853). Type: Singapore, *Lobb* s.n. (holo. BO).
D.conostalix Rchb.f. in Ann. Bot. Syst. 6:292 (1861). Type: Singapore, *Cuming* s.n., cult. Loddiges no.158 (holo. W).

HABITAT: kerangas forest with dominant *Baeckea* and *Ploiarium*; wet sandy areas.
ALTITUDINAL RANGE: sea level – 600m.
DISTRIBUTION: Borneo (B, K, SA, SR). Also Peninsular Malaysia, Singapore, Vietnam, Sumatra, Bangka, New Guinea, Australia (Queensland).
GENERIC SECTION: *Conostalix.*

D.lobulatum *Rolfe & J.J.Sm.* in Orch. Java: 336 (1905). Types: Java, South Preangen, *Raciborski* s.n. (syn. BO); Soekaboemi, *Smith* s.n. (syn. BO); Groeda, *Smith* s.n.(syn. BO). Cultivated material received from Bogor under numbers 67 and 139, possibly from the type collections, are at K.

HABITAT: lowland forest.
ALTITUDINAL RANGE: lowlands.
DISTRIBUTION: Borneo (B, SA). Also Sumatra, ?Bangka, Java, Maluku (Ambon).
GENERIC SECTION: *Aporum.*

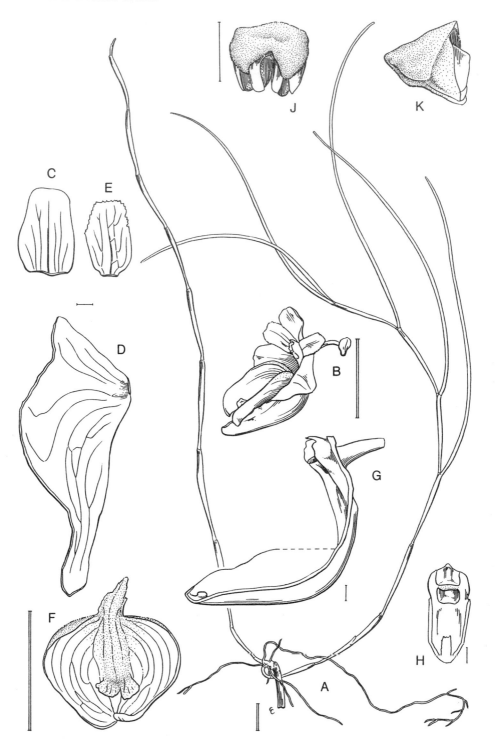

D.lowii *Lindl.* in Gard. Chron.:1046 (1861). Type:Borneo, imported by *Low* (holo. K-LINDL).

D.lowii Rchb.f. var. *pleiotrichum* Rchb.f. in Gard. Chron.:424 (1885). Type: Borneo, *Low* s.n. (holo. W, possible iso. K).

HABITAT: unknown.
ALTITUDINAL RANGE: unknown.
DISTRIBUTION: Borneo (?K, SA, SR). Endemic.
GENERIC SECTION: *Formosae.*

D.lucens *Rchb.f.* in Bot. Zeitung (Berlin) 21:128 (1863). Type: Borneo, *Low* s.n., cult. *Bullen* (holo. W).

HABITAT: unknown.
ALTITUDINAL RANGE: unknown.
DISTRIBUTION: Borneo (unspecified). Endemic.
GENERIC SECTION: *Calcarifera.*

D.maraiparense *J.J.Wood & C.L. Chan* in Orch. Borneo 1:119, fig.25, Pl.5E (1994). Type: Sabah, Mt. Kinabalu, Bembangan River, *Chew & Corner* RSNB 4962 (holo. K, iso. SING).

HABITAT: lower montane *Leptospermum recurvum* forest on ultramafic substrate; lower montane oak forest.
ALTITUDINAL RANGE: 1200 – 2000m.
DISTRIBUTION: Borneo (SA). Endemic.
GENERIC SECTION: *Distichophyllum.*

D.microglaphys *Rchb.f.* in Gard. Chron.:1014 (1868). Type: Borneo, ex *Low*, cult. *Bullen* (holo. W).

D.callibotrys Ridl. in J. Linn. Soc., Bot. 32:258 (1896). Type: Singapore, Toas, Sungai Mora, *Ridley* s.n. (holo. SING).

HABITAT: podsol forest, including *Dacrydium* with *Rhododendron malayanum* understorey; low, rather open swampy forest on ultramafic substrate; coastal *Agathis* forest; mixed hill forest.
ALTITUDINAL RANGE: sea level – 800m.
DISTRIBUTION: Borneo (K, SA, SR). Also Peninsular Malaysia, Singapore.
GENERIC SECTION: *Amblyanthus.*

D.minimum *Ames & C.Schweinf.*, Orch. 6:107, Pl.91 (1920). Type: Sabah, Mt. Kinabalu, Marai Parai, *Clemens* 287 (holo. AMES, iso. K, SING).

HABITAT: lower montane forest; oak-laurel forest; low stature forest on ultramafic substrate.
ALTITUDINAL RANGE: 1200 – 2000m.
DISTRIBUTION: Borneo (SA). Endemic.
GENERIC SECTION: *Rhopalanthe.*

FIG. 31. *Dendrobium limii.* **A,** habit; **B,** flower (side view); **C,** dorsal sepal; **D,** lateral sepal; **E,** petal; **F,** lip (front view); **G,** pedicel with ovary, column and part of mentum (side view); **H,** column (front view); **J,** anther cap with pollinia (front view); **K,** anther cap (side view). A-K from *Lim Weng Hee* TT 5.4. Scale: single bar = 1 mm; double bar = 1 cm. Drawn by Eleanor Catherine.

D.modestissimum *Kraenzl.* in Engl., Pflanzenr. Orch. Monand. IV. 50. II. B. 21:206 (1910). Type: Sarawak, *Beccari* 446 (holo. B,destroyed, iso. FI).

> HABITAT: unknown.
> ALTITUDINAL RANGE: unknown.
> DISTRIBUTION: Borneo (SR). Endemic.
> GENERIC SECTION: *Aporum.*

D.moquetteanum *J.J.Sm.* in Bull. Jard. Bot. Buitenzorg, ser. 2, 25:50 (1917). Type: Borneo, locality unknown, cult. *Moquette* (holo. BO).

> HABITAT: podsol forest.
> ALTITUDINAL RANGE: 400 – 500m.
> DISTRIBUTION: Borneo (K, SA). Endemic.
> GENERIC SECTION: *Distichophyllum.*

Dendrobium nabawanense *J.J.Wood et A.Lamb* **sp.nov.** *D.hosei* Ridl., species malayana atque borneensis, valde affinis, sed lobis partis labii proximi leniter rotundatis, disco crista centrali carnosa elevata plusminusve plana tricostata fere a basi partis labii proximi fere usque ad apicem partis distalis extendenti instructo, mento recto differt. Typus: East Malaysia, Sabah, Keningau District, near Nabawan, S.E. of Keningau, 21.5.1985, *Wood* 750 (holotypus K, herbarium and spirit material).

Terrestrial or epiphytic herb. Stems 24-45(-60) cm high, erect. Leaves linear-lanceolate, acutely narrowly unequally bilobed, blade (3.5-)6-8 x 0.4-0.6 cm, sheaths 1.2-2.5 cm long, black-ramentaceous, older sheaths becoming less so. Inflorescences 1-flowered, emerging from base of sheath opposite leaf blade, horizontal to porrect; peduncle and rachis 0.8-1 cm long, filiform, stiff; floral bracts 3, 1 cm long, ovate, acute, minuteley ramentaceous. Flower 2 cm across, unscented, resupinate. Pedicel with ovary 1.3 cm long, narrowly clavate. Sepals and petals stiff, spreading, white. Dorsal sepal 9 x 5 mm, ovate-elliptic, apiculate, reflexed. Lateral sepals obliquely triangular-ovate, acute, lower margin 1.5-1.6 cm long, upper margin 0.9 -1 cm long, 0.7 cm wide near base. Mentum 4-5 mm long, straight, obtuse, white with pale salmon-pink veins. Petals 8-9 x 2.5-3 mm, ligulate to spathulate, subacute, margin minutely erose above, somewhat reflexed, white. Lip 1.4-1.6 cm long, pandurate, minutely papillose, fleshy, waxy in appearance, white; hypochile 6 -7 mm long, (6 -)7-8 mm wide unflattened, with erect rounded margins, veined ochre-brown inside, with scattered papillose hairs at base each side of central ridge; epichile 4-5 x 6-9 mm, reniform, very shallowly bilobulate, spreading, flushed apricot-yellow; disc with a central fleshy, raised, three-ribbed, shiny pale ochre-brown, rather flat, glabrous ridge extending from near base of hypochile to just below apex of epichile, the three ribs more distinct and minutely papillose at the apex, the three nerves each side sometimes thickened. Column 3 mm long, oblong, with short, obtuse apical wings and an acicular anther connective, white, foot 4-5 mm long, brown, margins of stigmatic cavity green. Anther cap 2 x 2 mm, ovate, cucullate, minutely papillose, white. Fig.32.

Other material examined: Sabah, near Nabawan, 540 m, 19.2.1981, *Collenette* 2292 (K, herbarium and spirit material); Bornion (old Rashna) road 16 km from junction of Sook to Nabawan road, 600 m, 31.7.1977, *Lamb* SAN 87110 (K, SAN, herbarium material only); Lahad Datu District, Gunung Nicola, 700 m, 24.2.1992, *Lamb* AL 1452/92 (K, herbarium and spirit material).

D. nabawanense is closely related to *D. hosei* Ridl. (section Distichophyllum), from Peninsular Malaysia and Borneo, but differs in having gently rounded lip hypochile lobes, a disc with a central fleshy, raised, rather flat, 3-ribbed elongate ridge and a straight mentum. It is found in the same habitat as *D. trullatum* where it is common on the lower branches of trees, but seems to avoid *Dacrydium.*

D.nieuwenhuisii *J.J.Sm.* in Icon. Bogor. 3:25, t.211 (1906). Type: Kalimantan, Sungai Merase and Bukit Mili, *Nieuwenhuis* s.n. (holo. BO).

HABITAT: unknown.
ALTITUDINAL RANGE: unknown.
DISTRIBUTION: Borneo (K). Endemic.
GENERIC SECTION: *Calcarifera.*

D.nycteridoglossum *Rchb.f.* in Gard. Chron., ser. 2, 26:616 (1866). Type: 'of Papuan origin', *Linden* s.n. (holo. W).
   D.platyphyllum Schltr. in Bull. Herb. Boissier, ser. 2, 6:457 (1906). Type: Labuan, *Schlechter* s.n.(holo. B, destroyed).

HABITAT: unknown.
ALTITUDINAL RANGE: sea level.
DISTRIBUTION: Borneo (SA). Endemic.
GENERIC SECTION: *Aporum.*

NOTE: The origin of the type is most likely false.

D.oblongum *Ames & C.Schweinf.*, Orch. 6:108 (1920). Type: Sabah, Mt. Kinabalu, *Haslam* s.n. (holo. AMES).

HABITAT: hill and lower montane forest; oak-laurel forest with *Agathis*; sometimes on ultramafic substrate.
ALTITUDINAL RANGE: 900 – 1700m.
DISTRIBUTION: Borneo (K, SA, SR). Endemic.
GENERIC SECTION: *Oxystophyllum.*

D.olivaceum *J.J.Sm.* in Bull. Jard. Bot. Buitenzorg, ser. 2, 8:41 (1912). Type: Kalimantan, Liang Gagang, *Hallier* s.n., cult. Bogor under number 540a (holo. BO).

HABITAT: lower montane forest, epiphytic on *Phyllocladus;* very low and open podsol forest; recorded as a terrestrial among shrubs on a sandstone road cutting.
ALTITUDINAL RANGE: 900 – 1600m.
DISTRIBUTION: Borneo (K, SA).Endemic.
GENERIC SECTION: *Distichophyllum.*

D.orbiculare *J.J.Sm.* in Bull. Jard. Bot. Buitenzorg, ser. 2, 13:23 (1914). Types: Kalimantan, Damoes, *Hallier* 551 (syn. BO); Semedoem, *Hallier* 673 (syn. BO, iso. K).
   D.fuscopilosum Ames & C. Schweinf., Orch. 6:101 (1920). Type:Sabah, Mt. Kinabalu, Gurulau Ridge, *Clemens* 300 (holo. AMES, iso. K, SING).

HABITAT: lower montane forest; open mossy low forest; low stature forest on ultramafic substrate; lower montane dipterocarp forest; sandy areas near rivers and streams.
ALTITUDINAL RANGE: 300 – 1700m.
DISTRIBUTION: Borneo (K, SA, SR). Endemic.
GENERIC SECTION: *Conostalix.*

NOTE: *D.fuscopilosum* is cited in error as '*Dendrochilum fuscopilosum*' by Masamune (1942).

D.osmophytopsis *Kraenzl.* in Engl.,Pflanzenr. Orch. Monand. IV.50. II.21:172(1910). Type: Sarawak, 'Berg Mattau' (probably Mattan), *Beccari* 1684 (as "1686") (holo. FI).

HABITAT: unknown.
ALTITUDINAL RANGE: unknown.
DISTRIBUTION: Borneo (SR). Endemic.

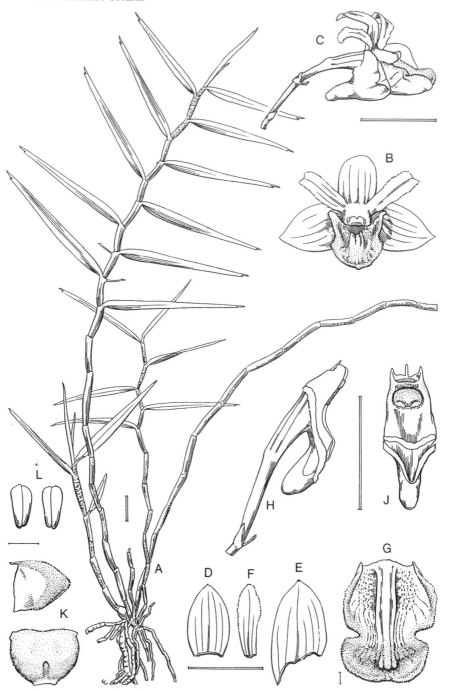

FIG. 32. *Dendrobium nabawanense.* **A**, habit; **B**, flower (front view); **C**, flower (side view); **D**, dorsal sepal; **E**, lateral sepal; **F**, petal; **G**, lip (front view); **H**, pedicel with ovary, column and mentum (side view); **J**, column and mentum (front view); **K**, anther cap (back and side views); **L**, pollinia. **A** from *Collenette* 2292; **B** from *Wood* 750; **C-K** from *Collenette* 2292. Scale: single bar = 1 mm; double bar = 1 cm. Drawn by Eleanor Catherine.

GENERIC SECTION: *Grastidium.*

D.ovatifolium *Ridl.* in J. Linn. Soc., Bot. 31:271 (1896). Type: Sarawak, Beutang Lupar, *Hose* s.n. (holo. SING).

HABITAT: lowland dipterocarp forest.
ALTITUDINAL RANGE: 100m.
DISTRIBUTION: Borneo (SR). Endemic.
GENERIC SECTION: *Distichophyllum.*

D.ovipostoriferum *J.J.Sm.* in Bot. Jahrb. Syst. 48:100 (1912). Type: Kalimantan, between Sungai Tarik and Kwaru, *Winkler* 3062 (holo. HBG).

HABITAT: unknown.
ALTITUDINAL RANGE: unknown.
DISTRIBUTION: Borneo (K, SA). Endemic.
GENERIC SECTION: *Formosae.*

D.paathii *J.J.Sm.* in Orchidee (Bandoeng) 4:183 (1935). Type: Kalimantan, Westerafdeeling, Gunung Raja near Singkawang, *Paath* s.n. (holo. L).

HABITAT: hill forest.
ALTITUDINAL RANGE: 300m.
DISTRIBUTION: Borneo (K). Endemic.
GENERIC SECTION: *Calcarifera.*

D.pachyanthum *Schltr.* in Feddes Repert. 9:290 (1911). Type: Sarawak, Kuching, *Schlechter* 15852 (holo. B,destroyed).

HABITAT: epiphytic on *Gymnostoma sumatrana* in hill forest; heath forest with *Dacrydium, Nepenthes veitchii,* etc. on podsolic soils.
ALTITUDINAL RANGE: sea level – 1200m.
DISTRIBUTION: Borneo (SA, SR). Endemic.
GENERIC SECTION: *Distichophyllum.*

D.pachyphyllum *(Kuntze) Bakh.f.* in Blumea 12, 1:69 (1963). Type: Java, Nusa Kambanga, *Blume* s.n. (holo. BO).
Desmotrichum pusillum Blume, Bijdr.:331, fig.35 (1825).
Dendrobium pumilum Roxb., Fl. Ind., ed.2, 3:479 (1832). Type: Bangladesh, Chittagong, *Roxburgh* drawing (holo. K).
Callista pachyphylla Kuntze, Rev. Gen. Pl.2:654 (1891).
Dendrobium borneense Finet in Bull. Soc. Bot. France 50:373, Pl.11 (1903). Type: Borneo, locality unknown, *Beccari* 442 (holo. ?P,iso. K).

HABITAT: very open kerangas forest with *Dacrydium;* hill and lower montane forest, sometimes on ultramafic substrate.
ALTITUDINAL RANGE: sea level – 2100m.
DISTRIBUTION: Borneo (K, SA, SR). Also Bangladesh, India, Burma, Peninsular Malaysia,
Singapore, Vietnam, Thailand, Sumatra, Java, Riau Archipelago, Mentawai, Siberut.

GENERIC SECTION: *Bolbidium.*

D.pandaneti *Ridl.* in J. Linn. Soc., Bot. 32:257 (1896). Type: Singapore, Bukit Mandai, *Ridley* 5029 (holo. BM).

HABITAT: lowland forest; often epiphytic on wild sago palm.
ALTITUDINAL RANGE: sea level – 400m.
DISTRIBUTION: Borneo (SA). Also Peninsular Malaysia, Singapore, Sumatra,

Lingga Archipelago, Java.
GENERIC SECTION: *Distichophyllum.*

D.panduriferum *Hook.f.*, Fl. Brit. Ind. 6:186 (1890). Type: Burma, Pegu, Rangoon, *Gilbert* s.n. (holo. K).

HABITAT: hill forest on ultramafic substrate.
ALTITUDINAL RANGE: 500-600m.
DISTRIBUTION: Borneo (SA). Also Burma, Peninsular Malaysia, Thailand.
GENERIC SECTION: *Calcarifera.*

D.parthenium *Rchb.f.* in Gard. Chron., ser. 2, 24:489 (1885). Type: Borneo, *Bull* s.n. (holo. W).

HABITAT: hill forest on ultramafic substrate; reported epiphytic on *Gymnostoma sumatrana.*
ALTITUDINAL RANGE: 400 – 900m.
DISTRIBUTION: Borneo (SA). Endemic.
GENERIC SECTION: *Formosae.*

D.patentilobum *Ames & C.Schweinf.*, Orch. 6:110 (1920). Type: Sabah, Mt. Kinabalu, Marai Parai, *Clemens* 366 (holo. AMES, iso. BO, K, SING).

HABITAT: hill and lower montane forest; low stature forest on ultramafic substrate; recorded as epiphytic on *Gymnostoma sumatrana.*
ALTITUDINAL RANGE: 800 – 1500m.
DISTRIBUTION: Borneo (SA). Endemic.
GENERIC SECTION: *Aporum.*

D.pensile *Ridl.* in J. Linn. Soc., Bot. 32:253 (1896). Types: Singapore, Selitar, *Ridley* s.n.; Peninsular Malaysia, Johore, *native collector* s.n.; Indonesia, Riau (Rhio), *native collector* s.n. (syn. SING).

HABITAT: hill forest.
ALTITUDINAL RANGE: 700-900m.
DISTRIBUTION: Borneo (SA). Also Peninsular Malaysia, Sumatra.
GENERIC SECTION: *Grastidium.*

D.pictum *Lindl.* in Gard. Chron.:548 (1862). Type: Borneo, *Low* s.n. (holo. K-LINDL).

HABITAT: unknown.
ALTITUDINAL RANGE: unknown.
DISTRIBUTION: Borneo (unspecified). Endemic.
GENERIC SECTION: *Calcarifera.*

D.pinifolium *Ridl.* in J. Linn. Soc., Bot. 31:269 (1896). Type: Sabah, Sandakan, *Pryer* s.n. (holo. SING).
D.squarrosum J.J.Sm. in Bull. Jard. Bot. Buitenzorg, 13:22 (1914). Type: Kalimantan, Samenggaris, *Amdjah* s.n., cult. Bogor number 157 (holo. BO).

HABITAT: podsol forest with *Dacrydium*; low stature forest on ultramafic substrate; lowland and hill forest.
ALTITUDINAL RANGE: 300 – 600m.
DISTRIBUTION: Borneo (K, SA, SR).
GENERIC SECTION: *Conostalix.*

D.piranha *C.L. Chan & P.J. Cribb* in Orch. Borneo 1:127, fig.28, Pl.6B (1994). Type:

Sabah, Mt. Kinabalu, Marai Parai, *Bailes & Cribb* 815 (holo. K).

> HABITAT: lower montane forest; low scrubby ridge forest on serpentine.
> ALTITUDINAL RANGE: 1400 – 2400m.
> DISTRIBUTION: Borneo (SA). Endemic.
> GENERIC SECTION: *Distichophyllum.*

D.planibulbe *Lindl.* in Bot. Reg. 29:70, misc. 54 (1843). Type: Peninsular Malaysia, *Cuming* s.n., said to have come from 'Manila' (holo. K-LINDL).

> HABITAT: unknown.
> ALTITUDINAL RANGE: unknown.
> DISTRIBUTION: Borneo (K). Also Peninsular Malaysia, Thailand, Sumatra.
> GENERIC SECTION: *Rhopalanthe.*

> NOTE: The origin of Cuming's type is probably false.

D.pogoniates *Rchb.f.* in Gard. Chron., ser. 2, 26:199 (1886). Type: Sabah, locality unknown, cult. *O'Brien* (holo. W).

> HABITAT: lowland dipterocarp forest.
> ALTITUDINAL RANGE: lowlands.
> DISTRIBUTION: Borneo (K, SA). Endemic.
> GENERIC SECTION: *Breviflores.*

D.prostratum *Ridl.* in J. Linn. Soc., Bot., 32:248 (1896). Types: Singapore, Kranji, *Ridley* s.n. (syn. SING); Selitar, *Ridley* s.n. (syn. SING); Sungai Blukang, *Ridley* s.n. (syn. SING); Peninsular Malaysia, Selangor, Seppan, *Ridley* s.n. (syn. SING).

> HABITAT: lowland and hill forest on ultramafic substrate; podsol forest.
> ALTITUDINAL RANGE: sea level – 600m.
> DISTRIBUTION: Borneo (SA, SR). Also Peninsular Malaysia, Singapore.
> GENERIC SECTION: *Aporum.*

D.pseudoaloifolium *J.J.Wood* in Kew Bull. 39(1):82, fig.7 (1984). Type: Sarawak, Gunung Mulu National Park, between Sungai Berar and Sungai Mentawai, *Nielsen* 715 (holo. AAU).

> HABITAT: mixed dipterocarp forest with kerangas element.
> ALTITUDINAL RANGE: 100 – 300m.
> DISTRIBUTION: Borneo (SR). Endemic.
> GENERIC SECTION: *Aporum.*

D.pseudodichaea *Kraenzl.* in Engl., Pflanzenr. Orch. Monand. IV. 50. II. B. 21:171 (1910). Type: Sarawak, *Beccari* 659 (holo. FI).

> HABITAT: unknown.
> ALTITUDINAL RANGE: unknown.
> DISTRIBUTION: Borneo (SR). Endemic.
> GENERIC SECTION: *Grastidium.*

D.puberilingue *J.J.Sm.* in Mitt. Inst. Allg. Bot. Hamburg, 7:53, t.8, fig.43 (1927). Type: Kalimantan, Bukit Raja, *Winkler* 900 (holo. HBG).

> HABITAT: lower montane ridge forest with *Dacrydium, Leptospermum, Phyllocladus, Podocarpus, Tristania,* etc.
> ALTITUDINAL RANGE: 1250 – 1800m.
> DISTRIBUTION: Borneo (K, SA). Endemic.
> GENERIC SECTION: *Rhopalanthe.*

D.quadrangulare *C.S.P.Parish & Rchb.f.* in Flora 69:553 (1886). Type: Burma, Tenasserim, *Parish* 120 (holo. K).

HABITAT: unknown.
ALTITUDINAL RANGE: unknown.
DISTRIBUTION: Borneo (SA). Also Burma, Peninsular Malaysia, Thailand.
GENERIC SECTION: *Bolbidium.*

NOTE: The specimen *Creagh* s.n.(K) matches *D.quadrangulare* in habit, but lack of flowers prevents exact determination. References to this species are probably referable to *D.pachyphyllum* (Kuntze) Bakh.f.

D.radians *Rchb.f.,* Xenia Orch. 2, t.146 (1867). Type: Borneo, *Low* s.n. (holo. W).

HABITAT: unknown.
ALTITUDINAL RANGE: unknown.
DISTRIBUTION: Borneo (unspecified). Endemic.
GENERIC SECTION: *Formosae.*

D.reflexibarbatulum *J.J.Sm.* in Mitt. Inst. Allg. Bot. Hamburg, 7:54, t.8, fig.44 (1927). Type: Kalimantan, Nanga Kruab, *Winkler* 89 (holo. HBG).

HABITAT: lowland forest.
ALTITUDINAL RANGE: lowlands.
DISTRIBUTION: Borneo (K). Endemic.
GENERIC SECTION: *Aporum.*

D.rhodostele *Ridl.* in Trans. Linn. Soc. London, Bot. 2, ser. 3:360 (1893). Type: Peninsular Malaysia, Pahang, Tahan Valley, *Ridley* s.n. (holo. SING).

HABITAT: unknown.
ALTITUDINAL RANGE: unknown.
DISTRIBUTION: Borneo (SA). Also Peninsular Malaysia, Thailand, Sumatra.
GENERIC SECTION: *Aporum.*

D.rosellum *Ridl.* in J. Linn. Soc., Bot., 31:268 (1896). Type: Sarawak, Selabat, *Haviland* s.n. (holo. SING, iso. BO).

HABITAT: lowland forest on white sandy soil and limestone.
ALTITUDINAL RANGE: lowlands.
DISTRIBUTION: Borneo (B, K, SA, SR). Endemic.
GENERIC SECTION: *Aporum.*

D.salaccense *(Blume.) Lindl.,* Gen. Sp. Orch. Pl.86 (1830). Type: Java, Salak, *Blume* s.n. (holo. BO).
Grastidium salaccense Blume., Bijdr.:333 (1825).
Dendrobium gemellum auct., non Lindl.

HABITAT: riparian forest.
ALTITUDINAL RANGE: 100-500m.
DISTRIBUTION: Borneo (SA, SR). Also China, Burma, Peninsular Malaysia, Thailand, Laos, Vietnam, Sumatra, Java.
GENERIC SECTION: *Grastidium.*

D.sambasanum *J.J.Sm.* in Bull. Dép. Agric. Indes Néerl. 22:25 (1909). Type: Kalimantan, Sungai Sambas, *Hallier* 1146 (holo. BO, iso. K).

HABITAT: unknown.
ALTITUDINAL RANGE: unknown.

DISTRIBUTION: Borneo (K). Endemic.
GENERIC SECTION: *Strongyle.*

D.sanderianum *Rolfe* in Kew Bull.:155 (1894). Type: Borneo, imported by *Sander &
Co.* (holo. K).

HABITAT: unknown.
ALTITUDINAL RANGE: unknown.
DISTRIBUTION: Borneo (unspecified). Endemic.
GENERIC SECTION: *Formosae.*

D.sandsii *J.J.Wood & C.L.Chan* in Orch. Borneo 1:129, fig.29, Pl.6C (1994). Type:
Sabah, Kinabatangan District, Maliau River valley, *Sands* 4064 (holo. K).

HABITAT: mixed hill forest.
ALTITUDINAL RANGE: 300-400m.
DISTRIBUTION: Borneo (SA). Endemic.
GENERIC SECTION: *Distichophyllum.*

D.sanguinolentum *Lindl.* in Bot. Reg. 28:62, misc. 73 (1842). Type: 'Ceylon', but
probably Peninsular Malaysia, (holo. K-LINDL).

HABITAT: lowland and hill forest.
ALTITUDINAL RANGE: lowlands and hills.
DISTRIBUTION: Borneo (SA). Also Peninsular Malaysia, Thailand, Sumatra,
Java, Tambelan Islands, Philippines (Sulu Archipelago).
GENERIC SECTION: *Calcarifera.*

D.sarawakense *Ames* in Merr., Bibl. Enum. Born. Pl.:164 (1921). Type: Sarawak,
Quop, *Hewitt* s.n. (holo. SING, iso. K).
   D.multiflorum Ridl. in J. Straits. Branch. Roy. Asiat. Soc. 50:134 (1908), non C.S.P.
Parish & Rchb.f.

HABITAT: unknown.
ALTITUDINAL RANGE: unknown.
DISTRIBUTION: Borneo (SR). Endemic.
GENERIC SECTION: *Calcarifera.*

D.sculptum *Rchb.f.* in Bot. Zeitung (Berlin) 21:128 (1863). Type: Borneo, imported
by *Low*, cult. B*ullen* (holo. W).

HABITAT: rather low, small-crowned lower montane forest on podsolic soil
formations over sandstone.
ALTITUDINAL RANGE: 1200 – 1500m.
DISTRIBUTION: Borneo (SA, SR). Endemic.
GENERIC SECTION: *Formosae.*

D.secundum *(Blume) Lindl.* in Bot. Reg. 15: t.1291 (1829). Type: Java, Tjikao, *Blume*
s.n. (holo. BO).
   Pedilonum secundum Blume, Bijdr.:322 (1825).
   Dendrobium bursigerum Lindl. in J. Linn. Soc., Bot. 32:258 (1896). Type:
Philippines, locality unknown, *Cuming* 2066 (holo. K-LINDL).

HABITAT: lowland and hill forest; rubber plantations.
ALTITUDINAL RANGE: sea level – 1200m.
DISTRIBUTION: Borneo (K, SA, SR). Widespread in Burma, Thailand, Laos,
Cambodia, Vietnam, Peninsular Malaysia, Sumatra, Riau Archipelago, Krakatau,
Java, Flores, Sulawesi, Philippines, Irian Jaya (Waigeo Is.).

GENERIC SECTION: *Pedilonum.*

D.setifolium *Ridl.* in J. Linn. Soc., Bot. 31:270 (1896). Type: Sarawak, *Haviland* s.n. (holo. SING, iso. K).

HABITAT: lower montane forest; observed on *Dacrydium.*
ALTITUDINAL RANGE: 1300 – 1600m.
DISTRIBUTION: Borneo (SA, SR). Also Peninsular Malaysia, Singapore, Thailand, Riau Archipelago.
GENERIC SECTION: *Rhopalanthe.*

D.singaporense *A.D.Hawkes & A.H.Heller* in Lloydia 20, 2:124 (1957). Type: Singapore, *Loddiges* s.n. (holo. K-LINDL).
Dendrobium teres Lindl. in Bot. Reg. 29, misc. 111 (1840), non Roxb.

HABITAT: unknown.
ALTITUDINAL RANGE: lowlands.
DISTRIBUTION: Borneo (K, SR). Also Peninsular Malaysia, Thailand, Sumatra, Natuna Islands.
GENERIC SECTION: *Strongyle.*

D.singkawangense *J.J.Sm.* in Gard. Bull. Straits Settlem. 9:91 (1935). Type: Kalimantan, Wester Afdeeling, Singkawang, *Paath* s.n.(holo. L).

HABITAT: riverine hill and lower montane forest on sandstone ridges.
ALTITUDINAL RANGE: 500 – 1700m.
DISTRIBUTION: Borneo (K, SA). Endemic.
GENERIC SECTION: *Formosae.*

D.singulare *Ames & C.Schweinf.*, Orch. 6:112 (1920). Type: Sabah, Mt. Kinabalu, Lobong, *Clemens* 118 (holo. AMES).

HABITAT: lower montane forest.
ALTITUDINAL RANGE: 1500m.
DISTRIBUTION: Borneo (SA). Endemic.
GENERIC SECTION: *Grastidium.*

D.sinuatum *(Lindl) Lindl. ex Rchb.f.* in Ann. Bot. Syst. 6:280 (1861). Type: Singapore, *Cuming* s.n. (holo. K-LINDL).
Aporum sinuatum Lindl. in Bot. Reg. 27, misc.1 (1841).

HABITAT: low and open mossy ridge forest with a dense undergrowth of bamboo and rattans.
ALTITUDINAL RANGE: 1100 – 1700m.
DISTRIBUTION: Borneo (B, SA, SR). Also Peninsular Malaysia, Singapore, Thailand, Mentawai Islands (Siberut).
GENERIC SECTION: *Oxystophyllum.*

D.smithianum *Schltr.* in Feddes Repert. 10:74 (1911). Types: Sulawesi, Minahassa, Ayermadidi, *Schlechter* 20522 (syn. B, destroyed, isosyn. K); Toli Toli District, Lampasioe, *Schlechter* 20681 (syn. B,destroyed, isosyn. K).

HABITAT: hill forest on limestone; podsol forest.
ALTITUDINAL RANGE: 300 – 1300m.
DISTRIBUTION: Borneo (SA). Also Sulawesi.
GENERIC SECTION: *Aporum.*

D.spathipetalum *J.J.Sm.* in Bull. Jard. Bot. Buitenzorg., ser. 2, 13:20 (1914). Type: Kalimantan, Kota Waringin, *van Nouhuys* 9 (holo. BO).

HABITAT: unknown.
ALTITUDINAL RANGE: unknown.
DISTRIBUTION: Borneo (K). Endemic.
GENERIC SECTION: *Distichophyllum.*

D.spectatissimum *Rchb.f.* in Linnaea 41:41 (1877). Type: Borneo, *Lobb* s.n. (holo. W).
  D.speciosissimum Rolfe in Orchid Rev. 3:119 (1895). Type: Sabah, Mt. Kinabalu, *Low* s.n. (holo. K).
  D.reticulatum J.J.Sm. in Bull. Jard. Bot. Buitenzorg, ser. 2, 13:18 (1914). Type: Sabah, Mt. Kinabalu, *Dumas* s.n. (holo. BO).

HABITAT: lower montane *Leptospermum* scrub on ultramafic substrate; most commonly epiphytic on *Leptospermum.*
ALTITUDINAL RANGE: 1600 – 1700m.
DISTRIBUTION: Borneo (SA). Endemic.
GENERIC SECTION: *Formosae.*

D.spegidoglossum *Rchb.f.* in Bonplandia 2:88 (1854). Type: ?Burma, 'Ostindien', imported by *Schiller* s.n. (holo. W).

HABITAT: hill and lower montane forest.
ALTITUDINAL RANGE: 600 – 1500m.
DISTRIBUTION: Borneo (K, SA, SR). Also Burma, Peninsular Malaysia, Singapore, Thailand, Riau Archipelago, Sumatra, Java.
GENERIC SECTION: *Breviflores.*

D.spurium *(Blume) J.J.Sm.*, Fl. Buit.:343 (1905). Type: Java, Seribu, *Blume* s.n. (holo. BO).
  Dendrocolla spuria Blume, Bijdr.:290 (1825).

HABITAT: peat swamp forest; hill forest on ultramafic substrate.
ALTITUDINAL RANGE: 200 – 900m.
DISTRIBUTION: Borneo (K, SA, SR). Also Peninsular Malaysia, Sumatra, Java, Bali, Philippines.
GENERIC SECTION: *Euphlebium.*

D.stuartii *F.M. Bailey* in Proc. Roy. Soc. Queensland 1:12 (1884). Type: Australia, cult. Brisbane Botanic Gardens ex 'Herberton', *Stuart* s.n. (holo. BRI, not located, neo. Fitzgerald's illustration in Aust. Orch. 2(3): t.6 (1888) CBG).

HABITAT: riverine forest on ultramafic substrate.
ALTITUDINAL RANGE: 100 – 300m.
DISTRIBUTION: Borneo (K). Also Peninsular Malaysia, Thailand, Sumatra, Java, Australia.
GENERIC SECTION: *Dendrobium.*

D.subulatoides *Schltr.* in Feddes Repert. 9:290 (1911). Types: Sarawak, Kuching, *Beccari* 439 (syn. B, destroyed, isosyn. FI); Kuching, *Hewitt* s.n. (syn. B, destroyed); Mt. Matang, *Beccari* 3657 (syn. B, destroyed, isosyn. FI).

HABITAT: lowland and hill forest.
ALTITUDINAL RANGE: sea level – 900m.
DISTRIBUTION: Borneo (SA, SR). Endemic.
GENERIC SECTION: *Strongyle.*

D.subulatum *(Blume) Lindl.*, Gen. Sp. Orch. Pl.:91 (1830). Type: Java, locality unknown, *Blume* s.n. (holo. BO).
    Onychium subulatum Blume, Bijdr.:328 (1825).

HABITAT: lowlands.
ALTITUDINAL RANGE: sea level – lowlands.
DISTRIBUTION: Borneo (K, SA, SR). Also Peninsular Malaysia, Singapore, Thailand, Sumatra, Mentawai, Bangka, Java.
GENERIC SECTION: *Strongyle.*

D.swartzii *A.D.Hawkes and A.H.Heller* in Lloydia 20, 2:124 (1957). Type: Borneo, *Low* s.n. (holo. W).
    D.lilacinum Rchb.f. in Gard. Chron.:674 (1865), non Teijsm. & Binn.
    Callista lilacina (Rchb.f.) Kuntze, Rev. Gen. Pl.2:654 (1891).

HABITAT: unknown.
ALTITUDINAL RANGE: unknown.
DISTRIBUTION: Borneo (unspecified). Endemic.
GENERIC SECTION: *Calcarifera.*

D.takahashii *Carr* in Orchid Rev. 42:14 (1934). Type: Kalimantan, Martapura near Banjermasin, *Takahashi* s.n. (holo. SING).

HABITAT: stunted trees on exposed ridges.
ALTITUDINAL RANGE: 100m.
DISTRIBUTION: Borneo (K). Endemic.
GENERIC SECTION: *Formosae.*

D.tenue *J.J.Sm.* in Bull. Jard. Bot. Buitenzorg, ser. 2, 25:31 (1917). Types: Kalimantan, locality unknown, *Nieuwenhuis* s.n., cult. Bogor under number 1412 (syn. BO); Martapura (Martapoera), *Korthals* s.n., Herb. Lugd. Bat. number 903, 348-29-30 (syn. BO).

HABITAT: unknown.
ALTITUDINAL RANGE: 100m.
DISTRIBUTION: Borneo (K). Endemic.
GENERIC SECTION: *Rhopalanthe.*

D.tetrachromum *Rchb.f.* in Gard. Chron., 13:712 (1880). Type: Borneo, *Curtis* s.n.(holo. W).

HABITAT: riverine forest; hill forest.
ALTITUDINAL RANGE: 500 – 1100m.
DISTRIBUTION: Borneo (K, SA). Endemic.
GENERIC SECTION: *Dendrobium.*

D.tetralobum *Schltr.* in Bull. Herb. Boissier, ser. 2, 6:458 (1906). Type: Kalimantan, West Koetai, Samarinda, *Schlechter* 3329 (holo. B, destroyed).

HABITAT: unknown.
ALTITUDINAL RANGE: unknown.
DISTRIBUTION: Borneo (K). Endemic.
GENERIC SECTION: *Aporum.*

D.torquisepalum *Kraenzl.* in Engler, Pflanzenr. Orch. Monand. IV. 50. II. B. 21:187 (1910). Type: Sarawak, *Beccari* 1872 (holo. B, destroyed, iso. FI).

HABITAT: unknown.
ALTITUDINAL RANGE: unknown.

DISTRIBUTION: Borneo (SR). Endemic.
GENERIC SECTION: *Distichophyllum.*

D.tricuspe *(Blume) Lindl.*, Gen. Sp. Orch. Pl.:88 (1830). Type: Java, Salak, *Blume* s.n. (holo. L).
Onychium tricuspe Blume, Bijdr.:326 (1825).

HABITAT: lowland dipterocarp forest on sandstone and mudstone.
ALTITUDINAL RANGE: lowlands.
DISTRIBUTION: Borneo (SA): Also Sumatra, Java.
GENERIC SECTION: *Rhopalanthe*

D.tridentatum *Ames & C.Schweinf.*, Orch. 6:115 (1920). Type: Sabah, Mt. Kinabalu, Marai Parai, *Clemens* 257(holo. AMES).

HABITAT: lower and upper montane forest; montane oak forest; on ultramafic substrate.
ALTITUDINAL RANGE: 1200 – 2200m.
DISTRIBUTION: Borneo (SA). Endemic.
GENERIC SECTION: *Rhopalanthe.*

D.trullatum *J.J.Wood et A.Lamb* **sp.nov.** *D.pachyglosso* Par. et Rchb.f., species burmanica, laotica, annamensis, siamensis atque malayana et *D.pinifolio* Ridl., species borneensis, sed foliis brevioribus impariter bilobatis, vaginis glabris, labio trullato ecalloso lobo medio anguste triangulari acuto, cuius marginibus sunt undulato-convolutis atque papillosis, differt. Typus: East Malaysia, Sabah, Keningau District, near Nabawan, S.E. of Keningau, c.420m, 13.10.1985, *Wood* 598 (holotypus K, herbarium and spirit material).

Epiphytic herb. Stems 12-14 cm long, 0.3 cm wide, caespitose, slightly sinuate-fractiflex. Leaves linear-ligulate, slightly acutely unequally bilobed, stiff, blade 1.2-2 x 0.25-0.4 cm, sheaths 0.5-0.7 cm long, glabrous. Inflorescences 1(-2)-flowered, emerging from sheath base opposite leaf blade, sessile; floral bracts 3-4, 1 mm long, ovate, acute. Flowers c.1.3 cm across, non-resupinate, scented. Pedicel with ovary 3 mm long, clavate. Sepals stiff, spreading, pale green or bright yellow. Dorsal sepal 6 x 2-2.1 mm, triangular-ovate, acute. Lateral sepals 7.5-8 x 2.5 mm, obliquely triangular-ovate, acute. Mentum 2 mm long, conical, subacute. Petals 5.5-6 x 1 mm, linear, acute, pale green or bright yellow. Lip stiff, fleshy, 3-lobed, 6.5-7 mm long, 3 mm wide across side lobes when flattened, trullate in outline, cream; disc ecallose, farinose; mid-lobe 2 – 2.5 mm long, narrowly triangular-ovate, acute, papillose, margin strongly undulate-convolute and meeting in the middle; side lobes 4.5 mm long, 0.5 mm wide at middle, erect, broadly triangular, rounded. Column 2 mm long, oblong, foot 2 mm long, straight. Anther cap 1 x 1 mm, ovate, cucullate. Fig.33.

Other material examined: Sabah, near Nabawan, 540 m, 19.2.1981, *Collenette* 2287 (K, herbarium and spirit material); near Nabawan, 450 m, 22.9.1987, *Lamb* AL 841/87 (K, herbarium material only).

D.trullatum belongs to section *Conostalix* and is related to *D.pachyglossum* C.S.P. Parish and Rchb.f. from Burma, Laos, Vietnam, Thailand and Peninsular Malaysia, and *D.pinifolium* Ridl. from Borneo. It differs from both in having shorter, unequally bilobed leaves with glabrous sheaths and a trullate ecallose lip with a narrowly triangular, acute mid-lobe having an undulate-convolute, papillose margin.

This species has so far only been found growing in podsol forest on sandstone soils, composed of *Dacrydium*, *Eugenia* and *Garcinia*, etc. with an understorey of *Rhododendron longiflorum* and *R.malayanum*. The field layer consists of ferns, orchids

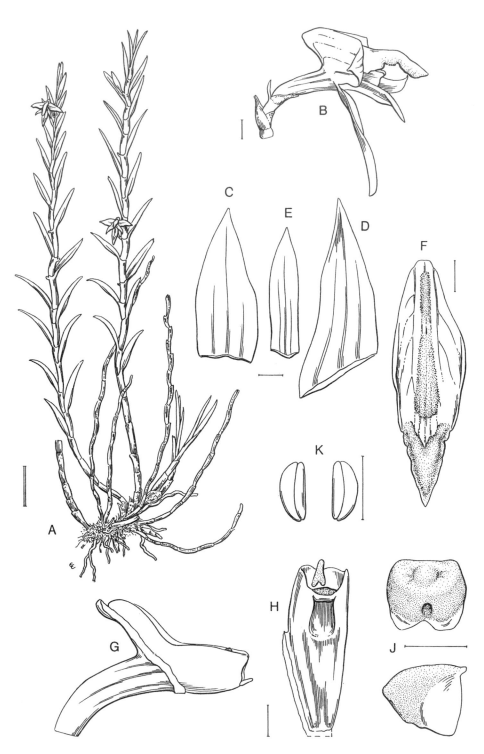

(eg. *Bromheadia finlaysoniana*, *Bulbophyllum* spp., *Dendrochilum simplex*), *Nepenthes ampullaria*, rattans and Araceae.

D.truncatum *Lindl.* in J. Linn. Soc., Bot. 3:15 (1859). Types: Java, *Reinwardt* s.n. (syn. K-LINDL); *Lobb* s.n. (syn. K-LINDL).

HABITAT: lowland forest; rubber plantations; podsol forest; lower montane forest.
ALTITUDINAL RANGE: sea level – 800m.
DISTRIBUTION: Borneo (K, SA). Also Peninsular Malaysia, Thailand, Sumatra, Bangka, Java.
GENERIC SECTION: *Rhopalanthe.*

D.uncatum *Lindl.* in J. Linn. Soc., Bot. 3:5 (1859). Type: Java, *Lobb* 156 (holo. K-LINDL).

HABITAT: lower montane forest.
ALTITUDINAL RANGE: 1400m.
DISTRIBUTION: Borneo (SA). Also Java, Krakatau.
GENERIC SECTION: *Strongyle.*

D.uniflorum *Griff.*, Not. 3:305 (1851). Type: Peninsular Malaysia, Mt. Ophir, *Griffith* s.n. (holo. K, iso. P).
D.quadrisulcatum J.J.Sm. in Bull. Jard. Bot. Buitenzorg, ser. 2, 25:49 (1917). Type: Kalimantan, Gunung Damoes, *Hallier* 563 (holo. L).

HABITAT: hill dipterocarp and lower montane forest; podsol forest with *Dacrydium* and *Tristania.*
ALTITUDINAL RANGE: lowlands – 1800m.
DISTRIBUTION: Borneo (K, SA, SR). Also Peninsular Malaysia, Vietnam, Thailand, Philippines.
GENERIC SECTION: *Distichophyllum.*

D.ustulatum *Carr* in Gard. Bull. Straits Settlem. 7:13 (1932). Type: Peninsular Malaysia, Pahang, Fraser's Hill, Ulu Jeriau, *Carr* s.n. (holo. SING).

HABITAT: hill forest.
ALTITUDINAL RANGE: 900m.
DISTRIBUTION: Borneo (SA, SR). Also Peninsular Malaysia.
GENERIC SECTION: *Bolbidium.*

D.ventripes *Carr* in Gard. Bull. Straits Settlem. 8:103 (1935). Type: Sarawak, Dulit ridge, *Synge* S.422 (holo. SING, iso. K).

HABITAT: lower montane forest; oak-Ericaceae moss forest; upper montane forest on ultramafic substrate.
ALTITUDINAL RANGE: 1200 – 2100m.
DISTRIBUTION: Borneo (SA, SR). Endemic.
GENERIC SECTION: *Rhopalanthe.*

FIG. 33. *Dendrobium trullatum.* **A**, habit; **B**, flower (side view); **C**, dorsal sepal; **D**, lateral sepal; **E**, petal; **F**, lip (front view); **G**, pedicel with ovary and column (side view); **H**, column (front view); **J**, anther cap (back and side views); **K**, pollinia. A from *Lamb* AL 841/87; **B-K** from *Wood* 598. Scale: single bar = 1 mm; double bar = 1 cm. Drawn by Eleanor Catherine.

D.villosulum *Lindl.*, Gen. Sp. Orch. Pl.:86 (1830). Type: Peninsular Malaysia, Penang, *Wallich* catalogue number 2006 (holo. K-WALL).
D.melanochlamys Holttum in Gard. Bull. Straits Settlem. 11:280 (1947).

HABITAT: podsol forest with *Dacrydium* and *Tristania*; ridge forest dominated by *Quercus*; lower montane forest; moss forest; damp rocky places near waterfalls.
ALTITUDINAL RANGE: 400 – 1700m.
DISTRIBUTION: Borneo (K, SA, SR). Also Peninsular Malaysia, Singapore.
GENERIC SECTION: *Conostalix.*

D.viriditepalum *J.J.Sm.* in Bull. Jard. Bot. Buitenzorg, ser. 2, 25:54 (1917). Types: Sumatra, Lampong, Gunung Soegi, *van Andel* s.n., cult. Bogor number 32 (syn. BO); Benkoelen, Boekit Kaba, Soeban Ajam, *Ajoeb* s.n., cult. Bogor (syn. BO).

HABITAT: unknown.
ALTITUDINAL RANGE: unknown.
DISTRIBUTION: Borneo (K). Also Sumatra.
GENERIC SECTION: *Calcarifera.*

D.xanthoacron *Schltr.* in Bull. Herb. Boissier, ser. 2, 6:459 (1906). Type: Kalimantan, West Koetai, Samarinda, *Schlechter* 13332 (holo. B,destroyed).

HABITAT: unknown.
ALTITUDINAL RANGE: lowlands.
DISTRIBUTION: Borneo (K, SA). Endemic.
GENERIC SECTION: *Aporum.*

D.xiphophyllum *Schltr.* in Feddes Repert., 9:291 (1911). Types: Sarawak, Gunung Matang, *Beccari* 2049 (syn. B,destroyed, isosyn. FI); Penrisen ('Pennisen'), *Brooks* s.n. (syn. B, destroyed).

HABITAT: lower montane forest.
ALTITUDINAL RANGE: 700 – 800m.
DISTRIBUTION: Borneo (SA, SR). Endemic.
GENERIC SECTION: *Aporum.*

DIPLOCAULOBIUM (Rchb.f.) Kraenzl.

in Engl., Pflanzenr. Orch. Monand. IV. 50. II. B. 21:331 (1910).

Epiphytic herbs. Pseudobulbs clustered, of one internode. Leaves solitary, terminal, subcoriaceous, erect, linear to elliptic, lacking a petiole. Inflorescences 1 to several, one-flowered, usually borne in succession. Flowers lasting less than a day, borne on pedicels from within a large bract at the apex of the pseudobulb; sepals and petals slender, often filiform for the most part but the bases of the lateral sepals rather broad and forming, with the column-foot, a distinct mentum; lip attached to the apex only of the column-foot, trilobed with rather small side lobes, or almost entire, the mid-lobe usually rather long, the callus of longitudinal keels which may be undulate; column with a distinct but not very long foot; pollinia 4.

About 100 species distributed from Peninsular Malaysia to New Guinea, Australia and the Pacific islands. The majority of species are endemic to New Guinea.

D.longicolle *(Lindl.) Kraenzl.* in Engl., Pflanzenr. Orch. Monand. IV. 50. II. B. 21:340 (1910). Type: Singapore, *Cuming* s.n., cult. Loddiges (holo. K-LINDL).

Dendrobium longicolle Lindl. in Bot. Reg. 26, 74,misc. 172 (1840).

HABITAT: lowland and hill forest.
ALTITUDINAL RANGE: under 300m.
DISTRIBUTION: Borneo (SR). Also Peninsular Malaysia, Singapore.

D.vanleeuwenii *(J.J.Sm.) P.F.Hunt & Summerh.* in Taxon 10(4):109 (1961). Type: Irian Jaya, Rouffaer River, *Docters van Leeuwen* 10336 (holo. L).
Dendrobium vanleeuwenii J.J.Sm. in Nova Guinea 18:40, t.8, fig.22 (1936).

HABITAT: podsol forest with *Dacrydium*.
ALTITUDINAL RANGE: 600m.
DISTRIBUTION: Borneo (B, SA, SR). Also Irian Jaya.

EPIGENEIUM Gagnep.

Bull. Mus. Hist. Nat. (Paris), ser. 2, 4:592 (1932).

Epiphytic or terrestrial herbs. Aerial shoots terminating in either a 1-, 2- or 3-leaved pseudobulb consisting of 1 internode, the new shoot arising at the base of the pseudobulb, sometimes pendulous, covered with conspicuous imbricate brown sheaths when young. Pseudobulbs usually short and conical, or ovoid. Leaves oblong or obovate, coriaceous. Inflorescences slender, arising between or just below the leaves, racemose, 1- to several-flowered. Flowers medium-sized to large, resupinate, long lasting; sepals and petals narrowly elliptic, subequal; dorsal sepal enclosing column; lateral sepals larger, adnate to column-foot forming a short mentum; petals longly decurrent on mentum; lip pandurate-oblong to 3-lobed, disc with a callus which is lobulate or ridged at base; column short to rather long, with or without short stelidia, foot long; pollinia 2, or 4 in 2 groups.

About 35 species distributed in mainland Asia, Taiwan, Malaysia, Indonesia and the Philippines.

E.geminatum *(Blume) Summerh.* in Kew Bull. 12:262 (1957). Type: Java, Gede & Salak, *Blume* s.n. (holo. BO, iso. L).
Desmotrichum geminatum Blume, Bijdr.:332 (1825).

HABITAT: lower montane forest.
ALTITUDINAL RANGE: 1500 – 1900m.
DISTRIBUTION: Borneo (SA, SR). Also Peninsular Malaysia, Java.

E.kinabaluense *(Ridl.) Summerh.* in Kew Bull.12:262 (1957). Type: Sabah, Mt. Kinabalu, *Haviland* 1253 (holo. K).
Dendrobium kinabaluense Ridl. in Stapf in Trans. Linn. Soc. London, Bot., ser. 2, 4:234(1894).
Sarcopodium kinabaluense (Ridl.) Rolfe in Orchid Rev. 18:239 (1910).
S.suberectum Ridl. in Kew Bull.:211 (1914). Type: Sarawak, Gunung Rumput, *Anderson* 172 (holo. K).
Katherinea kinabaluense (Ridl.) A.D.Hawkes in Lloydia 19:96 (1956).
Epigeneium suberectum (Ridl.) Summerh. in Kew Bull. 12:265 (1957).

HABITAT: lower montane oak-laurel forest; upper montane forest; mossy rocks on exposed ridges; exposed branches in full sunlight; mostly in low ericaceous – myrtaceous scrub on ultramafic substrate.
ALTITUDINAL RANGE: 1200 – 3400m.

DISTRIBUTION: Borneo (SA, SR). Endemic.

E.labuanum *(Lindl.) Summerh.* in Kew Bull. 12:262 (1957). Type: Borneo, *Lobb* s.n. (holo. K-LINDL).
  Dendrobium labuanum Lindl. in J. Linn. Soc., Bot. 3:6 (1858).
  Sarcopodium labuanum (Lindl.) Rolfe in Orchid Rev. 18:239 (1910).
  Katherinea labuana (Lindl.) A.D.Hawkes in Lloydia 19:96 (1956).

HABITAT: lower montane ridge forest.
ALTITUDINAL RANGE: 1200 – 1400m; the type locality is presumably Labuan Island which is at around sea level.
DISTRIBUTION: Borneo (SA, SR). Also Vietnam.

E.longirepens *(Ames & C.Schweinf.) Seidenf.* in Dansk Bot. Ark. 34 (1):18 (1980). Type: Sabah, Mt. Kinabalu, Marai Parai, *Clemens* 245 (holo. AMES, iso. K, SING).
  Dendrobium longirepens Ames & C.Schweinf., Orch. 6:105 (1920).
  Desmotrichum longirepens (Ames & C.Schweinf.) A.D.Hawkes in Lloydia 20:126 (1957).
  Flickingeria longirepens (Ames & C.Schweinf.) A.D.Hawkes in Orchid Weekly 2, 46:456 (1961).
  Ephemerantha longirepens (Ames & C.Schweinf.) P.F.Hunt & Summerh. in Taxon 10:105 (1961).

HABITAT: lower montane forest, frequently on ultramafic substrate; oak, chestnut and *Dacrydium* forest on podsolic soil.
ALTITUDINAL RANGE: 900 – 1700m.
DISTRIBUTION: Borneo (SA). Endemic.

E.radicosum *(Ridl.) Summerh.* in Kew Bull. 12:264 (1957). Type: Sarawak, Tiang Lagu, *Hewitt* s.n. (holo. SING).
  Dendrobium radicosum Ridl. in J. Straits Branch Roy. Asiat. Soc., 49:29 (1908).
  Sarcopodium radicosum (Ridl.) Rolfe in Orchid Rev. 18:239 (1910).

HABITAT: unknown.
ALTITUDINAL RANGE: unknown.
DISTRIBUTION: Borneo (SR). Endemic.

E.speculum *(J.J.Sm.) Summerh.* in Kew Bull. 12:264 (1957). Type: Kalimantan, Bukit Kasian, *Nieuwenhuis* 274 (holo. L).
  Dendrobium speculum J.J.Sm. in Bull. Dép. Agric. Indes Néerl. 5:34 (1907).
  Sarcopodium speculum (J.J.Sm.) Carr in Gard. Bull. Straits Settlem. 8:109 (1935).
  Katherinea specula (J.J.Sm.) A.D.Hawkes in Lloydia 19:97 (1956), as '*K.speculum*'.

HABITAT: transitional moss forest; podsol forest with *Dacrydium, Tristania*, etc.; often terrestial.
ALTITUDINAL RANGE: 400 – 900m.
DISTRIBUTION: Borneo (K, SA, SR). Endemic.

E.treacherianum *(Rchb.f.ex Hook.f.) Summerh.* in Kew Bull. 12:265 (1957). Type: Borneo, *Low* s.n. (holo. K).
  Dendrobium treacherianum Rchb.f.ex Hook.f. in Bot. Mag. 107, t.6591 (1881).
  Sarcopodium treacherianum (Rchb.f. ex Hook.f.) Rolfe in Orchid Rev. 18:239 (1910).
  Katherinea treacheriana (Rchb.f.ex Hook.f.) A.D.Hawkes in Lloydia 19:97 (1956).

HABITAT: lowland and mixed hill dipterocarp forest.
ALTITUDINAL RANGE: 100 – 400m.
DISTRIBUTION: Borneo (K, SA, SR). Also Philippines.

E.tricallosum *(Ames & C.Schweinf.)* *J.J.Wood* in Lindleyana 5(2):99 (1990). Type: Sabah, Mt. Kinabalu, *Clemens* s.n. (holo. AMES).
   Dendrobium tricallosum Ames & C.Schweinf., Orch. 6:114 (1920).

HABITAT: lower montane forest; ridge forest with *Dacrydium, Leptospermum, Podocarpus, Tristania* and *Phyllocladus,* associated with *Dilochia rigida, Chelonistele* spp. and *Gahnia* spp.
ALTITUDINAL RANGE: 800 – 1800m.
DISTRIBUTION: Borneo (SA). Endemic.

E.verruciferum *(J.J.Sm.) Summerh.* in Kew Bull. 12:266 (1957). Type: Kalimantan, *Moquette* s.n. (holo. BO).
   Dendrobium verruciferum J.J.Sm. in Bull. Dép. Agric. Indes Néerl. 15:12 (1908).
   Sarcopodium verruciferum (J.J.Sm.) Rolfe in Orchid Rev. 18:239 (1910).
   Dendrobium interruptum J.J.Sm. in Bull. Dép. Agric. Indes Néerl. 45:18 (1911).
   Sarcopodium interruptum (J.J.Sm.) Ames in Merr., Bibl. Enum. Born. Pl.:167 (1921).
   Katherinea interrupta (J.J.Sm.) A.D.Hawkes in Lloydia 19:96 (1956).
   K.verrucifera (Rolfe) A.D.Hawkes in Lloydia 19:98 (1956).

HABITAT: unknown.
ALTITUDINAL RANGE: unknown.
DISTRIBUTION: Borneo (K). Endemic.

E.zebrinum *(J.J.Sm.) Summerh.* in Kew Bull. 12:266 (1957). Type: Kalimantan, Kelam, *Hallier* s.n. (holo. BO).
   Dendrobium zebrinum J.J.Sm. in Icon. Bogor. 2:72, t.113c (1903).
   D.sulphuratum Ridl.in J. Straits Branch Roy. Asiat. Soc. 49:239 (1908). Type: Sarawak, Sajingkat, *Hewitt* s.n. (holo. SING).
   Sarcopodium sulphuratum (Ridl.) Rolfe in Orchid Rev. 18:239 (1910).
   S.zebrinum (J.J.Sm.) Kraenzl. in Engl.,Pflanzenr. Orch. Monand. IV. 50. II. B. 21:324 (1910).
   Dendrobium citrinocastaneum Burkill in Gard. Bull. Straits Settlem. 3:12 (1923). Type: Peninsular Malaysia, Johor, near Johor Baharu, *Feddersen* s.n. (holo. SING).
   Sarcopodium citrinocastaneum (Burkill) Ridl., Fl. Mal. Pen. 4:29 (1924).
   Katherinea citrinocastanea (Burkill) A.D.Hawkes in Lloydia 19:95 (1956).
   K.zebrina (J.J.Sm.) A.D.Hawkes in Lloydia 19:98 (1956).

HABITAT: unknown.
ALTITUDINAL RANGE: unknown.
DISTRIBUTION: Borneo (K, SR). Also Peninsular Malaysia, Sumatra.

FLICKINGERIA A.D.Hawkes

Orchid Weekly 2:451 (Jan.1961).

Desmotrichum Blume, non Kutz.
Ephemerantha P. F.Hunt & Summerh.

Epiphytic herbs. Aerial shoots terminating in a unifoliate pseudobulb consisting of 1 internode, erect and bushy or drooping and laxly branched, new branches usually arising at base of pseudobulb. Leaves narrowly- to oblong-elliptic, coriaceous. Inflorescences borne from chaffy bracts in front or behind the leaf base, 1 – 2-flowered. Flowers fragile, ephemeral, lasting less than a day; sepals and petals acute; lateral sepals adnate to column-foot forming a rather long mentum; petals narrower than sepals; lip 3-lobed, 2 – 3-keeled, side lobes erect, mid-lobe variable in shape, straight, curved or very undulate-plicate and transversely bilobed, often narrow at base forming a mesochile; column short, with a long foot; pollinia 4, naked, i.e. without a stipes or caudicle.

Between 65 and 70 species distributed in mainland Asia, Malaysia, Indonesia, the Philippines, New Guinea, Australia and the Pacific islands.

F.bancana *(J.J.Sm.) A.D.Hawkes* in Orchid Weekly 2(46):452 (1961). Type: Bangka, cult. *Bogor* (holo. L, iso. K).
Dendrobium bancanum J.J.Sm. in Bull. Dép. Agric. Indes Néerl. 22:23 (1909).

HABITAT: sclerophyllous ridge forest on limestone.
ALTITUDINAL RANGE: 200 – 300m.
DISTRIBUTION: Borneo (K, SR). Also Peninsular Malaysia, Singapore, Thailand, Sumatra, Bangka.

F.bicarinata *(Ames & C.Schweinf.) A.D.Hawkes* in Orchid Weekly 2(46): 452 (1961). Type: Sabah, Mt. Kinabalu, Kiau, *Clemens* 335 (holo. AMES).
Dendrobium bicarinatum Ames & C. Schweinf., Orch. 6:98 (1920).
Desmotrichum bicarinatum (Ames & C.Schweinf.) A.D.Hawkes & A.H.Heller in Lloydia 20:125 (1957).
Ephemerantha bicarinata (Ames & C.Schweinf.) P.F.Hunt & Summerh. in Taxon 10(4):102 (1961).

HABITAT: hill forest.
ALTITUDINAL RANGE: 800 – 900m.
DISTRIBUTION: Borneo (SA). Endemic.

F.bicostata *(J.J.Sm.) A.D.Hawkes* in Orchid Weekly 2(46):452 (1961). Types: Kalimantan, *Nieuwenhuis* s.n. (syntype: BO); Pontianak, *Romburgh* s.n. (syn. BO).
Dendrobium bicostatum J.J.Sm. in Icon. Bogor. 3:19, Pl.209 (1906).
D.mattangianum Kraenzl. in Engl.,Pflanzenr. Orch. Monand. IV. 50. II. B. 21:159 (1910). Type: Sarawak, Mt. Mattang, *Beccari* 1346 (holo. B, destroyed, iso. FI).
Desmotrichum bicostatum (J.J.Sm.) Kraenzl.in Engl., Pflanzenr. Orch. Monand. IV. 50. II. B. 21:353 (1910).
Ephemerantha bicostata (J.J.Sm.)P.F.Hunt & Summerh. in Taxon 10(4):102 (1961).

HABITAT: unknown.
ALTITUDINAL RANGE: unknown.
DISTRIBUTION: Borneo (K, SR). Endemic.

F.comata *(Blume) A.D.Hawkes* in Orchid Weekly 2(46):453 (1961). Type: Java, Buitenzorg, *Blume* s.n. (holo. BO).
Desmotrichum comatum Blume, Bijdr.:330 (1825).
Dendrobium criniferum Lindl. in Bot. Reg. 30, 41, misc. 45 (1844). Type: Sri Lanka, *Power* s.n., cult. Duke of Northumberland (holo. K-LINDL).

HABITAT: unknown.
ALTITUDINAL RANGE: unknown.
DISTRIBUTION: Borneo (unspecified). Widespread throughout most of S.E. Asia, north to Taiwan, south and east to Australia and New Caledonia.

F.crenicristata *(Ridl.) J.J.Wood* in Lindleyana 5(2):101 (1990). Type: Sarawak, Quop, *Hewitt* s.n.(holo. SING, iso. SAR).
Dendrobium crenicristatum Ridl. in J. Straits Branch Roy. Asiat. Soc. 54:48 (1910).
Ephemerantha crenicristata (Ridl.) P.F.Hunt & Summerh. in Taxon 10(4):103 (1961).

HABITAT: unknown.
ALTITUDINAL RANGE: unknown.
DISTRIBUTION: Borneo (SR). Endemic.

F.denigrata *(J.J.Sm.) J.J.Wood* in Lindleyana 5(2):101 (1990). Type: Kalimantan, Nieuwenhuis Expedition, *Jaheri* s.n., cult. Bogor number 166 (holo. BO).
Dendrobium denigratum J.J.Sm. in Bull. Jard. Bot. Buitenzorg, ser. 3, 2:60 (1920).
Ephemerantha denigrata (J.J.Sm.) P.F.Hunt & Summerh. in Taxon 10:103 (1961).

HABITAT: unknown.
ALTITUDINAL RANGE: unknown.
DISTRIBUTION: Borneo (K). Endemic.

F.dura *(J.J.Sm.) A.D.Hawkes* in Orchid Weekly 2(46):454 (1961). Type: Java, *Gede*, HLB 903, 348-22 (holo. BO).
Dendrobium durum J.J.Sm., Orch. Java: 320 (1905).

HABITAT: lower montane forest.
ALTITUDINAL RANGE: 1600m.
DISTRIBUTION: Borneo (SA). Also Java.

F.fimbriata *(Blume) A.D.Hawkes* in Orchid Weekly 2 (46):454 (1961). Type: Java, Nusa Kambanga Islands, *Blume* s.n. (holo. BO).
Desmotrichum fimbriatum Blume, Bijdr.:329 (1825).
Dendrobium flabellum Rchb.f. in Bonplandia 5:56 (1857). Type: Java, Salak, *Zollinger* 1294 (holo. W).
D.kunstleri Hook.f., Fl. Brit. Ind. 5:714 (1890). Types: Peninsular Malaysia, Perak, *Scortechini* 253b (syn. K); Perak, Larut, *King's collector* 1877 & 6897 (syn. K).
D.flabellum Rchb.f. var. *validum* J.J.Sm. in Bull. Dép. Agric. Indes Néerl. 45:17 (1911). Type: Kalimantan, Pontianak, cult. B*ogor* (holo. BO).

HABITAT: lowland forest on limestone; hill forest; low stature forest on ultramafic substrate.
ALTITUDINAL RANGE: sea level – 800m.
DISTRIBUTION: Borneo (K, SA, SR). Widespread over virtually all of S.E. Asia.

F.flabelloides *(J.J.Sm.) J.J.Wood* in Lindleyana 5(2):101 (1990). Type: Kalimantan, Pontianak, cult. B*ogor* number b133 & d53 (holo. L).
Dendrobium flabelloides J.J.Sm. in Bull. Jard. Bot. Buitenzorg, ser. 3, 2:58 (1920).
Ephemerantha flabelloides (J.J.Sm.) P.F.Hunt & Summerh. in Taxon 10:104 (1961).

HABITAT: unknown.
ALTITUDINAL RANGE: lowlands.
DISTRIBUTION: Borneo (K). Endemic.

F.labangensis *(J.J.Sm.) J.J.Wood* in Lindleyana 5(2):101 (1990). Type: Kalimantan, Gunung Labang, *Amdjah* s.n., cult. Bogor (holo. BO).
Dendrobium labangense J.J.Sm. in Bull. Jard. Bot. Buitenzorg, ser. 3, 2:61 (1920).
Ephemerantha labangensis (J.J.Sm.) P.F.Hunt & Summerh. in Taxon 10:104 (1961).

HABITAT: unknown.
ALTITUDINAL RANGE: unknown.
DISTRIBUTION: Borneo (K). Endemic.

F.luxurians *(J.J) Sm.) A.D.Hawkes* in Orchid Weekly 2(46):457 (1961). Type: Java, cult. B*ogor* (holo. L).
Dendrobium luxurians J.J.Sm. in Bull. Jard. Bot. Buitenzorg, ser. 3, 3:288, t.4, fig.2 (1921).

HABITAT: riverine forest.
ALTITUDINAL RANGE: lowlands – 1000m.
DISTRIBUTION: Borneo (K, SA, SR). Also Peninsular Malaysia, Java, Philippines.

F.pseudoconvexa *(Ames) A.D.Hawkes* in Orchid Weekly 2(46):458 (1961). Type: Philippines, Luzon, Laguna Province, San Antonio, *Maximo Ramos* Bur. Sci. 15075 (holo. AMES).
Dendrobium pseudoconvexum Ames, Orch. 5:135 (1915).

HABITAT: hill and lower montane forest; Fagaceae, *Dacrydium, Leptospermum* forest on sandstone; ultramafic and dioritic rock; oak-laurel forest; lower montane ridge forest with *Dacrydium, Podocarpus, Phyllocladus,*etc.
ALTITUDINAL RANGE: 900 – 1800m.
DISTRIBUTION: Borneo (SA). Also Philippines.

F.scopa *(Lindl.) Brieger* in Schltr., Die Orchideen 1 (11-12):742 (1981). Type: Philippines, Manila, *Loddiges* 273 (holo. K-LINDL).
Dendrobium scopa Lindl. in Bot. Reg. 28:55, misc. 55 (1842).

HABITAT: unknown.
ALTITUDINAL RANGE: lowlands.
DISTRIBUTION: Borneo (SA, SR). Also Philippines, distribution elsewhere uncertain, but Valmayor (1984) quotes Taiwan, Sulawesi, New Guinea, Guam and Samoa.

F.xantholeuca *(Rchb.f.) A.D.Hawkes* in Orchid Weekly 2(46):460 (1961). Type: Java, locality unknown, *Kuhl & van Hasselt* s.n. (holo. W, iso. L).
Dendrobium xantholeucum Rchb.f., Xenia Orch. 2:73, t.118, fig.1 (1862).
D.xantholeucum Rchb.f. var. *obtusilobum* J.J.Sm. in Bull. Jard. Bot. Buitenzorg, ser. 3, 2:66 (1920). Type: Kalimantan, Gunung Labang, *Amdjah* s.n., cult. Bogor under number 41 (holo. L).

HABITAT: unknown.
ALTITUDINAL RANGE: unknown.
DISTRIBUTION: Borneo (K). Also Peninsular Malaysia, Singapore, Thailand, Sumatra, Riau Archipelago, Java.

SUBTRIBE BULBOPHYLLINAE

BULBOPHYLLUM Thouars

Hist. Orchid., tabl. esp. 3 sub u.(1822).

Vermeulen, J.J. (1991). Orchids of Borneo Vol. 2 *Bulbophyllum.* Bentham-Moxon Trust, Royal Botanic Gardens Kew and Toihaan Publishing Company, Kota Kinabalu.

Epiphytic, lithophytic or rarely terrestrial herbs. Rhizome short to long, creeping or pendulous. Pseudobulbs clustered or remote, 1- to several-leaved. Leaves thin-textured to coriaceous. Inflorescences lateral, arising from base of pseudobulbs, or from nodes on the rhizome, racemose or pseudoumbellate, 1- to many-flowered. Flowers usually resupinate, minute to large; dorsal sepal usually free; lateral sepals connate at base to column-foot forming a saccate mentum; petals usually free, shorter than sepals; lip simple to 3-lobed, often sigmoid and recurved, versatile, fleshy, sometimes ciliate or pubescent; column short, with apical aristate teeth or wings (stelidia), and a foot. Pollinia 2 or 4, occasionally with a viscidium or viscidia, or a stipes.

A cosmopolitan genus containing around 1000 species, particularly well represented in Africa, S.E. Asia and New Guinea. The largest orchid genus in Borneo.

B.aberrans *Schltr.* in Feddes Repert. 10:177 (1911). Type: Sulawesi, Lampasioe, Djangdjang, Toli-Toli, *Schlechter* 20666 (holo. B, destroyed).

HABITAT: unknown.
ALTITUDINAL RANGE: unknown.
DISTRIBUTION: Borneo (unspecified). Also Sulawesi.
GENERIC SECTION: *Polyblepharon.*

B.acuminatum *Ridl.,*Mat. Fl. Mal. Pen. 1:81 (1907). Type: Singapore, Chua Chu Kang, *Ridley* s.n. (holo. SING).

HABITAT: podsol forest with *Dacrydium, Tristania, Rhododendron,* etc.; usually epiphytic on tree trunks near forest floor.
ALTITUDINAL RANGE: sea level – 500m.
DISTRIBUTION: Borneo (B, K, SA, SR). Also Peninsular Malaysia, Singapore, Thailand.
GENERIC SECTION: *Cirrhopetalum.*

B.acutum *J.J.Sm.,* Orch. Java: 466 (1905). Type: Java, *Blume* s.n., HLB 902, 322-404 (holo. L).

HABITAT: lower montane forest, either low and open or high with a closed canopy.
ALTITUDINAL RANGE: 1200 – 1800m.
DISTRIBUTION: Borneo (SA). Also China, India (Arunachal Pradesh), Thailand, Sumatra, Java.
GENERIC SECTION: *Micromonanthe.*

B.alatum *J.J.Verm.,* Orch. Borneo 2:167, fig.54, Pl.12B (1991). Type: Sabah, Mt. Kinabalu, Pinosuk Plateau, *Lamb* AL726/86 (holo. K).

HABITAT: lower montane forest; Fagaceae, *Dacrydium, Leptospermum* forest.
ALTITUDINAL RANGE: 1200 – 1800m.
DISTRIBUTION: Borneo (SA). Endemic.

GENERIC SECTION: *Monilibulbus.*

B.anaclastum *J.J.Verm.* in Blumea 38(1):149, fig.5 (1993). Type: Sabah, Mt. Kinabalu, *Carr* SFN 27902 (holo. L, iso. AMES, LAE, SING).

HABITAT: montane forest.
ALTITUDINAL RANGE: 1700 – 1900m.
DISTRIBUTION: Borneo (SA). Endemic.
GENERIC SECTION: *Sestochilus.*

B.anceps *Rolfe* in Lindenia 8: t.351 (1892). Type: Borneo, cult. *Linden* (holo. K).
   B.racemosum Rolfe in Kew Bull.:61 (1893). Type: Borneo, cult. *Sir T.Lawrence* (holo. K).

HABITAT: unknown.
ALTITUDINAL RANGE: unknown.
DISTRIBUTION: Borneo (unspecified). Endemic.
GENERIC SECTION: *Sestochilus.*

B.angulatum *J.J.Sm.* in Bull. Dép. Agric. Indes Néerl. 15:19 (1908). Type: Kalimantan, Martapoera, *Korthals* s.n., HLB 902, 322-427-429 (holo. BO).

HABITAT: unknown.
ALTITUDINAL RANGE: unknown.
DISTRIBUTION: Borneo (K). Endemic.
GENERIC SECTION: *Leptopus.*

B.anguliferum *Ames & C.Schweinf.,* Orch. 6:164 (1920). Type: Sabah, Mt. Kinabalu, Lobong Cave, *Clemens* 133 (holo. AMES).
   B.muluense J.J.Wood in Kew Bull. 39:95, fig.14 (1984). Type: Sarawak, Gunung Mulu National Park, summit area, *Hansen* 521 (holo. C, iso. K).

HABITAT: high and wet lower and upper montane forest; dry, low and open, often scrubby montane forest, sometimes in very exposed places.
ALTITUDINAL RANGE: 1500 – 2300m.
DISTRIBUTION: Borneo (SA, SR). Endemic.
GENERIC SECTION: *Monilibulbus.*

B.antenniferum *(Lindl.) Rchb.f.* in Ann. Bot. Syst. 6:250 (1861). Type: Philippines, *Cuming* s.n. (holo. K-LINDL).
   Cirrhopetalum antenniferum Lindl. in Bot. Reg. 29: sub. t.49 (1843).
   Bulbophyllum leysianum Burb. in J.Hort. Soc. London, 17:134, fig.19 (1895). Type: Sabah, Sinaroup, *Burbidge* s.n. (holo. not located).

HABITAT: lowland, hill and lower montane forest.
ALTITUDINAL RANGE: 400 – 1500m.
DISTRIBUTION: Borneo (SA). Also Java, Philippines, New Guinea, Solomon Islands (Guadalcanal).
GENERIC SECTION: *Ephippium.*

B.apheles *J.J.Verm.,* Orch. Borneo 2:245, fig.85, Pl.17A (1991). Type: Sabah, Mt. Kinabalu, Pinosuk Plateau, *Vermeulen* 487 (holo. L, iso. K).

HABITAT: low and open podsol forest dominated by Fagaceae and *Dacrydium*; lower montane forest.
ALTITUDINAL RANGE: 1300 – 1700m.
DISTRIBUTION: Borneo (SA). Endemic.
GENERIC SECTION: *Sestochilus.*

B.apodum *Hook.f.*, Fl. Brit. Ind. 5:766 (1890). Types: Peninsular Malaysia, Malacca, *Maingay*, Kew Distrib. 1619 (syn. K); Perak, Batu Kurau, *Scortechini* s.n. (syn. K).
B.saccatum Kraenzl. in Bot. Jahrb. Syst. 34:250 (1905). Type: Borneo, *Beccari* 486 (holo. FI).

HABITAT: lowland mixed dipterocarp forest; lowland podsol forest; coastal shrub vegetation; hill and lower montane forest on ultramafic substrate.
ALTITUDINAL RANGE: sea level – 2000m.
DISTRIBUTION: Borneo (K, SA, SR). Also Thailand, Vietnam, Peninsular Malaysia, Sumatra, Java, Sulawesi, Philippines.
GENERIC SECTION: *Aphanobulbon.*

B.armeniacum *J.J.Sm.* in Bull. Jard. Bot. Buitenzorg, ser. 2, 25:70 (1917). Type: Sumatra, cult. Bogor 746, *Jacobson* s.n. (holo. BO).

HABITAT: lower montane forest; *Agathis* dominated podsol forest; low and open forest on steep ridges; usually in well exposed places.
ALTITUDINAL RANGE: 700 – 1700m.
DISTRIBUTION: Borneo (K, SA, SR). Also Peninsular Malaysia, Sumatra.
GENERIC SECTION: *Aphanobulbon.*

B.asperulum *J.J.Sm.* in Bull. Dép. Agric. Indes Néerl. 22:38 (1909). Type: Kalimantan, Sungai Kenepai, *Hallier* 2089 (holo. BO).

HABITAT: unknown.
ALTITUDINAL RANGE: unknown.
DISTRIBUTION: Borneo (K). Endemic.
GENERIC SECTION: *Cirrhopetalum.*

B.attenuatum *Rolfe* in Kew Bull.:45 (1896). Type: Borneo, introduced by *Linden* (holo. K).
B.pugioniforme J.J.Sm. in Bull. Dép. Agric. Indes Néerl. 45:21 (1911). Type: Kalimantan, Sungai Kenepai, *Hallier* 2036 (holo. BO).

HABITAT: unknown.
ALTITUDINAL RANGE: unknown.
DISTRIBUTION: Borneo (K). Endemic.
GENERIC SECTION: *Intervallatae.*

B.auratum *(Lindl.) Rchb.f.* in Ann. Bot. Syst. 6:261(1861). Type: Philippines, Manila, *Cuming* s.n. (holo. K-LINDL).
Cirrhopetalum auratum Lindl. in Bot. Reg. 26: misc.50, 107 (1840).
C.borneense Schltr. in Bull. Herb. Boissier, ser. 2, 6:464 (1906). Type: Kalimantan, Koetai, Moera Maratu, *Schlechter* 13401 (holo. B, destroyed).

HABITAT: mixed dipterocarp forest; very open kerangas forest.
ALTITUDINAL RANGE: 400 – 1100m.
DISTRIBUTION: Borneo (B, K, SA). Also Thailand, Peninsular Malaysia, Sumatra.
GENERIC SECTION: *Cirrhopetalum.*

B.beccarii *Rchb.f.* in Gard. Chron., ser. 2, 11:41 (1879). Type: Sarawak, Kuching, *Beccari* 3515 (holo. FI).

HABITAT: lowland forest; podsol forest with *Dacrydium, Tristania, Rhododendron,* etc.; peat swamp forest, with *Shorea albida.*
ALTITUDINAL RANGE: sea level – 600m.
DISTRIBUTION: Borneo (B, K, SA, SR). Endemic.

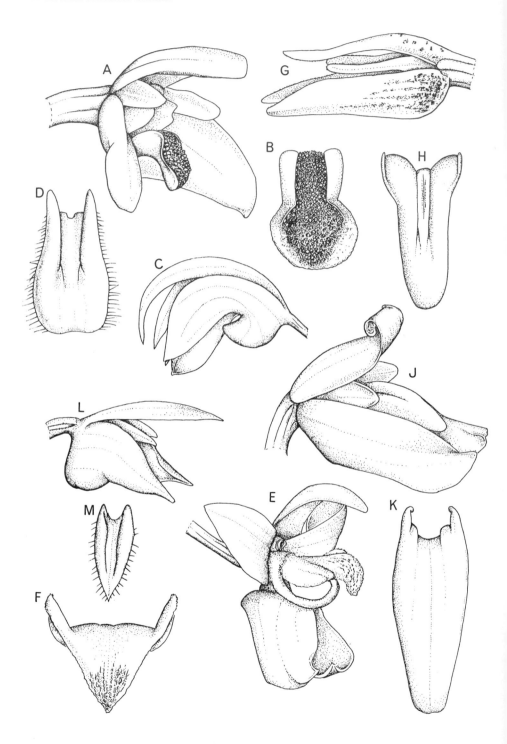

GENERIC SECTION: *Sestochilus.*

B.biflorum *Teijsm. & Binn.* in Natuurk. Tijdschr. Ned.-Indië, 5:488 (1853). Type: Java, Salak, fide *Teijsmann & Binnendijk* (holo. ?BO, iso. ?L).

HABITAT: open secondary forest; hill forest on limestone and ultramafic substrate.
ALTITUDINAL RANGE: 600 – 1200m.
DISTRIBUTION: Borneo (SA). Also Thailand, Peninsular Malaysia, Sumatra, Java, Bali, Lombok, Philippines.
GENERIC SECTION: *Cirrhopetalum.*

B.binnendijkii *J.J.Sm.*, Orch. Java: 442 (1905). Type: Java, Salak, *Teijsmann & Binnendijk* s.n.,HLB 904, 44-31 (holo. BO, iso. L).

HABITAT: moss forest.
ALTITUDINAL RANGE: 1300 – 1400m.
DISTRIBUTION: Borneo (SR). Also Sumatra, Java.
GENERIC SECTION: *Sestochilus.*

B.blumei *(Lindl.) J.J.Sm.*, Orch. Java:459 (1905). Type: Java, Salak, *Blume* s.n. (holo. L).
Ephippium ciliatum Blume, Bijdr.:309 (1825), non *Bulbophyllum ciliatum* (Blume) Lindl.
Cirrhopetalum blumei Lindl., Gen. Sp. Orch. Pl.:59 (1830). Type: based on same type as *Ephippium ciliatum* Blume.
Bulbophyllum ephippium Ridl. in Mat. Fl. Mal. Pen. 1:78 (1907). Type: based on same type as *Ephippium ciliatum* Blume.

HABITAT: lowland forest; forest on limestone.
ALTITUDINAL RANGE: lowlands.
DISTRIBUTION: Borneo (K, SA). Also Peninsular Malaysia, Sumatra, Java, Philippines, Kei Islands, New Guinea, Australia, Solomon Islands.
GENERIC SECTION: *Ephippium.*

B.botryophorum *Ridl.* in J. Linn. Soc., Bot. 32:275 (1896). Types: Peninsular Malaysia, Pahang, near Pekan, *Ridley* s.n. (syn. SING); Singapore, Sungai Buloh, Toas, Kranji, *Ridley* s.n. (syn. SING).

HABITAT: lowland forest; low and open podsol forest; forest on ultramafic substrate.
ALTITUDINAL RANGE: sea level – 800m.
DISTRIBUTION: Borneo (SA). Also Peninsular Malaysia, Singapore.
GENERIC SECTION: *Globiceps.*

B.brevicolumna *J.J.Verm.*, Orch. Borneo 2:37, fig.8 (1991). Type: Sabah, Sipitang District, Ulu Padas, *Vermeulen* 565 (holo. L).

HABITAT: lower montane forest; riverine forest.
ALTITUDINAL RANGE: 1300 – 1500m.
DISTRIBUTION: Borneo (SA). Endemic.
GENERIC SECTION: *Aphanobulbon.*

FIG. 34. *Bulbophyllum* flowers and lips. **A & B**, *B. armeniacum*, flower, x 12, lip, x 32; **C & D**, *B. ceratostylis*, flower, x 6, lip, x 12; **E & F**, *B. ecornutum*, flower, x 3, lip, x 6; **G & H**, *B. fulvibulbum*, flower, x 3, lip, x 10; **J & K**, *B. ionophyllum*, flower, x 10, lip, x 20; **L & M**, *B. vermiculare*, flower, x 12, lip, x 12. Drawn by Sarah Thomas.

B.breviflorum *Ridl.* in Stapf in Trans. Linn. Soc. London, Bot. 4:236 (1894). Type: Sabah, Mt. Kinabalu, *Haviland* s.n. (holo. ?BM).

HABITAT: lower montane forest.
ALTITUDINAL RANGE: 1200 – 1800m.
DISTRIBUTION: Borneo (SA, SR). Endemic.
GENERIC SECTION: *Aphanobulbon.*

B.brienianum *(Rolfe) J.J.Sm.* in Feddes Repert. 32:306 (1933). Type: Borneo, introduced by *Messrs. Linden* (holo. K).
Cirrhopetalum brienianum Rolfe in Kew Bull.:62 (1893).
C.brunnescens Ridl. in J. Linn. Soc., Bot. 31:279 (1896). Type: Borneo, locality uncertain, cult. *Nanson* (holo. SING).
C.makoyanum Rchb.f. var. *brienianum* (Rolfe) Ridl. in J. Linn. Soc., Bot. 32:285 (1896).
C.adenophorum Schltr. in Bull. Herb. Boissier, ser. 2, 6:463 (1906). Type: Kalimantan, Koetai, Samarinda, *Schlechter* 13343 (holo. B,destroyed).
Bulbophyllum makoyanum (Rchb.f.) Ridl. var. *brienianum* (Ridl) Ridl., Mat. Fl. Mal. Pen. 1:81 (1907).
B.adenophorum (Schltr.) J.J.Sm. in Bull. Jard. Bot. Buitenzorg, ser. 2, 8:22 (1912).
B.brunnescens (Ridl.) J.J.Sm., loc. cit.:23 (1912).

HABITAT: unknown.
ALTITUDINAL RANGE: lowlands.
DISTRIBUTION: Borneo (K). Also Peninsular Malaysia, Sumatra, Riau Archipelago.
GENERIC SECTION: *Cirrhopetalum.*

B.calceolus *J.J.Verm.*, Orch. Borneo 2:157, fig.51, Pl.11D & E (1991). Type: Sabah, Mt. Kinabalu, Pinosuk Plateau near Kundassang, *Vermeulen* 489 (holo. L, iso. K).

HABITAT: rather open lower montane forest; *Dacrydium – Leptospermum* forest; riverine forest; podsol forest; ridge top forest with *Agathis alba*, small rattans, etc.
ALTITUDINAL RANGE: 1300 – 1700m.
DISTRIBUTION: Borneo (SA). Endemic.
GENERIC SECTION: *Leptopus.*

B.capitatum *(Blume) Lindl.*, Gen. Sp. Orch. Pl.:56 (1830). Type: Java, Gede and Salak, *Blume* s.n. (holo. ?BO).
Diphyes capitata Blume, Bijdr.:314 (1825).

HABITAT: unknown.
ALTITUDINAL RANGE: unknown.
DISTRIBUTION: Borneo (SR). Also Java.
GENERIC SECTION: *Desmosanthes.*

B.carinilabium *J.J.Verm.*, Orch. Borneo 2:233, fig.81, Pl.16C (1991). Type: Sabah/Sarawak border, N.W. of Long Pa Sia to Long Samado trail, *Vermeulen* 641a (holo. L, iso. K).

HABITAT: lower montane mossy forest; low very dense forest on steep sandstone ridges; rather open podsol forest.
ALTITUDINAL RANGE: 1500 – 2000m.
DISTRIBUTION: Borneo (SA). Endemic.
GENERIC SECTION: *Hirtula.*

B.catenarium *Ridl.* in Trans. Linn. Soc. London, Bot. 4:235 (1894). Type: Sabah, Mt. Kinabalu, *Haviland* 1164 (lecto. K).

    B.carunculaelabrum Carr in Gard. Bull. Straits Settlem. 7:25 (1932). Type: Peninsular Malaysia, Johore, Sedili River, *Corner* s.n. (holo. SING).

HABITAT: lower and upper mossy montane forest; Ericaceae dominated forest; dry montane low and open *Dacrydium – Leptospermum* forest.
ALTITUDINAL RANGE: 1500 – 2300m.
DISTRIBUTION: Borneo (K, SA, SR). Also Peninsular Malaysia.
GENERIC SECTION: *Monilibulbus.*

B.caudatisepalum *Ames & C.Schweinf.*, Orch. 6:166 (1920). Type: Sabah, Mt. Kinabalu, Lobong Cave, *Clemens* 113A (holo. AMES).

    B.cuneifolium Ames & C.Schweinf., loc. cit.:172 (1920). Type: Sabah, Mt. Kinabalu, Kiau, *Clemens* 195 (holo. AMES).

    B.pergracile Ames & C.Schweinf., loc. cit.:190 (1920). Type: Sabah, Mt. Kinabalu, Kiau, *Clemens* 326 (holo. AMES).

    B.koyanense Carr in Gard. Bull. Straits Settlem. 8:112 (1935). Type: Sarawak, Mt. Dulit, Ulu Koyan, *Synge* S.541 (holo. SING, iso. K).

    B.dulitense Carr in loc. cit.:114 (1935). Type: Sarawak, Mt. Dulit, Dulit ridge, *Synge* S.516 (holo. SING, iso. K, L).

HABITAT: lowland and montane forests of different composition including mixed forest, lowland dipterocarp forest, low and open shrubby forest on ridges, high montane forest, moss forest, low and open *Dacrydium – Leptospermum* forest; often on ultramafic substrate.
ALTITUDINAL RANGE: 100 – 2400m.
DISTRIBUTION: Borneo (K, SA, SR). Also Peninsular Malaysia.
GENERIC SECTION: *Aphanobulbon.*

B.ceratostylis *J.J.Sm.* in Recueil Trav. Bot. Néerl. 1:154 (1904). Type: Sumatra, *Korthals* s.n., HLB904, 44-108 (holo. L).

    B.eximium Ames & C.Schweinf., Orch. 6:178 (1920). Type: Sabah, Mt. Kinabalu, Kiau, *Clemens* 317 (holo. AMES, iso. BM, BO, K, SING).

HABITAT: hill and lower montane forest; low and open podsol forest.
ALTITUDINAL RANGE: 900 – 1600m.
DISTRIBUTION: Borneo (SA). Also Sumatra.
GENERIC SECTION: *Aphanobulbon.*

B.cernuum *(Blume) Lindl.*, Gen. Sp. Orch. Pl.:48 (1930). Type: Java, Salak, *Blume* s.n., HLB902, 322-400 (holo. L).

    Diphyes cernua Blume, Bijdr.:318 (1825).

    Bulbophyllum gibbilingue J.J.Sm. in Bull. Jard. Bot. Buitenzorg, ser. 3, 11:142 (1931). Type: Kalimantan, West Koetai, Long Petak, *Endert* 3223 (holo. BO, sterile, drawing in ms. J.J.Sm. in L).

HABITAT: very dry, low and open ridge forest.
ALTITUDINAL RANGE: 1300m.
DISTRIBUTION: Borneo (K, SA). Also Thailand, Java.
GENERIC SECTION: *Monilibulbus.*

B.chanii *J.J.Verm. & A.Lamb*, Orch. Borneo 2:177, fig.58, Pl.12E (1991). Type: Sabah, *Vermeulen & Chan* 381 (holo. L).

HABITAT: high non-mossy lower montane forest.
ALTITUDINAL RANGE: 1400 – 1900m.

DISTRIBUTION: Borneo (SA). Endemic.
GENERIC SECTION: *Monilibulbus*.

B.cheiri *Lindl.* in Bot. Reg. 30, misc. 66 (1844). Type: Philippines, *Loddiges* s.n. (holo. K-LINDL).

HABITAT: lowland forest.
ALTITUDINAL RANGE: sea level – 100m.
DISTRIBUTION: Borneo (SA). Also Peninsular Malaysia, Sulawesi, Philippines.
GENERIC SECTION: *Sestochilus*.

B.cheiropetalum *Ridl.* in Kew Bull.:477 (1926). Type: Peninsular Malaysia, Kedah Peak, *Flippance* s.n. (holo. SING).

HABITAT: unknown.
ALTITUDINAL RANGE: unknown.
DISTRIBUTION: Borneo (unspecified). Also Peninsular Malaysia.
GENERIC SECTION: *Epicrianthes*.

B.cleistogamum *Rolfe* in J. Linn. Soc., Bot. 31:277 (1896). Type: Riau Archipelago, *native collector* s.n. (holo. K).

HABITAT: podsol forest with *Dacrydium*, etc.
ALTITUDINAL RANGE: 400 – 500m.
DISTRIBUTION: Borneo (SA, SR). Also Peninsular Malaysia, Singapore, Riau Archipelago and Sumatra.
GENERIC SECTION: *Intervallatae*.

B.coelochilum *J.J.Verm.*, Orch. Borneo 2:179, fig.59 (1991). Type: Sabah, Sipitang District, Ulu Padas, *Vermeulen* 631 (holo. L).

HABITAT: dry,low, Open forest on steep ridges; low, open podsol forest with dense undergrowth.
ALTITUDINAL RANGE: 1300 – 1500m.
DISTRIBUTION: Borneo (SA). Endemic.
GENERIC SECTION: *Monilibulbus*.

B.comberi *J.J.Verm.* in Comber, Orch. Java: 267, line drawing on p.268 (1990). Type: Sabah, Mt. Kinabalu, *Vermeulen & Lamb* 356 (holo. L).
B.ciliatum sensu Carr in Gard. Bull. Straits Settl. 7:29 (1932), non (Blume) Lindl.

HABITAT: lower montane forest.
ALTITUDINAL RANGE: unknown.
DISTRIBUTION: Borneo (SA). Also Peninsular Malaysia, Java, Flores.
GENERIC SECTION: *Micromonanthe*.

B.compressum *Teijsm. & Binn.* in Natuurk. Tijdschr. Ned.-Indië 24:307 (1862). Type: Sumatra, Palembang, *Teijsmann* s.n. (holo. BO).

HABITAT: lowland and hill mixed dipterocarp forest.
ALTITUDINAL RANGE: 100 – 600m.

FIG. 35. *Bulbophyllum* flowers and lips. **A** & **B**, *B. calceolus*, flower, x 2½, lip, x 6; **C** & **D**, *B. longimucronatum*, flower, x 5, lip, x 8; **E** & **F**, *B. obtusipetalum*, flower, x 6, lip, x 20; **G** & **H**, *B. sigmoideum*, flower, x 6, lip, x 12; **J** & **K**, *B. undecifilum*, flower, x 6, lip, x 12; **L** & **M**, *B. uniflorum*, flower, x 2, lip, x 2½. Drawn by Sarah Thomas.

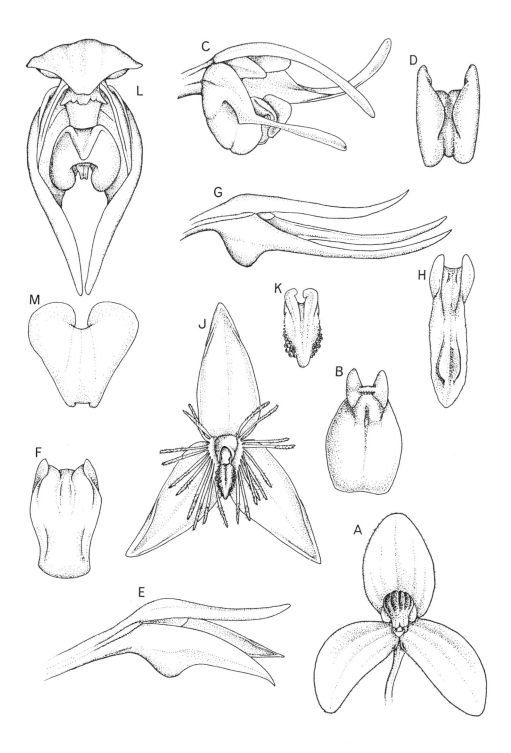

DISTRIBUTION: Borneo (SA). Also in Sumatra, Java, Sulawesi.
GENERIC SECTION: *Desmosanthes.*

B.concinnum *Hook.f.*, Fl. Brit. Ind. 6:187 (1890). Type: Singapore, Chang Chu Rang, *Ridley* s.n. (holo. K).

HABITAT: heath forest; podsol forest with *Dacrydium* etc.
ALTITUDINAL RANGE: 100 – 500m.
DISTRIBUTION: Borneo (SA, SR). Also Thailand, Vietnam, Peninsular Malaysia, Singapore, Mentawai.
GENERIC SECTION: *Desmosanthes.*

B.coniferum *Ridl.* in J. Fed. Malay States Mus. 4:67 (1909). Type: Peninsular Malaysia, Telom, *Ridley* s.n.(holo. SING).
B.reflexum Ames & C. Schweinf., Orch. 6:192 (1920). Type: Sabah, Mt. Kinabalu, Kiau, *Clemens* 384 (holo. AMES).

HABITAT: lower montane forest.
ALTITUDINAL RANGE: 1000 – 2200m.
DISTRIBUTION: Borneo (K, SA, SR). Also Peninsular Malaysia, Sumatra, Java.
GENERIC SECTION: *Globiceps.*

B.coriaceum *Ridl.* in Stapf in Trans. Linn. Soc. London, Bot., ser. 2, 4:235 (1894). Type: Sabah, Mt. Kinabalu, *Haviland* 1100 (holo. SING, iso. K).
B.kinabaluense Rolfe in Gibbs in J. Linn. Soc., Bot. 42:148 (1914). Type: Sabah, Mt. Kinabalu, Lobong, Paka-paka Cave, *Gibbs* 4252 (holo. BM).
B.venustum Ames & C.Schweinf., Orch. 6:198 (1920). Type: Sabah, Mt. Kinabalu, Pakka, *Clemens* 113 (holo. AMES, iso. K).

HABITAT: lower montane Fagaceae – *Dacrydium* – *Leptospermum* forest; upper montane *Leptospermum* – *Dacrydium* scrub on granite; granite rock crevices; sometimes on ultramafic substrate.
ALTITUDINAL RANGE: 1200 – 3500m.
DISTRIBUTION: Borneo (SA). Endemic.
GENERIC SECTION: *Aphanobulbon.*

B.cornutum *(Blume) Rchb.f.* in Ann. Bot. Syst. 6:247 (1861). Types: Java, *Blume* 1966, HLB 902, 322-925 (lecto. L); HLB 902, 926-928 (syn. L).
Ephippium cornutum Blume, Bijdr.:310 (1825).
Bulbophyllum concavum Ames & C.Schweinf., Orch. 6:168 (1920). Type: Sabah, Mt. Kinabalu, Kiau, *Clemens* 94 (holo. AMES).

HABITAT: hill and lower montane forest; Fagaceae – *Dacrydium* – *Leptospermum* forest on podsolic soil.
ALTITUDINAL RANGE: 900 – 1700m.
DISTRIBUTION: Borneo (SA). Also Java, Philippines.
GENERIC SECTION: *Sestochilus.*

B.corolliferum *J.J.Sm.* in Bull. Jard. Bot. Buitenzorg, ser. 2, 25:80 (1917). Type: Kalimantan, Moeara Tewe, cult. Bogor (holo. BO).

HABITAT: unknown.
ALTITUDINAL RANGE: unknown.
DISTRIBUTION: Borneo (K). Also Thailand, Peninsular Malaysia, Sumatra.
GENERIC SECTION: *Cirrhopetalum.*

B.crepidiferum *J.J.Sm.* in Bull. Jard. Bot. Buitenzorg, ser. 3, 2:88 (1920). Type: Sumatra, Agam, Gunung Singgalang, *Jacobson* s.n., cult. under numbers 1290 and 1291 (holo. BO).

HABITAT: lower montane forest.
ALTITUDINAL RANGE: 1700m.
DISTRIBUTION: Borneo (unspecified). Also Sumatra.
GENERIC SECTION: *Leptopus.*

B.cumingii *(Lindl.) Rchb.f.* in Ann. Bot. Syst. 6:261 (1861). Types: Philippines, Panay, *Cuming* s.n., *Loddiges* 198 (syn. K-LINDL).
Cirrhopetalum cumingii Lindl. in Paxton's Mag. Bot. 8:165 (1841).

HABITAT: lowland mixed dipterocarp forest.
ALTITUDINAL RANGE: lowlands.
DISTRIBUTION: Borneo (SA). Also Philippines.
GENERIC SECTION: *Cirrhopetalum.*

B.cuspidipetalum *J.J.Sm.* in Bull. Dép. Agric. Indes Néerl. 15:16 (1908). Type: Kalimantan, locality unknown, *Nieuwenhuis* s.n.(holo. BO).

HABITAT: unknown.
ALTITUDINAL RANGE: unknown.
DISTRIBUTION: Borneo (K). Endemic.
GENERIC SECTION: *Sestochilus.*

B.dearei *(Hort.) Rchb.f.* in Flora 71:156 (1888). Type: Origin unknown, cult. *Deare* (holo. W).
Sarcopodium dearei Hort. in Gard. Chron., ser. 2, 20:108 (1883).
Bulbophyllum reticosum Ridl. in J. Linn. Soc., Bot. 31:273 (1896). Type: Borneo, *Durnford* s.n. (holo. SING).
B.punctatum Ridl. in J. Straits Branch Roy. Asiat. Soc. 50:129 (1908). Type: Sarawak, Matang, *Hewitt* s.n. (holo. SING).

HABITAT: hill and lower montane forest; forest on limestone.
ALTITUDINAL RANGE: 700 – 1200m.
DISTRIBUTION: Borneo (probably K, SA, SR,). Also Peninsular Malaysia, Philippines.
GENERIC SECTION: *Sestochilus.*

B.decatriche *J.J.Verm.*, Orch. Borneo 2:117, fig.37 (1991). Type: Sarawak, locality unknown, cult. Semenggoh, *Vermeulen* 1170 (holo. L).

HABITAT: unknown.
ALTITUDINAL RANGE: unknown.
DISTRIBUTION: Borneo (SR). Endemic.
GENERIC SECTION: *Epicrianthes.*

B.deltoideum *Ames & C.Schweinf.*, Orch. 6:174 (1920). Type: Sabah, Mt. Kinabalu, Lobong Cave, *Clemens* 115 (holo. AMES, iso. SING).
B.angustatifolium J.J.Sm. in Mitt. Inst. Allg. Bot. Hamburg 7:63, t.10, fig.52 (1927). Type: Kalimantan, Bukit Raja, *Winkler* 941A (holo. HBG).

HABITAT: dry, rather low and open lower montane forest on steep ridges; very wet, mossy, low podsol forest on sandy soil.
ALTITUDINAL RANGE: 1200 – 2000m.
DISTRIBUTION: Borneo (K, SA). Endemic.
GENERIC SECTION: *Aphanobulbon.*

B.devogelii *J.J.Verm.*, Orch. Borneo 2:45, fig.11, Pl.5B (1991). Type: Sabah, Crocker Range, Gunung Alab, *Vermeulen* 514 (holo. L, iso. K, SNP).

HABITAT: moss forest.
ALTITUDINAL RANGE: 1600 – 2000m.
DISTRIBUTION: Borneo (SA). Endemic.
GENERIC SECTION: *Aphanobulbon.*

B.dibothron *J.J.Verm.* & *A.Lamb* ined. Type: Sabah, Crocker Range, near Tambunan, cult. *Jongejan* 1956 (holo. L).

HABITAT: montane forest.
ALTITUDINAL RANGE: 1500 – 2000m.
DISTRIBUTION: Borneo (SA). Endemic.
GENERIC SECTION: *Monilibulbus.*

B.disjunctum *Ames* & *C.Schweinf.*, Orch. 6:176 (1920). Type: Sabah, Mt. Kinabalu, Marai Parai, *Clemens* 254 (holo. AMES, iso. BM, BO, K, SING).
    B.densissimum Carr in Gard. Bull. Straits Settlem. 8:116 (1935). Type: Sarawak, Mt. Dulit, *native collector* in *Synge* S.98 (holo. SING, iso. K).

HABITAT: lower montane mossy forest.
ALTITUDINAL RANGE: 900 – 2500m.
DISTRIBUTION: Borneo (K, SA, SR). Also Thailand.
GENERIC SECTION: *Aphanobulbon.*

B.dransfieldii *J.J.Verm.*, Orch. Borneo 2:181, fig.60, Pl.12F (1991). Type: Sabah, Crocker Range, Sinsuron Road, *Dransfield* 5456 (holo. K, iso. L).

HABITAT: montane forest
ALTITUDINAL RANGE: 2000m.
DISTRIBUTION: Borneo (SA). Endemic.
GENERIC SECTION. *Monilibulbus.*

B.dryas *Ridl.* in J. Fed. Malay States Mus. 6:175 (1915). Type: Peninsular Malaysia, Pahang, Gunung Tahan, *Ridley* 18 (holo. K).

HABITAT: lower and upper montane Fagaceae – *Dacrydium* -*Leptospermum* forest.
ALTITUDINAL RANGE: 1500 – 2000m.
DISTRIBUTION: Borneo (SA). Also Peninsular Malaysia, Sumatra.
GENERIC SECTION: *Aphanobulbon.*

B.ecornutum *(J.J.Sm.)* *J.J.Sm.* in Bull. Jard. Bot. Buitenzorg, ser. 2, 13:32 (1914). Types: Java, Salak, *Blume* s.n., HLB 902, 322-938 (lecto. L); *Waitz* s.n., HLB 902, 322-924 (syn. L).
    B.cornutum (Blume) Rchb.f. var. *ecornutum* J.J.Sm., Orch. Java:445(1905).

HABITAT: podsol forest; shrubby forest on limestone.
ALTITUDINAL RANGE: 400 – 1300m.
DISTRIBUTION: Borneo (SA). Also Thailand, Sumatra, Java and possibly Maluku (Banda).
GENERIC SECTION: *Sestochilus.*

B.elachanthe *J.J.Verm.*, Orch. Borneo 2:49, fig.13 (1991). Type: Sabah, Sipitang District, Ulu Padas, *Vermeulen* & *Duistermaat* 902 (holo. L).

HABITAT: podsol forest.
ALTITUDINAL RANGE: 1400m.

DISTRIBUTION: Borneo (SA). Endemic.
GENERIC SECTION: *Aphanobulbon.*

B.elongatum *(Blume) Hassk.*, Cat. Hort. Bog.:39 (1844). Type: Java, Salak, *Blume* s.n.(holo. BO).
Ephippium elongatum Blume, Bijdr.:309 (1825).

HABITAT: unknown.
ALTITUDINAL RANGE: unknown.
DISTRIBUTION: Borneo (B). Also Peninsular Malaysia, Java.
GENERIC SECTION: *Altisceptrum.*

B.epicrianthes *Hook.f.*, Fl. Brit. Ind. 5:753 (1890). Type: based on type of *Epicranthes javanica* Blume.
Epicranthes javanica Blume, Bijdr.:307 (1825) – changed by Blume to '*Epicrianthes*' in Fl. Java n.s.1:praef. VI(1858). Type: Java, *Blume* 191, HLB 904, 44-40 (holo. L).

HABITAT: lowland forest; rather open podsol forest.
ALTITUDINAL RANGE: 400m.
DISTRIBUTION: Borneo (K, SA). Also Sumatra, Java and Maluku(Ambon).
GENERIC SECTION: *Epicrianthes.*

B.fissibrachium *J.J.Sm.* in Bull. Jard. Bot. Buitenzorg, ser. 3, 9:166 (1927). Type: Sumatra, Bengkoeloe, Gunung Dempo, *Ajoeb* 506 (holo. BO).

HABITAT: unknown.
ALTITUDINAL RANGE: unknown.
DISTRIBUTION: Borneo (SA). *B.fissibrachium* is native to Sumatra.
GENERIC SECTION: *Globiceps.*

NOTE: The specimen *Lamb* AL660/86 is listed as *B.aff.fissibrachium* by Vermeulen(1991). No further details are provided.

B.flammuliferum *Ridl.* in J. Bot. 36:211 (1898). Type: Peninsular Malaysia, Selangor, Gua Batu, *Ridley* s.n. (holo. SING).

HABITAT: hill forest with *Gymnostoma* on ultramafic substrate.
ALTITUDINAL RANGE: 600m.
DISTRIBUTION: Borneo (SA). Also Peninsular Malaysia.
GENERIC SECTION: *Desmosanthes.*

B.flavescens *(Blume) Lindl.*, Gen. Sp. Orch. Pl.:54 (1830). Types: Java, Salak, *Blume* s.n., HLB 902, 322-412 (lecto. L); HLB 902, 322-415 (syn. L).
Diphyes flavescens Blume, Bijdr.:313 (1825).
Bulbophyllum montigenum Ridl. in Trans. Linn. Soc. London, Bot., ser. 2, 4:235(1895). Type: Sabah, Mt. Kinabalu, *Haviland* 1252 (holo. K).
B.puberulum Ridl. in J. Linn. Soc., Bot. 31:275 (1896). Type: Sarawak, *Haviland* 2324 (holo. SING, iso. K).
B.barrinum Ridl. in Sarawak Mus. J. 1:32 (1912). Type: Sarawak, Mt. Penrissen, *Moulton* s.n.(holo. SING).
B.lanceolatum Ames & C.Schweinf., Orch. 6:180 (1920). Type: Sabah, Mt. Kinabalu, Gurulau Spur, *Clemens* 305 (holo. AMES,iso. BO).
B.exiliscapum J.J.Sm. in Mitt. Inst. Allg. Bot. Hamburg 7:64, t.10, fig 53 (1927). Type: Kalimantan, Bukit Mehipit, *Winkler* 1146 (holo. BO).
B.flavescens (Blume) Lindl. var. *temelense* J.J.Sm. in Bull. Jard. Bot. Buitenzorg, ser. 3, 11:147 (1931). Type: Kalimantan, West Koetai, Long Temelen, *Endert* 2926 (holo. L).

HABITAT: a wide range of lowland and montane primary forest types, on sandstone, ultramafic and limestone soil.
ALTITUDINAL RANGE: 100 – 3800m.
DISTRIBUTION: Borneo (K, SA, SR). Also Peninsular Malaysia, Sumatra, Java, Philippines.
GENERIC SECTION: *Aphanobulbon.*

B.flavofimbriatum *J.J.Sm.* in Bull. Jard. Bot. Buitenzorg, ser. 3, 11:143 (1931). Type: Kalimantan, West Koetai, Long Petak, *Endert* 3222 (holo. L, iso. BO).
Epicranthes flavofimbriata (J.J.Sm.) Garay & W.Kittr. in Bot. Mus. Leafl. 30:49 (1985).

HABITAT: dense and high lower montane forest; low and open podsol forest.
ALTITUDINAL RANGE: 800 – 1500m.
DISTRIBUTION: Borneo (K, SA). Endemic.
GENERIC SECTION: *Epicrianthes.*

B.foetidolens *Carr* in Gard. Bull. Straits Settlem. 5:135, t.3, fig.2 (1930). Type: Peninsular Malaysia, Gunung Tahan, Wray's Camp, *Carr* s.n. (holo. SING).

HABITAT: mixed hill dipterocarp forest; lower montane forest; riverine forest; often growing on sandstone and other boulders in forest, associated with *Nepenthes leptochila* at one locality.
ALTITUDINAL RANGE: 300 – 1500m.
DISTRIBUTION: Borneo (K, SA, SR). Also Peninsular Malaysia.
GENERIC SECTION: *Sestochilus.*

B.fulvibulbum *J.J.Verm.*, Orch. Borneo 2:23, fig.2(1991). Type: Sabah, Sipitang District, Ulu Padas, *Vermeulen* 603 (holo. L, iso. K).

HABITAT: low and open podsol forest with dense undergrowth of rattans and *Pandanus.*
ALTITUDINAL RANGE: 1500m.
DISTRIBUTION: Borneo (SA). Endemic.
GENERIC SECTION: *Altisceptrum.*

B.gibbosum *(Blume) Lindl.*, Gen. Sp. Orch. Pl.:54 (1830). Type: Java, Salak, *Blume* 1873, HLB 902, 322-417 (holo. L).
Diphyes gibbosa Blume, Bijdr.:312 (1830).
Bulbophyllum magnivaginatum Ames & C.Schweinf., Orch. 6:186 (1920). Type: Sabah, Mt. Kinabalu, Kiau, *Clemens* 36 (holo. AMES, iso. BO, K).

HABITAT: hill forest.
ALTITUDINAL RANGE: 800 – 900m.
DISTRIBUTION: Borneo (SA). Also Peninsular Malaysia, Sumatra, Java, Bali.
GENERIC SECTION: *Aphanobulbon.*

B.gibbsiae *Rolfe* in Gibbs in J. Linn. Soc., Bot. 42:149 (1914). Type: Sabah, Mt. Kinabalu, Penibukan Ridge, *Gibbs* 4059 (holo. K).
B.minutiflorum Ames & C.Schweinf., Orch. 6:188 (1920). Type: Sabah, Mt. Kinabalu, *Haslam* s.n. (holo. AMES).

HABITAT: lower montane oak-laurel forest; 'heath forest'; moss forest; Fagaceae – *Dacrydium* – *Leptospermum* forest on sandstone and shale; very wet, mossy, low podsol forest on sandy soil.
ALTITUDINAL RANGE: 800 – 2200m.
DISTRIBUTION: Borneo (SA, SR). Endemic.
GENERIC SECTION: *Aphanobulbon.*

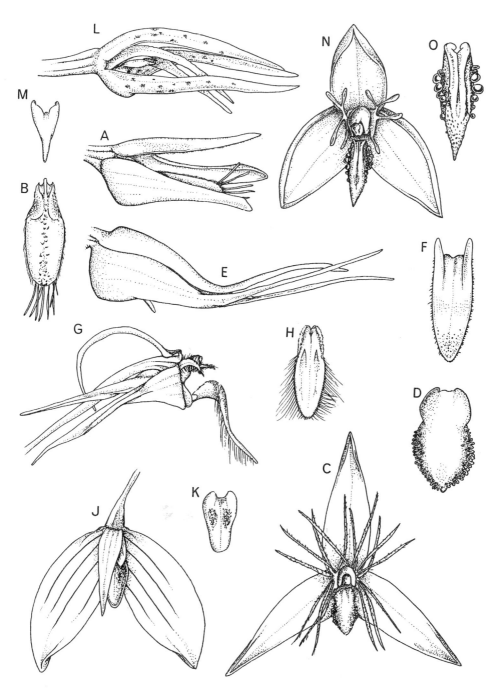

FIG. 36. *Bulbophyllum* flowers and lips. **A & B**, *B. carinilabium*, flower, x 5, lip, x 5; **C & D**, *B. decatriche*, flower, x 3, lip, x 6; **E & F**, *B. flavescens*, flower, x 8, lip, x 10; **G & H**, *B. jolandae*, flower, x 3, lip, x 3; **J & K**, *B. leproglossum*, flower, x 4, lip, x 2½; **L & M**, *B. penduliscapum*, flower, x 6, lip, x 6; **N & O**, *B. vesiculosum*, flower, x 4, lip, x 8. Drawn by Sarah Thomas.

B.gilvum *J.J.Verm. & A.Lamb* ined. Type: Sabah, Sipitang District, Ulu Padas, near Long Pa Sia, *Vermeulen* 569 (holo. L).

> HABITAT: low, open, mossy podsol forest.
> ALTITUDINAL RANGE: 1300 – 1600m.
> DISTRIBUTION: Borneo (SA). Endemic.
> GENERIC SECTION: *Leptopus.*

B.glaucifolium *J.J.Verm.*, Orch. Borneo 2:61, fig.16 (1991). Type: Sabah, Crocker Range, cult. *Leiden* 26437 (holo. L).

> HABITAT: high mixed montane forest.
> ALTITUDINAL RANGE: 1700m.
> DISTRIBUTION: Borneo (SA). Endemic.
> GENERIC SECTION: *Aphanobulbon.*

B.grandilabre *Carr* in Gard. Bull. Straits Settlem. 8:117 (1935). Type: Sarawak, Mt. Dulit, *Synge* S.135 (holo. K).

> HABITAT: low and rather open montane mossy forest; *Agathis* dominated ridge-top forest with small rattans.
> ALTITUDINAL RANGE: 1400 – 1600m.
> DISTRIBUTION: Borneo (SA, SR). Endemic.
> GENERIC SECTION: *Aphanobulbon.*

B.groeneveldtii *J.J.Sm.* in Bull. Jard. Bot. Buitenzorg, ser. 3, 2:95 (1920). Type: Sumatra, Agam, Bukit Batoe Banting, *Groeneveldt* s.n., cult.*Jacobson* 951 (holo. BO, iso. L).

> HABITAT: low and open podsol forest; high montane forest.
> ALTITUDINAL RANGE: 1100 – 1300m.
> DISTRIBUTION: Borneo (SA). Also Sumatra.
> GENERIC SECTION: *Hirtula.*

B.grudense *J.J.Sm.*, Orch. Java: 458 (1905). Type: Java, Gunung Grudo (Groeda), *Smith* s.n. (holo. BO).

> HABITAT: unknown.
> ALTITUDINAL RANGE: unknown.
> DISTRIBUTION: Borneo (unspecified). Also Java.
> GENERIC SECTION: *Micromonanthe.*

B.habrotinum *J.J.Verm. & A.Lamb* ined. Type: Kalimantan Timur, Apo Kayan, cult. *Leiden* 913245 (holo. L).

> HABITAT: lower montane forest.
> ALTITUDINAL RANGE: 1000 – 1500m.
> DISTRIBUTION: Borneo (K, SA). Endemic.
> GENERIC SECTION: *Cirrhopetalum.*

B.hcldiorum *J.J.Verm.*, Orch. Borneo 2:143, fig.46, Pl.10E (1991). Type: Sabah, Mt. Kinabalu, summit trail, *Vermeulen & Duistermaat* 547 (holo. L, iso. K, SNP).

> HABITAT: lower and upper montane mossy *Dacrydium – Leptospermum* forest; sometimes on ultramafic substrate.
> ALTITUDINAL RANGE: 1800 – 2500m.
> DISTRIBUTION: Borneo (SA). Endemic.
> GENERIC SECTION: *Globiceps.*

B.hemiprionotum *J.J.Verm. & A.Lamb* ined. Type: Sabah, Mt. Kinabalu, *Vermeulen* 467 (holo. L).

HABITAT: mixed montane forest; low, open forest with rattans.
ALTITUDINAL RANGE: 1700 – 2000m.
DISTRIBUTION: Borneo (SA). Endemic.
GENERIC SECTION: uncertain.

B.hirtulum *Ridl.* in J. Bot., 38:71 (1900). Type: Peninsular Malaysia, Pinang, West Hill, *Ridley & Curtis* s.n. (holo. BM, iso. K).
   B.trichoglottis Ridl. in J. Fed. Malay States Mus. 4:66 (1909). Type: Peninsular Malaysia, Telom, *Ridley* s.n. (holo. SING).

HABITAT: shrubby forest on limestone hills; low, very open forest on steep sandstone ridges; high montane forest.
ALTITUDINAL RANGE: 400 – 1300m.
DISTRIBUTION: Borneo (SA, SR). Also Peninsular Malaysia, Sumatra.
GENERIC SECTION: *Hirtula.*

B.hymenanthum *Hook.f.*, Fl. Brit. Ind. 5:767 (1890). Types: India, Assam, Khasia Hills, Myrung, *Hooker & Thompson* 1523 & 1525 (syn. K).

HABITAT: open mossy forest.
ALTITUDINAL RANGE: 1800m.
DISTRIBUTION: Borneo (SA). Also India, Thailand.
GENERIC SECTION: *Aphanobulbon.*

B.hymenochilum *Kraenzl.* in Bot. Jahrb. Syst. 34:252 (1904). Type: Sarawak, Mt. Mattang (Mattan), *Beccari* 1536 (holo. FI).
   Cirrhopetalum rhombifolium Carr in Gard. Bull. Straits Settlem. 8:119 (1935). Type: Sarawak, Mt. Dulit, near Long Kapa, *Richards* 2362 (holo. K).
   Bulbophyllum rhombifolium (Carr) Masam., Enum. Phan. Born.:132 (1942).

HABITAT: unknown.
ALTITUDINAL RANGE: under 300m.
DISTRIBUTION: Borneo (SR). Endemic.
GENERIC SECTION: *Cirrhopetalum.*

B.inunctum *J.J.Sm.* in Icon. Bogor. 3:37, t.215 (1906). Type: Kalimantan, Pontianak, cult. B*ogor* (holo. BO).

HABITAT: unknown.
ALTITUDINAL RANGE: unknown.
DISTRIBUTION: Borneo (B, K, SR). Endemic.
GENERIC SECTION: *Sestochilus.*

B.ionophyllum *J.J.Verm.*, Orch. Borneo 2:65, fig.18, Pl.6B (1991). Type: Sabah, *Chan* s.n. (holo. L).

HABITAT: lower montane forest; podsolic Fagaceae – *Dacrydium* forest.
ALTITUDINAL RANGE: 1200 – 1700m.
DISTRIBUTION: Borneo (SA). Endemic.
GENERIC SECTION: *Aphanobulbon.*

B.jolandae *J.J.Verm.*, Orch. Borneo 2:241, fig.84, Pl.16F (1991). Type: Sabah, Sipitang District, Ulu Padas, *Vermeulen* 602 (holo. L).

HABITAT: open montane podsol forest.

ALTITUDINAL RANGE: 1500m.
DISTRIBUTION: Borneo (SA). Endemic.
GENERIC SECTION: *Hirtula.*

B.kemulense *J.J.Sm.* in Bull. Jard. Bot. Buitenzorg, ser. 3, 11:144 (1931). Type: Kalimantan, West Koetai, Gunung Kemoel, *Endert* 4278 (holo. BO).

HABITAT: lower and upper montane forest.
ALTITUDINAL RANGE: 1500 – 1800m.
DISTRIBUTION: Borneo (K, SA). Endemic.
GENERIC SECTION: *Globiceps.*

B.kestron *J.J.Verm. & A.Lamb* in Malayan Orchid Rev. 22:45 (1988). Type: Sabah, Gunung Alab, *Lamb* AL 566/86 (holo. K).

HABITAT: lower montane forest.
ALTITUDINAL RANGE: 1400 – 2000m.
DISTRIBUTION: Borneo (SA). Endemic.
GENERIC SECTION: *Monilibulbus.*

B.lacinulosum *J.J.Sm.* in Mitt. Inst. Allg. Bot. Hamburg 7:61, t.9, fig.49 (1927). Type: Kalimantan, Bukit Raja, *Winkler* 939 (holo. HBG).

HABITAT: lower montane forest.
ALTITUDINAL RANGE: 1200 – 1300m.
DISTRIBUTION: Borneo (K). Endemic.
GENERIC SECTION: *Monilibulbus.*

B.lambii *J.J.Verm.*, Orch. Borneo 2:185, fig.62, Pl.13B (1991). Type: Sabah, Mt. Kinabalu, Pinosuk Plateau, *Lamb* AL 577/86 (holo. K).

HABITAT: lower montane forest.
ALTITUDINAL RANGE: 1500 – 2100m.
DISTRIBUTION: Borneo (SA). Endemic.
GENERIC SECTION: *Monilibulbus.*

B.lasianthum *Lindl.* in Gard. Chron.:53 (1855). Type: origin unknown, cult. (holo. K-LINDL).

HABITAT: lowland forest; lower montane forest.
ALTITUDINAL RANGE: lowlands – 1200m.
DISTRIBUTION: Borneo (SR). Also Peninsular Malaysia, Sumatra.
GENERIC SECTION: *Sestochilus.*

B.latisepalum *Ames & C.Schweinf.*, Orch. 6:182 (1920). Type: Sabah, Mt. Kinabalu, *Clemens* s.n. (holo. AMES).

HABITAT: lower montane forest.
ALTITUDINAL RANGE: 1300 – 1900m.
DISTRIBUTION: Borneo (SA). Endemic.
GENERIC SECTION: *Monilibulbus.*

B.laxiflorum *(Blume) Lindl.*, Gen. Sp. Orch. Pl.:57 (1830). Type: Java, Tjiapus, *Blume* s.n. (holo. L, iso. P).
Diphyes laxiflora Blume, Bijdr.:316(1825).
Bulbophyllum pedicellatum Ridl., J. Linn. Soc., Bot. 31:278, t.13 (1896). Type: Sarawak, *Haviland* 2346 (holo. K).

HABITAT: riverine forest; lowland, hill and lower montane forest.
ALTITUDINAL RANGE: lowlands – 1400m.
DISTRIBUTION: Borneo (K, SA, SR). Also Burma, Laos, Thailand, Peninsular Malaysia, Sumatra, Java, Sulawesi, Philippines.
GENERIC SECTION: *Desmosanthes.*

B.lemniscatoides *Rolfe* in Gard. Chron., ser. 3, 8:672 (1890). Type: Java, locality unknown, cult. *van Lansberge* (holo. K).

HABITAT: unknown.
ALTITUDINAL RANGE: unknown.
DISTRIBUTION: Borneo (K). Also Thailand, Vietnam, Sumatra, Java.
GENERIC SECTION: *Pleiophyllus.*

B.leniae *J.J.Verm.*, Orch. Borneo 2:189, fig.64 (1991). Type: Sabah, Sipitang District, Ulu Padas, *Vermeulen & Duistermaat* 1033 (holo. L).

HABITAT: low and open, dry forest on steep ridges; open podsol forest with dense undergrowth of rattans and bamboo.
ALTITUDINAL RANGE: 1300 – 1500m.
DISTRIBUTION: Borneo (?K, SA). Endemic.
GENERIC SECTION: *Monilibulbus.*

B.lepidum *(Blume) J.J.Sm.*, Orch. Java: 471 (1905). Type: Java, Pantjar, *Blume* 1934 (holo. L).
   Ephippium lepidum Blume, Bijdr.:310 (1825).
   Bulbophyllum lepidum (Blume) J.J.Sm. var. *insigne* J.J.Sm. in Bull. Jard. Bot. Buitenzorg, ser. 3, 2:101 (1920). Type: Kalimantan, Kota Waringin, *van Nouhuys*, Bogor cult. no. 4, 6 (holo. L).
   B.lepidum (Blume) J.J.Sm. var. *angustum* Ridl in J. Linn. Soc., Bot. 32:282 (1896). Type: Sabah, Sandakan, *Pryer* s.n. (holo. SING).
   B.gamosepalum (Griff.) J.J.Sm. in Bull. Jard. Bot. Buitenzorg, ser. 2, 8:24 (1912).

HABITAT: hill forest; riverine forest.
ALTITUDINAL RANGE: lowlands – 900m.
DISTRIBUTION: Borneo (B, K, SA, SR). Also in Burma, Andaman Islands, Thailand, Laos, Cambodia, Vietnam, Peninsular Malaysia, Sumatra, Mentawai, Bangka, Java, Maluku.
GENERIC SECTION: *Cirrhopetalum.*

B.leproglossum *J.J.Verm. & A.Lamb* in Malayan Orchid Rev. 22:46 (1988). Type: Sabah, Sipitang District, Ulu Padas, *Vermeulen* 584 (holo. L).

HABITAT: high montane forest.
ALTITUDINAL RANGE: 1300 – 1500m.
DISTRIBUTION: Borneo (SA). Endemic.
GENERIC SECTION: *Monilibulbus.*

B.limbatum *Lindl.* in Bot. Reg. 26, misc. 74 (1840). Type: Singapore, *Cuming* s.n., cult.*Loddiges* (holo. K-LINDL).

HABITAT: unknown.
ALTITUDINAL RANGE: unknown.
DISTRIBUTION: Borneo (unspecified). Also Burma, Thailand, Peninsular Malaysia, Singapore, Sumatra, Riau Archipelago, Bangka.
GENERIC SECTION: *Hirtula.*

B.lissoglossum *J.J.Verm.* in Orch. Borneo 2:25, fig.3, Pl.4B (1991). Type: Sabah, Mt. Kinabalu, near Park Headquarters, *Vermeulen & Chan* 390 (holo. L, iso. K).

HABITAT: montane podsol forest; lower montane forest.
ALTITUDINAL RANGE: 1000 – 2000m.
DISTRIBUTION: Borneo (SA, SR). Endemic.
GENERIC SECTION: *Altisceptrum.*

B.lobbii *Lindl.* in Bot. Reg. 33, sub t.29 (1847). Type: Java, *Lobb* s.n. (holo. K-LINDL).
B.claptonense Rolfe in Orchid Rev. 8:198 (1905). Type: Borneo, cult. *Glasnevin Bot.Gard.* (holo. K).

HABITAT: lowland, hill and lower montane forest.
ALTITUDINAL RANGE: sea level – 2000m.
DISTRIBUTION: Borneo (K, SA, SR). Widespread in India, Burma, Thailand, Cambodia, Peninsular Malaysia, Sumatra, Java, Bali, Flores, Philippines.
GENERIC SECTION: *Sestochilus.*

B.lohokii *J.J.Verm. & A.Lamb* ined. Type: Sabah, *Lamb* AL 569/86 (holo. L).

HABITAT: upper montane forest.
ALTITUDINAL RANGE: 1700 – 2000m.
DISTRIBUTION: Borneo (SA). Endemic.
GENERIC SECTION: *Hapalochilus.*

B.longerepens *Ridl.* in J. Straits Branch Roy. Asiat. Soc. 49:28 (1907). Type: Sarawak, Santubong, *Hewitt* s.n. (holo. SING).

HABITAT: podsol forest.
ALTITUDINAL RANGE: 400m.
DISTRIBUTION: Borneo (SA, SR). Endemic.
GENERIC SECTION: *Desmosanthes.*

B.longhutense *J.J.Sm.* in Bull. Jard. Bot. Buitenzorg, ser. 3, 11:141 (1931). Type: Kalimantan, Long Hoet, *Endert* 2833 (holo. L).

HABITAT: hilly lowland forest; mossy high lower montane forest.
ALTITUDINAL RANGE: 200 – 1700m.
DISTRIBUTION: Borneo (K, SA). Endemic.
GENERIC SECTION: *Monilibulbus.*

B.longiflorum *Thouars*, Hist. Orch.:III & t.98 (1822). Type: Mauritius, *Thouars* s.n. (holo. P, iso. L).

HABITAT: hill forest.
ALTITUDINAL RANGE: 600 – 900m.
DISTRIBUTION: Borneo (SA). Widespread from Africa, Madagascar and Mauritius to Borneo, Maluku, New Guinea, Vanuatu, New Caledonia, Fiji, Tahiti, Guam, Austral Islands, Society Islands and Australia.
GENERIC SECTION: *Cirrhopetalum.*

B.longimucronatum *Ames & C.Schweinf.*, Orch. 6:184 (1920). Type: Sabah, Mt. Kinabalu, Kiau, *Clemens* 56 (holo. AMES).

HABITAT: hill forest; mossy lower montane forest; also recorded on a steep, cleared roadside close to primary forest on sandstone rock.
ALTITUDINAL RANGE: 900 – 1900m.

DISTRIBUTION: Borneo (SA). Endemic.
GENERIC SECTION: *Aphanobulbon.*

B.lordoglossum *J.J.Verm. & A.Lamb* ined. Type: Sabah, Sandakan Zone, Tawai Plateau, near Telupid, *Vermeulen* 900 (holo. L).

HABITAT: forest on ultramafic substrate.
ALTITUDINAL RANGE: 700m.
DISTRIBUTION: Borneo (SA). Endemic.
GENERIC SECTION: *Hybochilus.*

B.lygeron *J.J.Verm.*, Orch. Borneo 2:69, fig.20, Pl.6D (1991). Type: Sabah, *Vermeulen & Lamb* 353 (holo. L, iso. K).

HABITAT: lower montane forest; *Agathis* or *Lithocarpus* dominated forest; very low and open podsol forest.
ALTITUDINAL RANGE: 1200 – 2000m.
DISTRIBUTION: Borneo (SA). Endemic.
GENERIC SECTION: *Aphanobulbon.*

B.macranthum *Lindl.* in Bot. Reg. 30, t.13 (1844). Type: Singapore, probably plate in Bot. Reg. according to Vermeulen (1991).

HABITAT: lowland and hill forest; occasionally in lower montane forest; sometimes on ultramafic substrate.
ALTITUDINAL RANGE: sea level – 700m (-1400m).
DISTRIBUTION: Borneo (B, K, SA, SR). Burma, Vietnam, Thailand, Peninsular Malaysia, Sumatra, Java, Philippines.
GENERIC SECTION: *Sestochilus.*

B.macrochilum *Rolfe* in Kew Bull.:45 (Feb. 1896). Type: Borneo, *Haviland* s.n. (holo. K).
 B.mirandum Kraenzl. in Bot. Jahrb. Syst. 34:255 (1905). Type: Sarawak (holo. *Beccari* illustration, FI).
 B.intervallatum J.J.Sm. in Bull. Jard. Bot. Buitenzorg, ser. 3, 9:167 (1927). Type: Kalimantan, Bandjermasin, *Delmaar* 14, cult. Bogor (holo L).

HABITAT: lowland and lower montane forest; high mixed dipterocarp forest; *Agathis* forest; *Shorea albida* forest; podsol forest; swamp forest.
ALTITUDINAL RANGE: sea level – 1200m.
DISTRIBUTION: Borneo (B, K, SA, SR). Also Peninsular Malaysia.
GENERIC SECTION: *Intervallatae.*

B.mahakamense *J.J.Sm.* in Bull. Dép. Agric. Indes Nécrl. 22:34 (1909). Type: Kalimantan, Mahakam, Sungai Bloe-oe, *Jaheri* s.n. (holo. BO).

HABITAT: unknown.
ALTITUDINAL RANGE: unknown.
DISTRIBUTION: Borneo (K). Endemic.
GENERIC SECTION: uncertain.

B.makoyanum *(Rchb.f.) Ridl.*, Mat. Fl. Mal. Pen. 1:81 (1907). Type: received from *Messrs. Makoy*, erroneously said to have originated from Brazil (holo. W).
 Cirrhopetalum makoyanum Rchb.f. in Gard. Chron. n.s. 11, 1:234 (1879).

HABITAT: lowland forest.
ALTITUDINAL RANGE: 100 – 200m.
DISTRIBUTION: Borneo (K). Also Singapore, Philippines.

GENERIC SECTION: *Cirrhopetalum.*

B.malleolabrum *Carr* in Gard. Bull. Straits Settlem. 7:24 (1932). Types: Peninsular Malaysia, Pahang, *Henderson* A458 (syn. SING) & 23474 (syn. SING).

HABITAT: low and open wet podsol forest with a dense undergrowth of *Pandanus* and rattan.
ALTITUDINAL RANGE: 1300 – 1500m.
DISTRIBUTION: Borneo (SA). Also Peninsular Malaysia.
GENERIC SECTION: *Monilibulbus.*

B.mandibulare *Rchb.f.* in Gard. Chron. 17:366 (1882). Type: Sabah, *Burbidge* s.n., cult. V*eitch* (holo. W).

HABITAT: riverine forest.
ALTITUDINAL RANGE: 300 – 1000m.
DISTRIBUTION: Borneo (SA). Endemic.
GENERIC SECTION: *Intervallatae.*

B.marudiense *Carr* in Gard. Bull. Straits Settlem. 8:111 (1935). Type: Sarawak, Marudi, *Synge* S.2 (holo. SING, iso. K).

HABITAT: on *Podocarpus* in sandy forest; 'heath woodland'; podsolic dipterocarp -*Dacrydium* forest on very wet sandy soil; rather open swampy forest.
ALTITUDINAL RANGE: sea level – 700m.
DISTRIBUTION: Borneo (SA, SR). Endemic.
GENERIC SECTION: *Micromonanthe.*

B.mastersianum *(Rolfe)* *J.J.Sm.* in Bull. Jard. Bot. Buitenzorg, ser. 2, 8:26 (1912). Type: 'Indes Néerlandaise', introduced by *Linden* (holo. K).
Cirrhopetalum mastersianum Rolfe in Lindenia 6:33, t.255 (1890).

HABITAT: hill forest.
ALTITUDINAL RANGE: 500m.
DISTRIBUTION: Borneo (SA, SR). Also Maluku.
GENERIC SECTION: *Cirrhopetalum.*

B.medusae *(Lindl.)* *Rchb.f.* in Ann. Bot. Syst. 6:262 (1861). Type: Borneo, *Lobb* s.n. (holo. K-LINDL).
Cirrhopetalum medusae Lindl. in Bot. Reg. 28: t.12 (1842).
Phyllorchis medusae (Lindl.) Kuntze, Rev. Gen. Pl.2.:677 (1891).

HABITAT: lowland forest, either high with a closed canopy or low, open and podsolic.
ALTITUDINAL RANGE: lowlands.
DISTRIBUTION: Borneo (B, K, SA, SR). Also Thailand, Peninsular Malaysia, Sumatra, Bangka, Lesser Sunda Islands.
GENERIC SECTION: *Desmosanthes.*

B.membranaceum *Teijsm. & Binn.* in Ned. Kruidk. Arch. 3:397 (1855). Type: Java, Salak, cult. *Bogor* (holo. BO).

HABITAT: lowland forest.
ALTITUDINAL RANGE: sea level – 400m.
DISTRIBUTION: Borneo (K, SA). Also Peninsular Malaysia, Thailand, Sumatra, Sulawesi, probably New Guinea.
GENERIC SECTION: *Polyblepharon.*

B.membranifolium *Hook.f.*, Fl. Brit. Ind. 5:756 (1890). Type: Peninsular Malaysia, Perak, Gunung Batu Pateh, *Wray* 1123 (holo. K).
  B.insigne Ridl. in J. Linn. Soc., Bot. 31:272 (1896). Type: Borneo, locality uncertain, *Durnford* s.n. (holo. SING).
  B.cryptophoranthoides Kraenzl. in Bot. Jahrb. Syst. 34:251 (1905). Type: Sarawak, Mattang, *Beccari* 1874 (holo. FI).
  B.scandens Kraenzl., loc. cit.:253 (1905). Type: Sarawak, Mattang, *Beccari* 1954 (holo. FI).

HABITAT: podsol forest; swampy forest; forest on ultramafic substrate; lower montane forest dominated by *Dacrydium* & *Leptospermum*.
ALTITUDINAL RANGE: 1300 – 2100m.
DISTRIBUTION: Borneo (K, SA, SR). Also Peninsular Malaysia, Sumatra, Philippines.
GENERIC SECTION: *Sestochilus*.

B.microglossum *Ridl.* in J. Linn. Soc., Bot. 38:325 (1908). Type: Peninsular Malaysia, Pahang, Gunung Tahan, *Wray & Robinson* 5397 (holo. BM, iso. K).

HABITAT: oak-laurel forest; lower montane forest.
ALTITUDINAL RANGE: 900 – 2200m.
DISTRIBUTION: Borneo (SA, SR). Also Peninsular Malaysia.
GENERIC SECTION: *Sestochilus*.

B.minutulum *Ridl.*, Fl. Mal. Pen. 4:62 (1924). Type: Peninsular Malaysia, Pahang, Bukit Fraser, *Burkill & Holttum* 7797 (holo. SING, iso. K).

HABITAT: lower montane forest of various kinds; less frequent in dry and open conditions.
ALTITUDINAL RANGE: 700 – 2000m.
DISTRIBUTION: Borneo (K, SA). Also Peninsular Malaysia.
GENERIC SECTION: *Monilibulbus*.

B.mirabile *Hallier f.* in Ann. Jard. Bot. Buitenzorg, 13:316, t.28, fig.1a-d (1896). Type: Kalimantan, locality unknown, *Jaheri* s.n., cult. Bogor (holo. BO).

HABITAT: unknown.
ALTITUDINAL RANGE: 200 – 300m.
DISTRIBUTION: Borneo (B, K). Endemic.
GENERIC SECTION: *Hirtula*.

B.mirum *J.J.Sm.* in Icon. Bogor. 3:41, Pl.216 (1906). Type: Sumatra, Padang Pandjang, *Storm van's Gravesande* s.n. (holo. BO).

HABITAT: unknown.
ALTITUDINAL RANGE: unknown.
DISTRIBUTION: Borneo (unspecified). Also Sumatra, Java, Bali.
GENERIC SECTION: *Cirrhopetalum*.

B.montense *Ridl.* in Trans. Linn. Soc. London, Bot., ser. 2, 4:234 (1894). Type: Sabah, Mt. Kinabalu, *Haviland* 1099 (holo. SING, iso. K).
  B.vinculibulbum Ames & C.Schweinf., Orch. 6:202 (1920). Type: Sabah, Mt. Kinabalu, Pakka Cave to Lobong, *Topping* 372 (holo. AMES).

HABITAT: open mixed upper montane forest; open *Dacrydium-Leptospermum* forest; frequently on ultramafic substrate.
ALTITUDINAL RANGE: 1300 – 3400m.
DISTRIBUTION: Borneo (SA). Endemic.

GENERIC SECTION: *Monilibulbus.*

B.multiflexum *J.J.Sm.* in Mitt. Inst. Allg. Bot. Hamburg 7:66, t.10, fig.55 (1927). Type: Kalimantan, Bukit Tilung, *Winkler* 1533 (holo. HBG).

HABITAT: primary forest.
ALTITUDINAL RANGE: 800m.
DISTRIBUTION: Borneo (K). Endemic.
GENERIC SECTION: *Aphanobulbon.*

B.multiflorum *(Breda) Kraenzl.* in Gard. Chron., ser. 3, 19:326 (1896). Type: Java, Sumbangeractabr., *Kuhl & van Hasselt* s.n. (holo. not located).
Odontostylis multiflora Breda in Kuhl & van Hasselt, Gen. Sp. Orch., sub.t.4 (1827).

HABITAT: unknown.
ALTITUDINAL RANGE: unknown.
DISTRIBUTION: Borneo (SA). Also Peninsular Malaysia, Sumatra, Java.
GENERIC SECTION: *Desmosanthes.*

B.muscohaerens *J.J.Verm. & A.Lamb* ined. Type: Sabah, Crocker Range, near Tambunan, *Vermeulen* 647 (holo. L).

HABITAT: upper montane forest.
ALTITUDINAL RANGE: 1800 – 2000m.
DISTRIBUTION: Borneo (SA). Endemic.
GENERIC SECTION: *Polyblepharon.*

B.mutabile *(Blume) Lindl.*, Gen. Sp. Orch. Pl.:48 (1830). Types: Java, Salak, *Blume* 257 (lecto. L), *Blume* 459 (syn. L), *Blume* 460 (syn. L).
Diphyes mutabilis Blume, Bijdr.:312 (1825).
Bulbophyllum altispex Ridl. in Trans. Linn. Soc. London, Bot. 4:236 (1894). Type: Sabah, Mt. Kinabalu, *Haviland* 1143 (holo. K).

var. **mutabile**

HABITAT: lowland and montane forest of different types.
ALTITUDINAL RANGE: 400 – 2700m.
DISTRIBUTION: Borneo (B, K, SA, SR). Also Thailand, Peninsular Malaysia, Sumatra, Java, Sulawesi, Philippines.
GENERIC SECTION: *Aphanobulbon.*

var. **obesum** *J.J.Verm.*, Orch. Borneo 2:74, fig.21D (1991). Type: Sabah, Mt. Kinabalu, Park Headquarters area, *Vermeulen* 468 (holo. L).

HABITAT: lower montane forest; *Dacrydium – Leptospermum* forest; montane podsol forest.
ALTITUDINAL RANGE: 1400 – 2400m.
DISTRIBUTION: Borneo (SA). Endemic.
GENERIC SECTION: *Aphanobulbon.*

B.nabawanense *J.J.Wood & A.Lamb*, Orch. Borneo 1:73, fig.9, Pl.2D (1994). Type: Sabah, near Nabawan, *Lamb* AL 131/83 (holo. K).

HABITAT: podsol forest.
ALTITUDINAL RANGE: 400 – 500m.
DISTRIBUTION: Borneo (SA). Endemic.
GENERIC SECTION: *Sestochilus.*

B.nematocaulon *Ridl.*, Fl. Mal. Pen. 4:63 (1924). Types: Peninsular Malaysia, Perak, *Scortechini* 614 (syn. SING, isosyn. K); Bukit Fraser, *Burkill & Holttum* 8955 (syn. SING, isosyn. K).
  B. johannis-winkleri J.J.Sm. in Mitt. Inst. Allg. Bot. Hamburg 7:59 , t.9, fig.47 (1927). Type: Kalimantan, Bukit Raja, *Winkler* 899 (holo. HBG).

HABITAT: lowland and montane forest, either in high forest or low, open podsol forest.
ALTITUDINAL RANGE: 400 – 1800m.
DISTRIBUTION: Borneo (K, SA). Also Peninsular Malaysia.
GENERIC SECTION: *Micromonanthe.*

B.nieuwenhuisii *J.J.Sm.* in Bull. Jard. Bot. Buitenzorg, ser. 3, 8:61 (1926). Type: Kalimantan, Bukit Kasian, *Nieuwenhuis* 1899 (holo. BO).

HABITAT: unknown.
ALTITUDINAL RANGE: unknown.
DISTRIBUTION: Borneo (K). Endemic.
GENERIC SECTION: *Intervallatae.*

B.nubinatum *J.J.Verm.*, Orch. Borneo 2:203, fig.70, Pl.14B (1991). Type: Sabah, Mt. Kinabalu, summit trail, *Chan* 63/87 (holo. L, iso. K, SNP).

HABITAT: lower montane forest; upper montane *Dacrydium – Leptospermum* forest; upper montane forest on ultramafic substrate.
ALTITUDINAL RANGE: 1800 – 3000m.
DISTRIBUTION: Borneo (SA). Endemic.
GENERIC SECTION: *Monilibulbus.*

B.obtusipetalum *J.J.Sm.*, Orch. Java: 424 (1905). Types: Sumatra, *Korthals* s.n., HLB 902, 322-410 (lecto. L); Java, HLB 904, 33-388-390 (syn. L); locality unknown, HLB 902, 322-409 & -411 (syn. L).
  B.spinulipes J.J.Sm. in Mitt. Inst. Allg. Bot. Hamburg 7:61, t.10, fig.50 (1927). Type: Kalimantan, Bukit Mehipit, *Winkler* 737 (holo. BO, iso. HBG).

HABITAT: high montane forest.
ALTITUDINAL RANGE: 1300 – 1500m.
DISTRIBUTION: Borneo (K, SA). Also Peninsular Malaysia, Sumatra, Java.
GENERIC SECTION: *Aphanobulbon.*

B.obtusum *(Blume) Lindl,* Gen. Sp. Orch. Pl.:56 (1830). Type: Java, Gede & Salak, *Blume* s.n. (holo. BO).
  Diphyes obtusa Blume, Bijdr.:315 (1825).

HABITAT: lower montane forest.
ALTITUDINAL RANGE: 1200 – 1800m.
DISTRIBUTION: Borneo (unspecified). Also Java.
GENERIC SECTION: *Desmosanthes.*

B.odoardii *Pfitzer* in Engl. & Prantl., Nat. Pflanzenfam. 2, 6:179 (1889). Type: Sarawak, *Beccari* 431 (holo. FI, iso. K).

HABITAT: unknown.
ALTITUDINAL RANGE: unknown.
DISTRIBUTION: Borneo (SR). Endemic.
GENERIC SECTION: *Monilibulbus.*

B.odoratum *(Blume) Lindl.*, Gen. Sp. Orch. Pl.:54 (1830). Type: Java, Tjiburrum, *Blume* s.n., HLB 904, 33-386 (holo. L).
  Diphyes odorata Blume, Bijdr.:312 (1825).
  Bulbophyllum elatius Ridl. in J. Linn. Soc., Bot. 31:275 (1896). Types: Sumatra, Ayer Mancior, Padang, *Beccari* 559 (lecto. BM, isolecto. FI, K); Sabah, Ulu Tawaran, *Haviland* 1371 (syn. BM, K, SING); Sarawak, *Hose* s.n. (syn. BM, SING).
  B.brookesii Ridl. in J. Straits Branch Roy. Asiat. Soc. 50:131 (1908). Type: Sarawak, Bidi, *Brookes* s.n. (holo. SING).
  B.hortense J.J.Sm. in Bull. Jard. Bot. Buitenzorg, ser. 2, 9:79 (1913). Type: ?Kalimantan, cult. *Bogor* s.n. (holo. BO, iso. L).
  B.crassicaudatum Ames & C.Schweinf., Orch. 6:170 (1920). Type: Sabah, Mt. Kinabalu, *Clemens* s.n. (holo. AMES, iso. BO, K).

  HABITAT: lowland, hill and lower montane forest; mixed dipterocarp forest.
  ALTITUDINAL RANGE: sea level – 2400m.
  DISTRIBUTION: Borneo (K, SA, SR). Also Peninsular Malaysia, Sumatra, Java, Lesser Sunda Islands, Sulawesi, Maluku, Philippines.
  GENERIC SECTION: *Aphanobulbon.*

B.osyriceroides *J.J.Sm.* in Bull. Jard. Bot. Buitenzorg, ser. 3, 2:97 (1920). Type: Sumatra, Pasiaman, Kajoe Tanam, *Groeneveldt* s.n., cult. *Jacobson* under nos. 1265 & 1380 (holo. BO).

  HABITAT: unknown.
  ALTITUDINAL RANGE: lowlands.
  DISTRIBUTION: Borneo (SA). Also Sumatra.
  GENERIC SECTION: *Globiceps.*

B.otochilum *J.J.Verm.*, Orch. Borneo 2:269, fig.94, Pl.18E (1991). Type: Kalimantan, Mt. Meratus, *de Vogel* 1195 (holo. L, iso. BO).

  HABITAT: high ridge forest.
  ALTITUDINAL RANGE: 700m.
  DISTRIBUTION: Borneo (K, SA, SR). Endemic.
  GENERIC SECTION: *Sestochilus.*

B.ovalifolium *(Blume) Lindl.*, Gen. Sp. Orch. Pl.:49 (1830). Type: Java, Gede, *Blume* s.n., HLB 902, 322-463 (holo. L).
  Diphyes ovalifolia Blume, Bijdr.:318 (1825).

  HABITAT: lower montane forest.
  ALTITUDINAL RANGE: 1200 – 2000m.
  DISTRIBUTION: Borneo (SA). Also Thailand, Peninsular Malaysia, Sumatra, Java, Flores, Sulawesi.
  GENERIC SECTION: *Monilibulbus.*

B.papillatum *J.J.Sm.* in Bull. Dép. Agric. Indes Néerl. 43:60 (1910). Type: Java, Salak, Tjigombong, *Smith* s.n. (holo. L).
  B.papillosum J.J.Sm. in Orch. Java: 464 (1905), non Finet.

  HABITAT: unknown.
  ALTITUDINAL RANGE: unknown.
  DISTRIBUTION: Borneo (unspecified). Also Java.
  GENERIC SECTION: *Micromonanthe.*

B.patens *King ex Hook.f.*, Fl. Brit. Ind. 6:187(1890). Type: Peninsular Malaysia, Perak, *Kunstler* s.n. (holo. drawing in CAL).

HABITAT: *Shorea albida* dominated swamp forest; peat swamp forest; lowland podsol forest; mixed dipterocarp forest.
ALTITUDINAL RANGE: sea level – 100m.
DISTRIBUTION: Borneo (B, K, SA, SR). Also Peninsular Malaysia, Sumatra.
GENERIC SECTION: *Sestochilus*.

B.pelicanopsis *J.J.Verm. & A.Lamb* in Malayan Orchid Rev. 22:47 (1988). Type: Sabah, *Vermeulen* 582 (holo. L).

HABITAT: high lower montane *Agathis – Lithocarpus* forest.
ALTITUDINAL RANGE: 1200 – 1500m.
DISTRIBUTION: Borneo (SA). Endemic.
GENERIC SECTION: *Monilibulbus*.

B.penduliscapum *J.J.Sm.*, Icon. Bogor. 2:101, t.119b (1903). Type: North Sumatra, *Heldt* s.n. (holo. BO).
B.macrophyllum Kraenzl. in Bot. Jahrb. Syst. 34:249 (1905). Type: Sarawak, *Beccari* 3113 (lecto.:FI); Sungai Mattan, *Beccari* 1344 (syn. FI).

HABITAT: low and open podsol forest; higher, denser, slightly podsolic forest; low forest on limestone hills.
ALTITUDINAL RANGE: 300 – 1100m.
DISTRIBUTION: Borneo (SA, SR). Also Sumatra, Philippines.
GENERIC SECTION: *Altisceptrum*.

B.perparvulum *Schltr.* in Beih. Bot. Centralbl. 33,2:419 (1915). Type: Sarawak, Bidi, *Brookes* s.n. (holo. SING, iso. K).
B.perpusillum Ridl. in J. Straits Branch Roy. Asiat. Soc., 50:130 (1908), non Kraenzl.

HABITAT: unknown.
ALTITUDINAL RANGE: lowlands.
DISTRIBUTION: Borneo (SR). Endemic.
GENERIC SECTION: *Micromonanthe*.

B.phaeoneuron *Schltr.* in Bot. Jahrb. Syst. 45, Beibl. 104:48 (1911). Type: Sumatra, kampung Tengah, *Schlechter* 16022 (holo. B, destroyed).

HABITAT: high, dry montane forest; dry, low and open *Leptospermum* forest.
ALTITUDINAL RANGE: 1400 – 2000m.
DISTRIBUTION: Borneo (SA). Also Sumatra.
GENERIC SECTION: *Monilibulbus*.

B.pileatum *Lindl.* in Bot. Reg. 30, misc. 73 (1844). Type: Singapore, *Loddiges* 178 (holo. K-LINDL).

HABITAT: mixed oak-chestnut forest on sandstone ridges.
ALTITUDINAL RANGE: 900 – 1000m.
DISTRIBUTION: Borneo (SA). Also Peninsular Malaysia, Singapore, Sumatra.
GENERIC SECTION: *Sestochilus*.

B.placochilum *J.J.Verm.*, Orch. Borneo 2:29, fig.5 (1991). Type: Sabah, Mt. Kinabalu, *Clemens* 34065 (holo. L, iso. BM).

HABITAT: lower montane forest.
ALTITUDINAL RANGE: 1500 – 1700m.
DISTRIBUTION: Borneo (SA). Endemic.
GENERIC SECTION: *Altisceptrum*.

B.planibulbe *(Ridl.) Ridl.*, Mat. Fl. Mal. Pen. 1:79 (1907). Type: Peninsular Malaysia, Pekan and Kwala Pahang, *Ridley* s.n. (holo. SING).
   Cirrhopetalum planibulbe Ridl. in Trans. Linn. Soc. London, Bot. 3:364, t.64 (1893).

HABITAT: hill forest on ultramafic substrate.
ALTITUDINAL RANGE: 700 – 900m.
DISTRIBUTION: Borneo (K). Also Thailand, Peninsular Malaysia, Sumatra.
GENERIC SECTION: *Desmosanthes.*

B.pocillum *J.J.Verm.*, Orch. Borneo 2:83, fig.24, Pl.7D (1991). Type: Sabah, *Vermeulen & Lamb* 354 (holo. L, iso. K).

HABITAT: lower montane riverine forest; *Dacrydium* – *Leptospermum* forest; low and open podsol forest.
ALTITUDINAL RANGE: 1200 – 2100m.
DISTRIBUTION: Borneo (SA). Endemic.
GENERIC SECTION: *Aphanobulbon.*

B.polygaliflorum *J.J.Wood* in Kew Bull. 39(1):96, fig.15 (1984). Type: Sarawak, Gunung Mulu National Park, *Hansen* 499 (holo. C, iso. K).

HABITAT: upper montane moss forest.
ALTITUDINAL RANGE: 1700m.
DISTRIBUTION: Borneo (SA, SR). Endemic.
GENERIC SECTION: *Hirtula.*

B.porphyrotriche *J.J.Verm.*, Orch. Borneo 2:127, fig.40 (1991). Type: Sabah, Sipitang District, east of Long Pa Sia – Long Miau trail, *Vermeulen & Duistermaat* 1106 (holo. L, iso. K).

HABITAT: very open kerangas forest.
ALTITUDINAL RANGE: 1100m.
DISTRIBUTION: Borneo (SA). Endemic.
GENERIC SECTION: *Epicrianthes.*

B.prianganense *J.J.Sm.* in Bull. Jard. Bot. Buitenzorg, ser. 2, 9:76 (1913). Type: Java, South Priangan, *Raciborski* s.n. (holo. BO).
   B.hamatifolium J.J.Sm. in Bull. Jard. Bot. Buitenzorg, 13:26 (1914). Type: Kalimantan, Kapoeas, *Nieuwenhuis* s.n., cult. Bogor number 1763 (holo. BO).

HABITAT: swamp forest.
ALTITUDINAL RANGE: lowlands.
DISTRIBUTION: Borneo (SA). Also Sumatra, Java.
GENERIC SECTION: *Aphanobulbon.*

B.pugilanthum *J.J.Wood*, Orch. Borneo 1:77, fig.10, Pl.2E (1994). Type: Sabah, Mt. Kinabalu, Park Headquarters, *Lamb* AL 56/83 (holo. K).

HABITAT: lower montane forest; lower montane riverine forest; upper montane forest; sometimes on ultramafic substrate.
ALTITUDINAL RANGE: 1300 – 2400m.
DISTRIBUTION: Borneo (SA). Endemic.
GENERIC SECTION: *Aphanobulbon.*

B.puguahaanense *Ames*, Orch. 5:185 (1915). Type: Philippines, Leyte, Puguahaan, Dagami, *Wenzel* 128 (holo. AMES).

HABITAT: unknown.
ALTITUDINAL RANGE: unknown.
DISTRIBUTION: Borneo (unspecified). Also Peninsular Malaysia, Philippines.
GENERIC SECTION: *Cirrhopetalum.*

B.pumilio *Ridl.* in J. Straits Branch Roy. Asiat. Soc., 50:129 (1908). Type:Sarawak, Bidi, *Brookes* s.n. (holo. SING, iso. K).

HABITAT: *Agathis – Lithocarpus* forest on sandstone.
ALTITUDINAL RANGE: 1400 – 1500m.
DISTRIBUTION: Borneo (SA, SR). Endemic.
GENERIC SECTION: *Aphanobulbon.*

B.puntjakense *J.J.Sm.* in Bull. Dép. Agric. Indes Néerl. 13:46 (1907). Type: Java, Poentjak, *Smith* s.n. (holo. BO).

HABITAT: lower montane forest.
ALTITUDINAL RANGE: 1500 – 2000m.
DISTRIBUTION: Borneo (SA). Also Java, Bali.
GENERIC SECTION: *Monilibulbus.*

B.purpurascens *Teijsm. & Binn.* in Natuurk. Tijdschr. Ned.-Indië 24:308 (1862). Type: Java, locality unknown, *Teijsmann* s.n. (holo. BO).
  Cirrhopetalum peyerianum Kraenzl. in Bot. Jahrb. Syst. 17:485 (1893). Type: Sumatra, locality unknown, *Peyer* s.n. (holo. B,destroyed).
  C.citrinum Ridl. in J. Linn. Soc., Bot. 31:279 (1896). Type: Sarawak, *Haviland* 2322 (holo. SING, iso. K).
  C.pallidum Schltr. in Bull. Herb. Boissier, ser. 2, 6:464 (1906). Type: Kalimantan, Samarinda, *Schlechter* 13344 (holo. B, destroyed).
  Bulbophyllum citrinum (Ridl.) Ridl., Mat. Fl. Mal. Pen. 1:75 (1907).
  B.peyerianum (Kraenzl.) Seidenf. in Dansk Bot. Ark. 29, 1:151 (1973).

HABITAT: lowland and hill forest.
ALTITUDINAL RANGE: sea level – 1200m.
DISTRIBUTION: Borneo (B, K, SA, SR). Also Thailand, Peninsular Malaysia, Sumatra, Bangka, Java, Krakatau.
GENERIC SECTION: *Cirrhopetalum.*

B.pustulatum *Ridl.* in Journ. Straits Branch Roy. Asiat. Soc. 39:74 (1903). Type: Peninsular Malaysia, Mt. Ophir, *Ridley* s.n. (holo. SING).

HABITAT: unknown.
ALTITUDINAL RANGE: unknown.
DISTRIBUTION: Borneo (unspecified). Also Peninsular Malaysia.
GENERIC SECTION: *Sestochilus.*

B.putidum *(Teijsm. & Binn.)* *J.J.Sm.* in Bull. Jard. Bot. Buitenzorg, ser. 2, 8:27 (1912). Type: Sumatra, Palembang, *Teijsmann* s.n. (holo. BO).
  Cirrhopetalum putidum Teijsm. & Binnend. in Natuurk. Tijdschr. Ned.-Indië 23:311 (1862).

HABITAT: unknown.
ALTITUDINAL RANGE: unknown.
DISTRIBUTION: Borneo (K, SR). Also India (Bengal, Sikkim), Thailand, Indonesia, Peninsular Malaysia, Sumatra, Bangka, Philippines.
GENERIC SECTION: *Cirrhopetalum.*

B.pyridion *J.J.Verm.*, Orch. Borneo 2:217, fig.75 (1991). Type: Sabah, Ulu Padas, *Vermeulen* 577 (holo. L).

HABITAT: high lower montane forest.
ALTITUDINAL RANGE: 1500m.
DISTRIBUTION: Borneo (SA). Endemic.
GENERIC SECTION: *Monilibulbus.*

B.rajanum *J.J.Sm.* in Mitt. Inst. Allg. Bot. Hamburg 7:62, t.10, fig.51 (1927). Type: Kalimantan, Bukit Raja, *Winkler* 1018 (holo. HBG).

HABITAT: lower montane forest.
ALTITUDINAL RANGE: 1400m.
DISTRIBUTION: Borneo (K). Endemic.
GENERIC SECTION: *Aphanobulbon.*

B.rariflorum *J.J.Sm.* in Bull. Dép. Agric. Indes Néerl. 15:20 (1908). Type: Kalimantan, Pontianak, cult. Bogor (holo. BO).

HABITAT: unknown.
ALTITUDINAL RANGE: lowlands.
DISTRIBUTION: Borneo (K). Endemic.
GENERIC SECTION: *Hirtula.*

B.refractilingue *J.J.Sm.* in Bull. Jard. Bot. Buitenzorg, ser. 3, 11:145 (1931). Type: Kalimantan, West Koetai, Long Hoet, *Endert* 2690 (holo. L, iso. BO).

HABITAT: lowland forest; podsol forest; forest on limestone.
ALTITUDINAL RANGE: 100 – 400m.
DISTRIBUTION: Borneo (B, K, SA, SR). Endemic.
GENERIC SECTION: *Sestochilus.*

B.restrepia *Ridl.* in Trans. Linn. Soc. London, Bot. 3:365 (1893). Type: Peninsular Malaysia, Pramau, near Pekan, *Ridley* s.n. (holo. SING).

HABITAT: unknown.
ALTITUDINAL RANGE: lowlands.
DISTRIBUTION: Borneo (K). Also Peninsular Malaysia, Singapore.
GENERIC SECTION: *Ephippium.*

B.reticulatum *Bateman ex Hook.f.* in Bot. Mag. 92, t.5605 (1866). Type: Borneo, *Lobb* s.n. (holo. Bot. Mag. t.5605).
   B.carinatum Cogn. in J. Orchidées 6:216 (1895), non (Teijsm. & Binn.) Naves. Type: Borneo, cult.*l'Horticulture Internationale* (holo. not located).
   B.katherinae A.D. Hawkes in Lloydia 19:93 (1956).

HABITAT: hill forest on limestone.
ALTITUDINAL RANGE: lowlands and hills.
DISTRIBUTION: Borneo (SR). Endemic.
GENERIC SECTION: *Sestochilus.*

B.rhizomatosum *Ames & C.Schweinf.*, Orch. 6:194 (1920). Type: Sabah, Mt. Kinabalu, Lobong Cave, *Clemens* 106 (holo. AMES, iso. BM, BO).

HABITAT: lower montane forest; once recorded from lowland forest.
ALTITUDINAL RANGE: (100-)1100 – 2000m.
DISTRIBUTION: Borneo (B, SA). Also Peninsular Malaysia, Philippines.
GENERIC SECTION: *Aphanobulbon.*

B.rhynchoglossum *Schltr.* in Feddes Repert. 8:569 (1910). Type: Sarawak, Kuching, *Schlechter* 15851 (holo. B, destroyed, iso. K).

HABITAT: unknown.
ALTITUDINAL RANGE: lowlands.
DISTRIBUTION: Borneo (SR). Endemic.
GENERIC SECTION: *Leptopus.*

B.romburghii *J.J.Sm.* in Bull. Dép. Agric. Indes Néerl. 5:17 (1907). Type: Sumatra, 'Padangsche Bovenlanden', *Romburgh* s.n. (holo. BO).

HABITAT: dry, rather low and open ridge forest; riverine forest; forest on ultramafic substrate.
ALTITUDINAL RANGE: 400 – 1300m.
DISTRIBUTION: Borneo (SA). Also Sumatra.
GENERIC SECTION: *Desmosanthes.*

B.rubiferum *J.J.Sm.* in Bull. Jard. Bot. Buitenzorg, ser. 2, 26:73 (1918). Types: Java, cult. *Winckel* 355 (syn. BO, isosyn. L) and *Bakhuisen van den Brink* s.n. (syn. BO).

HABITAT: lower montane mossy forest.
ALTITUDINAL RANGE: 1700m.
DISTRIBUTION: Borneo (SA). Also Java.
GENERIC SECTION: *Globiceps.*

B.ruficaudatum *Ridl.* in J. Straits Branch Roy. Asiat. Soc. 54:50 (1910). Type: Sarawak, Kuching, *Hewitt* s.n. (holo. SING).
  B.microbulbon auct., non Schltr.
  B.nanobulbon Seidenf. in Dansk Bot. Ark. 29 (1):87, fig.40 (1973), nom. superfl.

HABITAT: unknown.
ALTITUDINAL RANGE: lowlands.
DISTRIBUTION: Borneo (SR). Also Singapore.
GENERIC SECTION: *Cirrhopetalum*

B.rugosum *Ridl.* in J. Linn. Soc., Bot. 32:266 (1896). Type: Singapore, Chan Chew Kang, *Goodenough* s.n. (holo. SING).

HABITAT: lowland mixed dipterocarp forest.
ALTITUDINAL RANGE: 100m.
DISTRIBUTION: Borneo (K, SA, SR). Also Peninsular Malaysia, Singapore, Sumatra.
GENERIC SECTION: *Sestochilus.*

B.salaccense *Rchb.f.* in Bonplandia 5:57 (1857). Type: Java, Salak, *Zollinger* s.n. (holo. W).

HABITAT: lowland, hill and lower montane forest of various types including low and open lowland podsol forest; riverine forest; wet or dry open forest on hill and mountain ridges; moss forest; upper montane *Dacrydium – Leptospermum* forest; also on sandstone rocks.
ALTITUDINAL RANGE: 400- 3000m.
DISTRIBUTION: Borneo (SA, SR). Also Peninsular Malaysia, Sumatra, Java.
GENERIC SECTION: *Globiceps.*

B.scabrum *J.J.Verm.* and *A.Lamb* in Malayan Orchid Rev.22:46(1988). Type: Sabah, Mt. Kinabalu, *Vermeulen* 483 (holo. L).

HABITAT: lower montane Fagaceae – *Dacrydium* – *Leptospermum* forest.
ALTITUDINAL RANGE: 1500 – 1700m.
DISTRIBUTION: Borneo (SA). Endemic.
GENERIC SECTION: *Monilibulbus*.

B.schefferi *(Kuntze) Schltr.* in Beih. Bot. Centralbl. 28, 2:417 (1915). Types: Java, Gede, *Blume* s.n., HLB 904, 44-52 (lecto. L) and 902, 322-424 (syn. L).
Diphyes gracilis Blume, Bijdr.:319 (1825).
Bulbophyllum gracile (Blume) Lindl., Gen. Sp. Orch. Pl.:50 (1830), non Thouars, nec C.S.P. Parish. & Rchb.f.
Phyllorchis schefferi Kuntze, Rev. Gen. Pl.2: 676 (1891).
Bulbophyllum corticicola Schltr. in Feddes Repert. 8:568 (1910). Type: Sarawak, Kuching, *Schlechter* 15844 (holo. B, destroyed, iso. BM, BO, K, L, S).
B.corticicola Schltr. var. *minor* Schltr. in Feddes Repert. 8:569 (1910). Type: Sarawak, Kuching, *Schlechter* 15844 (holo. B, destroyed).
B.bilobipetalum J.J.Sm. in Mitt. Inst. Allg. Bot. Hamburg 7:60, t.9, fig.48 (1927). Type: Kalimantan, Bukit Raja, *Winkler* 975 (holo. HBG).

HABITAT: high montane *Agathis* – *Lithocarpus* ridge forest; also recorded from fruit trees in the lowlands.
ALTITUDINAL RANGE: sea level – 1500m.
DISTRIBUTION: Borneo (K, SA, SR). Also Sumatra, Java, Lombok, Philippines.
GENERIC SECTION: *Monilibulbus*.

B.scintilla *Ridl.* in J. Straits Branch Roy. Asiat. Soc. 50:129 (1908). Type: Sarawak, Kuching, *Hewitt* s.n. (holo. SING).

HABITAT: unknown.
ALTITUDINAL RANGE: lowlands.
DISTRIBUTION: Borneo (SR). Endemic.
GENERIC SECTION: *Monilibulbus*.

B.sessile *(J.König) J.J.Sm.*, Orch. Java: 448 (1915). Type: Origin unknown, probably Thailand, *König* s.n. (holo. K).
Epidendrum sessile J.König in Retz. Observ. Bot. 6:60 (1791).

HABITAT: swamp forest; hill and lower montane forest; forest on ultramafic substrate.
ALTITUDINAL RANGE: sea level – 1500m.
DISTRIBUTION: Borneo (K, SA, SR). Widespread in Burma, Thailand, Indochina, Malaysia and Indonesia east to New Guinea, Solomon Islands and Fiji.
GENERIC SECTION: *Oxysepalum*.

B.sigmoideum *Ames & C.Schweinf.*, Orch. 6:196 (1920). Type: Sabah, Mt. Kinabalu, Gurulau Spur, *Clemens* 316 (holo. AMES).

HABITAT: hill and lower montane forest.
ALTITUDINAL RANGE: 900 – 1700m.
DISTRIBUTION: Borneo (SA). Endemic.
GENERIC SECTION: *Aphanobulbon*.

B.similissimum *J.J.Verm.*, Orch. Borneo 2:225, fig.78, Pl.15C & D (1991). Type: Sabah, Ulu Padas, *Vermeulen* 558 (holo. L).

HABITAT: *Agathis* – *Dacrydium* forest; *Castanopsis* – *Lithocarpus* ridge forest.
ALTITUDINAL RANGE: 1300 – 1500m.
DISTRIBUTION: Borneo (SA). Endemic.

GENERIC SECTION: *Monilibulbus.*

B.singaporeanum *Schltr.* in Feddes Repert. 9:165 (1911). Types: Singapore, Selitar, *Ridley* s.n. (syn. SING); Bukit Mandai, *Ridley* s.n. (syn. SING); Bukit Timah, *Ridley* s.n. (syn. SING, isosyn. K); Kranji, *Ridley* s.n. (syn. SING); Choa Chu Kang, *Ridley* s.n. (syn. SING).
  B.densiflorum Ridl. in J. Linn. Soc., Bot. 32:277 (1896), non Rolfe.

HABITAT: unknown.
ALTITUDINAL RANGE: lowlands.
DISTRIBUTION: Borneo (unspecified). Also Peninsular Malaysia, Singapore.
GENERIC SECTION: *Sestochilus.*

B.sopoetanense *Schltr.* in Feddes Repert. 10:181 (1912). Type: Sulawesi, Gunung Sopoetan, *Schlechter* 20614 (holo. B, destroyed, iso. K, L).
  B.conspectum J.J.Sm. in Mitt. Inst. Allg. Bot. Hamburg 7:65, t.10, fig.54 (1927). Type: Kalimantan, Bukit Raja, *Winkler* 1029 (holo. BO, iso. HBG).
  B.balapiuense sensu J.J.Wood in Jermy, Studies Fl. G. Mulu Nat. Park, Sarawak: 12 (1984), non J.J.Sm.

HABITAT: lower montane forest; moss forest; low open ridge forest.
ALTITUDINAL RANGE: 1200 – 2100m.
DISTRIBUTION: Borneo (K, SA, SR). Also Sulawesi.
GENERIC SECTION: *Aphanobulbon.*

B.stipitatibulbum *J.J.Sm.* in Bull. Jard. Bot. Buitenzorg, ser. 3, 11:148 (1931). Type: Kalimantan, West Koetai, Gunung Kemoel, *Endert* 4402 (holo. L).

HABITAT: lower montane forest.
ALTITUDINAL RANGE: 1800 – 2100m.
DISTRIBUTION: Borneo (K, SA). Endemic.
GENERIC SECTION: *Aphanobulbon.*

B.stormii *J.J.Sm.* in Bull. Dép. Agric. Indes Néerl. 5:20 (1907). Type: Sumatra, Padang Pandjang, *Storm van's Gravesande* s.n., cult. Bogor (holo. BO).

HABITAT: wet or dry, dense or open montane forest; montane podsol forest.
ALTITUDINAL RANGE: 1300 – 1900m.
DISTRIBUTION: Borneo (SA). Also Peninsular Malaysia, Sumatra.
GENERIC SECTION: *Monilibulbus.*

B.streptotriche *J.J.Verm.*, Orch. Borneo 2:129, fig.41, Pl.9F (1991). Type: Sabah, Ulu Padas, *Vermeulen* 646 (holo. L, iso. K).

HABITAT: high montane ridge forest on sandstone.
ALTITUDINAL RANGE: 1300m.
DISTRIBUTION: Borneo (SA). Endemic.
GENERIC SECTION: *Epicrianthes.*

B.striatellum *Ridl.* in Ann. Bot.(J.König & Sims) 4:335, Pl.22, fig.7 (1890). Type: Singapore, Char Chu Rang (= Choa Chu Kang), *Ridley* s.n. (holo. SING).

HABITAT: very wet, mossy, low podsol forest; mossy low ridge forest with climbing bamboo and rattans; ridge top forest with *Agathis*, small rattans, etc.; high montane Fagaceae – *Dipterocarpus* – *Agathis* forest.
ALTITUDINAL RANGE: lowlands – 2000m.
DISTRIBUTION: Borneo (SA, SR). Also Peninsular Malaysia, Singapore.
GENERIC SECTION: *Micromonanthe.*

B.subclausum *J.J.Sm.* in Bull. Dép. Agric. Indes Néerl. 22:35 (1909). Type: Sumatra, Bukit Gompong, *Piepers* s.n. (holo. BO).

> HABITAT: lower montane forest.
> ALTITUDINAL RANGE: 1200 – 2100m.
> DISTRIBUTION: Borneo (SA). Also Sumatra.
> GENERIC SECTION: *Aphanobulbon.*

B.subumbellatum *Ridl.* in J. Linn. Soc., Bot. 31:274 (1896). Type: Sarawak, near Kuching, *Haviland* 964 (holo. K).

> HABITAT: unknown.
> ALTITUDINAL RANGE: lowlands.
> DISTRIBUTION: Borneo (SR). Endemic.
> GENERIC SECTION: *Sestochilus.*

B.succedaneum *J.J.Sm.* in Mitt. Inst. Allg. Bot. Hamburg 7:67, t.10, fig.56 (1927). Type: Kalimantan, Bukit Raja, *Winkler* 861 (holo. HBG).

> HABITAT: high montane forest; low and open montane podsol forest.
> ALTITUDINAL RANGE: 1000 – 1700m.
> DISTRIBUTION: Borneo (K, SA). Endemic.
> GENERIC SECTION: *Intervallatae.*

B.supervacaneum *Kraenzl.* in Feddes Repert. 26:173 (1929). Type: Borneo, 'Mada bei Telang', *Grabowsky* s.n. (holo. not located).

> HABITAT: unknown.
> ALTITUDINAL RANGE: unknown.
> DISTRIBUTION: Borneo (unspecified). Endemic.
> GENERIC SECTION: *Aphanobulbon.*

B.taeniophyllum *C.S.P.Parish & Rchb.f.* in J. Bot., N.S. 3:198 (1874). Type: Burma, Moulmein, *Parish* 346 (holo. W).
> *B.fenestratum* J.J.Sm. in Bull. Dép. Agric. Indes Néerl. 13:48 (1907). Type: Bangka, *Smith* s.n. (holo. ?BO).

> HABITAT: unknown.
> ALTITUDINAL RANGE: unknown.
> DISTRIBUTION: Borneo (K). Also Burma, Thailand, Laos, Peninsular Malaysia, Sumatra, Bangka, Java.
> GENERIC SECTION: *Cirrhopetalum.*

B.tardeflorens *Ridl.* in J. Linn. Soc., Bot. 31:276 (1896). Type: Borneo, cult. Singapore Bot. Gard. (holo. SING).

> HABITAT: unknown.
> ALTITUDINAL RANGE: unknown.
> DISTRIBUTION: Borneo (unspecified).
> GENERIC SECTION: *Intervallatae.*

B.tenompokense *J.J.Sm.* in Feddes Repert. 36:116 (1934). Type: Sabah, Mt. Kinabalu, Tenompok, *Clemens* 29674 (holo.L).

> HABITAT: lower montane forest.
> ALTITUDINAL RANGE: 1500m.
> DISTRIBUTION: Borneo (SA). Endemic.
> GENERIC SECTION: *Polyblepharon.*

B.tenuifolium *(Blume) Lindl.*, Gen. Sp. Orch. Pl.:50 (1830). Type: Java, Salak, *Blume* s.n. (holo. L).
  Diphyes tenuifolia Blume, Bijdr.:316 (1825).
  Bulbophyllum microstele Schltr. in Feddes Repert. 8:569 (1910). Type: Sarawak, Kuching, *Schlechter* 15835 (holo. B, destroyed, iso. K).

  HABITAT: podsol forest; lower montane forest.
  ALTITUDINAL RANGE: sea level – 1500m.
  DISTRIBUTION: Borneo (K, SA, SR). Also Thailand, Peninsular Malaysia, Java.
  GENERIC SECTION: *Leptopus*.

B.teres *Ridl.* in Gard. Bull. Straits Settlem. 8:115 (1935). Type: Sarawak, Dulit ridge, *Synge* S.535 (holo. SING, iso. K).

  HABITAT: moss forest; lower montane Fagaceae/*Dacrydium*/*Leptospermum* forest.
  ALTITUDINAL RANGE: 1300 – 1800m.
  DISTRIBUTION: Borneo (SA, SR). Endemic.
  GENERIC SECTION: *Aphanobulbon*.

B.thymophorum *J.J.Verm. & A.Lamb* in Malayan Orchid Rev. 22:47 (1988). Type: Sabah, Crocker Range, *Lamb* AL 564/86 (holo. K).

  HABITAT: high, wet or dry montane forest; transitional zone between montane Fagaceae/Lauraceae forest and Ericaceae dominated forest.
  ALTITUDINAL RANGE: 1600 – 1900m.
  DISTRIBUTION: Borneo (SA) Endemic.
  GENERIC SECTION: *Monilibulbus*.

B.tortuosum *(Blume) Lindl.*, Gen. Sp. Orch. Pl.:50 (1830). Type: Java, Salak, *Blume* s.n. (holo. BO).
  Diphyes tortuosa Blume, Bijdr.:311 (1825).

  HABITAT: dry, rather low and open ridge forest; hill forest on ultramafic substrate.
  ALTITUDINAL RANGE: 1300m.
  DISTRIBUTION: Borneo (SA). Also Bhutan, Thailand, Laos, Vietnam, Peninsular Malaysia, Sumatra, Java.
  GENERIC SECTION: *Polyblepharon*.

B.trifolium *Ridl.* in J. Linn. Soc., Bot. 32:278 (1896). Type: Singapore, Sungai Morai, *Goodenough* s.n. (holo. SING).
  B.tacitum Carr in Gard. Bull. Straits Settlem. 8:110 (1935). Type: Sarawak, Marudi, *Synge* S.26 (holo. SING, iso. K).

  HABITAT: rather low and open vegetation including podsol forest, forest on ultramafic substrate and open lower montane forest.
  ALTITUDINAL RANGE: sea level – 2000m.
  DISTRIBUTION: Borneo (SA, SR). Also Peninsular Malaysia.
  GENERIC SECTION: *Globiceps*.

B.trulliferum *J.J.Verm. & A.Lamb* ined. Type: Sabah, Tenom District, Crocker Range, near Tenom, cult. *Tenom Orchid Centre* T. O. C. 2321 (holo. L).

  HABITAT: hill forest.
  ALTITUDINAL RANGE: 900m.
  DISTRIBUTION: Borneo (SA). Endemic.
  GENERIC SECTION: *Hirtula*.

B.tryssum *J.J.Verm. & A.Lamb* ined. Type: Sabah, Crocker Range, near Kimanis, *Vermeulen* 508 (holo. L).

HABITAT: hill and lower montane forest.
ALTITUDINAL RANGE: 400 – 1600m.
DISTRIBUTION: Borneo (K, SA). Endemic.
GENERIC SECTION: *Hybochilus.*

B.tumidum *J.J.Verm.*, Orch. Borneo 2:101, fig.32, Pl.8D (1991). Type: Sabah, Ulu Padas, *Vermeulen* 639 (holo. L).

HABITAT: very open, dry montane forest on steep sandstone ridges.
ALTITUDINAL RANGE: 1300m.
DISTRIBUTION: Borneo (SA). Endemic.
GENERIC SECTION: *Brachypus.*

B.turgidum *J.J.Verm.*, Orch. Borneo 2:103, fig.33, Pl.8E (1991). Type: Sabah, Gunung Alab, *Lamb* AL 571/86 (holo. K).

HABITAT: rather low or open lower montane forest; moss forest; podsol forest.
ALTITUDINAL RANGE: 1300 – 1900m.
DISTRIBUTION: Borneo (SA). Endemic.
GENERIC SECTION: *Brachypus.*

B.undecifilum *J.J.Sm.* in Bull. Jard. Bot. Buitenzorg, ser. 3, 9:51, t.4,v (1927). Type: Java, Gunung Wiroe, *Bakhuizen van den Brink* s.n., cult. *J.J.Smith* 389 (holo. L).

HABITAT: high lower montane mossy forest; low and open montane podsol forest.
ALTITUDINAL RANGE: 1200 – 1700m.
DISTRIBUTION: Borneo (SA). Also Java.
GENERIC SECTION: *Epicrianthes.*

B.unguiculatum *Rchb.f.* in Linnaea 22:864 (1849). Type: Java, *Schierbrand* s.n. (holo. W).

HABITAT: unknown.
ALTITUDINAL RANGE: unknown.
DISTRIBUTION: Borneo (unspecified). Also Sumatra, Java.
GENERIC SECTION: *Aphanobulbon.*

B.uniflorum *(Blume) Hassk.*, Cat. Bog.:39 (1844). Type: Java, Gede, *Blume* s.n. (holo. L).
Ephippium uniflorum Blume, Bijdr.:309 (1825).
Bulbophyllum hewittii Ridl. in J. Straits Branch Roy. Asiat. Soc. 54:49 (1909). Type: Sarawak, Mt. Poe, *Hewitt* s.n. (holo. SING).

HABITAT: hill and lower montane forest; dense riverine forest; open podsol forest.
ALTITUDINAL RANGE: 600 – 1800m.
DISTRIBUTION· Borneo (SA, SR). Also Peninsular Malaysia, Sumatra, Java, Philippines.
GENERIC SECTION: *Sestochilus.*

B.vaginatum *(Lindl.) Rchb.f.* in Ann. Bot. Syst. 6:261 (1861). Type: Singapore, *Wallich* s.n., cat. no. 1979 (holo. K-LINDL).
Cirrhopetalum vaginatum Lindl., Gen. Sp. Orch. Pl.:59 (1830).

HABITAT: mangrove forest; lowland forest; open secondary forest; mixed dipterocarp forest; forest on limestone; peat swamp forest.
ALTITUDINAL RANGE: sea level – 600m.
DISTRIBUTION: Borneo (B, K, SA, SR). Also Thailand, Peninsular Malaysia, Singapore, Sumatra, Bangka, Java, Maluku.
GENERIC SECTION: *Cirrhopetalum*.

B.vermiculare *Hook.f.*, Fl. Brit. Ind. 6:188 (1890). Type: Singapore, Kranji, *Ridley* s.n. (holo. BM, iso. K).
    B.brookeanum Kraenzl. in Bot. Jahrb. Syst. 34:250 (1905). Type: Sarawak, *Beccari* 306 (holo. FI, iso. K).

HABITAT: lowland mixed dipterocarp forest; mangrove forest.
ALTITUDINAL RANGE: sea level – 500m.
DISTRIBUTION: Borneo (B, K, SA, SR). Also Peninsular Malaysia, Philippines.
GENERIC SECTION: *Aphanobulbon*.

B.vesiculosum *J.J.Sm.* in Bull. Jard. Bot. Buitenzorg, ser. 2, 25:63 (1917). Type: Sumatra, Lampong near Menggala, *Gusdorf* s.n., cult. Bogor 116 (holo. L).

HABITAT: open podsol forest.
ALTITUDINAL RANGE: 400m.
DISTRIBUTION: Borneo (SA). Also Peninsular Malaysia, Sumatra.
GENERIC SECTION: *Epicrianthes*.

B.vinaceum *Ames & C.Schweinf.*, Orch. 6:200 (1920). Type: Sabah, Mt. Kinabalu, Marai Parai Spur, *Clemens* 240 (holo. AMES).

HABITAT: hill and lower montane forest on ultramafic substrate.
ALTITUDINAL RANGE: 600 – 1000m.
DISTRIBUTION: Borneo (K, SA). Endemic.
GENERIC SECTION: *Sestochilus*.

B.viridescens *Ridl.* in J. Linn. Soc., Bot. 38:325 (1908). Type: Peninsular Malaysia, Gunung Tahan, *Robinson & Wray* 5313 (holo. SING).

HABITAT: unknown.
ALTITUDINAL RANGE: unknown.
DISTRIBUTION: Borneo (unspecified). Also Peninsular Malaysia.
GENERIC SECTION: *Aphanobulbon*.

TRIAS Lindl.

Gen. Sp. Orch. Pl.:60 (1830).

Small epiphytic herbs. Rhizome creeping. Pseudobulbs globular, 1-leaved. Leaf oblong to elliptic. Inflorescences 1-flowered, arising from base of pseudobulbs, rarely as long as leaf. Flowers resupinate. Sepals nearly uniform, spreading in a regular triangle, broad, triangular-ovate. Petals much smaller, usually linear. Lip with small auriculate basal lobes or entire, ecallose. Column usually with insignificant stelidia. Anther cap with a prolongation in front, of varying shape. Pollinia 4.

Nine species distributed in India, Burma, Laos, Thailand, Vietnam and Borneo. The centre of distribution lies in Thailand.

T.antheae *J.J.Verm.* & *A.Lamb* ined. Type: Sabah, Nabawan, cult. *Tenom Orchid Centre* T. O. C.2600 (holo L).

HABITAT: podsol forest.
ALTITUDINAL RANGE: 400m.
DISTRIBUTION: Borneo (SA). Endemic.

T.tothastes *(J.J.Verm.) J.J.Wood,* **comb. nov.**
Bulbophyllum tothastes J.J.Verm., Orch. Borneo 2:277, fig.98, Pl.19C (1991). Type: Sabah, Ulu Padas, *Vermeulen* 659 (holo L).

HABITAT: very dry and open forest on steep exposed ridges; also recorded from *Agathis* forest on sandy ridge in peat swamp forest.
ALTITUDINAL RANGE: sea level – 1300m.
DISTRIBUTION: Borneo (K, SA). Endemic.

TRIBE VANDEAE

SUBTRIBE AERIDINAE

GROUP 1. Pollinia 4, more or less equal, globular, free from each other. See p.321 for Group 2.

ADENONCOS Blume

Bijdr. 8:381 (1825).

Small monopodial epiphytes. Stems 10 – 30cm long. Leaves short, narrow, to 1 x 7cm, very fleshy, acute, arranged in 2 rows. Inflorescences short, 1 – 5 flowered. Flowers small, green or yellowish, lasting a long time; sepals and petals free; dorsal sepal 1.5 – 5mm; petals narrower than sepals; lip entire or lobed, concave, somewhat saccate, with a papillose basal keel; column short and erect; stipes linear or clavate, 2 times the diameter of the pollinia; viscidium narrowly elliptic; pollinia 4, ± equal.

About 16 species distributed from Thailand and Indochina to Indonesia and New Guinea.

A.borneensis *Schltr.* in Bull. Herb. Boissier, ser. 2, 6:465 (1906). Type: Kalimantan, Koetai, Sungai Penang, Samarinda, *Schlechter* 13352 (holo. B, destroyed).

HABITAT: mixed dipterocarp forest on sandstone and mudstone.
ALTITUDINAL RANGE: 300 – 400m.
DISTRIBUTION: Borneo (K, SA). Endemic.

A.parviflora *Ridl.* in J. Linn. Soc., Bot. 32:350 (1896). Type: Peninsular Malaysia, Kuala Lumpur, *Kelsall* s.n. (holo. drawing by Ridley, K).
Saccolabium adenoncoides Ridl. in J. Straits Branch Roy. Asiat. Soc., 54:53 (1910). Type: Sarawak, Kuching, *Hewitt* 500 (holo. SING, iso. K).
Adenoncos adenoncoides (Ridl.) Garay in Bot. Mus. Leafl. 23, 4:156 (1972).

HABITAT: podsol forest; mixed dipterocarp forest; high lower montane forest; hill dipterocarp/lower montane transitional forest.
ALTITUDINAL RANGE: 300 – 1300m.
DISTRIBUTION: Borneo (SA, SR). Also Thailand, Peninsular Malaysia, Sumatra.

A.saccata *J.J.Sm.* in Bot. Jahrb. Syst. 48:104 (1912). Type: Kalimantan, Sungei Tarik, *Winkler* 3042 (holo. BO).

HABITAT: swamp forest.
ALTITUDINAL RANGE: sea level.
DISTRIBUTION: Borneo (K). Endemic.

A.sumatrana *J.J.Sm.* in Bull. Dép. Agric. Indes Néerl. 22:44 (1909). Type: Sumatra, Djambi, *Holten* s.n. (holo. BO).

HABITAT: secondary forest; peat swamp forest.
ALTITUDINAL RANGE: lowlands.
DISTRIBUTION: Borneo (SR). Also Thailand, Peninsular Malaysia, Bangka, Sumatra.

A.triloba *Carr* in Gard. Bull. Straits Settlem. 8:123 (1935). Type: Sarawak, Niah, *Synge* S.603 (holo. SING, iso. K).

HABITAT: lowland forest on limestone.
ALTITUDINAL RANGE: below 300m.
DISTRIBUTION: Borneo (SR). Endemic.

A.virens *Blume*, Bijdr.:381 (1825). Type: Java, Pantjar, *Blume* s.n. (holo. BO).

HABITAT: lowland and hill forest.
ALTITUDINAL RANGE: lowlands and hills.
DISTRIBUTION: Borneo (K). Also Peninsular Malaysia, Sumatra, Java.

DORITIS Lindl.

Gen. Sp. Orch. Pl.:178 (1833).

Small to medium sized monopodial terrestrials or lithophytes. Stems erect, usually less than 10cm high. Leaves 5 – 15 x 1.5 – 3cm, obtuse to acute. Inflorescences erect, 20 – 70cm long, laxly many flowered, sometimes branching. Flowers 2 – 3cm across, mauve to deep purple, lip with white lines on disc; sepals and petals spreading; dorsal sepal narrowly elliptic; lateral sepals oblong, falcate, the base connate to the column-foot forming a mentum; petals obovate; lip clawed, geniculate, claw with 2 tooth-like lateral appendages and a callus, blade 3-lobed; column erect, to 0.5cm long, with a foot up to 1cm long, stigmatic cavity large, rostellar projection and anther very elongate; stipes long, slender; viscidium ovate; pollinia 4, equal, globular.

Two species have been described, viz. *D.pulcherrima* Lindl. which is distributed on mainland Asia and Hainan, south to Sumatra and Borneo, and *D.regnieriana* (Rchb.f.)Holttum, probably endemic to Thailand.

D.pulcherrima *Lindl.*, Gen. Sp. Orch. Pl.:178 (1833). Type: Vietnam, Turon, *Finlayson* 521, Wallich Cat. 7348 (holo. K-LINDL).

HABITAT: unknown.
ALTITUDINAL RANGE: unknown.
DISTRIBUTION: Borneo (SA). Also China, India, Burma, Thailand, Laos, Vietnam, Cambodia, Peninsular Malaysia, Sumatra.

MICROSACCUS Blume

Bijdr. 6, tab. 3, & 8:367 (1825).

Small monopodial epiphytes. Stems up to 20cm long, often curved, rooting at the base. Leaves fleshy, laterally flattened, arranged in 2 rows, usually imbricate at the base. Inflorescences very short, arising from leaf axils, few flowered, usually 2 borne opposite one another. Flowers white, somewhat fleshy; sepals free, lateral sepals sometimes decurrent along the spur; lip spurred, immobile, entire, emarginate or slightly 2-lobed; column short, stout, without a foot; stipes linear – clavate, 1.5 – 3 times the diameter of the pollinia; viscidium narrowly elliptic; pollinia 4, equal.

About 12 or 13 species are known, distributed from Burma to Indonesia and the Philippines.

M.ampullaceus *J.J.Sm.* in Bull. Jard. Bot. Buitenzorg, ser. 3, 5:99 (1922). Type: Sumatra, Palembang, *Endert* s.n., cult. Bogor no. 32 (holo. BO).

HABITAT: podsol forest.
ALTITUDINAL RANGE: 1200 – 1300m.
DISTRIBUTION: Borneo (SA). Also Peninsular Malaysia, Sumatra.

M.borneensis *J.J.Sm.* in Bull. Jard. Bot. Buitenzorg, ser. 3, 11:156 (1931). Type: Kalimantan, West Koetai, near Benoewa Toewa, *Endert* 1619 (holo. L, iso. BO).

HABITAT: swamp forest.
ALTITUDINAL RANGE: sea level.
DISTRIBUTION: Borneo (K). Endemic.

M.griffithii *(C.S.P.Parish & Rchb.f.) Seidenf.* in Opera Bot. 95:25 (1988). Type: Burma, Tenasserim, *Parish* 334 (holo. W).
Saccolabium griffithii C.S.P. Parish & Rchb.f. in Trans. Linn. Soc. London, Bot. 30:145 (1874).

HABITAT: lower montane forest.
ALTITUDINAL RANGE: 1500 – 2000m; also recorded from around sea level in Kalimantan.
DISTRIBUTION: Borneo (K, SA). Also Burma, Thailand, Cambodia, Peninsular Malaysia, Singapore, Sumatra, Java, Philippines.

M.longicalcaratus *Ames & C.Schweinf.*, Orch. 6:232(1920). Type: Sabah, Mt. Kinabalu, Kiau, *Clemens* 342 (holo. AMES, iso. BO, K, SING).

HABITAT: lower montane forest.
ALTITUDINAL RANGE: 1500m.
DISTRIBUTION: Borneo (SA). Endemic.

TAENIOPHYLLUM Blume

Bijdr. 6, tab.3, & 8:355 (1825).

Small monopodial epiphytes or lithophytes with short stems bearing many long, spreading, flattened or terete greyish-green roots which are usually adpressed to the substrate. Leaves reduced to tiny brown scales. Inflorescences lateral, with a short peduncle, rachis slowly elongating, bearing flowers in succession, 1 or 2 at a time, bracts alternate, in 2 ranks. Flowers small; sepals and petals free and widely spreading or connate at the base and not opening widely; lip fixed to the base of the column, spurred or saccate, entire or 3-lobed, often with an apical tooth or bristle; column short and stout, foot absent, rostellar projection very variable; viscidium usually narrowly elliptic or ovoid; pollinia 4, equal.

The function of the leaves is carried out by the numerous green roots.

A genus of between 120 and 180 species distributed from tropical Africa (1 sp.) through tropical Asia to Australia and the Pacific islands. The centre of distribution is New Guinea with about 100 species.

T.affine *Schltr.* in Feddes Repert. 9:294 (1911). Type: Sarawak, Kuching, *Schlechter* 15848 (holo. B, destroyed).

HABITAT: unknown.
ALTITUDINAL RANGE: under 300m.
DISTRIBUTION: Borneo (SR). Endemic.

T.borneense *Schltr.* in Bull. Herb. Boissier, ser. 2, 6:466 (1906). Type: Kalimantan, Koetai, Moreara Bangkal, *Schlechter* 13563 (holo. B, destroyed).

HABITAT: unknown.
ALTITUDINAL RANGE: unknown.
DISTRIBUTION: Borneo (K). Endemic.

T.esetiferum *J.J.Sm.* in Bull. Jard. Bot. Buitenzorg, ser. 3, 11:157 (1931). Type: Kalimantan, West Koetai, Gunung Kemoel, *Endert* 3549 (holo. BO).

HABITAT: hill forest; lower montane ridge forest.
ALTITUDINAL RANGE: 900 – 1200m.
DISTRIBUTION: Borneo (K, SA).

T.filiforme *J.J.Sm.* in Bull. Inst. Bot. Buitenzorg, 7:4 (1900). Type: Sulawesi, Bone, Gorontalo, *Smith* s.n. (holo. BO).

HABITAT: unknown.
ALTITUDINAL RANGE: unknown.
DISTRIBUTION: Borneo (unspecified). Also Andaman Islands, Thailand, Peninsular Malaysia, Singapore, Sumatra, Java, Bali, Sulawesi, ? New Guinea.

T.gracillimum *Schltr.* in Bull. Herb. Boissier, ser. 3, 6:466 (1906). Type: Kalimantan, Koetai, Long Dett, *Schlechter* 13556 (holo. B, destroyed).

HABITAT: lower montane forest.
ALTITUDINAL RANGE: 1200m.
DISTRIBUTION: Borneo. Also Thailand, Peninsular Malaysia.

T.hirtum *Blume*, Bijdr.:356 (1825). Type: Java, Gegar Bentang, *Blume* s.n. (holo. ?L).

HABITAT: lower montane forest.
ALTITUDINAL RANGE: 1500m.
DISTRIBUTION: Borneo (SA). Also Java.

T.kapahense *Carr* in Gard. Bull. Straits Settlem. 8:125 (1935). Type: Sarawak, Kapah, *Synge* s.n. (holo. K).

HABITAT: unknown.
ALTITUDINAL RANGE: unknown.
DISTRIBUTION: Borneo (SR). Endemic.

T.obtusum *Blume*, Bijdr.:357 (1825). Type: Java, Burangrang, Krawang, *Blume* s.n. (holo. BO).

HABITAT: mangrove forest; coffee plantations.
ALTITUDINAL RANGE: sea level.
DISTRIBUTION: Borneo (SA, SR). Also Thailand, Cambodia, Peninsular Malaysia, Singapore, Sumatra, Java.

T.proliferum *J.J.Sm.* in Bull. Jard. Bot. Buitenzorg, ser. 2, 26:121 (1918). Type: Java, Priangan, Tjidadap, Tjibeber, *Bakhuizen van den Brink* s.n. (holo. BO).

HABITAT: hill forest; lower montane forest.
ALTITUDINAL RANGE: 900m.
DISTRIBUTION: Borneo (SA). Also Sumatra, Java.

T.rubrum *Ridl.* in J. Linn. Soc., Bot. 32:364 (1896). Type: Peninsular Malaysia, Sungei Ujong, Linsum estate, *Ridley* s.n. (holo. SING).

HABITAT: hill forest.
ALTITUDINAL RANGE: 800m.
DISTRIBUTION: Borneo (SA). Also Peninsular Malaysia.

T.rugulosum *Carr* in Gard. Bull. Straits Settlem. 7:72 (1932). Type: Peninsular Malaysia, Pahang, Sat River, *Ridley* s.n. (holo. SING).

HABITAT: lowland forest.
ALTITUDINAL RANGE: sea level.
DISTRIBUTION: Borneo (K). Also Peninsular Malaysia.

T.stella *Carr* in Gard. Bull. Straits Settlem. 7:69, Pl.8B (1932). Type: Peninsular Malaysia, Pahang, Tembeling, *Carr* s.n. (holo. SING).

HABITAT: hill forest on ultramafic substrate.
ALTITUDINAL RANGE: 1100m.
DISTRIBUTION: Borneo (SA). Also Peninsular Malaysia.

GROUP 2. Pollinia 4, appearing as 2 pollen masses, each completely divided into either rather unequal, or more or less equal, semiglobular free halves. See p.347 for Group 3.

ABDOMINEA J.J.Sm.

Bull. Jard. Bot. Buitenzorg, ser. 2, 14:52 (1914).

Tiny short stemmed monopodial epiphytes. Leaves obovate, 3 – 5cm long. Inflorescences many flowered racemes with persistent bracts. Flowers very small, about 1.2mm across; sepals and petals similar, pale greenish-yellow or cinnamon-orange with black spots; lip sac-shaped with a pointed apex, epichile pure white with a red-brown spot, spur or sac translucent green; column short and stout; anther cap purple; rostellar projection longer and broader than the rest of the column, from a narrow base widening into a cordate, acuminate blade; stipes linear-clavate, acute, more than 3 times the diameter of the pollinia; viscidium very inconspicuous if at all present; pollinia 4, appearing as 2 pollen masses.

A monotypic genus distributed from Thailand to the Philippines.

A.minimiflora *(Hook.f.) J.J.Sm.* in Bull. Jard. Bot. Buitenzorg, ser. 2, 25:98 (1917). Type: Peninsular Malaysia, Perak, *Scortechini* 635 (holo. K).
Saccolabium minimiflorum Hook.f., Fl. Brit. Ind. 6:59 (1890).

HABITAT: lowland and hill forest on limestone; also recorded from ultramafic substrate.

ALTITUDINAL RANGE: 300 – 900m.
DISTRIBUTION: Borneo (SA). Also Thailand, Peninsular Malaysia, Java, Philippines.

ARACHNIS Blume

Bijdr., tab.3, & 8:365 (1825).

Tan, K.W. (1975-1976). Taxonomy of *Arachnis, Armodorum, Esmeralda* & *Dimorphorchis*, Orchidaceae, Part I and Part II. Selbyana 1(1):1-15 (1975) & 1(4):365-373.

Large robust monopodial terrestrials or epiphytes. Stems often scrambling, occasionally branching, sometimes several metres long. Leaves strap-shaped, rigid, to 30cm long (in Asiatic species), apex bilobed. Inflorescences rigid, very often long and branched, few to many flowered. Flowers often large and showy, to about 7cm across, fragrant, green or yellow with maroon blotches or bars; sepals and petals narrowly oblong to linear, spreading; lip much shorter, articulated to the short column-foot by a short strap, 3-lobed, mid-lobe with a raised central ridge or callus, basally saccate or with a short spur; column short and stout; stipes short, broad; viscidium broadly ovate; pollinia 4, appearing as 2 unequal masses.

Thirteen species and one natural hybrid widely distributed from India(Sikkim) in the west, China (Yunnan) in the north, Java and Bali in the south, eastwards to the Solomon Islands. Borneo is the centre of speciation.

A.breviscapa *(J.J.Sm.) J.J.Sm.* in Natuurk. Tijdschr. Ned.-Indië 72:74 (1912). Type: Sarawak, Quop, *Hewitt* s.n. (holo. BO, iso. K).
Arachnanthe breviscapa J.J.Sm. in Bull. Dép. Agric. Indes Néerl., 22:48 (1909).
Vandopsis breviscapa (J.J.Sm.) Schltr. in Feddes Repert. 10:196 (1911).

HABITAT: riverine forest.
ALTITUDINAL RANGE: 300 – 600m.
DISTRIBUTION: Borneo (SA, SR). Endemic.

A.calcarata *Holttum* in Sarawak Mus. J. 5:172 (1949). Type: Sarawak, Gunung Rumput (Mt. Poi), *Anderson* 171 (holo. SING, iso. SAR).

HABITAT: lower montane forest.
ALTITUDINAL RANGE: 700 – 1600m.
DISTRIBUTION: Borneo (SA, SR). Endemic.

A.flosaeris *(L.) Rchb.f.* in Bot. Centralbl. 28:343 (1886). Type: Locality unknown (holo – LINN, IDC Microfiche – Savage, 1945, reference 1062.2).
Epidendrum flosaeris L., Sp. Plant.:952 (1753).
Arachnis flosaeris (L.) Rchb.f. var. *gracilis* Holttum in Malayan Orchid Rev. 2:65 (1935).

HABITAT: lowland and hill forest on limestone and sandstone.
ALTITUDINAL RANGE: 300 – 900m.
DISTRIBUTION: Borneo (K, SA). Also ? India, Thailand, Peninsular Malaysia, Sumatra, Java, Bali, Philippines.

A.grandisepala *J.J.Wood* in Orchid Rev. 89, 1050:113 (1981). Type: Sabah, Crocker Range, Keningau to Kimanis road, *Lamb* 105 in SAN 91558 (holo. K, iso. SAN).

HABITAT: lower montane forest.

ALTITUDINAL RANGE: 900m.
DISTRIBUTION: Borneo (SA). Endemic.

A.hookeriana *(Rchb.f.) Rchb.f.* in Bot. Centralbl. 28:343 (1886). Type: Labuan, *Motley* s.n. (holo. K).
Renanthera hookeriana Rchb.f., Xenia Orch. 2:42, t.113 (1862).

HABITAT: coastal rocks and scrub.
ALTITUDINAL RANGE: sea level.
DISTRIBUTION: Borneo (B, SA, SR). Also Peninsular Malaysia, Singapore, Riau, Archipelago.

A.longisepala *(J.J.Wood) Shim & A.Lamb* in Orchid Digest 46:178 (1982). Type: Sabah, Lahad Datu District, Mt. Tribulation, near Segama, *Cockburn* in SAN 84949 (holo. K, iso. L, SAN).
A.calcarata Holttum subsp. *longisepala* J.J.Wood in Orchid Rev. 89, 1050:113, fig.96 (1981).

HABITAT: hill forest on ultramafic substrate.
ALTITUDINAL RANGE: 600 – 800m.
DISTRIBUTION: Borneo (SA). Endemic.

A.x maingayi *(Hook.f.) Schltr.* in Feddes Repert. 10:197 (1911). Type: Peninsular Malaysia, Malacca, *Maingay* 1645 (holo. K).
Arachnanthe maingayi Hook.f., Fl. Brit. Ind. 6:28 (1890).

HABITAT: unknown.
ALTITUDINAL RANGE: 900m.
DISTRIBUTION: Borneo (SA). Also Peninsular Malaysia, Singapore, Riau, Archipelago.

NOTE: A natural hybrid between *A.flosaeris* var. *gracilis* and *A.hookeriana.*

BOGORIA J.J.Sm.

Orch. Java 6:566 – 567 (1905).

Small monopodial branch epiphytes. Stems very short with thick, flattened, greenish roots. Leaves flat, strap-shaped, apex unequally bilobed. Inflorescences borne from the stem below the leaves, simple, peduncle narrow and terete, rachis deeply sulcate, angular, with 3 – 4 flowers open simultaneously. Flowers small, greenish-yellow, lip white marked crimson; sepals and petals free, spreading; lip 3-lobed, deeply saccate, immobile, side lobes converging, mid-lobe thickened and merely forming the rim of the pouch; column with a foot; anther cap large; pollinia 4, appearing as 2 unequal masses.

Four species, distributed in Java and Borneo, Sumatra, the Philippines and New Guinea.

B.raciborskii *J.J.Sm.*, Orch. Java.: 566 (1905). Type: Java, Kota Batoe near Buitenzorg, *Raciborski* s.n. (holo. BO).

HABITAT: hill dipterocarp/oak/chestnut forest.
ALTITUDINAL RANGE: 900 – 1100m.
DISTRIBUTION: Borneo (SA). Also Java.

CERATOCHILUS Blume

Bijdr. 6: tab.3, & 8:358 (1825).

Very small monopodial epiphytes. Stems to 8cm. Leaves very short, fleshy, laterally flattened. Inflorescences arising laterally from near the stem apex, 1-flowered. Flowers large in proportion to the rest of the plant, transparent white, sometimes fading to scarlet, 1 – 2.2cm across; sepals and petals free, spreading; dorsal sepal 0.5 – 1.3 x 0.5cm; lip immobile, spurred, with very small side lobes which clasp the column, mid-lobe tiny (*C.biglandulosus*) or expanded into a broadly oblong-elliptic, emarginate blade (*C.jiewhoei*), with a hairy pale green callus on either side of the spur entrance (*C.biglandulosus*); column short, foot absent; stipes linear-clavate; pollinia 4, appearing as 2 unequal masses.

One species, *C.biglandulosus* Blume, endemic to Java, the other, *C.jiewhoei*, endemic to Borneo. A further undescribed species is recorded from Borneo (Sabah).

C.jiewhoei *J.J.Wood & Shim* in Orch. Borneo 1:93, fig.15, Pl.3D (1994). Type: Sabah, Mt. Kinabalu, Pinosuk Plateau, *Lamb* AL 58/83 (holo. K).

HABITAT: lower montane oak-laurel forest.
ALTITUDINAL RANGE: 900 – 1800m.
DISTRIBUTION: Borneo (SA). Endemic.

CLEISOCENTRON Brühl

Guide Orchids Sikkim : 136 (1926).

Monopodial epiphytes. Stems long, usually pendent, up to 1m. Leaves strap-shaped to terete, unequally bilobed or acute. Inflorescences axillary, simple or branched, short, few to many-flowered. Flowers pinkish-white to lavender-blue; sepals and petals spreading; lip immobile, 3-lobed, with a gently curving cylindrical spur, inside which is either an upward pointing central protuberance on the back wall (*C.merrillianum*), or a decurved shelf-like back wall callus, front wall callus flap-like, median septum absent; column erect, cylindrical, with a foot decurrent on the back wall of the lip, or free; stipes long and slender; viscidium relatively large; pollinia 4, appearing as 2 pollen masses, each completely divided into free halves.

Five species distributed in the Himalayan region, Burma, Vietnam and Borneo.

C.merrillianum *(Ames) Christenson* in Amer. Orchid Soc. Bull. 61, 3:246 (1992). Type: Sabah, Mt. Kinabalu, Marai Parai Spur, *Clemens* s.n. (holo. AMES, iso. K).
Sarcanthus merrillianus Ames, Orch. 6:230, Pl.97 (1920).
Robiquetia merrilliana (Ames) Lückel, M.Wolff & J.J.Wood in Die Orchidee 40, 3:109 (1989).

HABITAT: lower and upper montane forest; lower montane riverine forest; recorded as epiphytic on *Agathis*; sometimes on ultramafic substrate.
ALTITUDINAL RANGE: 1100 – 3000m.
DISTRIBUTION: Borneo (SA). Endemic.

A second species, material of which we have not seen, has been recorded from Mount Kinabalu in Sabah (Lamb, pers.comm.).

CLEISOMERIA Lindl. ex G. Don

in Loud., Encycl. ed.4, suppl. 2:1447 (1855).

Monopodial epiphytes. Stems short. Leaves strap-shaped, apex unequally bilobed. Inflorescences racemose, or with a small branch, densely many-flowered, pendulous; rachis densely pubescent; floral bracts much longer than ovary, densely pubescent. Flowers densely pubescent; sepals and petals free, spreading; sepals keeled; lip 3-lobed, spurred, with a narrow longitudinal median flange in the spur, otherwise without ornaments, spur entrance nearly closed by fleshy cushions; column short, foot absent; viscidium small; stipes Y-shaped, more than twice length of diameter of pollinia; pollinia 4, appearing as 2 nearly equal halves.

Two species distributed from Burma, Peninsular Malaysia through Thailand and Indochina to Borneo.

C.lanatum *(Lindl.) Lindl. ex G.Don* in Loud., Encycl., ed.4, suppl. 2:1447 (1855). Type: Burma, Moulmein, *Waily* s.n. (holo. K-LINDL).
Cleisostoma lanatum Lindl. in J. Hort. Soc. London 4:164 (1849).

HABITAT: lowland dipterocarp forest.
ALTITUDINAL RANGE: lowlands.
DISTRIBUTION: Borneo (SA). Also Burma, Thailand, Peninsular Malaysia, Laos, Vietnam, Cambodia.

CLEISOSTOMA Blume

Bijdr. 6: tab.3, & 8:362 (1825).

Sarcanthus Lindl., Coll. Bot. t.39B (1826), non Lindl. (1824).

Small to medium sized monopodial epiphytes. Stems up to 60cm long. Leaves strap-shaped or terete, to 30 x 4.5cm, apex usually unequally bilobed. Inflorescences usually branched, many flowered, erect, horizontal or pendulous. Flowers small, subtended by rather small bracts; sepals and petals free, spreading, usually the same size; lip 3-lobed, saccate or spurred, always with a callus on the back wall and often with outgrowths on the front wall closing the entrance, sac or spur usually having a longitudinal internal septum; column short and stout, with a distinct, although sometimes short foot; viscidium ranging from small and subglobose to broad and horseshoe-shaped; pollinia 4, appearing as 2 unequal masses.

Between 80 and 100 species widespread from India throughout Asia to N.E. Australia, New Guinea and the Pacific islands.

C.bicrure *(Ridl.) Garay* in Bot. Mus. Leafl. 23, 4:170 (1972). Type: Sarawak, Matang, *Ridley* s.n. (holo. SING).
Saccolabium bicrure Ridl. in J. Straits Branch Roy. Asiat. Soc. 44:190 (1905).

HABITAT: coffee plantations.
ALTITUDINAL RANGE: unknown.
DISTRIBUTION: Borneo (SR). Endemic.

C.brachystachys *(Ridl.) Garay* in Bot. Mus. Leafl. 23, 4:170 (1972). Type: Sarawak, Tambusan, cult. *Singapore Bot.Gard.* (holo. SING).

Saccolabium brachystachys Ridl. in Journ. Straits Branch Roy. Asiat. Soc. 49:36 (1907).

HABITAT: unknown.
ALTITUDINAL RANGE: unknown.
DISTRIBUTION: Borneo (SR). Endemic.

C.discolor *Lindl.* in Bot. Reg. 31, misc. 59 (1845). Type: India, locality unknown, *Loddiges* s.n. (holo. K-LINDL).

HABITAT: hill forest on ultramafic substrate; riverine forest.
ALTITUDINAL RANGE: 500 – 600m.
DISTRIBUTION: Borneo (K, SA, SR). Also India, Thailand, Cambodia, Peninsular Malaysia, Sumatra, Java.

C.duplicilobum *(J.J.Sm.) Garay* in Bot. Mus. Leafl. 23(4):171 (1972). Type: Java, Dieng, *Kamerling* s.n. (holo. BO).
Sarcanthus duplicilobus J.J.Sm. in Bull. Dép. Agric. Indes Néerl. 13:64 (1907).

HABITAT: unknown.
ALTITUDINAL RANGE: 400 – 1400m.
DISTRIBUTION: Borneo (SA). Also Thailand, Laos, Vietnam, Sumatra, Java, Philippines (Palawan).

C.flexum *(Rchb.f.) Garay* in Bot. Mus. Leafl. 23, 4:171 (1972). Type: Borneo, *Veitch* s.n. (holo. W).
Sarcanthus flexus Rchb.f. in Gard. Chron., 16:492 (1881).

HABITAT: mixed hill dipterocarp forest on sandstone ridges.
ALTITUDINAL RANGE: 300 – 600m.
DISTRIBUTION: Borneo (K, SA). Endemic.

C.halophilum *(Ridl.) Garay* in Bot. Mus. Leafl. 23(4):171 (1972). Types: Singapore, Kranji, *Ridley* s.n.; Sungai Morai, *Ridley* s.n.; Sungai Tengeh, *Ridley* s.n.; Pulau Tekong, *Ridley* s.n. (all syn. SING); Peninsular Malaysia, Johore, Batu Pahat, *Ridley* s.n.; Tana Runto, *Ridley* s.n. (syn. SING).
Sarcanthus halophilus Ridl. in J. Linn. Soc., Bot. 32:367(1876).

HABITAT: mangrove forest.
ALTITUDINAL RANGE: sea level.
DISTRIBUTION: Borneo (SA, SR). Also Peninsular Malaysia.

C.inflatum *(Rolfe) Garay* in Bot. Mus. Leafl. 23(4):172 (1972). Type: Vietnam, Annam, *Micholitz* s.n. (holo. K).
Sarcanthus inflatus Rolfe in Kew Bull.:115 (1906).
Saccolabium laxum Ridl. in J. Straits Branch Roy. Asiat. Soc. 50:140 (1908). Type: Sarawak, Matang, *Hewitt* s.n. (holo. K).
Sarcanthus ridleyi J.J.Sm. in Natuurk. Tijdschr. Ned.-Indië 72:92 (1912).

HABITAT: lower montane forest.
ALTITUDINAL RANGE: 500 – 600m.
DISTRIBUTION: Borneo (SA, SR). Also Vietnam.

C.koeteiense *(Schltr.) Garay* in Bot. Mus. Leafl. 23(4):172 (1972). Type: Kalimantan, Koetei, Samarinda, *Schlechter* 13336 (holo. B,destroyed).
Saccolabium koeteiense Schltr. in Feddes Repert. 3:280 (1907).
Sarcanthus koeteiensis (Schltr.) J.J.Sm. in Natuurk. Tijdschr. Ned.-Indië 72:88 (1912).

HABITAT: unknown.
ALTITUDINAL RANGE: sea level.
DISTRIBUTION: Borneo (K). Endemic.

C.nieuwenhuisii *(J.J.Sm.) Garay* in Bot. Mus. Leafl. 23(4):172 (1972). Type: Kalimantan, Bukit Kasian, *Nieuwenhuis* s.n. (holo. BO).
Sarcanthus nieuwenhuisii J.J.Sm. in Icon. Bogor. 3:57, t.222 (1906).

HABITAT: unknown.
ALTITUDINAL RANGE: unknown.
DISTRIBUTION: Borneo (K). Endemic.

C.pinifolium *(Ridl.) Garay* in Bot. Mus. Leafl. 23(4):173 (1972). Type: Sarawak, Bidi, *Brookes* s.n., comm. *Hewitt* (holo. SING).
Saccolabium pinifolium Ridl. in J. Straits Branch Roy. Asiat. Soc. 50:141 (1908).

HABITAT: unknown.
ALTITUDINAL RANGE: lowlands.
DISTRIBUTION: Borneo (SR). Endemic.

C.ridleyi *Garay* in Bot. Mus. Leafl. 23(4):174 (1972). Type: Sarawak, Bidi, *Ridley* s.n. (holo. K).
Saccolabium ramosum Ridl. in J. Straits Branch Roy. Asiat. Soc. 44:190 (1905), non *Cleisostoma ramosum* (Lindl.)Hook.f.

HABITAT: lowland and hill forest.
ALTITUDINAL RANGE: lowlands – 600m.
DISTRIBUTION: Borneo (SA, SR). Endemic.

C.sagittatum *Blume*, Bijdr.:363, fig.27 (1825). Type: Java, Pantjar, *Blume* s.n. (holo. BO).

HABITAT: lower montane forest on ultramafic substrate, particularly along rivers.
ALTITUDINAL RANGE: 600 – 800m.
DISTRIBUTION: Borneo (SA). Also Sumatra, Java.

C.samarindae *(Schltr.) Garay* in Bot. Mus. Leafl. 23(4):174 (1972). Type: Kalimantan, Koetei, Samarinda, *Schlechter* 13331 (holo. B, destroyed).
Saccolabium samarindae Schltr. in Feddes Repert. 3:280 (1907).
Sarcanthus samarindae (Schltr.) J.J.Sm. in Natuurk. Tijdschr. Ned.-Indië 72:93 (1912).

HABITAT: unknown.
ALTITUDINAL RANGE: lowlands.
DISTRIBUTION: Borneo (K). Endemic.

C.scortechinii *(Hook.f.) Garay* in Bot. Mus. Leafl. 23(4):174 (1972). Types: Peninsular Malaysia, Perak, Gunung Arang Para, *Scortechini* 585B (syn. K); Batu Togoh, *Wray* 2179 (syn. K).
Sarcanthus scortechinii Hook.f., Fl. Brit. Ind. 6:68 (1890).

HABITAT: lowland forest; peat swamp forest; podsol forest.
ALTITUDINAL RANGE: sea level – 500m.
DISTRIBUTION: Borneo (K, SA, SR). Also Thailand, Peninsular Malaysia, Singapore, Sumatra, Java.

C.striatum *(Rchb.f.) Garay* in Bot. Mus. Leafl. 23(4):175 (1972). Type: India, Sikkim, locality unknown, cult. *Mackay* (holo. W).
Echioglossum striatum Rchb.f. in Gard. Chron. 12:390 (1879).

HABITAT: beside streams in hill forest.
ALTITUDINAL RANGE: 400 – 500m.
DISTRIBUTION: Borneo (SA). Also India (Assam, Sikkim), China, Vietnam, Peninsular Malaysia.

C.strongyloides *(Ridl.) Garay* in Bot. Mus. Leafl. 23(4):175 (1972). Type: Sarawak, Kuching, *Lewis* s.n., Comm. *Hewitt* (holo. SING, iso. K).
Saccolabium strongyloides Ridl. in J. Straits Branch Roy. Asiat. Soc. 50:141 (1908).
Sarcanthus strongyloides (Ridl.) J.J.Sm. in Natuurk. Tijdschr. Ned.-Indië 72:94 (1912).

HABITAT: unknown.
ALTITUDINAL RANGE: lowlands.
DISTRIBUTION: Borneo (SR). Endemic.

C.suaveolens *Blume,* Bijdr.:363 (1825). Type: Java, Kambangan Island, *Blume* s.n. (holo. BO).
Sarcanthus robustum O'Brien in Gard. Chron., ser. 3, 55:21 (1914), non *S.robustus* Schltr. Type: Borneo, ex *Rothschild* (holo. not located).
Cleisostoma borneense J.J.Wood in Jermy, Studies Fl. G. Mulu Nat. Park:17 (1984), **syn. nov.**

HABITAT: hill forest on limestone.
ALTITUDINAL RANGE: 200m.
DISTRIBUTION: Borneo (SA, SR). Also Sumatra, Java, Bali.

C.subulatum *Blume,* Bijdr.:363 (1825). Type: Java, Tjilele, Gunung Parang, Tjanjor, *Blume* s.n. (holo. BO).
Sarcanthus subulatum (Blume) Rchb.f. in Bonplandia 5:41 (1857).
Saccolabium secundum (Griff.) Ridl., Mat. Fl. Mal. Pen. 1:168 (1907). Type: India, Assam, Suddyah, *Griffith* s.n. (holo. Griffith's drawing).

HABITAT: unknown.
ALTITUDINAL RANGE: lowlands.
DISTRIBUTION: Borneo (K, SR). Also India, Burma, Thailand, Cambodia, Peninsular Malaysia, Riau Archipelago, Sumatra, Java, Buru, Sulawesi, Maluku, Philippines.

C.tenuirachis *(J.J.Sm.) Garay* in Bot. Mus. Leafl. 23(4):175 (1972). Type: Kalimantan, Gunung Damoes, *Hallier* 409 (holo. BO, iso. K).
Sarcanthus tenuirachis J.J.Sm. in Bull. Dép. Agric. Indes Néerl. 22:50 (1909).

HABITAT: hill forest on limestone.
ALTITUDINAL RANGE: 600m.
DISTRIBUTION: Borneo (K, SR). Endemic.

C.teretifolium *Teijsm. & Binn.* in Natuurk. Tijdschr. Ned.-Indië 27:20 (1864). Type: Sumatra, Palembang, *Teysmann* s.n. (holo. BO).

HABITAT: hill forest.
ALTITUDINAL RANGE: 300m.
DISTRIBUTION: Borneo (K, SR). Also Thailand, Peninsular Malaysia, Sumatra.

C.uraiense *(Hayata) Garay & H.R.Sweet*, Orch. S.Ryukyu Is.:156 (1974). Type: Taiwan, Urai, *Hayata* s.n. (holo. not located).
Sarcanthus uraiensis Hayata, Icon. Pl. Formos. 8:130 (1919).
S.micranthus Ames, Orch. 5:248 (1915), non *Cleisostoma micranthum* (Lindl.) King & Pantl. Type: Philippines, Leyte, Jaro, *Wenzel* 181 (holo. AMES, iso. K).

HABITAT: unknown.
ALTITUDINAL RANGE: unknown.
DISTRIBUTION: Borneo (SA). *C.uraiense* is native to Taiwan and the Philippines, and was probably introduced to the Ryukyu Islands.

NOTE: Specimen *Castro & Melegrito* 1704 (K) matches *C.uraiense* in habit but lack of flowers does not allow further determination.

C.williamsonii *(Rchb.f.) Garay* in Bot. Mus. Leafl. 23(4):176 (1972). Type: India, Assam, *Williamson* s.n. (holo. W).
Sarcanthus williamsonii Rchb.f. in Hamburger Garten-Blumenzeitung 21:353 (1865).

HABITAT: coastal *Eugenia* forest.
ALTITUDINAL RANGE: sea level.
DISTRIBUTION: Borneo (SA). Also Bhutan, India (Assam), China, Thailand, Vietnam, Peninsular Malaysia, Sumatra, Java.

CORDIGLOTTIS J.J.Sm.

Bull. Jard. Bot. Buitenzorg, ser. 3, 5:95-96 (1922).

Cheirorchis Carr in Gard. Bull. Straits Settlem. 7:46 (1932).

Small monopodial epiphytes. Stems very short, usually pendulous. Leaves terete or laterally compressed. Inflorescences usually with a peduncle and a much shorter rachis, few to many flowered. Flowers small, lasting one day; dorsal sepal to 1.4 x 0.5mm; lip mobile, slightly saccate, mid-lobe somewhat fleshy, powdery or hairy; column short, with a distinct foot; stipes broad and spathulate; viscidium large; pollinia 4, appearing as 2 unequal masses; fruits long and slender as in *Thrixspermum*.

Seven species distributed from S. Thailand to Sumatra and Borneo, absent from Java. The centre of speciation is Peninsular Malaysia.

C.major *(Carr) Garay* in Bot. Mus. Leafl. 23(4):176 (1972). Type: Peninsular Malaysia, Pahang, Krambit, *Carr* s.n. (holo. SING).
Cheirorchis major Carr in Gard. Bull. Straits Settlem. 7:43 (1932).

HABITAT: lowland mixed dipterocarp forest.
ALTITUDINAL RANGE: 200m.
DISTRIBUTION: Borneo (SA). Also Peninsular Malaysia.

C.multicolor *(Ridl.) Garay* in Bot. Mus. Leafl. 23(4):176 (1972). Type: Sarawak, Kuching, *Hewitt* s.n. (holo. SING).
Dendrocolla multicolor Ridl. in J. Straits Branch Roy. Asiat. Soc. 54:54 (1909).
Thrixspermum multicolor (Ridl.) Ames in Merr., Bibl. Enum. Born. Pl.:195 (1921).

HABITAT: unknown.
ALTITUDINAL RANGE: lowlands.
DISTRIBUTION: Borneo (SR). Endemic.

C.pulverulenta *(Carr) Garay* in Bot. Mus. Leafl. 23(4):176 (1972). Type: Peninsular Malaysia, Pahang, Tembeling, *Carr* 310 (holo. SING, iso. K).
Cheirorchis pulverulenta Carr in Gard. Bull. Straits Settlem. 7:45 (1932).

HABITAT: podsol forest.
ALTITUDINAL RANGE: 400 – 500m.
DISTRIBUTION: Borneo (SA). Also Peninsular Malaysia.

C.westenenkii *J.J.Sm.* in Bull. Jard. Bot. Buitenzorg, ser. 3, 5:95 (1922). Type: Sumatra, Benkoelen, Lebong, Rimbo Pengadang, Westenenk, *Ajoeb* 223 (holo. BO).

HABITAT: kerangas forest on ridges above rivers.
ALTITUDINAL RANGE: 900m.
DISTRIBUTION: Borneo (K). Also Sumatra.

DIMORPHORCHIS Rolfe

Orchid Rev. 27:149 (1919).

Tan, K.W. (1975-1976). Taxonomy of *Arachnis, Armodorum, Esmeralda* & *Dimorphorchis*, Orchidaceae, Part I and Part II. Selbyana 1(1):1-15 (1975) & 1(4):365-373 (1976).

Large monopodial epiphytes. Stems usually pendent, leafy, up to 200 cm long. Leaves strap-shaped, apex unequally bilobed, arcuate, 30 – 70 x 1.7 – 6 cm. Inflorescences pendent, 28 – 300 cm long, laxly few to many flowered, peduncle and rachis flexuous, tomentose, bracts 0.7 – 3 cm long. Flowers large, showy, dimorphic, resupinate, the basal 2 always strongly scented, 5 – 6.5 cm across, sepals and petals yellow or orange-yellow, spotted purple or with a few small red basal spots or a few red spots only on the lateral sepals, the apical flowers unscented, 5 – 6.5 cm across, sepals and petals yellow or white with either large irregular purple blotches or small red or purple spots; sepals free, spreading, usually acute, margin often undulate in apical flowers, outer surface stellate-pubescent, 2.4 – 3.5 x 1.3 – 2 cm; lip mobile, 3-lobed, very fleshy, sometimes L-shaped in side view, 0.8 – 1.3 cm long, side lobes erect, margins often incurved, to 6 mm long, mid-lobe at an obtuse or right angle to the base of the lip, sometimes bilaterally compressed, with a keel-like callus; column 0.5 – 1.2 cm long, with a short foot; anther pubescent; stipes broad; pollinia 4, appearing as 2 pollen masses.

Two species endemic to Borneo.

D.lowii *(Lindl.) Rolfe* in Orchid Rev. 27:149 (1919). Type: Sarawak, *Low* s.n. (holo. K-LINDL).
Vanda lowii Lindl. in Gard. Chron. 1:239 (1847).
Renanthera lowii (Lindl.) Rchb.f., Xenia Orch. 1:87 (1858).
Arachnanthe lowii (Lindl.) Benth. & Hook.f., Gen. Plant. 3:573 (1883).
Arachnis lowii (Lindl.) Rchb.f. in Beih. Bot. Centralbl. 28:344 (1886).
Vandopsis lowii (Lindl.) Schltr. in Feddes Repert. 10:196 (1911).

var. **lowii**

HABITAT: gulley and riverine forest, often overhanging water; swamp forest.
ALTITUDINAL RANGE: sea level – 1300m.
DISTRIBUTION: Borneo (K, SA, SR). Endemic.

var. **rohaniana** *(Rchb.f.) K.W.Tan* in Taxon: 116 (1974). Type: Borneo, *Hupe* s.n. (holo. W).

Renanthera rohaniana Rchb.f., Xenia Orch. 1:89 (1855).

Arachnis lowii (Lindl.) Rchb.f. var. *rohaniana* (Rchb.f.) J.J.Sm. in Natuurk. Tijdschr. Ned.-Indië 72:73 (1913).

HABITAT: unknown.
ALTITUDINAL RANGE: unknown.
DISTRIBUTION: Borneo (K, SR). Endemic.

D.rossii *Fowlie* in Orchid Digest 53:14 (1989). Type: Sabah, Lohan River, cult. *Los Angeles Arboretum, Fowlie & Ross* 83P912 (holo. LA).

var. **rossii**

HABITAT: riverine and hill forest on ultramafic substrate.
ALTITUDINAL RANGE: 500 – 1200m.
DISTRIBUTION: Borneo (SA). Endemic.

var. **graciliscapa** *A.Lamb & Shim*, Orch. Borneo 1:143, fig.34a, Pl.8C & D (1994). Type: Sabah, Ulu Moyog, *Hepburn* in *Lamb* AL 598/86 (holo. K).

HABITAT: lower montane oak-chestnut forest.
ALTITUDINAL RANGE: 300m.
DISTRIBUTION: Borneo (SA). Endemic.

var. **tenomensis** *A.Lamb*, Orch. Borneo 1:145, fig.34b, Pl.8E & F (1994). Type: Sabah, Ulu Mentailung, *Lamb* T26 (holo. K).

HABITAT: hill and lower montane forest on sandstone ridges.
ALTITUDINAL RANGE: 800 – 1000m.
DISTRIBUTION: Bornco (SA). Endemic.

GASTROCHILUS D.Don

Prodr. Fl. Nepal :32 (1825).

Christenson in Amer. Orchid Soc. Bull. 54(9):1111-1116 (1985) and Indian Orch. Journ. 2(1):19-29 (1987).

Small to medium-sized monopodial epiphytes. Stems usually short, up to 30 cm in a few species. Leaves narrowly elliptic or strap-shaped, apex unequally bilobed, acute or, rarely, with 3 setae, rather leathery. Inflorescences short and densely many flowered, ± umbellate, usually sessile, a few species having a short peduncle. Flowers with greenish yellow to yellow sepals and petals often spotted red and a white lip, the hypochile often spotted red and the epichile with a central yellow patch, spotted red; sepals and petals narrowly obovate; lip adnate to the column, immobile, divided into a semi-globose saccate hypochile and a fan-shaped, often broadly triangular epichile, epichile often hairy or papillose, margin entire to fimbriate; column short and stout, foot absent, rostellar projection oblong; stipes strap-shaped, longer than twice the diameter of the pollinia; pollinia 2, porate, or 4, unequal (*G.patinatus*).

Around 38 species distributed from Sri Lanka and India through East Asia to Japan, south to Indonesia.

G.patinatus *(Ridl.) Schltr.* in Feddes Repert. 12:314 (1913). Type: Peninsular Malaysia, Pahang, Kota Glanggi, *Ridley* s.n. (holo. SING).

Saccolabium patinatum Ridl.in J. Straits Branch Roy. Asiat. Soc. 39:84 (1903).

HABITAT: lowland mixed dipterocarp forest on limestone, sandstone and shales, often beside rivers.
ALTITUDINAL RANGE: lowlands – 300m.
DISTRIBUTION: Borneo (SA). Also Peninsular Malaysia, Sumatra.

KINGIDIUM P.F. Hunt

Kew Bull. 24:97 (1970).

Monopodial epiphytes resembling a small *Phalaenopsis*. Stems very short. Leaves oblong to oblong-elliptic, apex not unequally bilobed. Inflorescences with a long peduncle, rachis much shorter, not flattened, the whole as long as or longer than the leaves. Flowers small, about 1.25 cm across, white and pinkish-violet; sepals and petals free, spreading; lateral sepals attached to the sides of the column-foot, not to the lip; lip 3-lobed, saccate, or shallowly spurred; column-foot short; pollinia 4, appearing as 2 unequal masses.

Five species distributed from Sri Lanka to N.E. India and the Himalayas, through Burma, Thailand and Indonesia to S. China, south to Peninsular Malaysia, Java and the Philippines.

K.deliciosum *(Rchb.f.) H.R.Sweet* in Amer. Orchid Soc. Bull. 39:1095 (1970). Type: Origin unknown (holo. W).
Phalaenopsis deliciosa Rchb.f. in Bonplandia 2:93 (1854).
P.bella Teijsm. & Binn. in Natuurk. Tijdschr. Ned. -Indië 24:321 (1862). Type: Java, Salal, *Teijsmann* s.n. (syntype: BO); Sumatra, Radja Basa, Lompons, *Teijsmann* s.n. (syntype: BO).
Kingidium deliciosum (Rchb.f.) H.R. Sweet var. *bellum* (Teijsm. & Binn.)) O. Gruss & Röllke in Die Orchidee 44(5): 225 (1993), **syn.nov**.

HABITAT: riverine forest.
ALTITUDINAL RANGE: lowlands – 300m.
DISTRIBUTION: Borneo (K, SA). Also China, India, Sri Lanka, Nepal, Bhutan, Burma, Thailand, Indochina, Peninsular Malaysia, Sumatra, Java, Sulawesi, Maluku, Philippines.

MICROPERA Lindl.

Bot. Reg. 18:sub t.1522 (1832).

Camarotis Lindl., Gen. Sp. Orch. Pl.:219 (1833).

Small climbing monopodial epiphytes. Stems with internodes to about 2 cm, bearing many support roots at intervals. Leaves linear, to 17 x 2 cm but usually much less. Inflorescences simple, to about 15 cm long. Flowers small, non-resupinate, fleshy, usually yellow with purple markings, or pink; lateral sepals connate for a short distance at the base; lip sac-like, with a back wall callus, a longitudinal septum and usually a 2-lobed front wall callus, apex 3-lobed, mid-lobe much smaller than the sac or spur; column variable, usually without a foot, rostellar projection elongated, slender, usually twisted to one side, recalling *Ludisia*; stipes linear; viscidium small, ovate; pollinia 4, appearing as 2 unequal masses.

About 14 – 15 species distributed from India (Sikkim) and Indochina east to Australia and the Solomon Islands.

M.callosa *(Blume) Garay* in Bot. Mus. Leafl. 23(4):186 (1972). Type: Java, Pantjar, *Blume* s.n. (holo. BO).
   Cleisostoma callosum Blume, Bijdr.:364 (1825).
   Camarotis callosa (Blume) J.J.Sm. in Natuurk. Tijdschr. Ned.-Indië 72:97 (1912).

HABITAT: mixed lowland forest on ultramafic substrate; hill forest.
ALTITUDINAL RANGE: sea level – 300m.
DISTRIBUTION: Borneo (K, SA, SR). Also Java.

M.fuscolutea *(Lindl.) Garay* in Bot. Mus. Leafl. 23(4):186 (1972). Type: Borneo, imported by *Lowe*, comm. *Cox* (holo. K-LINDL).
   Sarcochilus fuscoluteus Lindl. in Bot. Reg. 33, sub t.18 (1847).
   Camarotis latisaccata J.J.Sm. in Bull. Jard. Bot. Buitenzorg, ser. 3, 9:187 (1927). Type: Kalimantan, Kapoeas, *Teysmann* 8445 (holo. BO).

HABITAT: citrus trees; hill forest on ultramafic substrate; riverine forest.
ALTITUDINAL RANGE: sea level – 600m.
DISTRIBUTION: Borneo (SA, SR). Also Peninsular Malaysia, Singapore.

M.pallida *(Roxb.) Lindl.* in Bot. Reg. 18, sub t.1522 (1832). Type: Bangladesh, Chittagong, *Roxburgh* drawing 2349 (holo. K).
   Aerides pallida Roxb., Fl. Ind. 3:475 (1832).
   Camarotis apiculata Rchb.f. in Bonplandia 5:39 (1857). Type: Java, Paradana, *Zollinger* 1359 (holo. W).

HABITAT: swamp forest.
ALTITUDINAL RANGE: sea level.
DISTRIBUTION: Borneo (K). Also Bangladesh, India, Burma, Thailand, Indochina, Peninsular Malaysia, Sumatra, Java.

ORNITHOCHILUS (Lindl.) Wall. ex Benth.

J. Linn. Soc., Bot. 18:335 (1881).

Small monopodial epiphytes. Stems to 40 cm long. Leaves fleshy, to 12 x 4 cm. Inflorescences branching, many flowered, as long as or longer than the leaves. Flowers to c.1.3 cm across; petals narrower than sepals; lip with a cylindrical spur extending from the distal part of the hypochile, epichile 3-lobed, apex ciliate or crenulate; column to 4 mm long, foot absent, rostellar projection elongated, fleshy, obtuse, anther cap truncate; stipes obovate – cuneiform, more than twice the diameter of the pollinia; viscidium obtriangular; pollinia 4, appearing as 2 pollen masses.

Three species distributed from India and Nepal to S. China, south to Indonesia.

O.difformis *(Wall. ex Lindl.) Schltr.* in Feddes Repert., Beih. 4:277 (1919). Type: Nepal, *Wallich* drawing (holo. K).
   Aerides difforme Wall. ex Lindl., Gen. Sp. Orch. Pl.:242 (1833).

var. **difformis**

HABITAT: hill and lower montane forest on sandstone ridges.

ALTITUDINAL RANGE: 900 – 1500m.
DISTRIBUTION: Borneo (K, SA). Also India, China, Burma, Thailand, Indochina, Peninsular Malaysia, Sumatra.

var. **kinabaluensis** *J.J.Wood, A.Lamb & Shim* in Orch. Borneo 1:191, fig.53a, Pl.12B (1994). Type: Sabah, Mt. Kinabalu, Pinosuk Plateau, *Jukian & Lamb* in *Lamb* AL 4/82 (holo. K).

HABITAT: lower montane forest.
ALTITUDINAL RANGE: 1300 – 1500m.
DISTRIBUTION: Borneo (SA). Endemic.

POMATOCALPA Breda

Gen. Sp. Orchid. Asclep. t.15 (1827).

Small to medium-sized monopodial epiphytes or terrestrials. Stems up to 40 cm. Leaves strap-shaped, to 20 x 5 cm. Inflorescences branched, often longer than the rest of the plant, on a long peduncle, many flowered. Flowers small, non-resupinate, usually yellow marked red; sepals and petals free, spreading; lip fleshy, 3-lobed, spurred or saccate, with a tongue-like, often bifurcate callus projecting from the back wall; column short and stout, rostellar projection hammer-shaped; pollinia 4, appearing as 2 unequal masses.

Some 35 – 40 species distributed from India throughout tropical Asia to New Guinea, Australia and the Pacific islands.

P.fusca *(Lindl.) J.J.Sm.* in Natuurk. Tijdschr. Ned.-Indië 72:104 (1912). Type: "Received from East India Company in 1846". (holo. K-LINDL).
Cleisostoma fuscum Lindl. in J. Hort. Soc. London 5:80 (1850).

HABITAT: lowland forest.
ALTITUDINAL RANGE: lowlands.
DISTRIBUTION: Borneo (K, SA, SR). Also Peninsular Malaysia.

P.kunstleri *(Hook.f)* *J.J.Sm.* in Natuurk. Tijdschr. Ned.-Indië 72:104 (1912). Type: Peninsular Malaysia, Perak, drawing by *Kunstler* (holo. CAL, copy in K).
Cleisostoma kunstleri Hook.f., Icon. Plant. 24:t. 2335 (1894).
Saccolabium pubescens Ridl., J. Linn. Soc., Bot. 31:295 (1896). Type: Sarawak, Kuching, *Haviland* 2333 (holo. SING, iso. K.)

HABITAT: podsol forest.
ALTITUDINAL RANGE: sea level – 400m.
DISTRIBUTION : Borneo (B, K, SA, SR). Also Thailand, Peninsular Malaysia, Sumatra, Mentawai Islands, Java, Philippines.

P.latifolia *(Lindl.) J.J.Sm.* in Natuurk. Tijdschr. Ned.-Indië 72:105 (1912). Type: Singapore, *Loddiges* s.n. (holo. K-LINDL).
Cleisostoma latifolium Lindl. in Bot. Reg. 26:60, misc. 127 (1840).
Pomatocalpa hortensis (Ridl.) J.J.Sm. in Natuuk. Tijdschr. Ned.-Indië 72:104 (1912).

HABITAT: coral rocks; strand vegetation; secondary forest; mixed lowland dipterocarp forest.
ALTITUDINAL RANGE: sea level – 300m.

DISTRIBUTION: Borneo (K, SA, SR). Also Thailand, Peninsular Malaysia, Singapore, Sumatra, Mentawai Islands, Java, Bali and possibly Sulawesi, Philippines.

P.sphaerophorum *(Schltr.) J.J.Sm.* in Natuurk. Tijdschr. Ned. -Indië 72:106 (1913). Type: Kalimantan, Koetei, Samarinda, *Schlechter* 13333 (holo. B, destroyed).
Saccolabium sphaerophorum Schltr. in Bull. Herb. Boissier, ser. 2, 6:472 (1906).

HABITAT: unknown.
ALTITUDINAL RANGE: lowlands.
DISTRIBUTION: Borneo (K). Endemic.

P.spicata *Breda* in Kuhl & van Hasselt, Gen. Sp. Orch. t.15 (1827). Type: Java, locality unknown, *Kuhl & van Hasselt* s.n. (holo. ?L).
Cleisostoma crassum Ridl. in J. Linn. Soc., Bot. 31:295 (1896). Type: Borneo, locality unknown, *native collector* s.n., cult. Singapore (holo. SING).

HABITAT: lowland and hill dipterocarp forest.
ALTITUDINAL RANGE: sea level – 400m.
DISTRIBUTION: Borneo (K, SA, SR). Also India, Andaman Islands, Burma, Thailand, Laos, Vietnam, China (Hainan), Peninsular Malaysia, Sumatra, Java, Bali, Kangean Islands, Philippines.

P.truncata *(J.J.Sm.) J.J.Sm.* in Natuurk. Tijdschr. Ned. -Indië 72:107 (1913). Types: Kalimantan, Sungai Landak, *Teysmann* s.n.; Samak, *Teysmann* s.n.; Pontianak, cult. Bogor, Moeara Tewe, cult. *Bogor* (syn. BO).
Cleisostoma truncatum J.J.Sm. in Bull. Jard. Bot. Buitenzorg, ser. 2, 3:68 (1912).

HABITAT: unknown.
ALTITUDINAL RANGE: lowlands.
DISTRIBUTION: Borneo (K). Endemic.

RENANTHERA Lour.

Fl. Cochinch.:516, 521 (1790).

Robust monopodial epiphytes or, rarely, terrestrials with long, climbing stems, often up to several metres. Leaves oblong, leathery, apex bilobed, up to 20cm long. Inflorescences branched, up to 80cm long, many flowered. Flowers resupinate, usually red, sometimes orange or yellow; lateral sepals usually broader than the dorsal, edges undulate; lip much smaller than the sepals and petals, immobile, 3-lobed, saccate or spurred, mid-lobe small, often recurved, with basal calli, side lobes erect; column short and stout, to 7mm long, rostellar projection short; viscidium transverse, fleshy; pollinia 4, appearing as 2 pollen masses.

About 14 species distributed from E. India (Assam) through China to the Philippines and south to Malaysia, Indonesia, New Guinea and the Solomon Islands.

R.angustifolia *Hook.f.*, Fl. Brit. Ind. 6:49 (1891). Type: Peninsular Malaysia, Perak, Gunung Bata Pateh, *Wray* 439 (holo. K).
Aerides matutina Blume, Bijdr.:366 (1825), non Willd. (1805).
Renanthera matutina (Blume) Lindl., Gen. Sp. Orch. Pl.:218 (1833).

HABITAT: podsol forest; lowland and hill dipterocarp forest.
ALTITUDINAL RANGE: 100 – 600m.

DISTRIBUTION: Borneo (SA). Also Peninsular Malaysia, Sumatra, Java, Philippines.

R.bella *J.J.Wood* in Orchid Rev. 89:116, fig.97-99 (1981). Type: Sabah, Bukit Hempuen (Ampuan), *Lamb* in SAN 89640 (holo. K, iso. SAN).

HABITAT: hill forest on ultramafic substrate.
ALTITUDINAL RANGE: 800 – 1100m.
DISTRIBUTION: Borneo (SA). Endemic.

R.elongata *(Blume) Lindl.*, Gen. Sp. Orch. Pl.:218 (1833). Type: Java, near Kuripan, *Blume* 1052 (holo. L).
Aerides elongatum Blume, Bijdr.:366 (1825).

HABITAT: rocks near the sea; riverine lowland and hill dipterocarp forest; swamp forest.
ALTITUDINAL RANGE: sea level – 1000m.
DISTRIBUTION: Borneo (K, SA, SR). Also Thailand (Terutao Island), Peninsular Malaysia, Singapore, Sumatra, Mentawai Islands, Java, Philippines.

R.isosepala *Holttum* in Seidenf. & Smitinand, Orch. Thailand 4:825 (1964). Type: Thailand, Prachuap, *Sagarik* 201 (holo. K).

HABITAT: swamp forest.
ALTITUDINAL RANGE: sea level.
DISTRIBUTION: Borneo (SA). Also Thailand.

RENANTHERELLA Ridl.

J. Linn. Soc., Bot. 32:354 (1896).

Similar to *Renanthera*, but plants smaller. Leaves semiterete, acute. Inflorescences usually unbranched, up to 10cm long. Flowers resupinate; lateral sepals and mid-lobe of lip recurved; column slender, curved forwards, about two thirds the length of the dorsal sepal; pollinia 4, appearing as 2 pollen masses.

A monotypic genus comprising *R.histrionica* (Rchb.f.) Ridl. subsp. *histrionica* from S. Thailand and Peninsular Malaysia, and subsp. *auyongii* (Christenson) Senghas from Borneo.

R.histrionica *(Rchb.f.) Ridl.* in J. Linn. Soc., Bot. 32:355 (1896).

subsp. **auyongii** *(Christenson) Senghas* in Schlechter, Die Orchideen 1(21):1327 (1988). Type: Sarawak, near Mt.Santubong, *Au Yong Nan Yip* s.n., cult. *J. Levy* 1277 (holo. SEL).
Renanthera auyongii Christenson in Orchid Digest 50:169 (1986).

HABITAT: trees near the sea.
ALTITUDINAL RANGE: sea level.
DISTRIBUTION: Borneo (SR). Endemic.

SARCOGLYPHIS Garay

Bot. Mus. Leafl. 23(4):200 (1972).

Small monopodial epiphytes resembling *Cleisostoma*. Stems to 10cm long, often much less, internodes rather short. Leaves strap shaped, to 15 x 2.5cm. Inflorescences branched, many flowered, to 30cm long. Flowers yellowish-green, sometimes with purplish or lilac markings, lip often lilac; sepals and petals free, spreading; dorsal sepal to 5mm long; lip spurred, with a back wall callus and longitudinal septum; column short and stout, both the fleshy rostellar projection and the anther apex are long and strongly incurved, rostellum raised, fleshy, laterally compressed, with a longitudinal furrow along its edge into which the stipes and dorsally placed pollinia recline; stipes long, slender, arcuate, horseshoe-shaped in side view; viscidium small, ovate; pollinia 4, appearing as 2 unequal masses.

Ten species distributed from Burma to S. China (Yunnan), south to Sumatra, Java and Borneo.

S.fimbriatus *(Ridl.) Garay* in Bot. Mus. Leafl. 23(4):773 (1974). Type: Sarawak, Quop, *Hewitt* 104 (holo. SING, iso. K).
  Saccolabium fimbriatum Ridl. in J. Straits Branch Roy. Asiat. Soc. 54:52 (1909).
  Pennilabium fimbriatum (Ridl.) Garay in Bot. Mus. Leafl.23(4):189 (1972).

HABITAT: unknown.
ALTITUDINAL RANGE: unknown.
DISTRIBUTION: Borneo (SR). Endemic.

S.potamophila *(Schltr.) Garay & W.Kittr.* in Bot. Mus. Leafl. 30(3):58 (1985). Type: Kalimantan, Koetei, Long Sele, Long Wahan, *Schlechter* 13518 (holo. B, destroyed).
  Sarcanthus potamophilus Schltr. in Feddes Repert. 3:279 (1907).
  Cleisostoma potamophilum (Schltr.) Garay in Bot. Mus. Leafl. 23(4):173 (1972).

HABITAT: unknown.
ALTITUDINAL RANGE: unknown.
DISTRIBUTION: Borneo (K). Endemic.

SCHOENORCHIS Reinw.

in Hornschuh, Syll. Pl. nov. 2:4 (1825).

Christenson, E. (1985). The Genus *Schoenorchis* Blume. Amer. Orchid Soc. Bull. 54(7):851 – 854.

Small monopodial epiphytes. Stems with condensed or elongated internodes, up to 30cm long. Leaves fleshy, either flat or semiterete, to 13cm long. Inflorescences simple or branched, pendulous or horizontal, with many small white or red-purple flowers. Flowers usually not opening widely; sepals and petals free; sepals often dorsally keeled; dorsal sepal less than 2mm long; lip spurred, very fleshy, longer than the sepals, 3-lobed; column very short, without a foot, anther and rostellar projection long, pointed and geniculate; viscidium narrowly elliptic to ovate; pollinia 4, appearing as 2 unequal masses.

Christenson (1985) estimated some 24 species, others say around 15. These are distributed in Sri Lanka, S. India, from the Himalayan region to China (Hainan) and the Philippines, Thailand, south to Indonesia and east to New Guinea, Australia and the Pacific islands.

S.aurea *(Ridl.) Garay* in Bot. Mus. Leafl. 23(4):202 (1972). Type: Sarawak, Kuching, *Ridley* s.n. (holo. SING, iso. K).
   Saccolabium aureum Ridl. in J. Straits Branch Roy. Asiat. Soc. 49:35 (1907).

   HABITAT: lowland primary and secondary forest.
   ALTITUDINAL RANGE: sea level – 200m.
   DISTRIBUTION: Borneo (SR). Endemic.

S.endertii *(J.J.Sm.) Christenson & J.J.Wood* in Lindleyana 5, 2:101 (1990). Type: Kalimantan, West Koetai, Long Temelen, *Endert* 2852 (holo. BO, iso. L).
   Robiquetia endertii J.J.Sm. in Bull. Jard. Bot. Buitenzorg, ser. 3, 11:153 (1931).

   HABITAT: low stature hill forest on ultramafic substrate.
   ALTITUDINAL RANGE: 400 – 600m.
   DISTRIBUTION: Borneo (K, SA). Endemic.

S.juncifolia *Blume ex Reinw.*, Cat. Gew. Buitenzorg:100 (1823). Type: Java, *Blume* s.n. (holo. BO).

   HABITAT: lower montane forest.
   ALTITUDINAL RANGE: 1500m.
   DISTRIBUTION: Borneo (SA). Also Sumatra, Java.

S.micrantha *Blume*, Bijdr.:362 (1825). Type: Java, Gede, Salak, etc., *Blume* s.n. (holo. BO).

   HABITAT: riverine forest; hill and lower montane forest.
   ALTITUDINAL RANGE: sea level – 1500m.
   DISTRIBUTION: Borneo (K, SA, SR). Also Thailand, Vietnam, Peninsular Malaysia, Sumatra, Java, Philippines, New Guinea east to Fiji.

S.pachyglossa *(Lindl.) Garay* in Bot. Mus. Leafl. 23(4):202 (1972). Type: Borneo, *Lobb* s.n. (holo. K-LINDL).
   Saccolabium pachyglossum Lindl. in J. Linn. Soc., Bot. 3:34 (1858).
   Gastrochilus pachyglossus (Lindl.) Kuntze, Rev. Gen. Pl.:661 (1891).

   HABITAT: unknown.
   ALTITUDINAL RANGE: unknown.
   DISTRIBUTION: Borneo (unspecified). Endemic.

S.paniculata *Blume*, Bijdr.:362 (1825). Type: Java, Buitenzorg, Salak and Seribu, *Blume* s.n. (holo. L)

   HABITAT: hill forest on ultramafic substrate.
   ALTITUDINAL RANGE: 400 – 500m.
   DISTRIBUTION: Borneo (SA). Also Sumatra, Java, Bali.

SMITINANDIA Holttum

Gard. Bull.Singapore 25:105 (1969).

   Monopodial epiphytes. Stems 25 – 40cm long. Leaves strap shaped, apex unequally bilobed, about 15 x 1.5cm. Inflorescences many flowered racemes as long as the leaves. Flowers rather fleshy, to 5mm across; petals much narrower than sepals; lip 3-lobed, with a distinct spur, the opening of which is ± closed by a high fleshy transverse wall at the base of the mid-lobe; column short and broad, without a foot, rostellar projection small; anther cap pointed; stipes broad below the pollinia, tapering towards the viscidium, twice the diameter of the pollinia; viscidium broad, ± triangular; pollinia 4, the pairs being completely split, the smaller half in each pair detaching from the larger as a free flat disc.

Three species distributed from N.W. Himalaya to Vietnam, south to Borneo, one species endemic to Sulawesi.

S.micrantha *(Lindl.) Holttum* in Gard. Bull.Singapore 25:106 (1969). Types: Nepal, *Wallich* 7300a (syn. K-WALL); Bangladesh, Sylhet, Chota Nagpur, *De Silva* in *Wallich* 7300b (syn. K-WALL).
Saccolabium micranthum Lindl., Gen. Sp. Orch. Pl.:220 (1833).

HABITAT: riverine forest on ultramafic substrate.
ALTITUDINAL RANGE: 200 – 300m.
DISTRIBUTION: Borneo (K). Also Bangladesh, India, Nepal, Bhutan, Burma, Thailand, Indochina, Peninsular Malaysia.

STAUROCHILUS Ridl. ex Pfitzer

Pflanzenr. Ergänz. 1:16 (1900).

Climbing monopodial epiphytes. Stems long, leafy. Leaves ligulate, obtusely unequally bilobed. Inflorescences often branching, longer than leaves, with several flowers on a long scape. Flowers similar to *Trichoglottis* but larger; lip with or without (in *S.fasciatus*) a distinct spur, 3-lobed mid-lobe (in *S.fasciatus*) and hairy basal tongue, disc hairy; column short and stout, without a foot, with small stelidia; pollinia 4, appearing as 2 unequal masses.

Between 12 and 14 species distributed from India, Burma and Thailand to Malaysia, Indonesia, with most species native to the Philippines.

S.fasciatus *(Rchb.f.) Ridl.* in J. Linn. Soc., Bot. 32:350 (1896). Type: 'Hinterindien', cult. *Veitch* & *Bull* (holo. W).
Trichoglottis fasciata Rchb.f. in Flora 55(9):137 (1872).

HABITAT: on limestone rocks.
ALTITUDINAL RANGE: 400 – 500m.
DISTRIBUTION: Borneo (SA). Also Peninsular Malaysia, Thailand, Laos, Cambodia, Vietnam, Sumatra, Anambas Islands, Batu Islands, Philippines.

THRIXSPERMUM Lour.

Fl. Cochinch.:516, 519 (1790).

Medium-sized monopodial epiphytes or, more rarely, terrestials. Stems varying from a few centimetres up to 1 metre in length. Leaves usually well spaced, less often a few close together, flattened, sometimes fleshy, never terete or laterally compressed as in *Cordiglottis*. Inflorescences short or long, sometimes dense, a few flowers opening at a time, the flowering of many lowland species initiated by a sudden afternoon rainstorm. Flowers ephemeral, often fully open for only half a day, very variable, from a few millimetres to several centimetres across; sepals and petals ± equal; lip immobile, 3-lobed, saccate but not truly spurred, usually with a partly hairy or papillose front wall callus, side lobes ± erect, mid-lobe usually fleshy; column short and stout, sometimes winged, with a long foot; stipes very short and broad, shorter than the diameter of the pollinia; pollinia 4, appearing as 2 unequal masses; fruits long and slender.

Around 100 species distributed from Sri Lanka and the Himalayan region east to the Pacific islands. The centre of distribution appears to be in Sumatra.

T.acuminatissimum *(Blume) Rchb.f.*, Xenia Orch. 2:121 (1867). Type: Java, Pantjar, *Blume* 1927 (holo. L).
   Dendrocolla acuminatissima Blume, Bijdr.:288 (1825).

HABITAT: recorded from rubber trees on edge of 'white sand forest'.
ALTITUDINAL RANGE: sea level.
DISTRIBUTION: Borneo (SA, SR). Also Thailand, Cambodia, Peninsular Malaysia, Singapore, Sumatra, Mentawai Islands, Java, Philippines.

T.affine *Schltr.* in Bull. Herb. Boissier, ser. 2, 6:468 (1906). Type: Kalimantan, Samarinda, *Schlechter* 13345 (holo. B, destroyed).

HABITAT: unknown.
ALTITUDINAL RANGE: lowlands.
DISTRIBUTION: Borneo (K). Endemic.

T.amplexicaule *(Blume) Rchb.f.*, Xenia Orch. 2:121 (1867). Type: Java, Salak and Kuripan, *Blume* s.n. (holo. L, iso. AMES).
   Dendrocolla amplexicaulis Blume, Bijdr.:288 (1825).

HABITAT: swamp forest; roadside vegetation; lower montane forest.
ALTITUDINAL RANGE: sea level – 1500m.
DISTRIBUTION: Borneo (K, SA, SR). Widespread from Andaman Islands, Thailand and Vietnam through Malaysia and Indonesia to the Philippines, New Guinea and the Solomon Islands.

T.borneense *(Rolfe) Ridl.* in J. Linn. Soc., Bot. 31:299 (1896). Type: Borneo, ex cult. *Linden* (holo. K).
   Sarcochilus borneensis Rolfe in Ill. Hort. 39:99, t.161 (1892).

HABITAT: lowland dipterocarp forest.
ALTITUDINAL RANGE: lowlands.
DISTRIBUTION: Borneo (unspecified). Endemic.

T.calceolus *(Lindl.) Rchb.f.*, Xenia Orch. 2:122 (1867). Type: Singapore, *Cuming* s.n., cult. *Loddiges* (holo. K-LINDL), originally said to have come from Manila.
   Sarcochilus calceolus Lindl. in Bot. Reg. 32, t.19 (1846).
   Aerides lobbii Teijsm.& Binn. in Natuurk. Tijdschr. Ned.-Indië 24:323 (1862). Type: Sarawak, *Lobb* s.n. (holo. ?L).

HABITAT: hill forest.
ALTITUDINAL RANGE: 200 – 300m.
DISTRIBUTION: Borneo (B, SA, SR). Also Thailand, Peninsular Malaysia, Singapore, Riau Archipelago, Sumatra, Belitung.

T.canaliculatum *J.J.Sm.* in Bull. Jard. Bot. Buitenzorg, ser. 2, 13:42 (1914). Type: Kalimantan, Gunung Labang, *Amdjah* s.n., cult. Bogor number 30 (holo. BO).

HABITAT: unknown.
ALTITUDINAL RANGE: unknown.
DISTRIBUTION: Borneo (K). Endemic.

T.centipeda *Lour.*, Fl. Cochinch.:520 (1790). Type: Vietnam, near Hue, fide Gagnepain, *Loureiro* s.n. (holo. BM).
T.arachnites (Blume) Rchb.f., Xenia Orch. 2:121 (1867). Type: Java, Salak and Seribu, *Blume* s.n. (holo. ?L).

HABITAT: lowland forest; lower montane oak-laurel forest; high kerangas forest; secondary forest; low,very open and dry forest on steep sandstone ridges; *Agathis* dominated heath podsol forest.
ALTITUDINAL RANGE: sea level – 2100m.
DISTRIBUTION: Borneo (B, K, SA, SR). Widespread from India, the Himalayan region and China through Burma, Thailand and Indochina to Malaysia, Indonesia and the Philippines.

T.crassifolium *Ridl.* in J. Straits Branch Roy. Asiat. Soc. 41:32 (1904). Type: Peninsular Malaysia, Johore, Castlewood, *Ridley* s.n. (holo. SING).

HABITAT: unknown.
ALTITUDINAL RANGE: unknown.
DISTRIBUTION: Borneo (SR). Also Peninsular Malaysia.

T.crescentiforme *Ames & C.Schweinf.*, Orch. 6:215 (1920). Type: Sabah, Mt. Kinabalu, Marai Parai Spur, *Clemens* 238 (holo. AMES).

HABITAT: lower montane forest.
ALTITUDINAL RANGE: 1200 – 1500m.
DISTRIBUTION: Borneo (SA). Endemic.

T.fimbriatum *(Ridl.) J.J.Wood*, **comb.nov.** Type: Sarawak, near the race course, *Ridley* & *Hewitt* s.n. (holo. SING).
Dendrocolla fimbriata Ridl. in J. Straits Branch Roy. Asiat. Soc. 49:39 (1907).

HABITAT: unknown.
ALTITUDINAL RANGE: lowlands.
DISTRIBUTION: Borneo (SR). Endemic.

T.fuscum *(Ridl.) Ames* in Merr., Bibl.Enum. Born. Pl.:194 (1921). Type: Borneo, locality unknown, *native collector* s.n., cult. Singapore Botanic Gardens (holo. SING).
Dendrocolla fusca Ridl. in J. Linn. Soc., Bot. 31:296 (1896).

HABITAT: unknown.
ALTITUDINAL RANGE: unknown.
DISTRIBUTION: Borneo (unspecified). Endemic.

T.infractum *Schltr.* in Bull. Herb. Boissier, ser. 2, 6:470 (1906). Type: Kalimantan, Bandjermassin, *Schlechter* 13324 (holo. B, destroyed).

HABITAT: unknown.
ALTITUDINAL RANGE: lowlands.
DISTRIBUTION: Borneo (K). Endemic.

T.inquinatum *J.J.Sm.* in Bull. Dép. Agric. Indes Néerl. 5:23 (1907). Type: Kalimantan, *Nieuwenhuis* s.n. (holo. BO).

HABITAT: unknown.

ALTITUDINAL RANGE: unknown.
DISTRIBUTION: Borneo (K). Endemic.

T.longicauda *Ridl.* in J. Linn. Soc., Bot. 31:299 (1896). Type: Sarawak, *Haviland* 2320 (holo. SING, iso. K).

HABITAT: limestone rocks; roadside banks with *Lycopodium*, *Nepenthes fusca*, *Rhododendron*, ferns, etc. on sandstone and shale; lower montane forest.
ALTITUDINAL RANGE: sea level – 1500m.
DISTRIBUTION: Borneo (SA, SR). Endemic.

T.maculatum *Schltr.* in Bull. Herb. Boissier, ser. 2, 6:470 (1906). Type: Kalimantan, Koetei, Long Wahan, *Schlechter* s.n. (holo. B, destroyed).

HABITAT: unknown.
ALTITUDINAL RANGE: unknown.
DISTRIBUTION: Borneo (K). Endemic.

T.pardale *(Ridl.) Schltr.* in Orchis 5, 4:56 (1911). Types: Peninsular Malaysia, Pahang River, *Ridley* s.n.; Pulau Chengei, *Ridley* 2365; Tulomalaty, *Ridley* s.n.; Kalambalai, *Ridley* s.n. (syn. SING, isosyn. of *Ridley* 2365 K).
Sarcochilus pardalis Ridl. in Trans. Linn. Soc. London, Bot., ser. 3:371 (1893).

HABITAT: hill forest.
ALTITUDINAL RANGE: 900m.
DISTRIBUTION: Borneo (SA). Also Peninsular Malaysia, Sumatra.

T.pensile *Schltr.* in Bot. Jahrb. Syst. 45, Beibl. 104:59 (1911). Type: Sumatra, Fort de Kock, *Schlechter* 15928 (holo. B, destroyed).

HABITAT: hill forest on ultramafic substrate.
ALTITUDINAL RANGE: 700 – 900m.
DISTRIBUTION: Borneo (SA). Also Thailand, Sumatra, Java.

T.ridleyanum *Schltr.* in Orchis 5, 4:57 (1911). Type: Singapore, Bukit Mandai, *Ridley* s.n. (holo. drawing, K).
Dendrocolla maculata Ridl. in J. Linn. Soc., Bot. 32:381 (1896).

HABITAT: mixed lowland dipterocarp forest.
ALTITUDINAL RANGE: lowlands.
DISTRIBUTION: Borneo (SA, SR). Also Thailand, Peninsular Malaysia, Singapore.

T.samarindae *Schltr.* in Bull. Herb. Boissier, ser. 2, 6:471 (1906). Type: Kalimantan, Koetei, Samarinda, *Schlechter* 13334 (holo. B, destroyed).

HABITAT: unknown.
ALTITUDINAL RANGE: lowlands.
DISTRIBUTION: Borneo (K). Endemic.

T.sarawakense *Ames* in Merr., Bibl.Enum. Born. Pl.:195 (1921). Type: Sarawak, Kuching, *Hewitt* 113 (holo. SING, iso. K).
Dendrocolla pulchella Ridl. in J. Straits Branch Roy. Asiat. Soc. 54:55 (1910), non Thwaites.

HABITAT: unknown.
ALTITUDINAL RANGE: lowlands.
DISTRIBUTION: Borneo (SR). Endemic.

T.scopa *(Rchb.f. ex Hook.f.) Ridl.* in J. Linn. Soc., Bot. 32:378 (1896). Type: Peninsular Malaysia, Perak, Larut, *King's collector* 5777 (holo. K).
Sarcochilus scopa Rchb.f. ex Hook.f., Fl. Brit. Ind. 6:40 (1890).

HABITAT: swampy forest.
ALTITUDINAL RANGE: sea level and lowlands.
DISTRIBUTION: Borneo (SA). Also Thailand, Peninsular Malaysia.

T.tortum *J.J.Sm.* in Bull. Jard. Bot. Buitenzorg, ser. 2, 13:40 (1914). Types: Kalimantan, Poetoes Sibou, *Nieuwenhuis* s.n., cult. Bogor (syn. BO); Sumatra, Djambi, *Grootings* s.n., cult. Bogor (syn. BO).

HABITAT: unknown.
ALTITUDINAL RANGE: unknown.
DISTRIBUTION: Borneo (K). Also Sumatra, Java.

T.triangulare *Ames & C.Schweinf.*, Orch. 6:217 (1920). Type: Sabah, Mt. Kinabalu, Kamborangah, *Haslam* 201 (holo. AMES, iso. K, SING).

HABITAT: lower montane forest; upper montane mossy forest; oak-laurel forest; often on ultramafic substrate.
ALTITUDINAL RANGE: 1200 – 3400m.
DISTRIBUTION: Borneo (SA). Endemic.

T.trichoglottis *(Hook.f.) Kuntze*, Rev. Gen. Pl.2:682 (1891). Types: Peninsular Malaysia, Perak, *Scortechini* s.n. and *King's collector* 5934 (syn. K); Singapore, Tanglin, *Ridley* s.n. (syn. K).
Sarcochilus trichoglottis Hook.f., Fl. Brit. Ind. 6:39 (1890).

HABITAT: lowland forest; low stature lower montane ridge forest on ultramafic substrate.
ALTITUDINAL RANGE: lowlands – 1500m.
DISTRIBUTION: Borneo (SA, SR). Also India, Andaman Islands, Burma, Thailand, Laos, ?Vietnam, Peninsular Malaysia, Singapore, Sumatra, Java.

TRICHOGLOTTIS Blume

Bijdr. 6: tab.3, fig.8, and 8:359 (1825).

Climbing monopodial epiphytes. Stems short or long, straggling, up to 80cm long, with elongated internodes. Leaves linear to elliptic, apex usually unequally bilobed. Inflorescences with a short peduncle, 1-4 flowered, but often more than one per node. Flowers rather small, opening widely, lasting about a week, resupinate, usually yellowish with light brown or purple markings; sepals and petals free; dorsal sepal up to 1.5cm long; lip spurred or saccate at base, 3-lobed, immobile, with a small, usually hairy basal tongue which emerges from the back wall, disc often hairy; column short and stout, without a foot, often with small roughly hairy stelidia; stipes linear-oblong; viscidium small, ovate or elliptic; pollinia 4, appearing as 2 unequal masses.

About 55 – 60 species are known, distributed from Sri Lanka and the Nicobar Islands east to New Guinea, Australia and the Solomon Islands, north to Thailand. The centre of distribution lies in Indonesia and the Philippines.

T.bipenicillata *J.J.Sm.* in Icon. Bogor. 2:125, t.125A (1903). Type: Kalimantan, Sungai Semitau, *Hallier* s.n. (holo. BO).

HABITAT: hill forest; mixed hill dipterocarp ridge forest.
ALTITUDINAL RANGE: 400 – 600m.
DISTRIBUTION: Borneo (K, SA). Endemic.

T.calcarata *Ridl.* in J. Linn. Soc., Bot. 31:292 (1896). Type: Sarawak, *Haviland* s.n. (holo. SING).

HABITAT: unknown.
ALTITUDINAL RANGE: unknown.
DISTRIBUTION: Borneo (SR). Endemic.

T.celebica *Rolfe* in Kew Bull.:130 (1899). Type: Sulawesi, Ranoeketan, Minahassa, *Koorders* 29505 (holo. K).

HABITAT: unknown.
ALTITUDINAL RANGE: unknown.
DISTRIBUTION: Borneo (K). Also Sumatra, Java, Sulawesi.

T.collenetteae *J.J.Wood, C.L. Chan & A.Lamb* in Wood, Beaman & Beaman, Plants of Mt. Kinabalu 2, Orchids: 327, fig.56, Pl.80A & B (1993). Type: Sabah, Crocker Range, Sinsuron Road, between Kota Kinabalu and Tambunan, *Collenette* 2295 (holo. K).

HABITAT: lower montane forest.
ALTITUDINAL RANGE: 1100 – 1500m.
DISTRIBUTION: Borneo (SA). Endemic.

T.cuneilabris *Carr* in Gard. Bull. Straits Settlem. 8:124 (1935). Type: Sarawak, Gunung Balapau, Ulu Tinjar, *native collector* in *Oxford Univ. Exp. 1932* no. 2407 (holo. SING, iso. K).

HABITAT: lower montane forest.
ALTITUDINAL RANGE: 600 – 900m.
DISTRIBUTION: Borneo (SR). Endemic.

T.geminata *J.J.Sm.*, Orch. Ambon :106 (1905). Type: Maluku, Ambon, coll. ? *Teysmann*, cult. *Bogor* (holo. BO).
Sarcanthus geminatus Teijsm. & Binn. in Natuurk. Tijdschr. Ned.-Indië 29:243 (1867).
Trichoglottis wenzelii Ames in Philipp. J. Sci. 8:440 (1913), **syn.nov.** Type: Philippines, Leyte, Dagami, *Wenzel* 15 (holo. AMES, iso. K).

HABITAT: scrub and andesitic volcanic rocks near the sea.
ALTITUDINAL RANGE: 200m.
DISTRIBUTION: Borneo (SA). Also ? Sumatra, Sulawesi, Maluku (Ambon, Saparua, Seram), Philippines.

T.kinabaluensis *Rolfe* in Gibbs in J. Linn. Soc., Bot. 42:157 (1914). Type: Sabah, Mt. Kinabalu, Kiau, *Gibbs* 3993 (holo. K).

HABITAT: hill and lower montane forest.
ALTITUDINAL RANGE: 900 – 1700m.
DISTRIBUTION: Borneo (SA). Endemic.

T.lanceolaria *Blume*, Bijdr.:360 (1825). Type: Java, Buitenzorg & Pantjar, *Blume* s.n. (holo. BO).

HABITAT: hill forest on ultramafic and other substrates.
ALTITUDINAL RANGE: 500 – 1200m.
DISTRIBUTION: Borneo (SA). Also Thailand, Vietnam, Peninsular Malaysia, Sumatra, Natuna Islands, Java.

T.lobifera *J.J.Sm.* in Bull. Jard. Bot. Buitenzorg, ser. 3, 11:155 (1931). Type: Kalimantan, West Koetai, Long Temelen, *Endert* 2864 (holo. BO).

HABITAT: ridge forest.
ALTYITUDINAL RANGE: 400 – 700m.
DISTRIBUTION: Borneo (K). Endemic.

T.maculata *(J.J.Sm.) J.J.Sm.* in Bull. Jard. Bot. Buitenzorg, ser. 2, 26:106 (1918). Type: Java, Bobodjong, *Smith* s.n. (syn. BO); Goenoeng Batoe, ?*Smith* s.n. (syn. BO).
T.lanceolaria Blume var. *maculata* J.J.Sm., Orch. Java: 619 (1905).

HABITAT: hill forest on ultramafic substrate.
ALTITUDINAL RANGE: 800 – 1100m.
DISTRIBUTION: Borneo (SA). Also Peninsular Malaysia, Sumatra, Java.

T.magnicallosa *Ames & C.Schweinf.*, Orch. 6:221 (1920). Type: Sabah, Mt. Kinabalu, *Haslam* s.n. (holo. AMES, iso. BO, K).

HABITAT: hill forest.
ALTITUDINAL RANGE: 500 – 900m.
DISTRIBUTION: Borneo (SA). Endemic.

T.odoratissima *Garay* in Bot. Mus. Leafl. 23(4):209 (1972). Type: Borneo, locality unknown, cult.*Dare* (holo. ?SING).
Saccolabium odoratissimum Ridl. in Sarawak Mus. J. 1:37 (1912), non J.J.Sm.

HABITAT: unknown.
ALTITUDINAL RANGE: unknown.
DISTRIBUTION: Borneo (unspecified). Endemic.

T.pantherina *J.J.Sm.* in Icon. Bogor. 2:123, t.124B (1903). Type: Kalimantan, Putus Sibau, *Nieuwenhuis* s.n. (holo. BO).

HABITAT: lower montane forest.
ALTITUDINAL RANGE: 1200 – 1300m.
DISTRIBUTION: Borneo (K, SA). Also Sumatra.

T.philippinensis *Lindl.* in Ann. Mag. Nat. Hist. 15:386 (1845). Type: Philippines, *Cuming* s.n. (holo. K-LINDL).

HABITAT: unknown.
ALTITUDINAL RANGE: unknown.
DISTRIBUTION: Borneo (SR). Also ? Sumatra, Philippines.

T.punctata *Ridl.* in J. Straits Branch Roy. Asiat. Soc. 49:37 (1908). Type: Sarawak, Lingga, *Hewitt* s.n. (holo. SING, iso. SAR).

HABITAT: unknown.
ALTITUDINAL RANGE: unknown.
DISTRIBUTION: Borneo (SR). Endemic.

T.retusa *Blume*, Bijdr.:360, Pl.2, fig.8 (1825). Type: Java, Buitenzorg & Bantam, *Blume* s.n. (holo. BO).

HABITAT: lowland and hill forest.
ALTITUDINAL RANGE: lowlands – 500m.
DISTRIBUTION: Borneo (SA, SR). Also Thailand, Indochina, Peninsular Malaysia, Sumatra, Mentawai Islands, Java, Sumbawa, Seram.

T.scapigera *Ridl.* in J. Linn. Soc., Bot. 32:357 (1896). Type: Peninsular Malaysia, Penang, Government Hill, *Curtis* 1964 (holo. SING, iso. K).

HABITAT: lowland dipterocarp forest; coffee plantations.
ALTITUDINAL RANGE: 300 – 400m.
DISTRIBUTION: Borneo (SA). Also Thailand, Peninsular Malaysia.

T.smithii *Carr* in Gard. Bull. Straits Settlem. 8:125 (1935). Type: Kalimantan, Koetai, cult. *van Gelder*, no. 24 (holo. BO).
T.quadricornuta J.J.Sm. in Bull. Jard. Bot. Buitenzorg, ser. 3, 9:188 (1927), non Kurz.
T.appendiculifera Holttum in Gard. Bull. Singapore 25:108 (1969). Type: Sabah, Sook Plain, *Alphonso* 1/110/66 (holo. K).

HABITAT: hill forest; low stature lower montane forest on ultramafic substrate; lower montane forest of *Agathis alba, Castanopsis and Lithocarpus*, etc.
ALTITUDINAL RANGE: lowlands – 1300m.
DISTRIBUTION: Borneo (SA, SR). Also ? Sumatra.

T.tenuis *Ames & C.Schweinf.*, Orch. 6:223 (1920). Type: Sabah, Mt. Kinabalu, Kiau, *Clemens* 60 (holo. AMES).

HABITAT: hill forest.
ALTITUDINAL RANGE: 900m.
DISTRIBUTION: Borneo (SA). Endemic.

T.uexkulliana *J.J.Sm.* in Icon. Bogor. 3, 60, t.223 (1906). Type: Kalimantan, Pontianak, cult. Bogor (holo. BO).

HABITAT: riverine forest.
ALTITUDINAL RANGE: 300m.
DISTRIBUTION: Borneo (K, SA, SR). Endemic.

T.valida *Ridl.* in J. Straits Branch Roy. Asiat. Soc. 44:192 (1905). Type: Kalimantan, Sambas River, *Micholitz* s.n. (holo. SING).

HABITAT: unknown.
ALTITUDINAL RANGE: unknown.
DISTRIBUTION: Borneo (K, SR). Endemic.

T.vandiflora *J.J.Sm.* in Bull. Dép. Agric. Indes Néerl. 22:49 (1909). Type: Kalimantan, Mahakam, Bloe-oe, *Jaheri* s.n. (holo. BO).

HABITAT: hill forest on ultramafic substrate.
ALTITUDINAL RANGE: 500m.
DISTRIBUTION: Borneo (K, SA). Endemic.

T.winkleri *J.J.Sm.* in Winkler in Bot. Jahrb. Syst. 48:105 (1912). Type: Kalimantan, Batu babi, *Winkler* 2796 (holo. HBG, iso. BO).

var. **winkleri**

HABITAT: unknown.
ALTITUDINAL RANGE: 200 – 700m.
DISTRIBUTION: Borneo (K). Also Peninsular Malaysia, Sumatra, Java.

var. **minor** *J.J.Sm.* in Bull. Jard. Bot. Buitenzorg, ser. 2, 26:102 (1918). Type: Java, Priangan, Tjisokan, Tjibeber, *Bakhuizen van den Brink* s.n., cult. *Winckel* (holo. BO).

HABITAT: hill forest on ultramafic substrate.
ALTITUDINAL RANGE: 200 – 700m.
DISTRIBUTION: Borneo (SA). Also Peninsular Malaysia, Java.

GROUP 3. Pollinia 2, sulcate, i.e. more or less, but not completely, cleft on split. See p.363 for Group 4.

AERIDES Lour.

Fl. Cochinch. 2:525 (1790).

Christenson, E. A. (1987). The Taxonomy of *Aerides* and Related Genera. Proc. 12th. World Orch. Conf.:35-40.

Medium sized, rather coarse monopodial epiphytes. Stems short to elongate, up to 1 metre long. Leaves oblong – ligulate to linear or terete, often very fleshy, apex usually distinctly bilobed, to 60cm long. Inflorescences variable, simple, usually densely many flowered, often pendent. Flowers quite showy, to 3cm or more across, white and rose-violet; sepals and petals broad, spreading; lateral sepals decurrent on column-foot; lip 3-lobed, immobile, with a basal, often forward-curving spur, side lobes decurrent on the column, mid-lobe often erose; column short, often broadened at the apex, foot short, rostellar projection usually long and pointed; stipes usually long and slender; pollinia 2, sulcate.

Between 17 and 20 species distributed from Sri Lanka, India and the Himalayan region to Thailand and Indochina south to Malaysia and Indonesia, north to the Philippines.

A.inflexa *Teijsm.& Binn.* in Natuurk. Tijdschr. Ned.-Indië 24:324 (1862). Type: Sulawesi, Goa, Makasar, *Tolson* s.n. (holo. ?L).
Aerides bernhardiana Rchb.f. in Gard. Chron., 24:650 (1885). Type: Borneo, ex *Veitch* (holo. W).

HABITAT: unknown.
ALTITUDINAL RANGE: unknown.
DISTRIBUTION: Borneo (unspecified). This record is based on the reported type locality of *A.bernhardiana*. All material examined has originated from Sulawesi.

A.odorata *Lour.*, Fl. Cochinch. 2:525 (1790). Type: probably Vietnam, *Loureiro* s.n. (holo. not located).

HABITAT: riverine forest; hill forest; lower montane forest.
ALTITUDINAL RANGE: lowlands – 1500m.

DISTRIBUTION: Borneo (K, SA, SR). Widespread from India, the Himalayan region and most of S.E. Asia east as far as Sulawesi and the Philippines.

ASCOCHILOPSIS Carr

Gard. Bull. Straits Settlem. 5:21 (1929).

Small monopodial epiphytes. Stems branching, to 7cm long. Leaves strap-shaped, apex unequally bilobed, 3-20 x 0.5-1.5cm, borne at apex of stem. Inflorescences erect, 2 or more together, peduncle 2-5cm long, rough-textured or hairy, rachis thickened, fleshy, to 4.5cm long, with 1 – 4 flowers open in succession. Flowers very small, lasting one day only, pale yellow with a white spur; sepals and petals free, spreading, to 3 x 1.2-1.5mm; lip immobile, entirely composed of a spur which is either laterally flattened (*A.myosurus*) or not (*A.lobata*), at the entrance of which are 3 small fleshy lobes (*A.myosurus*) or 2 distinct side lobes and a shallowly trilobed mid-lobe (*A.lobata*); column 1mm or less long, foot absent; *stipes* long and slender, widened below the pollinia; viscidium very small; pollinia 2, sulcate.

Two species, one from Peninsular Malaysia and Sumatra, the other, described below, endemic to Borneo.

Ascochilopsis lobata *J.J.Wood* et *A.Lamb* **sp.nov.** ab *A.myosuro* (Ridl.) Carr, species malayana et sumatrana, lobis lateralibus labii distinctis oblongis, lobo medio trilobulato, calcari angustiore distinguitur. Typus: East Malaysia, Sabah, Tenom District, Paling Paling Hills, Ulu Batu Tiningkang, 700-800m, April 1990, *Lamb & Surat* in *Lamb* AL 1252/90 (holotypus K, spirit material only).

Epiphytic herb. Stems 1-2.5 cm long. Leaves 4-7 per stem, 2-5 x 0.8-1.3 cm, fleshy, up to 2 or 3 mm thick, narrowly elliptic, ligulate, gradually narrowed at base, apex minutely unequally bilobed, lobules acute. Inflorescences erect or porrect, bearing many tiny flowers in succession, 1 or 2 open at a time; peduncle 1.5-2.3 cm long, scabrous-hairy, greenish-purple; rachis 0.5-1.5(-2.5) cm long, thickened, clavate, minutely scaberulus, green; floral bracts 0.4 mm long, triangular-ovate, acute, fawn coloured. Flowers 4-5 mm long, 5 mm across. Pedicel with ovary 2 mm long, pale yellow, flushed green. Sepals & petals spreading, translucent yellow, whitish at base. Dorsal sepal 2 x 1.2 mm, oblong-elliptic, acute. Lateral sepals 2.2 x 1.5 mm, ovate-elliptic, subacute, slightly asymmetrical. Petals 1.8 x 1 mm, oblong-elliptic, obtuse. Lip 3-lobed, c.3 mm long, fleshy, immobile; mid-lobe very small, 0.4-0.5 mm long, trilobulate, white, sidelobules rounded, midlobule triangular; side lobes c.1 mm long, oblong, apex very shallowly unequally retuse, longer than broad, white with a reddish-brown patch on the upper margins; spur 1.9-2 x 1.5 mm, saccate, obtuse, transversely flattened. Column 1 mm long, foot absent, white; rostellum elongate, apex upcurved. Anther cap yellow. Pollinia lost. Fig.37.

A.lobata was found growing on branches of dipterocarp trees in mixed hill dipterocarp forest on a sandstone ridge. It is distinguished from *A.myosurus* (Ridl.) Carr, from Peninsular Malaysia and Sumatra, by its distinct oblong lip side lobes, trilobulate mid-lobe and narrower spur.

The specific epithet is derived from the Latin *lobatus*, lobed, in reference to the lobing of the lip.

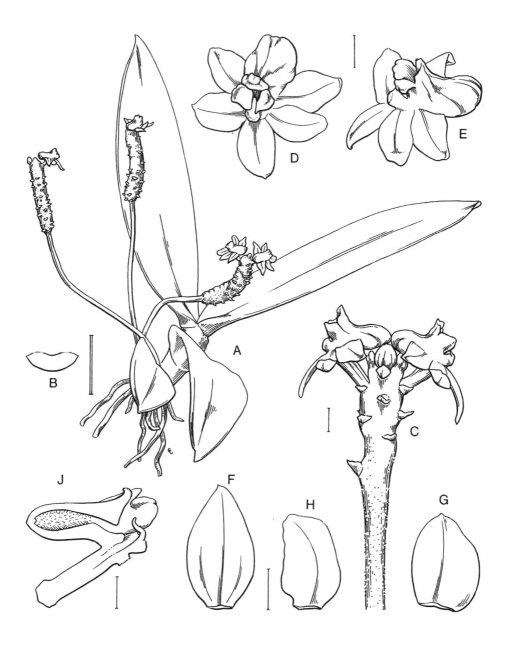

FIG. 37. *Ascochilopsis lobata*. **A**, habit; **B**, transverse section through leaf; **C**, inflorescence; **D**, flower, front view; **E**, flower, side view; **F**, dorsal sepal; **G**, lateral sepal; **H**, petal; **J**, longitudinal section through pedicel with ovary, lip and column. A-J from *Lamb & Surat* in *Lamb* AL 1252/90. Scale: single bar = 1 mm; double bar = 1 cm. Drawn by Eleanor Catherine.

BRACHYPEZA Garay

Bot. Mus. Leafl. 23(4):163 (1972).

Small monopodial epiphytes. Stems very short. Leaves few, fleshy, broad, to 15 (-25) x 5cm. Inflorescences few, spreading or pendent, as long as or shorter than leaves, peduncle long, rachis a few centimetres long, club-shaped, with only a few flowers open at a time. Flowers small; sepals and petals free, spreading, narrowly oblong, to about 1cm long; lip 3-lobed, pendent on column-foot, side lobes erect, ear-like, mid-lobe very small, with a conspicuous bag-like spur; column long, slender, arcuate, foot short, rostellar projection triangular; stipes oblong, as long as the pollinia; pollinia 2, sulcate.

About 7 species distributed from Thailand and Laos through Malaysia and Indonesia, north to the Philippines, east to New Guinea.

B.indusiata *(Rchb.f.) Garay* in Bot. Mus. Leafl. 23(4):164 (1972). Type: 'Insul. Sondaic', ex *Linden* (holo. W).
  Thrixspermum indusiatum Rchb.f. in Gard. Chron., 25:585 (1886).
  Sarcochilus indusiatus (Rchb.f.) Carr in Gard. Bull. Straits Settlem. 8:121 (1935).

  HABITAT: lowland forest; riverine forest; recorded from limestone.
  ALTITUDINAL RANGE: under 300m.
  DISTRIBUTION: Borneo (K, SA, SR). Endemic.

B.koeteiensis *(Schltr.) Garay* in Bot. Mus. Leafl. 23(4):164 (1972). Type: Kalimantan, Koetei, Long Sele, *Schlechter* 13464 (holo. B, destroyed).
  Sarcochilus koeteiense Schltr. in Bull. Herb. Boissier, ser. 2, 6:467 (1906).

  HABITAT: unknown.
  ALTITUDINAL RANGE: unknown.
  DISTRIBUTION: Borneo (K). Endemic.

B.stenoglottis *(Hook.f.) Garay* in Bot. Mus. Leafl. 23(4):164 (1972). Type: Peninsular Malaysia, ?Perak, *Scortechini* s.n. (holo. K, not located).
  Sarcochilus stenoglottis Hook.f., Fl. Brit. Ind. 6:34 (1890).

  HABITAT: hill forest.
  ALTITUDINAL RANGE: 200 – 300m.
  DISTRIBUTION: Borneo (SA). Also Peninsular Malaysia, Sumatra.

B.zamboangensis *(Ames) Garay* in Bot. Mus. Leafl. 23(4):164 (1972). Type: Philippines, Mindanao, Zamboanga Province, Flecha Point, *Merrill* 11640 (holo. Philippine Bureau of Science, iso. AMES).
  Sarcochilus zamboangensis Ames, Sched. Orch. 5:39 (1923).

  HABITAT: forest on limestone.
  ALTITUDINAL RANGE: 500m.
  DISTRIBUTION: Borneo (SA). Also Philippines.

MACROPODANTHUS L.O.Williams

Bot. Mus. Leafl. 6:103 (1938).

Medium-sized monopodial epiphytes. Stems to 10cm long. Leaves narrow, strap-shaped, apex unequally bilobed. Inflorescences often clustered, as long as the leaves, with up to 15 flowers in a loose raceme, rachis sometimes keeled. Flowers quite showy, white or yellow, marked red, lilac-pink or orange-brown; dorsal sepal free; lateral sepals & petals adnate to the column-foot; lip mobile, 3-lobed, lateral lobes linear or tooth-like, mid-lobe with a distinct saccate spur about 1cm long; column short, slightly broadened at the truncate apex, foot very long, rostellar projection elongated; stipes very long and slender, to 5mm long; viscidium obovate or oblong; pollinia 2, sulcate.

Around 6 species distributed from the Andaman Islands, Thailand and Peninsular Malaysia east to Indonesia and the Philippines.

M.membraniferus *(Carr)H.Ae. Peders.* in Opera Bot. 117: 56 (1993). Type: Sarawak, Dulit ridge, *native collector* in *Synge* S.455 (holo. K, not located).
　　Sarcochilus membraniferus Carr in Gard. Bull. Straits Settlem. 8:121 (1935).
　　Pteroceras membraniferum (Carr) Garay in Bot. Mus. Leafl. 23(4):193 (1972).

HABITAT: lower montane ridge forest.
ALTITUDINAL RANGE: 1200 – 1300m.
DISTRIBUTION: Borneo (SR). Endemic.

NOTE: A second undescribed species occurs in Sabah.

PAPILIONANTHE Schltr.

Orchis 9:78 (1915).

Scrambling epiphytes or lithophytes. Stems to 2 metres long. Leaves well spaced, terete, stiff, to 30cm long. Inflorescences 1 to few – flowered, as long as or shorter than the leaves. Flowers often large and showy, mauve or reddish; lip 3-lobed, up to 4cm wide, side lobes either parallel with or enclosing the column, spurred; column short and stout, with a foot, rostellar projection elongated; stipes broad; viscidium large; pollinia 2, sulcate.

About 8 species distributed from India to China, south to Indonesia, east as far as Borneo.

P.hookeriana *(Rchb.f.) Schltr.* in Orchis 9:80 (1915). Type: Labuan, *Motley* 347 (holo. K).
　　Vanda hookeriana Rchb.f. in Bonplandia 4:324 (1856).

HABITAT: swamps.
ALTITUDINAL RANGE: sea level.
DISTRIBUTION: Borneo (K, SA, SR). Also Thailand, Vietnam, Peninsular Malaysia, Sumatra, Bangka.

PARAPHALAENOPSIS A.D. Hawkes

Orquidea (Rio de Janeiro), 25:212 (1964 '1963').

Sweet, H.R. (1980). The genus *Phalaenopsis*. The Orchid Digest Inc.: 118 – 123.

Large monopodial epiphytes. Stems short. Leaves few, terete, canaliculate, 20 – 165cm long. Inflorescences few flowered, peduncle distinct, rachis congested. Flowers showy; dorsal sepal free; lateral sepals spreading, decurrent on the column-foot, lip immobile, 3-lobed, side lobes erect, mid-lobe narrow, with a broader forked or bilobed apex, callus conduplicate, plate-like, situated at the junction of the lobes, disc excavate behind the callus into a small sac-like nectary; column extended into a 3-fingered foot; pollinia 2, sulcate.

Four species endemic to Borneo.

P.denevei *(J.J.Sm.) A.D.Hawkes* in Orquidea (Rio de Janeiro) 25:212 (1964). Type: Kalimantan, Nanga Djetah Plantation near Pontianak, *de Neve* s.n., cult. Bogor under no. 484 (holo. BO).
Phalaenopsis denevei J.J.Sm. in Recueil Trav. Bot. Néerl. 22:264 (1925).

HABITAT: riverine forest.
ALTITUDINAL RANGE: lowlands.
DISTRIBUTION: Borneo (K). Endemic.

P.labukensis *Shim, A.Lamb & C.L.Chan* in Orchid Digest 45:139 (1981). Type: Sabah, Kuala Labuk near Pamol, *Lamb* SAN 91503 (holo. SAN, iso. K, LA).
Phalaenopsis labukensis (Shim, A. Lamb & C.L. Chan) Shim in Malayan Nat. J. 36(1):21 (1982).

HABITAT: hill forest; recorded from *Gymnostoma sumatrana* growing on ultramafic substrate.
ALTITUDINAL RANGE: 500 – 1000m.
DISTRIBUTION: Borneo (SA). Endemic.

P.laycockii *(M. R.Hend.) A.D.Hawkes* in Orquidea (Rio de Janeiro) 25:212(1964). Type: Kalimantan, cult. *Laycock* s.n. (holo. SING).
Phalaenopsis laycockii M. R. Hend. in Orchid Rev.43:108(1935).

HABITAT: lowland and hill forest.
ALTITUDINAL RANGE: lowlands – ?m.
DISTRIBUTION: Borneo (K). Endemic.

P.serpentilingua *(J.J.Sm.) A.D.Hawkes* in Orquidea (Rio de Janeiro) 25:212 (1964). Type: Kalimantan, Singkawang (Sintang), cult. B*ogor* (holo. L).
Phalaenopsis serpentilingua J.J.Sm. in Orchid Rev. 41:147 (1933).
P.denevei J.J.Sm. var. *alba* Price, Orchid Culture in Ceylon and the East, ed. 2:90 (1933). Type: cult. Sri Lanka, (specimen not preserved).
P.simonsei Simonse in Amer. Orchid Soc. Bull. 29:531(1960), nomen nudum.

HABITAT: wet mossy rocks.
ALTITUDINAL RANGE: lowlands.
DISTRIBUTION: Borneo (K). Endemic.

P.x thorntonii *(Holttum) A.D.Hawkes* in Phytologia 15:1(1967). Type: Kalimantan, *Thornton* s.n. (holo. SING).
Phalaenopsis x thorntonii Holttum in Orchid Rev. 74:290 (1966).

HABITAT: unknown.
ALTITUDINAL RANGE: unknown.
DISTRIBUTION: Borneo (K). Endemic.

NOTE: A natural hybrid between *P.denevei* and *P.serpentilingua*.

PHALAENOPSIS Blume

Bijdr. 6: tab.4, & 7:294 (1825).

Sweet, H.R. (1980). The genus *Phalaenopsis*. The Orchid Digest Inc.

Medium-sized to large monopodial epiphytes or lithophytes. Stems very short. Leaves few, narrowly oblong to broadly elliptic, normally widest in apical half, often fleshy, sometimes transversely striped, rather shiny, sometimes deciduous. Inflorescences one to many flowered, racemose or paniculate, peduncle long, rachis sometimes flattened. Flowers small to large, long lasting, waxy in texture, white, pink, mauve or violet to yellow with reddish-brown markings; sepals and petals spreading, free; dorsal sepal 1.5 – 4cm long; petals sometimes much broader than the sepals and with a clawed base; lip immobile, 3-lobed, side lobes erect, mid-lobe porrect, with a bifid or complex basal callus, apex sometimes extended into 2 recurved appendages or filiform 'antennae'; column erect, often expanded at apex, with a short foot, rostellar projection and stipes long and slender, pointing towards the base of the column; viscidium elliptic, shorter than the stipes; pollinia usually 2, sulcate (in Bornean species), rarely 4.

Between 40 and 45 species distributed from India to S. China, Thailand, Indochina, Malaysia and Indonesia to the Philippines and New Guinea. The majority of species occur in Indonesia and the Philippines.

P.amabilis *(L.) Blume*, Bijdr.:294 (1825). Type: Java, *Osbeck* s.n. (holo. LINN).
Epidendrum amabile L., Sp. Plant.:953 (1753).
Phalaenopsis amabilis (L.) Blume var. *fuscata* Rchb.f. in Bot. Zeitung (Berlin) 20:214 (1862). Type: Borneo, cult. *Low* (holo. W).
P.grandiflora Lindl. var. *aurea* Hort. in Proc. Roy. Hort. Soc. London 4:135 (1864). Type: Borneo, cult. *Warner* (neo. Select Orch. Pl., ser. 2, t.7, 1869).
P.grandiflora Lindl. var. *fuscata* (Rchb.f.) Burb. in Garden (London 1871 – 1927) 22:118 (1882).
P.amabilis (L.) Blume var. *aurea* (Hort.) Rolfe in Gard. Chron. ser. 2, 26:212 (1886).

HABITAT: low stature hill forest on ultramafic substrate; lower montane forest.
ALTITUDINAL RANGE: lowlands – 1500m.
DISTRIBUTION: Borneo (K, SA, SR). Also Sumatra, Java, Sulawesi and other Indonesian islands east to New Guinea (including Bismark Archipelago) and Australia (Queensland), north to the Philippines.

P.cochlearis *Holttum* in Orchid Rev. 72:408, fig.64 (1964). Type: Sarawak, *Kho* s.n., cult.Kew (holo. K).
Polychilos cochlearis (Holttum) Shim in Malayan Nat. J. 36:22 (1982).

HABITAT: forest on limestone.
ALTITUDINAL RANGE: 400 – 600m.
DISTRIBUTION: Borneo (SR). Endemic.

P.corningiana *Rchb.f.* in Gard. Chron. 11:620 (1879). Type: Borneo, cult. *Veitch* s.n. (holo. W).
P.sumatrana Korth.& Rchb.f. var. *sanguinea* Rchb.f. in Gard. Chron., ser. 2, 15:782 (1881). Type: Borneo, cult. *Veitch* s.n. (holo. W).
P.sumatrana Korth.& Rchb.f. subvar. *sanguinea* (Rchb.f.) Veitch, Man. Orch. Pl., pt. 7:40 (1891).

HABITAT: unknown.
ALTITUDINAL RANGE: unknown.
DISTRIBUTION: Borneo (SR). Endemic.

P.cornucervi *(Breda) Blume & Rchb.f.* in Hamburger Garten-Blumenzeitung 16:116 (1860). Type: Java, Bantam Province, *van Hasselt* s.n. (holo. L).
Polychilos cornucervi Breda in Kuhl and van Hasselt, Gen. Sp. Orch., t.1 (1827).

HABITAT: riverine forest; lowland forest; swamp forest.
ALTITUDINAL RANGE: sea level – 500m.
DISTRIBUTION: Borneo (K, SR). Also India, Nicobar Islands, Burma, Thailand, Laos, Peninsular Malaysia, Sumatra, Java.

P.fuscata *Rchb.f.* in Gard. Chron., ser. 2, 2:6 (1874). Type: Peninsular Malaysia, cult. *Bull* (holo. W).
Polychilos fuscata (Rchb.f.) Shim in Malayan Nat. J. 36:23 (1982).

HABITAT: lowland and hill dipterocarp forest; preferring shady places beside streams.
ALTITUDINAL RANGE: sea level – 1000m.
DISTRIBUTION: Borneo (K, SA). Also Peninsular Malaysia, Philippines.

P.x gersenii *(Teijsm.& Binn.) Rolfe* in Orchid Rev. 25:227 (1917). Type: Sumatra, *Gersen* s.n., icon. *Binnendijk* (holo. K).
P.zebrina Witte var. *gersenii* Teijsm. & Binn. in Natuurk. Tijdschr. Ned.-Indië 24:320 (1862).
P.x singuliflora J.J.Sm. in Feddes Repert. 31:80 (1932). Type: Kalimantan, Tajan, east of Pontianak, cult. *Bogor* (holo. L).

HABITAT: unknown.
ALTITUDINAL RANGE: lowlands.
DISTRIBUTION: Borneo (K, SR). Also Sumatra.

NOTE: A natural hybrid between *P.sumatrana* and *P.violacea*.

P.gigantea *J.J.Sm.* in Bull. Dép. Agric. Indes Néerl. 22:45 (1909). Type: Kalimantan, *Jaheri* s.n. (holo. L).
Polychilos gigantea (J.J.Sm.) Shim in Malayan Nat. J. 36:23 (1982).

HABITAT: lowland and hill dipterocarp forest.
ALTITUDINAL RANGE: sea level – 600m.
DISTRIBUTION: Borneo (K, SA). Endemic.

P.lamelligera *H.R.Sweet* in Amer. Orchid Soc. Bull. 38:516 (1969). Type: Borneo, cult. *Bull*, no. 83 (holo. K).
Polychilos lamelligera (H.R.Sweet) Shim in Malayan Nat. J. 36:24 (1982).

HABITAT: unknown.
ALTITUDINAL RANGE: unknown.
DISTRIBUTION: Borneo (SA). Endemic.

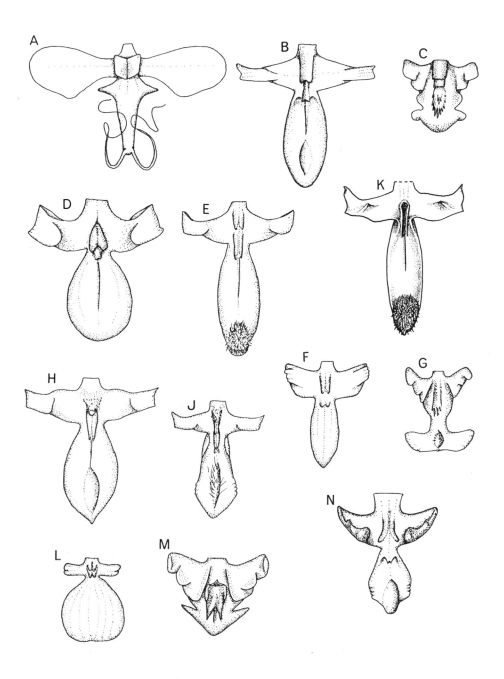

FIG. 38. *Phalaenopsis* lips. **A**, *P. amabilis*, x 1; **B**, *P. modesta*, x 3; **C**, *P. lamelligera*, x 2; **D**, *P. fuscata*, x 3; **E**, *P. sumatrana*, x 2; **F**, *P. maculata*, x 3; **G**, *P. pantherina*, x 1½; **H**, *P. violacea*, x 2; **J**, *P. mariae*, x 2; **K**, *P. corningiana*, x 2; **L**, *P. cochlearis*, x 2; **M**, *P. cornucervi*, x 2; **N**, *P.gigantea*, x 2. Drawn by Sarah Thomas.

355

P.maculata *Rchb.f.* in Gard. Chron., ser. 2, 16:134 (1881). Type: Borneo, *Curtis* s.n. (holo. W).
  P.cruciata Schltr. in Feddes Repert. 8:457 (1910). Type: Kalimantan, Koetai, Long Sele, *Schlechter* 13480 (holo. B, destroyed).
  Polychilos maculata (Rchb.f.) Shim in Malayan Nat. J. 36:24 (1982).

HABITAT: lowland and hill dipterocarp forest on limestone, ultramafics, etc.; riverine forest; rocky areas.
ALTITUDINAL RANGE: sea level – 1000m.
DISTRIBUTION: Borneo (K, SA, SR). Also Peninsular Malaysia.

P.mariae *Burb.* ex *R.Warner & H.Williams*, Orch. Album 2: t.80, sub. t.87 (1883). Type: Philippines, Sulu (Jolo) Island, Bunt-Doohan, *Burbidge* s.n. (holo. W).
  Polychilos mariae (Burb. ex R.Warner & H.Williams) Shim in Malayan Nat. J. 36:25 (1982).

HABITAT: unknown.
ALTITUDINAL RANGE: unknown.
DISTRIBUTION: Borneo (K, ?SA). Also Philippines.

P.modesta *J.J.Sm.* in Icon. Bogor. 3:47, t.218 (1906). Type: Kalimantan, *Nieuwenhuis* s.n., cult. Bogor (holo. BO).
  Polychilos modesta (J.J.Sm.) Shim in Malayan Nat. J. 36:25 (1982).

HABITAT: low stature hill forest on ultramafic substrate; mixed hill dipterocarp forest on sandstone.
ALTITUDINAL RANGE: sea level – 900m.
DISTRIBUTION: Borneo (K, SA). Endemic.

P.pantherina *Rchb.f.* in Bot. Zeitung (Berlin) 22:298 (1864). Type: Borneo, cult. *Low* (holo. W).
  P.luteola Burb., The Garden of the Sun: 258 (1880), nomen.
  Polychilos pantherina (Rchb.f.) Shim in Malayan Nat. J. 36:25 (1982).

HABITAT: riverine forest; hill forest on limestone; wet mossy rocks.
ALTITUDINAL RANGE: sea level – 800m.
DISTRIBUTION: Borneo (K, SA, SR). Endemic.

P.sumatrana *Korth. & Rchb.f.* in Hamburger Garten-Blumenzeitung 16:115 (1860). Type: Sumatra, *Korthals* drawing no.443 (holo. K-LINDL, iso. L).
  Polychilos sumatrana (Korth. & Rchb.f.) Shim in Malayan Nat. J. 36:26 (1982).

HABITAT: hill forest; riverine forest.
ALTITUDINAL RANGE: sea level – 700m.
DISTRIBUTION: Borneo (K, SA, SR). Also Burma, Thailand, Peninsular Malaysia, Sumatra, Mentawai, Bangka.

P.violacea *Witte* in Ann. Hort. Bot. 4:129 (1860). Type: Sumatra, Palembang, *Teijsmann* s.n., cult. Bogor and Leiden, icon. *Witte* (holo. L).
  P.violacea Witte var. *alba* Teijsm. & Binn. in Natuurk. Tijdschr. Ned.-Indië 24:320 (1862). Type: Sumatra, Palembang, *Teijsmann* s.n. (holo. ?L).
  Polychilos violacea (Witte) Shim in Malayan Nat. J. 36:27 (1982).

HABITAT: lowland riverine forest.
ALTITUDINAL RANGE: lowlands.
DISTRIBUTION: Borneo (K, SR). Also Peninsular Malaysia, Sumatra, Mentawai, Simeuluë.

PORPHYRODESME Schltr.

Feddes Repert., Beih. 1:982 (1913).

Scrambling monopodial epiphytes. Stems fragile, much branched, up to several metres long. Leaves borne on upper third of the stem, linear, acute, fleshy. Inflorescences bright red, branched, many flowered, up to 8cm long. Flowers small, 5-8mm across, scarlet-red; sepals and petals obovate-oblong; lip immobile, 3-lobed, bucket-shaped, laterally compressed, side lobes arcuately spreading, but converging at the base leaving only a narrow slit for the spur entrance; column short and stout, foot absent, rostellar projection ± sigmoid, its bifid apex turned abruptly upward; stipes slender, 3 times the diameter of the pollinia; pollinia 2, sulcate.

Two species from Borneo and New Guinea. Garay (1972) says that *Renanthera moluccana* Blume, from Maluku, may also possibly belong here. He also transferred *R.elongata* (Blume) Lindl. here, but studies by Seidenfaden (1988) have shown this to have four pollinia. The *Thrixspermum*-like appearance of the short-stemmed *P.hewittii* is far removed from the type of the genus, *P.papuana* (Schltr.) Schltr., and its position here is questionable.

P.hewittii *(Ames) Garay* in Bot. Mus. Leafl. 23(4):191 (1972). Type: Sarawak, Kuching, *Hewitt* s.n. (holo. SING, iso. K).
   Thrixspermum hewittii Ames in Merr., Bibl.Enum. Born. Pl.:194 (1921).
   Dendrocolla minima Ridl. in J. Straits Branch Roy. Asiat. Soc. 50:139 (1908), non Blume.

   HABITAT: unknown.
   ALTITUDINAL RANGE: unknown.
   DISTRIBUTION: Borneo (SR). Endemic.

P.papuana *(Schltr.) Schltr.* in Feddes Repert., Beih. 1:983 (1913). Type: Papua New Guinea, Ramu, *Schlechter* 14206 (holo. B, destroyed).
   Saccolabium porphyrodesme Schltr. in K.Schum. & Lauterb., Nachtr. Fl. Schutzgeb. Südsee: 229 (1905).

   HABITAT: hill forest.
   ALTITUDINAL RANGE: unknown.
   DISTRIBUTION: Borneo (SA). Also Papua New Guinea.

PTEROCERAS Hasselt ex Hassk.

Flora 25, Beibl.:6 (1842).

Pedersen, H.Ae. (1993). The genus *Pteroceras* (Orchidaceae) – a taxonomic revision. Opera Bot. 117:1-64.

Small, monopodial epiphytes. Stems short. Leaves few to about 10, apex usually unequally bilobed, to 10 x 1.5cm. Inflorescences often longer, few to many flowered. Flowers small, to about 1.5cm across, lasting only one day; sepals and petals free, spreading; lateral sepals may be distinctly broader than petals, not adnate to the column-foot; lip 3-lobed, fleshy, mobile, hinged to the column-foot, with a spur or sac usually pointing forward in line with the foot, without interior ornaments, although the front wall may be quite fleshy, side lobes usually large, erect, mid-lobe very short; column short and stout, foot long, rostellar projection oblong; stipes oblong; viscidium ± triangular; pollinia 2, sulcate.

Pedersen (1993) recognises 19 species distributed from N.E. India to Maluku.

P.biserratum *(Ridl.) Holttum* in Kew Bull. 14:269 (1960). Type: Peninsular Malaysia, Perak, Ipoh, *Ridley* 10156 (holo. SING).
  Sarcochilus biserratus Ridl. in J. Bot. 38:73 (1900).
  S.pachyrhachis Schltr. in Bull. Herb. Boissier, ser. 2, 6:468 (1906). Type: Kalimantan, Koetai, Sungai Penang, Samarinda, *Schlechter* 13348 (holo. B, destroyed).

  HABITAT: lowland dipterocarp forest.
  ALTITUDINAL RANGE: sea level – 500m.
  DISTRIBUTION: Borneo (K, SA). Also Peninsular Malaysia and Sumatra.

P.cladostachyum *(Hook.f.) H.Ae. Peders.* in Nord J. Bot. 12(4):387 (1992). Type: Peninsular Malaysia, Perak, colour plate by *Kunstler* (holo. CAL, copy at K).
  Sarcochilus cladostachyum Hook.f., Fl. Brit. Ind. 6:35 (1890).

  HABITAT: limestone rocks.
  ALTITUDINAL RANGE: sea level – 700m.
  DISTRIBUTION: Borneo (K, SA). Also Peninsular Malaysia, Java, Sulawesi, Philippines.

P.erosulum *H.Ae. Peders.* in Opera Bot. 117:45, fig.17 (1993). Type: Sabah, Sipitang District, 8km. N.W. of Long Pa Sia, *Wood* 676 (holo. K).

  HABITAT: lower montane forest.
  ALTITUDINAL RANGE: 1200 – 1300m.
  DISTRIBUTION: Borneo (SA). Endemic.

P.fragrans *(Ridl.) Garay* in Bot. Mus. Leafl. 23(4):193 (1972). Type: Sarawak, Matang estate, *Ridley* s.n. (holo. SING).
  Sarcochilus fragrans Ridl. in J. Straits Branch Roy. Asiat. Soc. 49:38 (1907).
  S.spathipetalus J.J.Sm. in Bull. Jard. Bot. Buitenzorg, ser. 3, 11:151 (1931). Type: Kalimantan, West Koetai, Long Temelen, prope Sungai Mahakam, *Endert* 2851 (holo. L, iso. BO).
  Pteroceras spathipetalum (J.J.Sm.) Garay in Bot. Mus. Leafl.23(4):194(1972).

  HABITAT: lowland and hill dipterocarp forest on ultramafic, sandstone and alluvial substrates.
  ALTITUDINAL RANGE: 400 – 1000m.
  DISTRIBUTION: Borneo (K, SA, SR). Endemic.

P.hirsutum *(Hook.f.) Holttum* in Kew Bull. 14:270 (1960). Type: Peninsular Malaysia, Perak, *Kunstler* s.n. (holo. CAL, original pencil drawing of floral details and copy of colour plate at K).
  Sarcochilus hirsutus Hook.f., Fl. Brit. Ind. 6:38 (1890).

  HABITAT: unknown.
  ALTITUDINAL RANGE: unknown.
  DISTRIBUTION: Borneo (unspecified). Also Peninsular Malaysia, Sumatra.

P.leopardinum *(C.S.P.Parish & Rchb.f.) Seidenf. & Smitinand*, Orch. Thail. 4:535, fig.395 (1963). Type: Burma, Tenasserim, Moulmein, *Parish* 269 (holo. K).
  Thrixspermum leopardinum C.S.P.Parish & Rchb.f. in Trans. Linn. Soc. London, Bot. 30:145 (1874).

  HABITAT: hill forest; lower montane forest.
  ALTITUDINAL RANGE: 500 – 600m.

DISTRIBUTION: Borneo (SA). Also Burma, Thailand, Vietnam, Philippines.

P.pallidum *(Blume) Holttum* in Kew Bull. 14:270 (1960). Type: Java, Nusa Kambangan Islands, *Blume* s.n. (holo. L).
Dendrocolla pallida Blume, Bijdr.:290 (1825).
Sarcochilus pallidus (Blume) Rchb.f. in Ann. Bot Syst. 6:500 (1863).

HABITAT: coastal forest.
ALTITUDINAL RANGE: sea level.
DISTRIBUTION: Borneo (K). Also Peninsular Malaysia, Thailand, Sumatra, Java.

P.spathibrachiatum *(J.J.Sm.) Garay* in Bot. Mus. Leafl. 23(4):194 (1972). Type: Sabah, Mt. Kinabalu, Penibukan ridge, upper Pani Taki river, *Clemens* s.n. (holo. L).
Sarcochilus spathibrachiatus J.J.Sm. in Blumea 5:310 (1943).

HABITAT: hill and lower montane forest.
ALTITUDINAL RANGE: 900 – 1700m.
DISTRIBUTION: Borneo (SA). Endemic.

P.teres *(Blume) Holttum* in Kew Bull. 14:271 (1960). Type: Java, Buitenzorg, *Blume* 744 (lecto. L, chosen by E. A. Christenson, not formally published).
Dendrocolla teres Blume, Bijdr.:289 (1825).
Sarcochilus suaveolens (Roxb.) Hook.f., Fl. Brit. Ind. 6:33 (1890). Type: Bangladesh, Chittagong, J. R. anno 1810, *Roxburgh* drawing no.2350 (holo. K).

HABITAT: lowland and hill mixed dipterocarp forest; lower montane forest.
ALTITUDINAL RANGE: sea level – 1500m.
DISTRIBUTION: Borneo (K, SA, SR). Also Bangladesh, India, Nepal, Burma, Thailand, Laos, Cambodia, Vietnam, Sumatra, Java, Lesser Sunda Islands, Sulawesi, Maluku, Philippines.

P.unguiculatum *(Lindl.) H.Ae. Peders.* in Nord. J. Bot. 12(4):388 (1992). Type: Philippines, Luzon, Manila, *Cuming* s.n. (holo. K-LINDL).
Sarcochilus unguiculatus Lindl. in Bot. Reg. 26:67 (1840).
S.pallidus Masam., Enum. Phan. Born.:215 (1942), non (Blume) Rchb.f.

HABITAT: mangrove swamps; rocky seashores, lowland forest.
ALTITUDINAL RANGE: sea level – 800m.
DISTRIBUTION: Borneo (SA). Also Peninsular Malaysia, Java, Timor, Sulawesi, Maluku.

P.vriesii *(Ridl.) Garay* in Bot. Mus. Leafl. 23(4):194 (1972). Type: Kalimantan, Pontianak, colour plate by *Ridley* (holo. K).
Sarcochilus vriesii Ridl. in J. Linn. Soc., Bot. 31:297 (1893).

HABITAT: unknown.
ALTITUDINAL RANGE: sea level.
DISTRIBUTION: Borneo (K). Endemic.

RHYNCHOSTYLIS Blume

Tab. Pl. Jav. Orchid., fig.49 (1825).

Bijdr., 7:285 (1825).

Robust monopodial epiphytes or occasionally lithophytes. Stems 10 – 25 cm long. Leaves linear, strap-shaped, apex unequally bilobed, with acute lobes, fleshy, with several longitudinal pale lines, 15 – 35 x 3 – 7 cm. Inflorescences 2 – 4, erect or drooping, densely many flowered cylindric racemes to 35cm long, peduncle rather short. Flowers medium-sized, showy, often whitish with pink or purple markings, 2 – 3.5 cm across; sepals and petals spreading; petals smaller than sepals; lip immobile, entire or slightly 3-lobed, deeply saccate or with a short backward-pointing laterally compressed spur without interior ornaments; column short and stout, with a short, indistinct foot, rostellar projection and anther elongated; stipes long and slender, 2.5 times the diameter of the pollinia; viscidium small; pollinia 2, sulcate.

Four species distributed from India to China, south to Java, east to the Philippines.

R.gigantea *(Lindl.) Ridl.* in J. Linn. Soc., Bot. 32:356 (1896). Type: Burma, Prome, *Wallich* 7306 (holo. K-LINDL).
Saccolabium giganteum Lindl., Gen. Sp. Orch. Pl.:221 (1833).

HABITAT: lowland forest on limestone.
ALTITUDINAL RANGE: lowlands.
DISTRIBUTION: Borneo (SA). Also China (Hainan), Burma, Thailand, Indochina, Singapore, Anambas and neighbouring islands.

R.retusa *(L.) Blume,* Bijdr.:286, fig.49 (1825). Type: India, Malabar coast, *coll.?* (lecto. Rheede's t.1 in Hort. Malab. 12 (1703)).
Epidendrum retusum L., Sp. Plant.:953 (1753).

HABITAT: unknown.
ALTITUDINAL RANGE: unknown.
DISTRIBUTION: Borneo: (occurence doubtful). Widespread in S. and S.E. Asia.

ROBIQUETIA Gaudich.

in Freycinet, Voy.Uranie:426, Pl.34(1829).

Medium-sized monopodial epiphytes. Stems pendent, to 50 cm long. Leaves oblong to narrowly elliptic, apex unequally bilobed, to 20 x 5cm. Inflorescences simple or branched, densely many flowered, pendent, to 25cm long. Flowers small, yellow marked red or brownish, purple-red or pink; sepals and petals spreading; lip immobile, 3-lobed, with an often apically inflated spur which occasionally has some callosities or scales inside on either the back or front, or both, side lobes small, sometimes fleshy; column short and stout, without a foot; stipes long, linear, spathulate, often uncinnate, rarely hamate; viscidium usually small; pollinia 2, sulcate.

About 40 species distributed from the Himalayan region to Australia and the Pacific islands, with most species in Indonesia.

R.crassa *(Ridl.) Schltr.* in Feddes Repert., Beih.1:983 (1913). Type: Sarawak, near Kuching, *Haviland* s.n. (holo. SING).
  Saccolabium crassum Ridl. in J. Linn. Soc., Bot. 31:294 (1896).

HABITAT: unknown.
ALTITUDINAL RANGE: lowlands.
DISTRIBUTION: Borneo (SR). Endemic.

R.crockerensis *J.J.Wood* & *A.Lamb* in Wood, Beaman & Beaman, Plants of Mt. Kinabalu 2, Orchids: 306, fig.51 (1993). Type: Sabah, Mt. Kinabalu, Mahandei River, *Carr* 3107 (SFN 26473) (holo. K, iso. A, C, L, LAE, SING).

HABITAT: hill forest; lower montane forest, sometimes on ultramafic substrate.
ALTITUDINAL RANGE: 800 – 1900m.
DISTRIBUTION: Borneo (SA). Endemic.

R.pinosukensis *J.J.Wood* & *A.Lamb* in Wood, Beaman & Beaman, Plants of Mt. Kinabalu 2, Orchids: 308, fig.52 (1993). Type: Sabah, Mt. Kinabalu, Mesilau Trail, *Chow & Leopold* SAN 74504 (holo. K, iso. A, L, SAN, SAR, SING).

HABITAT: lower montane forest.
ALTITUDINAL RANGE: 1500 – 2000m.
DISTRIBUTION: Borneo (SA). Endemic.

R.spatulata *(Blume) J.J.Sm.* in Natuurk. Tijdschr. Ned.-Indië 72:114 (1912). Type: Java, Pantjar, *Blume* 1929 (holo. L, iso. K).
  Cleisostoma spatulata Blume, Bijdr.:364 (1825).
  C.spicatum Lindl. in Bot. Reg. 33: sub. t.32 (1847). Type: Borneo, *Rollinson* s.n. (holo. K-LINDL).
  Saccolabium borneense Rchb.f. in Gard. Chron.:563 (1881). Type: Borneo, cult. *Bull* (holo. W).

HABITAT: riverine forest.
ALTITUDINAL RANGE: sea level – 1400m.
DISTRIBUTION: Borneo (K, SA, SR). Also China (Hainan), India, Burma, Thailand, Indochina, Peninsular Malaysia, Singapore, Sumatra, Java, Halmahera.

R.transversisaccata *(Ames & C.Schweinf.) J.J.Wood* in Wood, Beaman & Beaman, Plants of Mt. Kinabalu 2, Orchids: 310 (1993). Type: Sabah, Mt. Kinabalu, Kiau, *Clemens* 166 (holo. Ames, iso BO).
  Malleola transversisaccata Ames & C.Schweinf., Orch. 6:228 (1920).

HABITAT: hill & lower montane forest, sometimes on ultramafic substrate.
ALTITUDINAL RANGE: 900 – 1500m.
DISTRIBUTION: Borneo (SA). Endemic.

VANDA Jones ex R. Br.

Bot. Reg. 6: t.506 (1820).

Medium-sized to large monopodial epiphytes or lithophytes. Stems usually stiffly erect, 10 – 100cm long. Leaves linear, strap-shaped, apex praemorse, rigid, v-shaped in cross-section, arranged in 2 rows at an acute angle to the stem, usually decurved. Inflorescences large, usually simple, with rather few well-spaced flowers. Flowers showy, up to 5cm across; sepals and petals free, elliptic-obovate, narrowed at the base, margins often reflexed, twisted or undulate, variously coloured, often tessellated; dorsal sepal 2 – 4cm long; lip immobile, 3-lobed, usually divided into a hypochile and epichile, often shorter than sepals and petals, with a short spur, no adornments within the spur; column short and stout, without a distinct foot, rostellar projection broad, shelf-like; stipes and viscidium short and broad; pollinia 2, sulcate.

Between 40 and 50 species distributed from Sri Lanka and India north to S. China, south to Indonesia, eastwards to Australia, New Guinea and the Solomon Islands.

V.dearei *Rchb.f.* in Gard. Chron., ser. 2, 26:648 (1886). Type: Borneo, cult. *Schroeder* (holo. W).

> HABITAT: riverine forest; rocky places.
> ALTITUDINAL RANGE: sea level – 300m.
> DISTRIBUTION: Borneo (K, SA, SR). Endemic.

V.hastifera *Rchb.f.* in Linnaea 41:30 (1887). Type: Borneo, cult. *Linden* (holo. W).
Renanthera trichoglottis Ridl. in J. Linn. Soc., Bot. 31:293 (1896). Type: Sarawak, *Haviland* s.n. (holo. K).

> ### var. **hastifera**
>
> HABITAT: lowland, coastal and hill forest; mangrove forest; limestone rocks.
> ALTITUDINAL RANGE: lowlands.
> DISTRIBUTION: Borneo (K, SA, SR). Endemic.

> ### var. **gibbsiae** *(Rolfe) P.J. Cribb* in Wood, Beaman & Beaman, Plants of Mt. Kinabalu 2, Orchids: 339 (1993). Type: Sabah, Mt. Kinabalu, Kiau, bridle path to Kaung, *Gibbs* 3970 (holo. K, iso. BM).
> *V.gibbsiae* Rolfe in J. Linn. Soc., Bot. 42:158 (1914).
>
> HABITAT: hill dipterocarp forest; lower montane oak-chestnut forest.
> ALTITUDINAL RANGE: 800 – 1300m.
> DISTRIBUTION: Borneo (SA). Endemic.

V.helvola *Blume,* Rumphia 4:49 (1849). Type: Java, *Blume* s.n. (holo. L).

> HABITAT: hill and lower montane forest.
> ALTITUDINAL RANGE: 400 – 1500m.
> DISTRIBUTION: Borneo (SA). Also Peninsular Malaysia, Sumatra, Java.

V.lamellata *Lindl.* in Bot. Reg. 24, misc. 66 (1838). Type: Philippines, Manila, *Loddiges* s.n. (holo. K-LINDL).

> HABITAT: lowland forest on limestone; coastal scrub, sometimes on wooded sea cliffs.
> ALTITUDINAL RANGE: sea level – 100m.
> DISTRIBUTION: Borneo (SA). Also Philippines.

V.scandens *Holttum* in Sarawak Mus. J. 5:389 (1950). Type: Sarawak, Kuching, *Holttum* s.n., cult. Singapore (holo. K).

> HABITAT: lowland and hill dipterocarp forest on sandstone and limestone.
> ALTITUDINAL RANGE: sea level – 1000m.
> DISTRIBUTION: Borneo (SA, SR). Endemic.

GROUP 4. Pollinia 2, porate. See p.366 for Group 5.

ASCOCENTRUM Schltr.

Feddes Repert., Beih. 1:975 (1913).

Medium-sized monopodial epiphytes. Stems short. Leaves strap-shaped, rarely almost terete (not in Bornean species), apex usually unequally bilobed or praemorse, leathery or fleshy. Inflorescences racemose, many flowered, usually shorter than the leaves. Flowers showy, to 2cm across, red, reddish-yellow or orange; sepals and petals free, spreading, ovate; lip 3-lobed, immobile, with a slender spur longer than the mid-lobe, side lobes small, erect, mid-lobe entire, narrowly ovate or linear, strap-shaped; column short and stout, foot absent, rostellar projection broadly triangular; stipes strap-shaped, twice the diameter of the pollinia, viscidium large, as broad as long; pollinia 2, porate or, in one species, sulcate.

Some 10 species distributed from N.W. Himalaya to China and Taiwan, south through Thailand to Malaysia, Indonesia and the Philippines.

A.insularum *Christenson* in Lindleyana 7(2):89 (1992). Type: Kalimantan, Maratoea, *Mjoberg* s.n. (holo. AMES).

HABITAT: unknown.
ALTITUDINAL RANGE: sea level.
DISTRIBUTION: Borneo (K). Endemic.

BIERMANNIA King & Pantl.

J. As. Soc. Bengal 66:591 (1897).

Small monopodial epiphytes. Stems short. Leaves unequally bilobed, lobules usually acute. Inflorescences short. Flowers small; lip sessile, narrowly but firmly adnate to column-foot, its sides enveloping or parallel with column which is about half as long, spur absent, base of lip with a small, narrow slit-like opening leading to a very small, hidden gibbosity; column with a short, distinct foot; pollinia 2, slightly sulcate or with a small cavity, on a linear-oblong stipes.

Nine species distributed in India, Peninsular Malaysia, Thailand, Sumatra, Java, Bali and Borneo.

This genus, which is easily confused with *Chamaeanthus*, has been reported to occur in Sabah (possibly *B.sarcanthoides* (Ridl.) Garay from Peninsular Malaysia). We have as yet received no material to confirm this.

DYAKIA Christenson

Orchid Digest 50:63 (1986).

Medium-sized monopodial epiphyte. Stems short. Leaves flat, apex obtusely bilobed. Inflorescences many flowered. Flowers bright pinkish-red with a white lip, about 1cm across; sepals and petals spreading; lip 3-lobed, immobile, side lobes very small, with a laterally compressed spur and prominent back wall callus; column short and stout, foot absent, rostellar projection sigmoid, longer than the rest of the column; stipes elongated, said to have 2 lateral basal appendages, but this may be an error; pollinia 2, porate.

A monotypic genus endemic to Borneo.

D.hendersoniana *(Rchb.f.) Christenson* in Orchid Digest 50:63 (1986). Type: Borneo, cult. *Henderson*, Wellington Nursery, London (holo. W, iso. K).
   Saccolabium hendersonianum Rchb.f. in Gard. Chron., n.s.:356 (1875).
   Ascocentrum hendersonianum (Rchb.f.) Schltr., Die Orchideen, ed. 1:576 (1914).

HABITAT: rather open, swampy forest on ultramafic substrate; hill forest.
ALTITUDINAL RANGE: sea level – 700m.
DISTRIBUTION: Borneo (K, SA, SR). Endemic.

GASTROCHILUS D.Don

Prodr. Fl. Nepal : 32 (1825).

Small to medium-sized monopodial epiphytes. Stems usually short, up to 30cm in a few species. Leaves narrowly elliptic or strap-shaped, apex unequally bilobed, acute or, rarely, with 3 setae, rather leathery. Inflorescences short and densely many flowered, ± umbellate, usually sessile, a few species having a short peduncle. Flowers with greenish yellow or yellow sepals and petals, often spotted red and a white lip, the hypochile often spotted red and the epichile with a central yellow patch, spotted red; sepals and petals narrowly obovate; lip adnate to the column, immobile, divided into a semi-globose saccate hypochile and a fan-shaped, often broadly triangular epichile, epichile often hairy or papillose, margin entire to fimbriate; column short and stout, foot absent, rostellar projection oblong; stipes strap-shaped, longer than twice the diameter of the pollinia; pollinia 2, porate, or 4, unequal (*G.patinatus*).

Around 38 species distributed from Sri Lanka and India through E. Asia to Japan, south to Indonesia.

G.sororius *Schltr.* in Feddes Repert. 12:315 (1913). Types: Java, Koeripan; Salak, am Tjiapoes; Tjikoneng, *Smith* s.n.; Salabintana, *Smith* s.n.; Gede, Tjibodas, *Smith* s.n., *Hallier* s.n.; Tjikorai, *Ader* s.n.; Slamat, bei Djedjek; Groeda, *Smith* s.n.(syn. BO).

   Saccolabium calceolare sensu J.J.Sm., Orch. Java: 632 (1905), non (Sm.) Lindl.

HABITAT: hill and lower montane forest.
ALTITUDINAL RANGE: 900 – 1500m.
DISTRIBUTION: Borneo (SA). Also Sumatra, Java.

LUISIA Gaudich.

in Freycinet, Voy. Uranie: 427, t.37 (1829 "1826").

Seidenfaden, G. (1971). Notes on the Genus *Luisia*. Dansk Bot. Ark. 27(4):1-101.

Erect or climbing epiphytes or lithophytes. Stems often branching at the base, forming a tufted habit, others with a single shoot up to 40cm long. Leaves well spaced, terete, linear. Inflorescences dense, almost sessile, with less than 10 flowers, peduncle and rachis very short. Flowers usually small, about 1 cm across, occasionally up to 3 cm across, with yellow or yellowish green sepals and petals and a purple lip; sepals and petals free, spreading; petals often longer and narrower than sepals; lateral sepals often keeled; lip fleshy, immobile, usually divided into a ± concave hypochile and a ± wrinkled epichile; column short and stout, foot absent, rostellar projection and stipes short; viscidium short and broad; pollinia 2, porate.

About 40 species distributed from Sri Lanka and India to China south to Thailand, Indochina, east to Japan, New Guinea and the Pacific islands. The centre of distribution lies in Burma and Thailand.

L.antennifera *Blume*, Mus. Bot. Lugd. Bat. 1:64 (1849). Type: Kalimantan, Martapura, *Korthals* s.n. (holo. L).

HABITAT: coastal forest; rubber plantations.
ALTITUDINAL RANGE: sea level – 300m.
DISTRIBUTION: Borneo (K, SR). Also Thailand, Vietnam, Peninsular Malaysia, Sumatra, Java.

L.curtisii *Seidenf.* in Bot. Tidsskr. 68:83 (1973). Type: Peninsular Malaysia, Penang, Bukit Penara, *Curtis* 1176 (holo. K).
L.tristis sensu Hook.f., Fl. Brit. Ind. 6:25 (1890), non (G. Forst.) Hook. f.

HABITAT: lower montane forest.
ALTITUDINAL RANGE: 900 – 1500m.
DISTRIBUTION: Borneo (SA). Also Thailand, Vietnam, Peninsular Malaysia.

L.volucris *Lindl.*, Fol. Orch., Luisia:1 (1853). Type: India, Meghalaya State, Khasi (Khasia) Hills, *Lobb* s.n. (holo. K-LINDL).

HABITAT: unknown.
ALTITUDINAL RANGE: below 900m.
DISTRIBUTION: Borneo (SA). Also N.W. India.

L.zollingeri *Rchb.f.* in Ann. Bot. Syst. 6:622 (1863). Type: Java, Salak, *Zollinger* 1265 (holo. W).
L.brachystachys auct., non (Lindl.) Blume.

HABITAT: podsol forest; riverine forest.
ALTITUDINAL RANGE: 100 – 1500m.
DISTRIBUTION: Borneo (K, SA). Also Andaman Islands, Thailand, Vietnam, Peninsular Malaysia, Sumatra, Batu Islands, Java.

GROUP 5. Pollinia 2, entire, neither cleft, split or porate.

CHAMAEANTHUS Schltr.

in J.J.Sm., Orch. Java: 552 (1905).

Small monopodial epiphytes. Stems very short or up to a few centimeters long. Leaves few, somewhat fleshy, to 7 x 1 cm. Inflorescences simple, with up to 20 flowers, rachis usually club-shaped. Flowers small, ephemeral, only a few open at a time, greenish-yellow, looking superficially like a *Bulbophyllum*; sepals and petals similar, often linear and acuminate, to 4 – 6 x 3 mm; lateral sepals adnate to the column-foot; lip mobile, 3-lobed, side lobes ear-like, margins sometimes fimbriate, mid-lobe conical, fleshy; column short and stout with a 3 mm long foot, rostellar projection elongate; stipes strap-shaped, about twice the diameter of the pollinia; viscidium small, obovate; pollinia 2, entire, although some specimens have a very tiny notch.

About 3 species distributed from S. Thailand east to Java and New Guinea, north to the Philippines.

C.brachystachys *Schltr.* in J.J.Sm., Orch. Java: 552 (1905). Types: Java, Tegal, *Raciborski* s.n. (syn. B, destroyed); Djokjakarta, *Raciborski* s.n. (syn. B, destroyed).

HABITAT: swampy podsol forest; hill forest.
ALTITUDINAL RANGE: lowlands – 900m.
DISTRIBUTION: Borneo (K, SA). Also Thailand, Java.

NOTE: Bornean specimens differ in having entire rather than serrate lip side lobes. *C.wenzelii* Ames from the Philippines may be conspecific.

A second species, material of which we have not seen, has been recorded from lower montane forest on ultramafic substrate in Sabah by Lamb (pers. comm.).

CHRONIOCHILUS J.J.Sm.

Bull. Jard. Bot. Buitenzorg, ser. 2, 26:81 (1918).

Small monopodial epiphytes. Stems very short. Leaves few, flat. Inflorescences simple, few flowered, rachis sometimes flattened. Flowers small; sepals and petals free, spreading; lip mobile, sessile, arrow-head shaped, conical, with large ear-like side lobes, spur absent, the conical median part being solid; column with a distinct foot, rostellar projection prominent; stipes linear-oblong; viscidium large, oval, at least half as long as the stipes; pollinia 2, entire.

Four species distributed from S. Thailand through Malaysia to Indonesia.

C.ecalcaratus *(Holttum) Garay* in Bot. Mus. Leafl. 23(4):166 (1972). Type: Peninsular Malaysia, Pahang, Mentakab, *Carr* s.n. (holo. SING).
    Sarcochilus carrii Holttum in Gard. Bull. Straits Settlem. 11:288 (1947), non L.O.Williams.
    S.ecalcaratus Holttum, Rev. Fl Malay. 1:684 (1953).

HABITAT: hill forest on ultramafic substrate.

ALTITUDINAL RANGE: 700m.
DISTRIBUTION: Borneo (SA). *C.ecalcaratus* is native to Peninsular Malaysia.

NOTE: Bornean material differs from *C.ecalcaratus* in minor floral details.

C.minimus *(Blume) J.J.Sm.* in Bull. Jard. Bot. Buitenzorg, ser. 3, 8:366 (1927). Type: Java, Pantjar, *Blume* s.n. (holo. BO).
Dendrocolla minima Blume, Bijdr.:290 (1825).

HABITAT: low stature hill forest on ultramafic substrate.
ALTITUDINAL RANGE: 500 – 700m.
DISTRIBUTION: Borneo (SA, SR). Also Peninsular Malaysia, Java.

C.virescens *(Ridl.)Holttum* in Kew Bull. 14,2:273 (1960). Type: Peninsular Malaysia, Perak, Tapah, *Aeria* s.n. (holo. SING).
Sarcochilus virescens Ridl. in J. Straits Branch Roy. Asiat. Soc. 39:85 (1903).

HABITAT: low stature hill forest on ultramafic substrate.
ALTITUDINAL RANGE: 500 – 1000m.
DISTRIBUTION: Borneo (SA). Also Peninsular Malaysia, Thailand.

GROSOURDYA Rchb.f.

Bot.Zeitung (Berlin) 22:297 (1864).

Small monopodial epiphytes. Stems short. Leaves few, flat, to 10 x 2cm. Inflorescences usually shorter that the leaves, often many borne simultaneously on a plant, peduncle longer than the rachis, both prickly-hairy, with 1 or 2 flowers open at a time. Flowers ephemeral, to 1.5cm across, yellow marked red, lip white; sepals and petals free, spreading; lip mobile, 3-lobed, side lobes narrow, erect, recurved, mid-lobe with a distinct spur, blade apically bilobed with a small median tooth, giving a '4-lobed' appearance; column long, slender, bent forward at an obtuse angle to the base of the stigma, as long as or longer than the foot, rostellar projection elongated; stipes narrowly triangular; viscidium triangular; pollinia 2, entire.

About 10 species distributed in the Andaman Islands, Burma, Thailand and Indochina to Malaysia, Indonesia and the Philippines.

G.appendiculata *(Blume) Rchb.f.*, Xenia Orch. 2:123 (1867). Type: Java, Kuripan, *Blume* 1018 (holo. L).
Dendrocolla appendiculata Blume, Bijdr.:289 (1825).
Sarcochilus hirtulus Hook.f., Fl. Brit. Ind. 6:39 (1890). Type: Peninsular Malaysia, Perak, *Scortechini* 1559 (syn. K); Malacca, *Maingay* 3119 (syn. K).

HABITAT: riverine forest; lowland and hill forest.
ALTITUDINAL RANGE: sea level – 500m.
DISTRIBUTION: Borneo (K, SA, SR). Also Andaman Islands, Burma, Thailand, Vietnam, Peninsular Malaysia, Java, ?Sunda Islands, Philippines.

G.emarginata *(Blume) Rchb.f.*, Xenia Orch. 2:123 (1867). Type: Java, Pantjar and Meggamedong, *Blume* 1944 (holo. L).
Dendrocolla emarginata Blume, Bijdr.:290 (1825).
Sarcochilus emarginatus (Blume) Rchb.f. in Ann. Bot. Syst. 6:500 (1861).

HABITAT: unknown.

ALTITUDINAL RANGE: unknown.
DISTRIBUTION: Borneo (unspecified). Also Sumatra, Java.

MALLEOLA J.J.Sm. & Schltr.

Feddes Repert. Beih. 1:979 (1913).

Small monopodial epiphytes. Stems short or elongate and pendulous, to 30cm long. Leaves scattered along stem, often flushed purple-red, 7 – 4 x 1 – 3.5cm. Inflorescences many flowered, mostly shorter than the leaves, usually pendent. Flowers small, yellow, flushed purple, red or mauve; sepals and petals free, spreading; dorsal sepal to 4mm long; lip 3-lobed, immobile, with a variably shaped cylindrical spur with interior ornaments, side lobes short, broadly triangular, mid-lobe very small, conical or linear; column short and stout, hammer-shaped, foot absent; anther ± dorsal; stipes spathulate, very broad below the pollinia; viscidium very small; pollinia 2, entire.

Around 30 species distributed from Thailand and Vietnam to Malaysia and Indonesia, the Philippines, east to New Guinea and the Pacific islands. The centre of distribution lies in Indonesia. Several unidentified species occur in Borneo.

M.altocarinata *Holttum* in Gard. Bull.Singapore 11:283 (1947). Type: Peninsular Malaysia, Selangor, Ginting Simpah, *Mungo Park* s.n. (holo. SING).

HABITAT: riverine forest.
ALTITUDINAL RANGE: 200 – 300m.
DISTRIBUTION: Borneo (SA). *M.altocarinata* is native to Peninsular Malaysia and Thailand.

NOTE: The specimen *Lamb* AL 81/83(K) differs from *M.altocarinata* by its smaller flowers with longer spur and vertical callus at the base of the lip mid-lobe.

M.glomerata *(Rolfe)P.F.Hunt* in Kew Bull. 24:99 (1970). Type: Borneo, cult. *Rothschild* (holo. K).
Saccolabium glomeratum Rolfe in Kew Bull.:342 (1913).

HABITAT: unknown.
ALTITUDINAL RANGE: unknown.
DISTRIBUTION: Borneo (unspecified). Endemic.

M.kinabaluensis *Ames & C.Schweinf.*, Orch. 6:225, Pl.96 (1920). Type: Sabah, Mt. Kinabalu, Kiau, *Clemens* 330 (holo. AMES).

HABITAT: hill and lower montane forest.
ALTITUDINAL RANGE: 900 – 1500m.
DISTRIBUTION: Borneo (SA). Endemic.

M.penangiana *(Hook.f) J.J.Sm. & Schltr.* in Feddes Repert. Beih. 1:981 (1913). Type: Peninsular Malaysia, Perak, Sonkey River, *Curtis* 505 (holo. K).
Saccolabium penangianum Hook.f., Fl. Brit. Ind.6:57 (1890).

HABITAT: mixed hill dipterocarp forest.
ALTITUDINAL RANGE: 300m.
DISTRIBUTION: Borneo (SA). Also Thailand, Peninsular Malaysia.

M.serpentina *(J.J.Sm.) Schltr.* in Feddes Repert. Beih. 1:981 (1913). Type: Kalimantan, Pamatan, *Korthals* s.n. HLB 904, 84-115-116 (holo. L).
   Saccolabium serpentinum J.J.Sm. in Recueil Trav. Bot. Néerl. 1:157 (1904).
   HABITAT: unknown.
   ALTITUDINAL RANGE: unknown.
   DISTRIBUTION: Borneo (K). Endemic.

M.witteana *(Rchb.f.) J.J.Sm. & Schltr.* in Feddes Repert. Beih. 1:981 (1913). Type: Java, sent by *Lammers*, cult. R. B. G. Leiden (holo. W).
   Saccolabium witteanum Rchb.f. in Gard. Chron. 20:618 (1883).
   S.kinabaluense Rolfe in J. Linn. Soc., Bot. 42:158 (1914). Type: Sabah, Mt. Kinabalu, Lobong, *Gibbs* 4111 (holo. K).
   HABITAT: lower montane forest.
   ALTITUDINAL RANGE: 1500m.
   DISTRIBUTION: Borneo (SA). Also Peninsular Malaysia, Sumatra, Java.

MICROTATORCHIS Schltr.

in K.Schum. & Lauterb., Nachtr. Fl. Schutzgeb. Südsee:224 (1905).

Small monopodial epiphytes similar in habit to *Taeniophyllum*. Stems very short, with or without leaves. Inflorescences lateral, racemose, elongating gradually; rachis angled; floral bracts persistent, alternate, distichous. Flowers pale yellowish-green, not opening widely; sepals & petals similar, fused at base forming a short tube; lip entire or 3-lobed, with a bristle or tooth inside near apex, shortly spurred; column short, foot absent; pollinia 2, entire.

About 47 species distributed in Java, Borneo, Sulawesi and the Philippines eastward through New Guinea to the Solomon Islands, Vanuatu, New Caledonia, Fiji and other Pacific islands. The centre of speciation lies in New Guinea.

M.javanica *J.J.Sm.* in Bull. Jard. Bot. Buitenzorg, ser. 2, 26:115 (1918). Types: Java, Priangan, Tjidadap Tjibeber, *Bakhuizen van den Brink* s.n., *Winckel* s.n. (syn. BO).
   HABITAT: lower montane forest.
   ALTITUDINAL RANGE: 1500m.
   DISTRIBUTION: Borneo (SA). Also Java.
   NOTE: A new generic record for Borneo.

PENNILABIUM J.J.Sm.

Bull. Jard. Bot. Buitenzorg, ser. 2, 14:47 (1914).

Small monopodial epiphytes. Stems very short. Leaves few, clustered, rather fleshy, often twisted at the base, to 11 x 3cm. Inflorescences racemose, 3 – 8cm long, rachis somewhat thickened and complanate, with 1 or 2 flowers open at a time placed in 2 rows. Flowers lasting for a day or two, white, cream, yellow or orange; sepals and petals 1 – 2cm long; petals sometimes slightly toothed; lip 3-lobed, immobile, spurred, internal callosities absent, side lobes either well developed and truncate or reduced to small ear-like lobes, when present often fimbriate or toothed, mid-lobe large, fleshy and solid or reduced to a small fleshy lobe; column short, ± compressed dorsally, foot absent, rostellar projection elongate, prominent; stipes long, much broadened below the pollinia, often spathulate, 3 – 5 times the diameter of the pollinia; viscidium very small; pollinia 2, entire.

Some 10 – 12 species distributed from India (Assam) through Thailand and Malaysia to Indonesia and the Philippines.

P.angraecoides *(Schltr.) J.J.Sm.* in Bull. Jard. Bot. Buitenzorg, ser. 2, 13:47 (1914). Type: Kalimantan, Long Sele, *Schlechter* 13468 (holo. B, destroyed).
Saccolabium angraecoides Schltr. in Bull. Herb. Boissier, ser. 2, 6:472 (1906).

HABITAT: riverine forest; lower montane forest.
ALTITUDINAL RANGE: 400 – 1400m.
DISTRIBUTION: Borneo (K). Endemic.

P.lampongense *J.J.Sm.* in Bull. Jard. Bot. Buitenzorg, ser. 3, 10:74 (1928). Type: Sumatra, Lampoengs, Gunung Rate Telanggoran, *Iboet* 96 (holo. BO).

HABITAT: hill forest.
ALTITUDINAL RANGE: 200 – 300m.
DISTRIBUTION: Borneo (SA, SR). Also Sumatra.

PORRORHACHIS Garay

Bot. Mus. Leafl.23(4):191 (1972).

Small monopodial epiphytes. Stems short or long. Leaves narrow, rigid, 2.5 x 0.6 to 15 x 1 cm. Inflorescences simple, stiff, perpendicular to the stem, sometimes longer than the leaves, with up to 10 lax flowers. Flowers very small, greenish or yellow; sepals 2.5 – 7mm long; dorsal sepal incurved over the column; lateral sepals adpressed to the lip; lip to 3mm long, immobile, fleshy, narrow, laterally compressed, with a spur-like tubular cavity; column short and stout, foot absent, rostellar projection very short; stipes broadly rhomboid, about as long as the pollinia; viscidium small; pollinia 2, entire.

Two species only, viz. *P.galbina* (J.J.Sm.) Garay from Java and Borneo, and *P.macrosepala* (Schltr.) Garay from Sulawesi.

P.galbina *(J.J.Sm.) Garay* in Bot. Mus. Leafl. 23(4):191 (1972). Type: Java, Priangan, Pasir Angin (Tjadas Malang), Tjibeber, *Winckel* 339 (holo. L).
Saccolabium galbinum J.J.Sm. in Bull. Jard. Bot. Buitenzorg, ser. 2, 26:97 (1918).

HABITAT: lower montane forest; moss forest.
ALTITUDINAL RANGE: 1100 – 1500m.
DISTRIBUTION: Borneo (SA). Also Java.

SPONGIOLA J.J.Wood & A. Lamb

Orch. Borneo 1:283 (1994).

Small monopodial epiphyte. Stem short, to 7 cm long. Leaves 3 – 6, oblong or oblong-elliptic, apex obtuse and asymmetrical, 7.5 – 20 x 1.8 – 3 (– 4.2) cm.

Inflorescences pendent, racemose, occasionally branched, peduncle 12 – 15 cm long, rachis 18 – 20 cm long, with up to 25 flowers, usually 5 – 6 open at a time, sometimes opening in the middle of the raceme. Flowers ephemeral, lasting one day only, with semi-translucent pale yellow to yellowish-green sepals and petals and a white lip, spotted and flushed purple, spur yellow; sepals and petals free, spreading; dorsal sepal 9 – 10 x 3.5 mm; lip immobile, 3-lobed, ecallose, side lobes tiny, ear-like, mid-lobe resembling a small spongy pouch, hollow above, solid towards the apex, shallowly concave beneath, papillose-verrucose, 3 x 2 – 2.5 mm, with a conical spur 2mm long; column oblong, foot absent, 1.5 mm long, rostellar projection very short; stipes spathulate, broadened below pollinia; viscidium cucullate; pollinia 2, entire.

A monotypic genus endemic to Borneo.

S.lohokii *J.J.Wood & A.Lamb* in Orch. Borneo 1:285, fig.89, Pl.19 & 20A (1994). Type: Sabah, Batu Urun, *Lohok* in *Lamb* AL 426/85 (holo. K).

HABITAT: riverine forest on limestone and sandstone ridges.
ALTITUDINAL RANGE: 300 – 500m.
DISTRIBUTION: Borneo (SA). Endemic.

TUBEROLABIUM Yamam.

Bot. Mag. Tokyo 38:209 (1924).

Wood, J.J. (1990). Notes on *Trachoma, Tuberolabium* and *Parapteroceras* (Orchidaceae). Nord. J. Bot. 10 (5):481-486.

Trachoma Garay in Bot. Mus. Leafl. 23 (4):207 (1972).

Small monopodial epiphytes. Stems short. Leaves few, linear-falcate or strap-shaped, 7 – 14 x 2 – 3 cm. Inflorescences few to many flowered, a few flowers open at once or all open together, peduncle short, rachis fleshy, sulcate, terete, sometimes clavate. Flowers rather short-lived or lasting for about a week, up to 9 mm across, white, yellowish or greenish with various purple, brownish-purple or red markings; sepals and petals free, spreading; dorsal sepal 4.5 x 2.5 mm; petals narrower; lip 3-lobed, adnate to the base of the column, immobile, very fleshy, side lobes very small, tooth-like, mid-lobe very fleshy, laterally compressed, with incurved margins, 3 x 1.5 mm, with a conical spur as long as the blade, spur ornaments absent; column short and stout, foot absent, rostellar projection short; stipes linear, once or twice the diameter of the pollinia; viscidium small, ovate; pollinia 2, entire.

Eleven species distributed from India, Thailand and Peninsular Malaysia, north to Taiwan and the Philippines, south to Indonesia, east to New Guinea, Australia and the Pacific islands.

T.rhopalorrhachis *(Rchb.f.) J.J.Wood* in Nord. J. Bot.10, 5:482 (1990). Type: Java, Bondong Province, i.e. West Java Province, Bandung, *Zollinger* s.n. (holo. W).
Dendrocolla rhopalorrhachis Rchb.f., Xenia Orch. 1:214, t.86 (1856).
Trachoma rhopalorrhachis (Rchb.f.) Garay in Bot. Mus. Leafl. 23(4):208 (1972).

HABITAT: low stature hill forest on ultramafic substrate; lower montane forest.
ALTITUDINAL RANGE: 600 – 1500m.
DISTRIBUTION: Borneo (SA). Also Thailand, Peninsular Malaysia, Sumatra, Java, Maluku (Ambon).

POORLY KNOWN TAXA

Chelonistele tenuiflora *(Ridl.) Pfitzer* in Engl., Pflanzenr. 4,50:138 (1907). Type: Sarawak, *Haviland* s.n. (holo. SING).
Coelogyne tenuiflora Ridl. in J. Linn. Soc., Bot. 31:287 (1896).

De Vogel (1986) says that this taxon may possibly be conspecific with *Geesinkorchis phaiostele* (Ridl.) de Vogel.

Sarcochilus stellatus *Ridl.* in J. Straits Branch Roy. Asiat. Soc. 49:39 (1907). Type: Sarawak, locality unknown, cult. *Hose* (holo. ?SING).
Sarcochilus sensu stricto is confined to northern and eastern Australia, the centre of speciation, with a few outlying species in New Guinea, New Caledonia and Fiji. Most of the taxa formerly placed here have been transferred elsewhere, those from Borneo mainly to *Pteroceras*.

We have not seen authentic material of *S.stellatus*. Pedersen (pers. comm.) has doubts as to the generic position of this taxon, but suggests that it could belong in *Tuberolabium*. Ridley's description does not provide much information on pollinarium structure, but notes that the flowers appear to be regularly self pollinated. He describes the flowers as pale greenish-yellow, the lip having yellowish side lobes with dull red markings inside, a white callus with two violet spots and a white spur.

Tainia borneensis *Ridl.* in J. Straits Branch Roy. Asiat. Soc. 49:32 (1907). Type: Sarawak, *Hewitt* 15 (holo. K).

It is clear from the poorly preserved type material and from the original description that this is not conspecific with *Ania borneensis* (Rolfe) Senghas. The sterile shoots are one leaved and there is a cylindrical terminal internode as in *Tainia*, but the inflorescence is velutinous and arises from its base. The lip is saccate with an elongated acuminate tip suggesting a *Plocoglottis*.

EXCLUDED TAXA

Anoectochilus setaceus *Blume,* Bijdr.:412 (1825). Type: Java, Salak, Gede, Tankuwan Prahu, *Blume* s.n. (holo. ?L).

Records of this Javan species (Carr 1935) probably refer to *A. longicalcaratus* J.J.Sm.

Ascochilus mindanaensis *(Ames) Christenson* in Indian Orchid J. 1(4):151 (1985). Type: Philippines, Mindanao, Agusan Province, Butuan, *Weber* 139 (holo. AMES, iso. MO).

Seidenfaden (1988:234) quotes a distribution of 'Sumatra, Java, Borneo (?), Philippines' for *A. mindanaensis.* Although it still remains to be confirmed whether *Ascochilus* occurs on Borneo, we would expect this easily overlooked genus to be present there. Comber (1990) does not record it from Java.

Calanthe tunensis *J.J.Sm.* in Bull. Inst. Bot. Buitenzorg, 7:3(1900). Type: Maluku, Ambon, Tuna, *coll. unknown* (holo. BO).
C. tunensis appears in Masamune's enumeration (1942), presumably based on the quoted reference to Smith (1928) which in fact refers to Buru in Maluku (Moluccas) and not Borneo.

Corybas fornicatus *(Blume) Rchb.f.,* Beitr. Syst. Pfl.:42(1871). Type: Java, Salak, *Blume* s.n. (holo. L).

This species is listed from Sarawak by Ridley (1896) and Ames in Merrill (1921). It is a native of Peninsular Malaysia, Java and Bali and according to Dransfield et al. (1986) all records for Borneo are erroneous.

Cymbidium aloifolium *(L.) Sw.* in Nova Acta Regiae Soc. Sci. Upsal. 6:73 (1799). Type: India, Coromandel coast (lecto. *Roxburgh's* t.44 in Pl. Coast Coromandel 1(1795)).

A widespread species recorded from as far south as Java. Although reported from Borneo by J.J.Smith (1905), its occurence there remains unsubstantiated.

Dendrobium atropurpureum *(Blume) Miq.,* Fl. Ind. Bat. 3:644 (1859). Type: New Guinea, locality unknown, *Blume* s.n. (holo. ?L).

Records of this New Guinean species probably refer to either *D. concinnum* Miq. or *D. excavatum* (Blume) Miq.

Dendrobium bicaudatum *Reinw. ex Lindl.* in J. Linn. Soc., Bot. 3:20 (1859). Type: 'Java' (holo. *Reinwardt* illustration, ?L, copy at K).
D. burbidgei Rchb.f. in Gard. Chron. 2:300 (1878). Type: Philippines, Sulu Archipelago, *Burbidge* in *Veitch* s.n. (holo. W, iso. K).

This species is native to Sulawesi, Maluku (Ambon and Seram) and the Sulu Archipelago. References to its possible occurrence on Borneo are incorrect.

Dendrobium derryi *Ridl.,* Mat. Fl. Mal. Pen. 1:52 (1907). Type: Peninsular Malaysia, Perak, Larut Hills, *Derry* s.n. (holo. SING).

No material has been found to support Ridley's claim that this species occurs on Borneo.

Dendrobium scabrilingue *Lindl.* in J. Linn. Soc., Bot. 3:25 (1859). Type: Burma, Tenasserim, *Lobb* s.n., imported by *Veitch*, originally thought to have come from Borneo (holo. K-LINDL).

The original Bornean record of this species, a native of Burma, Laos and Thailand, was incorrect.

Dendrochilum microchilum *(Schltr.) Ames*, Orch. 2:87 (1908). Type: Philippines, imported from Manila, cult. Sandakan, Sabah, *Schlechter* s.n. (holo. B, destroyed).

A Philippine species described from material cultivated in Borneo.

Eria kingii *F.Muell.* in S. Sci. Rec. 2, 4:71 (1882). Type: Solomon Islands, Boneta, *Goldfinch* s.n. (holo. MEL).
E.moluccana Schltr. & J.J.Sm., Orch. Amb.:74 (1905). Types: Maluku, Ambon, Hila, *Treub* s.n. (syn. B, destroyed); between Toelehoe and Soeli; Larike; Alang,etc., *Smith* s.n. (syn. B, destroyed).

E.moluccana appears in Masamune's enumeration (1942), presumably based on the quoted reference to Smith (1928) which in fact refers to Buru in Maluku (Moluccas) and not Borneo.

Neuwiedia zollingeri *Rchb.f.* var. **zollingeri** in Bonplandia 5:58 (1857). Type: Java, *Zollinger* 2808 (holo. W).

The Low collection from Borneo, in the Kew herbarium, cited by Reichenbach in a note under the original description of *N.zollingeri* belongs to *N.veratrifolia*.

Paphiopedilum barbatum *(Lindl.)Pfitzer* in Jahrb. Wiss. Bot. 19:159 (1888). Type: Peninsular Malaysia, Mt. Ophir, *Cuming* s.n. (holo. K-LINDL).
Cypripedium nigritum Rchb.f. in Gard. Chron., 16:102 (1882). Type: 'Borneo', cult. *New Bulb Co.* (holo. W).
Cordula nigrita (Rchb.f.) Rolfe in Orchid Rev. 20:2 (1912).

C.nigritum is based on a plant said to have been imported from Borneo. Many early importations were distributed with false provenance to mislead competitors and it is probable that this taxon originated elsewhere. Rolfe (1896) equated plants imported by Messrs. Low & Co. from Borneo with *C.nigritum*. However, Low's specimens preserved at Kew are, according to Cribb (1987), small flowered *P.lawrenceanum* with warts on both upper and lower petal margins. *P.barbatum* is, so far as is known, restricted to Peninsular Malaysia, including Pinang (Penang) Island.

Pennilabium struthio *Carr* in Gard. Bull. Straits Settlem. 5:151, Pl.4.(1930). Type: Peninsular Malaysia, Pahang, Kuala Teku, *Carr* 174 (holo. K).

Seidenfaden (1988:330) includes Sabah (*Lamb* AL 382/85) in his distribution for this species. Comparison of *Lamb* AL 382/85 with type material of *P.struthio* shows that it differs in several respects. The lip side lobes are much shorter, the front of the callus has two teeth (in *P.struthio* it is longer and entire), the column is porrect and touches the callus (in *P.struthio* it is angled sharply back and placed far from the callus) and the rostellum is shorter.

The Thai plant figured by Seidenfaden (1988) & *Lamb* AL 382/85 appear to be identical and seem rather to be closer to the Sumatran *P.lampongense* J.J.Sm. than to *P.struthio*. The Bornean plant is therefore provisionally placed here under *P.lampongense*.

Phalaenopsis aphrodite *Rchb.f.* in Hamburger Garten-Blumenzeitung 18:35 (1862). Type: Philippines, Manila, *collector unknown* (holo. W).

Phalaenopsis equestris *(Schauer) Rchb.f.* in Linnaea 22:864 (1850). Type: Philippines, Luzon, Manila, *Meyens* s.n. (holo.:W).

Both these species occur in Taiwan and the Philippines and are included in an unpublished list of Bornean *Aeridinae* compiled by E. A. Christenson in 1986. We have no evidence that either occur on Borneo.

Pholidota pallida *Lindl.* in Bot. Reg. 21:1777 (1835). Type: Nepal, Gossain Kan, *Wallich* s.n. (lecto. K-LINDL).

References to this species, which is restricted to mainland Asia, are referable to *P.imbricata* Hook.

ADDITIONAL TAXA

Dendrobium excavatum *(Blume) Miq.*, Fl. Ind. Bat. 3:644 (1859). Type: Java, Salak, *Blume* s.n. (holo. L)
  Oxystophyllum excavatum Blume, Bijdr.:336 (1825).

HABITAT: unknown.
ALTITUDINAL RANGE: unknown.
DISTRIBUTION: Borneo (K). Also Peninsular Malaysia, Cambodia, Sumatra, Bangka, Mentawai, Java, Buru, ?Irian Jaya.
GENERIC SECTION: *Oxystophyllum.*

Dendrobium flexile *Ridl.* in J. Linn. Soc., Bot. 32:251 (1896). Type: Singapore, Bukit Timah, *Ridley* s.n. (holo. SING).

HABITAT: mangrove.
ALTITUDINAL RANGE: sea level.
DISTRIBUTION: Borneo (SR). Also Peninsular Malaysia, Singapore, Thailand, Sumatra.
GENERIC SECTION: *Strongyle.*

ACKNOWLEDGEMENTS

The preparation of this checklist has been generously supported by the Zürich Foundation for the Preservation and Study of Orchids (Stiftung zum Schutze und zur Erhaltung Wildwachsender Orchideen, Zürich) and we would particularly like to thank their Chairperson, Mrs. Meta Held, for her unstinted support over several years. Their patience in waiting for the incorporation of the data from the recent publication on the orchids of Mt. Kinabalu (Wood, Beaman & Beaman, 1993) has been much appreciated and will make this checklist a more useful and complete account.

Our collaborators on the *Orchids of Borneo* project include botanists in Sabah, Leiden, Singapore and Kew. We would particularly like to thank Chan Chew Lun, Tony Lamb, Phyau Soon Shim and William Wong in Sabah, Ed de Vogel and Jaap Vermeulen in Leiden, Kiat Tan in Singapore, Mark Clements in Canberra, Dr. Gunnar Seidenfaden in Denmark, Jim Comber of Southampton, Sarah Thomas at Kew and especially John and Reed Beaman of Michigan for their help and comments on this checklist.

We would like to thank Datuk Lamri Ali, the Director of Sabah National Parks, Malaysia and his staff, Dr.Suhirman and his staff in Bogor, Indonesia and the Director of the Forest Herbarium in Brunei for their assistance in the field. We thank the Director and staff of the Bogor Herbarium, the Curator of the Forest Herbarium in Kuching, Dr. Kiat Tan, Director of Singapore Botanic Gardens, the curators and staff of the herbaria at Edinburgh, the Natural History Museum in London, Leiden, Missouri, Paris and Sandakan for allowing us access to their collections. Professor Gren Lucas has given his wholehearted support to us in this venture.

The collation of the information and typing of the manuscript has been done by Mrs. Audrey Thorne and we are immensely grateful for all her hard work and patience over the past three years.

The illustrations have been drawn by Eleanor Catherine, Chan Chew Lun, Sarah Thomas and Mutsuko Nakajima. The colour photographs have been kindly lent by Reed Beaman, Sheila Collenette, Jim Comber, Phillip Cribb, John Dransfield, David Du Puy, Tony Lamb, Gwilym Lewis, Emil Lückel, Ray Oddy, Martin Sands and L. Vogelpoel. Dr. Mike Lock has overseen the production of the volume. The design is the work of the Media Resources Unit at Kew under the leadership of Milan Svanderlik.

LITERATURE CITED

Ames, O. in Merrill, E. (1921). A bibliographic enumeration of Bornean plants. J. Straits Branch Roy. Asiat. Soc., Special No.:134 – 204.

Ames, O. & Schweinfurth, C. (1920). The Orchids of Mount Kinabalu, British North Borneo. Merrymount Press, Boston.

Beccari, O. (1902). Nelle Foreste di Borneo – viaggi e ricerche di un naturalista. Salvadore Landi, Firenze.

Comber, J.B. (1990). Orchids of Java. Bentham-Moxon Trust. Royal Botanic Gardens, Kew.

Kores, P. (1989). A Precursory Study of Fijian Orchids. Allertonia 5(1):1 – 222.

Masamune, G. (1942). Enumeratio Phanerogamarum Bornearum :110 – 226. Tokyo.

Ridley, H.N. (1896). An Enumeration of all Orchidaceae hitherto recorded from Borneo. J. Linn. Soc., Bot. 31:261 – 305.

Russan, A. & Boyle, F. (1893). The Orchid Seekers – a story of adventure in Borneo. Chapman & Hall, London.

Seidenfaden, G. (1988). Orchid Genera in Thailand XIV. Fifty-nine vandoid Genera. Opera Bot. 95.

Smith, J.J. (1928). Orchidaceae Buruensis. Bull. Jard. Bot. Buitenzorg, ser. 3, 9:339 – 481.

Valmayor, H. (1984). Orchidiana Philippiniana Vol. 1 & 2. Eugenio Lopez Foundation Inc., Manila.

Vermeulen, J. (1991). Orchids of Borneo 2: *Bulbophyllum.* Royal Botanic Gardens, Kew and Toihaan Publishing Co., Kota Kinabalu.

Wood, J.J. (1984). A preliminary annotated checklist of the orchids of Gunung Mulu National Park, Sarawak, in Jermy, A.C. (editor). Studies on the Flora of Gunung Mulu National Park, Sarawak : 1 – 39. Kuching, Sarawak.

Wood, J.J., Beaman, R.S. & Beaman, J.H. (1993). The Plants of Mount Kinabalu 2, Orchids. Royal Botanic Gardens, Kew.

INDEX TO FIGURED TAXA

INDEX TO TAXA FIGURED IN COLOUR PLATES

INDEX TO SCIENTIFIC NAMES

Accepted names are in roman type. Synonyms are in *italics*. Numbers in **bold** refer to figures. Colour plates are designated Pl.

multiflexum J.J. Sm., 302
multiflorum (Breda) Kraenzl., 302
muluense J.J. Wood, 280
muscohaerens J.J. Verm. & A.Lamb
  ined., 302
mutabile (Blume) Lindl.
  var. mutabile, 302
  var. obesum J.J. Verm., 302
nabawanense J.J. Wood & A.Lamb,
  302
nanobulbon Seidenf., 309
nematocaulon Ridl., 303
nieuwenhuisii J.J. Sm., 303
nubinatum J.J. Verm., 303
obtusipetalum J.J. Sm., **287,** 303
obtusum (Blume) Lindl., 303
odoardii Pfitzer, 303
odoratum (Blume) Lindl., 304
osyriceroides J.J. Sm., 304
otochilum J.J. Verm., 304
ovalifolium (Blume) Lindl., 304
papillatum J.J. Sm., 304
papillosum J.J. Sm., non Finet, 304
patens King ex Hook.f., 304
pedicellatum Ridl., 296
pelicanopsis J.J. Verm. & A.Lamb, 305
penduliscapum J.J. Sm., **293,** 305
pergracile Ames & C. Schweinf., 285
perparvulum Schltr., 305
perpusillum Ridl., non Kraenzl., 305
peyerianum (Kraenzl.) Seidenf., 307
phaeoneuron Schltr., 305
pileatum Lindl., 305
placochilum J.J. Verm., 305
planibulbe (Ridl.) Ridl., 306
pocillum J.J. Verm., 306
polygaliflorum J.J. Wood, 306
porphyrotriche J.J. Verm., 306
prianganense J.J. Sm., 306
puberulum Ridl., 291
pugilanthum J.J. Wood, 306
pugioniforme J.J. Sm., 281
puguahaanense Ames, 306
pumilio Ridl., 307
punctatum Ridl., 289
puntjakense J.J. Sm., 307
purpurascens Teijsm. & Binn., 307
pustulatum Ridl., 307
putidum (Teijsm. & Binn.) J.J.Sm.,
  307
pyridion J. J. Verm., 308
racemosum Rolfe, 280
rajanum J.J. Sm., 308
rariflorum J.J. Sm., 308
reflexum Ames & C. Schweinf., 288

refractilingue J.J. Sm., 308, Pl. 3F
restrepia Ridl., 308
reticosum Ridl., 289
reticulatum Batem. ex Hook.f., 308
rhizomatosum Ames & C.Schweinf.,
  308
rhombifolium (Carr) Masam., 295
rhynchoglossum Schltr., 309
romburghii J.J. Sm., 309
rubiferum J.J. Sm., 309
ruficaudatum Ridl., 309
rugosum Ridl., 309
saccatum Kraenzl., 281
salaccense Rchb.f., 309
scabrum J.J. Verm. & A. Lamb, 309
scandens Kraenzl., 301
schefferi (Kuntze) Schltr., 310
scintilla Ridl., 310
sessile (J. König) J.J. Sm., 310
sigmoideum Ames & C. Schweinf.,
  287, 310
similissimum J.J. Verm., 310, Pl.4A
singaporeanum Schltr., 311
sopoetanense Schltr., 311
spinulipes J.J. Sm., 303
stipitatibulbum J.J. Sm., 311
stormii J.J. Sm., 311, Pl. 4B
streptotriche J.J. Verm., 311
striatellum Ridl., 311
subclausum J.J. Sm., 312
subumbellatum Ridl., 312
succedaneum J.J. Sm., 312
supervacaneum Kraenzl., 312
tacitum Carr, 313
taeniophyllum C.S.P. Parish &
  Rchb.f., 312
tardeflorens Ridl., 312
tenompokense J.J. Sm., 312
tenuifolium (Blume) Lindl., 313
teres Ridl., 313
thymophorum J.J. Verm. & A.Lamb,
  313
tortuosum (Blume) Lindl., 313
tothastes J.J. Verm., 316
trichoglottis Ridl., 295
trifolium Ridl., 313
trulliferum J.J. Verm. & A. Lamb
  ined., 313
tryssum J.J. Verm. & A. Lamb ined.,
  314
tumidum J.J. Verm., 314
turgidum J.J. Verm., 314
undecifilum J.J. Sm., **287,** 314
unguiculatum Rchb.f., 314
uniflorum (Blume) Hassk., **287,** 314

387

ovalifolia Blume, 304
tenuifolia Blume, 313
tortuosa Blume, 313
Diplocaulobium (Rchb.f.) Kraenzl., 272
  longicolle (Lindl.) Kraenzl., 272
  vanleeuwenii (J.J. Sm.) P.F. Hunt
    & Summerh., 273, Pl. 9F
Dipodium R. Br., 112
  paludosum (Griff.) Rchb.f., 112
  pictum (Lindl.) Rchb.f., 112
  purpureum J.J. Sm., 112
  scandens (Blume) J.J. Sm., 112
Doritis Lindl., 318
  pulcherrima Lindl., 318
  regnieriana (Rchb.f.) Holttum, 318
Dossinia E. Morren, 53
  marmorata E. Morren, 53
Dyakia Christenson, 364
  hendersoniana (Rchb.f.)
    Christenson, 364, Pl. 10A
Echioglossum Blume
  striatum Rchb.f., 328
Entomophobia de Vogel, 197
  kinabaluensis (Ames) de Vogel, 197
Ephemerantha P.F. Hunt & Summerh., 275
  bicarinata (Ames & C. Schweinf.)
    P.F. Hunt & Summerh., 276
  bicostata (J.J. Sm.) P.F. Hunt
    & Summerh., 276
  crenicristata (Ridl.) P.F. Hunt
    & Summerh., 276
  denigrata (J.J. Sm.) P.F. Hunt
    & Summerh., 277
  flabelloides (J.J. Sm.) P.F. Hunt
    & Summerh., 277
  labangensis (J.J. Sm.) P.F. Hunt
    & Summerh., 277
  longirepens (Ames & C. Schweinf.)
    P.F. Hunt & Summerh., 274
Ephippium Blume
  ciliatum Blume, 283
  cornutum Blume, 288
  elongatum Blume, 291
  lepidum Blume, 297
  uniflorum Blume, 314
Epicranthes Blume
  flavofimbriata (J.J. Sm.) Garay
    & W. Kittr., 292
  javanica Blume, 291
Epidendrum L.
  amabile L., 353
  concretum Jacq., 146
  equitans G. Forst., 106
  flosaeris L., 322
  retusum L., 360

sessile J. König, 310
Epigeneium Gagnep., 273
  geminatum (Blume) Summerh., 273
  kinabaluense (Ridl.) Summerh., 15,
    273
  labuanum (Lindl.) Summerh., 274
  longirepens (Ames & C. Schweinf.)
    Seidenf., 274
  radicosum (Ridl.) Summerh., 274
  speculum (J.J. Sm.) Summerh., 274,
    Pl. 10B
  suberectum (Ridl.) Summerh., 273
  treacherianum (Rchb.f. ex Hook.f.)
    Summerh., 6, 274
  tricallosum (Ames & C. Schweinf.)
    J.J. Wood, 275
  verruciferum (J.J. Sm.) Summerh.,
    275
  zebrinum (J.J. Sm.) Summerh., 275,
    Pl. 10C
Epiphanes Blume
  javanica Blume, 79
Epipogium J.G. Gmel. ex Borkh., 80
  roseum (D. Don) Lindl., 80
Eria Lindl., 6, 8, 205
  aeridostachya Rchb.f. ex Lindl., 219
  angustifolia Ridl., 205
  aporina Hook.f., 221
  atrovinosa Ridl., 205
  aurantia J.J. Sm., 206, **210**, Pl.10D & E
  aurantiaca Ridl., 206
  aurantiaca J.J.Sm., non Ridl., 206
  aurea Ridl., 221
  bancana J.J. Sm., 206
  berringtoniana Rchb.f., 206
  bifalcis Lindl., 206
  biflora Griff., 206
  bigibba (Benth. ex Kraenzl.) Rchb.f.,
    206
  biglandulosa J.J. Sm., 207
  borneensis Rolfe, 207
  bractescens Lindl., 207
  brevipedunculata Ames & C.Schweinf.,
    222
  brevirachis J.J. Sm., 222
  brookesii Ridl., 207
  canaliculata Blume, 222
  cinnabarina Rolfe, 213
  caricifolia J.J. Wood
    var. caricifolia, 207
    var. glabra J.J. Wood, 207
  carnea J.J. Sm., 207
  carnosissima Ames & C. Schweinf.,
    207
  cepifolia Ridl., 208

merrillianus Ames, 324
micranthus Ames, 329
nieuwenhuisii J.J. Sm., 327
potamophilus Schltr., 337
ridleyi J.J. Sm., 326
robustum O'Brien, 328
robustus Schltr., 328
samarindae (Schltr.) J.J. Sm., 357
scortechinii Hook.f., 327
strongyloides (Ridl.) J.J. Sm., 328
subulatum (Blume) Rchb.f., 328
tenuirachis J.J. Sm., 328
uraiensis Hayata, 329
williamsonii Rchb.f., 329
Sarcochilus R. Br.
biserratus Ridl., 358
borneensis Rolfe, 340
calceolus Lindl., 340
carrii Holttum, non L.O. Williams, 366
cladostachyum Hook.f., 358
ecalcaratus Holttum, 366
emarginatus (Blume) Rchb.f., 367
fragrans Ridl., 358
fuscoluteus Lindl., 333
hirsutus Hook.f., 358
hirtulus Hook.f., 367
indusiatus (Rchb.f.) Carr, 350
koeteiense Schltr., 350
membraniferus Carr, 351
pachyrhachis Schltr., 358
pallidus (Blume) Rchb.f., 359
pallidus Masam., non (Blume) Rchb.f., 359
pardalis Ridl., 342
scopa Rchb.f. ex Hook.f., 343
spathibrachiatus J.J. Sm., 359
spathipetalus J.J. Sm., 358
stellatus Ridl., 372
stenoglottis Hook.f., 350
suaveolens (Roxb.) Hook.f., 359
trichoglottis Hook.f., 343
unguiculatus Lindl., 359
virescens Ridl., 367
vriesii Ridl., 359
zamboangensis Ames, 350
Sarcoglyphis Garay, 337
fimbriatus (Ridl.) Garay, 337
potamophila (Schltr.) Garay & W.Kittr., 337
Sarcopodium Lindl. & Paxt.
beccarianum Kraenzl., 214
citrinocastaneum (Burkill) Ridl., 275
dearei Hort., 289
interruptum (J.J. Sm.) Ames, 275
kinabaluense (Ridl.) Rolfe, 273

labuanum (Lindl.) Rolfe, 274
radicosum (Ridl.) Rolfe, 274
speculum (J.J. Sm.) Carr, 274
suberectum Ridl., 273
sulphuratum (Ridl.) Rolfe, 275
treacherianum (Rchb.f. ex Hook.f.) Rolfe, 274
verruciferum (J.J. Sm.) Rolfe, 275
zebrinum (J.J. Sm.) Kraenzl., 275
Sarcostoma Blume, 220
borneense Schltr., 221
Schoenorchis Reinw., 337
aurea (Ridl.) Garay, 338
endertii (J.J. Sm.) Christenson & J.J. Wood, 338
juncifolia Blume ex Reinw., 338
micrantha Blume, 338
pachyglossa (Lindl.) Garay, 338
paniculata Blume, 338
Sigmatochilus Rolfe
kinabaluensis Rolfe, 148
Smitinandia Holttum, 338
micrantha (Lindl.) Holttum, 339
Spathoglottis Blume, 138
aurea Lindl., 138
confusa J.J. Sm., 138, Pl. 15F
gracilis Rolfe ex Hook.f., 138
kimballiana Hook.f., 138
microchilina Kraenzl., 139
plicata Blume, 139
Spiranthes Rich., 66
sinensis (Pers.) Ames, 66
Spongiola J.J. Wood & A. Lamb, 370
lohokii J.J. Wood & A. Lamb, 371
Staurochilus Ridl. ex Pfitzer, 339
fasciatus (Rchb.f.) Ridl., 339
Stereosandra Blume, 80
javanica Blume, 80
Stigmatodactylus Maxim. ex Makino, 69
Taeniophyllum Blume, 319
affine Schltr., 320
borneense Schltr., 320
esetiferum J.J. Sm., 320
filiforme J.J. Sm., 320
gracillimum Schltr., 320
hirtum Blume, 320
kapahense Carr, 320
obtusum Blume, 320
proliferum J.J. Sm., 320
rubrum Ridl., 321
rugulosum Carr, 321
stella Carr, 321
Tainia Blume, 139
beccarii (Schltr.) Gagnep., 133
bigibba Benth. ex Kraenzl., 206

wenzelii Ames, 36, 344
winkleri J.J. Sm.
  var. minor J.J. Sm., 347
  var. winkleri, 347
Trichotosia Blume, 221
annulata Blume, 221
aporina (Hook.f.) Kraenzl., 221
aurantiaca (J.J. Sm.) Kraenzl., 206
aurea (Ridl.) Carr, 221
brevipedunculata (Ames
    & C.Schweinf.) J.J. Wood, 221
brevirachis (J.J. Sm.) J.J. Wood, 35,
  222
canaliculata (Blume) Kraenzl., 222
carnea (J.J. Sm.) Kraenzl., 207
conifera (J.J. Sm.) J.J. Wood,35, 222
dajakorum Kraenzl., 224
ferox Blume, 222
fusca (Blume) Kraenzl., 222
gracilis (Hook.f.) Kraenzl., 222,
  Pl.16D
hallieri (J.J. Sm.) Kraenzl., 212
hispidissima (Ridl.) Kraenzl., 223
jejuna (J.J. Sm.) J.J. Wood, 35, 223
lacinulata (J.J. Sm.) Carr, 223
lawiensis (J.J. Sm.) J.J. Wood, 35, 223
microphylla Blume, 223
mollicaulis (Ames & C. Schweinf.)
    J.J. Wood, 223
odoardi Kraenzl., 223
pauciflora Blume, 224
pilosissima (Rolfe) J.J. Wood, 224
rubiginosa (Blume) Kraenzl., 224
sarawakensis Carr, 224
spathulata (J.J. Sm.) Kraenzl., 224
teysmannii (J.J. Sm.) Kraenzl., 224
unguiculata (J.J. Sm.) Kraenzl., 225
velutina (Lodd. ex Lindl.) Kraenzl.,
  225
vestita (Lindl.) Kraenzl., 225
Tropidia Lindl., 47
angulosa (Lindl.) Blume, 48
connata J.J. Wood & A. Lamb, 23, 34,
  47, **49**
curculigoides Lindl., 48
graminea Blume, 48
pedunculata Blume, 48
saprophytica J.J. Sm., 23, 48
Tuberolabium Yamam., 371
rhopalorrhachis (Rchb.f.) J.J.Wood,
  371
Tupistra Ker Gawl.
  singapureana Baker, 39
Vanda Jones ex R. Br., 6, 361
dearei Rchb.f., 362, Pl. 16E

gibbsiae Rolfe, 362
hastifera Rchb.f.
  var. gibbsiae (Rolfe) P.J. Cribb, 2,
    362
  var. hastifera , 362
helvola Blume, 362
hookeriana Rchb.f., 351
lamellata Lindl., 362
lowii Lindl., 330
scandens Holttum, 362
Vandopsis Pfitzer
breviscapa (J.J. Sm.) Schltr., 322
lowii (Lindl.) Schltr., 330
Vanilla Mill., 86
abundiflora J.J. Sm., 86
albida Blume, 86
borneensis Rolfe, 86
griffithii Rchb.f., 86
havilandii Rolfe, 86
kinabaluensis Carr, 6, 86
pilifera Holttum, 87
sumatrana J.J. Sm., 87
Vrydagzynea Blume, 61
albida (Blume) Blume, 62
angustisepala J.J. Sm., 62
argentistriata Carr, 62
beccarii Schltr., 62
bicostata Carr, 62
bractescens Ridl., 62
elata Schltr., 62
endertii J.J. Sm., 62
grandis Ames & C. Schweinf., 63
lancifolia Ridl., 63
nuda Blume, 63
pauciflora J.J. Sm.
  var. pauciflora, 63
  var. unistriata J.J. Sm., 63
semicordata J.J. Sm., 63
tilungensis J.J. Sm., 63
tristriata Ridl., 63
Wailesia Lindl.
  picta Lindl., 112
Wolfia Dennst.
  spectabilis Dennst., 113
Zeuxine Lindl., 64
biloba Ridl., 56
flava (Wall. ex Lindl.) Benth. ex
    Hook.f., 64
gracilis (Breda) Blume, 64
kutaiensis J.J. Sm., 64
linguella Carr, 64
papillosa Carr, 64
petakensis J.J. Sm., 65
purpurascens Blume, 65
strateumatica (L.) Schltr., 65, Pl.16F

violascens Ridl., 65
viridiflora (J.J. Sm.) Schltr., 65
Zosterostylis Blume
  arachnites Blume, 67